P9-DWG-589

D0037771

The Bacteria

VOLUME VIII: ARCHAEBACTERIA

THE BACTERIA

A TREATISE ON STRUCTURE AND FUNCTION

The Bacteria

A TREATISE ON STRUCTURE AND FUNCTION

CONSULTING EDITOR   *I. C. Gunsalus*

Department of Chemistry
School of Chemical Sciences
University of Illinois
Urbana, Illinois

VOLUME VIII

ARCHAEBACTERIA

EDITORS-IN-CHIEF

J. R. Sokatch

University of Oklahoma
Health Sciences Center
Oklahoma City, Oklahoma

L. Nicholas Ornston

Department of Biology
Yale University
New Haven, Connecticut

VOLUME EDITORS

Carl R. Woese

Department of Genetics and
Development
University of Illinois
Urbana, Illinois

Ralph S. Wolfe

Department of Microbiology
University of Illinois
Urbana, Illinois

1985

Academic Press, Inc.

(Harcourt Brace Jovanovich, Publishers)
ORLANDO   SAN DIEGO   NEW YORK   LONDON
TORONTO   MONTREAL   SYDNEY   TOKYO

QR
41
B131
—
v. 8

EB

COPYRIGHT © 1985, BY ACADEMIC PRESS, INC.
ALL RIGHTS RESERVED.
NO PART OF THIS PUBLICATION MAY BE REPRODUCED OR
TRANSMITTED IN ANY FORM OR BY ANY MEANS, ELECTRONIC
OR MECHANICAL, INCLUDING PHOTOCOPY, RECORDING, OR
ANY INFORMATION STORAGE AND RETRIEVAL SYSTEM, WITHOUT
PERMISSION IN WRITING FROM THE PUBLISHER.

ACADEMIC PRESS, INC.
Orlando, Florida 32887

United Kingdom Edition published by
ACADEMIC PRESS INC. (LONDON) LTD.
24–28 Oval Road, London NW1 7DX

Library of Congress Cataloging in Publication Data
(Revised for volume VIII)

Gunsalus, I. C. (Irwin Clyde), Date   ed.
    The bacteria; a treatise on structure and function.

    Includes bibliographies and index.
    Vol. 6-   edited by J. R. Sokatch and L. N. Ornston.
    Contents: v 1. Structure – v. 2. Metabolism – [etc.]
–   v. 8. Archaebacteria.
    1. Bacteriology–Collected works. I. Stanier,
Roger Y., joint ed. II. Sokatch, J. R. (John Robert),
Date.   III. Ornston, L. Nicholas.
QR41.G78   589.9   59-13831
ISBN 0–12–307208–5 (v. 8 : alk. paper)

PRINTED IN THE UNITED STATES OF AMERICA

85 86 87 88   9 8 7 6 5 4 3 2 1

5-21-86

CONTENTS OF VOLUME VIII

I. BIOCHEMICAL DIVERSITY AND ECOLOGY OF ARCHAEBACTERIA

4. Sedimentary Record and Archaebacteria 215

JÜRGEN HAHN AND PAT HAUG

II. TRANSLATION APPARATUS OF ARCHAEBACTERIA

5. The Structure and Evolution of Archaebacterial
Ribosomal RNA 257

GEORGE E. FOX

6. Transfer Ribonucleic Acids of Archaebacteria 311

RAMESH GUPTA

III. GENERAL MOLECULAR CHARACTERISTICS OF ARCHAEBACTERIA

CONTRIBUTORS TO VOLUME VIII

AUGUST BÖCK, *Institut für Genetik und Mikrobiologie der Universität München, D-8000 Munich 19, Federal Republic of Germany*

W. FORD DOOLITTLE, *Department of Biochemistry, Dalhousie University, Halifax, Nova Scotia, Canada B3H 4H7*

GEORGE E. FOX, *Department of Biochemical and Biophysical Sciences, University of Houston—University Park, Houston, Texas 77004*

RAMESH GUPTA,* *Department of Genetics and Development, University of Illinois, Urbana, Illinois 61801*

JÜRGEN HAHN, *Max-Planck-Institut für Chemie (Otto-Hahn-Institut), D-6500 Mainz, Federal Republic of Germany*

PAT HAUG, *Max-Planck-Institut für Chemie (Otto-Hahn-Institut), D-6500 Mainz, Federal Republic of Germany*

OTTO KANDLER, *Botanisches Institut der Universität München, D-8000 Munich 19, Federal Republic of Germany*

FRIEDRICH KLINK, *Biochemisches Institut, Christian-Albrechts-Universität, D-2300 Kiel, Federal Republic of Germany*

HELMUT KÖNIG, *Lehrstuhl für Mikrobiologie, Universität Regensburg, D-8400 Regensburg, Federal Republic of Germany*

D. J. KUSHNER, *Department of Biology, University of Ottawa, Ottawa, Ontario, Canada K1N 6N5*

THOMAS A. LANGWORTHY, *Department of Microbiology, School of Medicine, University of South Dakota, Vermillion, South Dakota 57069*

ALASTAIR T. MATHESON, *Department of Biochemistry and Microbiology, University of Victoria, Victoria, British Columbia, Canada V8W 2Y2*

R. SCHNABEL, *Max-Planck-Institut für Biochemie, D-8033 Martinsried, Federal Republic of Germany*

K. O. STETTER, *Lehrstuhl für Mikrobiologie, Universität Regensburg, D-8400 Regensburg, Federal Republic of Germany*

Present address: Department of Medical Biochemistry and Department of Chemistry and Biochemistry, Southern Illinois University, Carbondale, Illinois 62901.

M. THOMM, *Lehrstuhl für Mikrobiologie, Universität Regensburg, D-8400 Regensburg, Federal Republic of Germany*

WILLIAM B. WHITMAN,* *Department of Microbiology, University of Illinois, Urbana, Illinois 61801*

CARL R. WOESE, *Department of Genetics and Development, University of Illinois, Urbana, Illinois 61801*

RALPH S. WOLFE, *Department of Microbiology, University of Illinois, Urbana, Illinois 61801*

W. ZILLIG, *Max-Planck-Institut für Biochemie, D-8033 Martinsried, Federal Republic of Germany*

*Present address: Department of Microbiology, University of Georgia, Athens, Georgia 30605.

The Bacteria

A TREATISE ON STRUCTURE AND FUNCTION

VOLUME I: STRUCTURE

VOLUME II: METABOLISM

VOLUME III: BIOSYNTHESIS

VOLUME IV:

THE PHYSIOLOGY OF GROWTH

VOLUME V: HEREDITY

VOLUME VI: BACTERIAL DIVERSITY

VOLUME VII:

MECHANISMS OF ADAPTATION

INTRODUCTION
Archaebacteria: The Third Form of Life

Our concept of what a bacterium is has been strongly conditioned by our understanding of what it is not. It is not a eukaryote. Since the seminal realization by Chatton in the 1930s that all living (self-replicating) forms fall into one of two classes defined by cell morphology, our understanding of cells, their evolution, and their relationships has been fashioned in accord with a deceptively simple and remarkably subtle concept, the so-called prokaryote–eukaryote distinction. (It is also a concept whose etymological development was perhaps never recognized.) As is often the case, epoch-making ideas carry with them implicit, unanalyzed assumptions that ultimately impede scientific progress until they are recognized for what they are. So it is with the prokaryote–eukaryote distinction. Our failure to understand its true nature set the stage for the sudden shattering of the concept when a "third form of life" was discovered in the late 1970s, a discovery that actually left many biologists incredulous. Archaebacteria, as this third form has come to be known, have revolutionized our notion of the prokaryote, have altered and refined the way in which we think about the relationship between prokaryotes and eukaryotes (especially the role the former plays in the evolution of the latter), and will influence strongly the view we develop of the ancestor that gave rise to all extant life (particularly the conditions under which that entity arose). As a background against which to view the archaebacteria—whose physiological, morphological, ecological, and molecular aspects are discussed in this volume—we will trace the history of the prokaryote–eukaryote distinction and the parallel development of the concept of archaebacteria, which would prove its undoing.

Initially the definition of the prokaryotic cell was basically a negative one. The prokaryotic cell did not possess this or that eukaryotic feature—organelles, circumscribed nucleus, etc.—but no converse properties (those possessed by prokaryotes that were lacking in eukaryotes) were recognized. It is remarkable, therefore, that much of a concept of a prokaryote existed at all initially. (If a "car" were to be defined only as not having leaves or bark, etc., one would not have a very clear picture of what a car is.) Yet a definite concept of a "prokaryote" did exist, and it strongly influenced the course of microbiology.

For reasons that are not entirely clear, the prokaryote was taken to be a phylogenetically coherent unit, perhaps in part because the eukaryote was so conceived. The proof of this contention lies not so much in statements made by biologists over the years as in the fact that in defining the molecular phenotype of eukaryotes and prokaryotes, the biologist rarely felt a need to characterize other than the "typical prokaryote"—*Escherichia coli*—and if he did, it was for reasons unrelated to the matter of phylogenetic coherence of the group. Several generations of biologists have now been raised to believe that prokaryote–eukaryote is both a cytological and a phylogenetic dichotomy.

The term prokaryote also carried with it the implication (in the prefix "pro-") that it was a forerunner of the eukaryote. And so it continued to be regarded. Explanations for the origin of the eukaryotic cell in terms of endosymbioses are testimony to this. While such an idea seems a valid enough explanation for the origin of chloroplasts and mitochondria, it is at best a debatable speculation when applied to the origin of the eukaryotic nucleus and is near nonsense when invoked to account for the species that hosted the endosymbionts (it was a prokaryote that lost its cell wall). The general idea, too, embodies the more subtle implication that endosymbioses among prokaryotes are all there really is to eukaryotic cellular evolution, a notion which at best is misleading. Attempts by biologists to rationalize the origin of eukaryotic mechanisms (e.g., control of gene expression) in terms of their having arisen from their prokaryotic (i.e., eubacterial) counterparts have proved unsatisfying and stultifying exercises. Eukaryotic mechanisms, if anything, seem derived independently of, not from, their prokaryotic counterparts.

The final implication of the prokaryote–eukaryote distinction is that the eukaryote is somehow more advanced than the prokaryote. It is mainly because this view accords with anthropocentrism that it finds acceptance. Prokaryotes may have smaller genomes than do eukaryotes, but these are, if anything, more precisely organized, well defined, and efficiently functioning than are their eukaryotic counterparts. There are as many (or more) reasons to consider that the eukaryotic cell is the more primitive of the two than the reverse.

Conventional wisdom has always regarded the prokaryote, then, as a primitive, less advanced type of cell that arose before the eukaryotic cell and was capable of evolving into the latter. This notion is incorrect and pernicious.

With the onset of the molecular era in biology the prokaryote–eukaryote distinction in effect became redefined. Previously the distinction had been made in terms of noncomparable properties at the cytological level, e.g., whether the cell possessed a nucleus or organelles. Now the definition would be framed in terms of comparable properties at the molecular level. For any number of molecular properties there was a characteristic eukaryotic and a characteristic prokaryotic form—various enzyme quaternary structures, this or that biochemical pathway, the ribosome, control mechanisms, etc. While this drastic redefinition gave the notion of a prokaryote a definiteness it had previously lacked, while it made biologists aware of the true depth of the prokaryote–eukaryote distinction, and while it gave the idea some real evolutionary meaning, biologists approached the redefinition firm in their belief (mentioned above) that the prokaryote was a phylogenetically monolithic grouping, and so never saw fit to explore the phylogenetic diversity of its molecular phenotype. Had they done so, archaebacteria would certainly have been discovered ten to twenty years sooner.

Microbiology is a discipline that has developed virtually untouched by Darwin's grand idea. Despite the concerns and the efforts of the classical microbiologists, the natural, or evolutionary relationships among bacteria remained effectively unknown through the early 1970s, and so exerted no influence on the development of the field of microbiology. We cannot appreciate the profound effect the lack of a bacterial phylogeny has had on the course not only of microbiology but all of biology. Evolution provides the only unifying theme in an otherwise diverse and continually disintegrating discipline. And the evolutionary history of bacteria, because it transcends in time and to a large extent overlaps the evolution of eukaryotes, is the true base for the study of evolution.

In the past, evolutionary relationships were deduced largely from the morphological similarities and differences among the extant representatives of various (eukaryotic) lines and their fossilized ancestors. The success of this approach turned on the complexity of eukaryotic morphologies. Bacteria are not morphologically complex. And we now can see that attempts to use morphological (and other equally unreliable) characters to establish their natural relationships have created only a phylogenetic monstrosity: the current system of bacterial taxonomy. There is no point in attempting to construct a bacterial phylogeny in these terms.

While the bacterial phenotype (at least as classically defined) is too simple to serve as the basis on which to establish a phylogeny, this is not true of the (bacterial) genotype. For any given phenotypic character—the cytochrome c function, for example—myriad genotypic (i.e., sequence) equivalents exist. Thus, genotypic "phase space" is enormous compared to the "phase space" of the corresponding phenotypes. This in turn means that the majority of changes in genotypes that become fixed are selectively neutral—which gives the occurrences of such changes a chronometric quality—and that the biologist can use sequence comparisons as a rather precise, reliable, and powerful measure of evolutionary time (distance) and phylogenetic branch points.

The use of macromolecular sequence comparisons for determining evolutionary relationships and distances was developed in eukaryotic systems. However, its great impact will be in prokaryotic systems. Evolutionary distances in the bacterial world far exceed those among eukaryotes. This is readily seen in terms of the cytochrome c molecular chronometer. A phylogenetic tree constructed on the basis of cytochrome c sequence comparisons for the eukaryotes is no more than a branch of a larger eubacterial tree that itself covers only one of three of the sublines of purple bacteria, which in turn are one of about ten major divisions of the eubacteria. It is no wonder then that a molecule such as cytochrome c, which is a superb molecular chronometer for the eukaryotes, proves to be an inadequate chronometer in the case of the bacteria.

Ribosomal RNA (16 S) has proved to be an excellent molecular chronometer by which to measure the evolutionary distances encountered in the bacterial world. Ribosomal RNA sequence comparisons can measure not only the deepest divisions among organisms, but the more superficial ones as well. By this means it has been possible to construct a bacterial phylogeny, a structure that will serve as a framework within which to reconstruct and understand the evolutionary history of bacteria. It was this molecular chronometer that could show that there were three, not two, basic categories of living systems, three primary kingdoms: the eubacteria, the archaebacteria, and the eukaryotes.

Historically, archaebacteria have been known, but not recognized as such, for a long time. Methanogens trace their scientific lineage back to Volta, who concerned himself with the origin of the "combustible air" formed in the sediments of streams, bogs, and lakes rich in decaying vegetation. Only later, of course, was a microbial basis for this marsh gas demonstrated. H. A. Barker, more than any other microbiologist, was responsible for putting the study of methanogens on a sound scientific footing. It is through methanogenesis that we have been introduced to the biochemical uniqueness of archaebacteria. The impressive number of unique cofactors involved in methanogenesis would seem to represent only the tip of this biochemical iceberg.

A completely separate concern with archaebacteria began with the extreme halophiles, which produced reddening and rotting of salted meats and fish. Their incredible characteristics—growing in saturated brines and an extremely high internal salt concentration—made them objects of scientific fascination. It is through them that we have been introduced to both the unique lipids of archaebacteria (through the pioneering work of M. Kates) and the simplest of biological photosystems, the bacterial rhodopsin, or purple membrane, system.

The thermoacidophiles are a recent addition to the archaebacterial picture, several different strains having been discovered at about the same time, in the early 1970s, by several groups of workers. These organisms have introduced us to the remarkable conditions under which archaebacteria can grow; today we know they can grow even at temperatures above the normal boiling point of water. The thermoacidophiles were perhaps the first organisms, through their halobacteria-like lipid compositions, to provide evidence for a relationship among various of the bizarre organisms that would become grouped together as the archaebacteria. However, this early evidence was not interpreted in that way.

Initially the various archaebacterial idiosyncrasies were seen only as adaptations to unusual niches. Even today the majority of microbiologists see the thermoacidophiles, for example, as having adapted to their extreme temperature niches from an ancestral mesophilic condition. The same type of interpretation is applied to the extreme halophiles. Attempts to rationalize the similarities in lipid composition of these two types of organisms in this way are, in retrospect, rather amusing. It was only after the archaebacteria were shown to be a coherent and major grouping by a nonphenotypic, i.e., a genotypic, measure that the true significance of these many and common idiosyncrasies became apparent.

As revealed by the ribosomal RNA chronometer, the archaebacteria form a coherent phylogenetic unit that is equivalent in phylogenetic status to the eubacteria (and so is a deeper unit than the eukaryotes). That unit has two main branches: the methanogens and extreme halophiles, on the one hand, and the thermoacidophiles (sulfur-dependent archaebacteria), on the other. The isolated species *Thermoplasma acidophilum* does not group cleanly with either, but seems to cluster peripherally with the methanogens and extreme halophiles. The deep division within the kingdom is supported by a number of phenotypic differences, as the reader of this volume will see.

As one reads through this volume it will be useful to recognize from time to time that one is not reading about bacteria, as the name of this treatise might suggest. Archaebacteria resemble ordinary bacteria in only the most superficial of ways. They are now treated as "bacteria" principally because they are perceived as noneukaryotes. The more archaebacteria are perceived as a unique class of organisms that manifest "kingdom-level novelty" the more we will learn from them.

CARL R. WOESE AND RALPH S. WOLFE

PART I

Biochemical Diversity and Ecology of Archaebacteria

Chapter 1

Methanogenic Bacteria

William B. Whitman

Copyright © 1985 by Academic Press, Inc.
All rights of reproduction in any form reserved.
ISBN 0-12-307208-5

I. Introduction

Methanogenic bacteria are of special interest because they are the only arch-aebacteria not restricted to extreme environments. In fact, methanogens are found in a variety of anaerobic habitats including sediments, sludge digestors, the guts of insects, the large bowel of man and animals, and the rumen. They cohabitate these environments with both eubacteria and eukaryotes. Nevertheless, methanogens retain the characteristics unique to archaebacteria. Thus, these unique characteristics cannot be determined solely by an adaptation to the environment. Furthermore, the unique features of archaebacteria cannot be due to physical isolation from other organisms in extreme environments. For these reasons, methanogens may be the ideal archaebacterial subject for comparative studies.

So far, all known methanogens are obligate methane producers, and no secondary energy sources have been identified. This restriction to a narrow range of energy sources is not a general feature of archaebacteria. For example, *Halobacterium* can obtain energy from photosynthesis, respiration, and the fermentation of arginine. Moreover, in other respects the methanogens are themselves extremely diverse. The cell envelope structures, morphologies, and 16 S rRNAs of methanogens are as different or more different from each other as these characteristics of the eubacteria are different from each other. Thus, the narrow range of energy sources available to methanogens is anomalous. With our present knowledge, there is no definite solution to this apparent inconsistency. However, a reasonable explanation may be that methanogens, and archaebacteria in general, cannot compete successfully with eubacteria in temperate environments. Thus, archaebacteria are restricted to a niche unavailable to eubacteria, namely methane synthesis. In extreme environments which limit the growth of eubacteria and eukaryotes, archaebacteria can then exploit alternative energy sources. If facultative methanogens do exist, they should likewise be found in extreme environments.

Methanogens are not primitive microorganisms. They are highly specialized and possess elaborate systems unique to their way of life. Only our knowledge of methanogens is primitive. In spite of a wealth of information, answers to most of the compelling questions concerning these organisms are still elusive. This chapter will delineate some of the important questions and provide an overview of the ecology, physiology, and biochemistry of methanogens. Topics that are discussed in depth in subsequent chapters, including RNA and ribosomal characteristics, cell wall structure, lipids, and antibiotic sensitivity, will not be discussed here.

II. Ecology

The ecology of methanogenesis has been reviewed frequently (Mah *et al.*, 1977b; Zeikus, 1977; Bryant, 1979; Wolin, 1981; Wolin and Miller, 1982; Zehnder *et al.*, 1982; Mah, 1982). Therefore, the following discussion will emphasize only the major themes. The abundance of methane in the atmosphere is of particular interest because it demonstrates the global importance of methanogenesis. Methanogenic bacteria are the only archaebacteria that are truly cosmopolitan and play a major role in the cycling of elements in the biosphere.

Methane is an important trace gas in the earth's atmosphere. Although it represents only about 1% of the total atmospheric carbon (Ehhalt, 1974), methane may be a significant source of hydrogen compounds, water vapor, and carbon monoxide in the atmosphere (Ehhalt and Schmidt, 1978). The total atmospheric methane is on the order of 4×10^{15} g and may be increasing at a rate of approximately 2% per year (Ehhalt, 1974; Rasmussen and Khalil, 1981). Between 80 and 90% of the atmospheric methane is of biogenic origin (Sheppard *et al.*, 1982; Ehhalt, 1974). The methanogenic bacteria are the only significant source of biogenic methane. Nonbiogenic methane comes from fossil sources.

The contributions of various ecosystems to the atmospheric methane have been estimated from the methane fluxes of specific ecosystems, net primary productivities, and global inventories of land use (Table I; Sheppard *et al.*, 1982). These are not precise estimates because of incomplete measurements, high local variations, and uncertainties in the seasonal variation of methane fluxes. In addition, it is not clear whether or not all sources of methane have been taken into account. For instance, by one estimate methanogens in the guts of termites produce about 1.5×10^{14} g of methane per year (Zimmerman *et al.*, 1982). This contribution may not be included in field measurements of methane fluxes in ecosystems where termites are abundant. Furthermore, measurement of methane fluxes on the ground would not include arboreal methane generated in the wet wood of standing trees (Zeikus and Ward, 1974). Nevertheless, the present estimate of the total annual release of methane to the atmposhere, 1.21×10^{15} g methane per year, is similar to previous estimates, 0.55 to 1.1×10^{15} g methane per year (Ehhalt, 1974).

The sources of atmospheric methane provide a minimum estimate of the total biogenic methane produced as well as an indication of the relative importance of various ecosystems. It is only a minimum estimate because it neglects methane that is oxidized before reaching the atmosphere. Methane oxidation is likely to be substantial considering the ubiquity of aerobic methane-oxidizing bacteria. In

TABLE I

Sources of Atmospheric Methane[a]

Source	CH$_4$ flux (g/m^2 year)	Total annual emission rate (10^{14} g/year)
Terrestrial		
Forest		
Tropical rain	23.5	3.17
Seasonal	19.1	0.80
Temperate	14.6	0.79
Boreal	13.8	0.62
Woodland, scrubland	9.0	0.28
Savanna	9.6	1.37
Temperate grassland	4.1	0.15
Tundra, alpine meadow	8.1	0.07
Desert scrub	6.7	0.08
Cultivated land	6.1	0.34
Paddy field	55	0.39
Biomass burning	—[b]	0.6
Freshwater		
Marsh	78.8	0.39
Lakes, stream	102	0.51
Marine		
Open ocean	0.012	0.04
Continental shelf	0.012	0.003
Algal beds, reef	6.9	0.04
Estuaries	4.5	0.06
Ruminants	—	0.7
Human sewage	—	0.2
Human solid waste	—	0.5
Fossil sources	—	1.0
		12.1

[a] From Sheppard *et al.* (1982). Copyright by the American Geophysical Union.
[b] Not relevant.

addition, several authors have proposed that anaerobic methane oxidation by sulfate-reducing bacteria is a factor in methane evolution from marine sediments (Reeburgh and Heggie, 1977; Martens and Berner, 1977). In the Cariaco trench, as much as 85% of the total methane produced is oxidized in the anoxic water column (Reeburgh, 1976). Enrichments of anaerobic methane oxidizers have also been obtained from freshwater sediments (Panganiban *et al.,* 1979). A very small amount of methane may also be oxidized by the methanogenic bacteria themselves (Zehnder and Brock, 1979a, 1980). For these reasons, the atmospheric methane represents only a fraction of the total methane produced.

Interestingly, the major sources of atmospheric methane are terrestrial (Table I). Forests alone contribute almost one-half of the total biogenic methane released. These terrestrial systems are generally low flux systems and have not been studied in detail. Their significance comes from their enormous size. The high flux systems, which have been well studied, including paddy fields, marshes, and freshwater lakes, account for only 12% of the total methane released.

Marine environments contribute only a small portion of the atmospheric methane (Table I). Sulfate is abundant in seawater. In marine sediments, sulfate reduction predominates over methanogenesis from H_2 + CO_2 or acetate (Abram and Nedwell, 1978a,b; Lovley *et al.*, 1982; Mountfort and Asher, 1981; Oremland and Taylor, 1978; Sansone and Martens, 1981). Thus, when it occurs, methanogenesis is frequently limited to deeper sediments that are depleted in sulfate (Martens and Berner, 1974, 1977; Mountfort *et al.*, 1980; Hines and Buck, 1982) or less abundant substrates, such as methanol and trimethylamine (Oremland and Polcin, 1982; Senior *et al.*, 1982). Similarly, the addition of sulfate to freshwater sediments also inhibits methanogenesis (Cappenberg, 1974a,b, 1975; Cappenberg and Prins, 1974; Winfrey and Zeikus, 1977). The predominance of sulfate-reducing bacteria is probably explained by the more favorable thermodynamics for H_2 and acetate oxidation with sulfate as the electron acceptor than with carbon dioxide (McCarty, 1972; Thauer *et al.*, 1977; Schönheit *et al.*, 1982). Furthermore, some sulfate-reducing bacteria have a greater affinity for H_2 and acetate than do the methanogens (Kristjansson *et al.*, 1982; Schönheit *et al.*, 1982). Thus, the sulfate-reducing bacteria simply outcompete the methanogens in nutrient-limited environments.

In freshwater sediments, nitrate also inhibits methanogenesis (Balderston and Payne, 1976; Winfrey and Zeikus, 1979a). Like sulfate, the reduction of nitrate is thermodynamically more favorable than the reduction of carbon dioxide (Thauer *et al.*, 1977). In addition, intermediates of nitrate reduction, nitrite and nitrous oxide, inhibit the enzymes of methanogenesis (Gunsalus and Wolfe, 1978a). Therefore, nitrate reduction may block methanogenesis by substrate competition and direct inhibition.

In anaerobic environments where sulfate and nitrate are absent, methanogenesis is the terminal electron acceptor in the decomposition of organic matter. In anaerobic waste digestors, paddy soils, and freshwater sediments, about 70% of the methane is derived from acetate (Koyama, 1963; Jeris and McCarty, 1965; Smith and Mah, 1966; Mountfort and Asher, 1978; Winfrey and Zeikus, 1979b; Mackie and Bryant, 1981; King and Klug, 1982). Most of the acetate is in turn produced from the oxidation of propionate and other fatty acids by H_2-producing acetogenic bacteria (Bryant, 1979). Most of these acetogenic bacteria are probably obligate hydrogen producers (McInerny *et al.*, 1979, 1981a,b; Boone and Bryant, 1980), although facultative hydrogen producers are known to exist (Bryant *et al.*, 1977; McInerny and Bryant, 1981). Acetogenesis from H_2 + CO_2 may also play a role in certain systems (Mackie and Bryant, 1981). Hydrogen-producing acetogenic fermentations are only active at extremely low partial pressures of H_2 (Bryant *et al.*, 1967; Wolin, 1976). Methanogenesis from H_2 + CO_2 maintains the low partial pressure of H_2. Thus the reduction of CO_2 accounts for about 30% of the methane formed. This complex interaction between methanogens and the hydrogen-producing bacteria, called *interspecies hydrogen transfer,* demonstrates the intimate biochemical association of methanogens and eubacteria.

In the rumen, 5% or less of the total methane is derived from acetate (Oppermann *et al.*, 1961). Most of the methane is obtained from the reduction of carbon dioxide by H_2 or formate (Carroll and Hungate, 1955; Hungate, 1967; Hungate *et al.*, 1970). A small amount may also be derived from trimethylamine, which is formed from the degradation of choline (Patterson and Hespell, 1979). Acetate is a minor substrate because acetate and other fatty acids are absorbed directly by the animal. In addition, the short retention time of the rumen contents probably

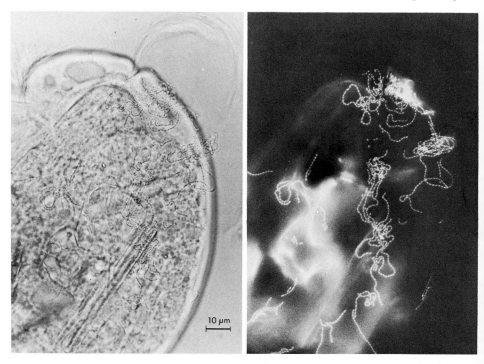

FIG. 1. Ectosymbiosis of a methanogen with the anaerobic rumen ciliate *Eudiplodinium maggi.* (Left) Bright field micrograph. (Right) Epifluorescence illumination of the identical field. Fluorescence microscopy was performed as Doddema and Vogels (1978). Photomicrographs by C. K. Stumm.

washes out the slow-growing obligate hydrogen-producing acetogenic bacteria before they can become established (Bryant, 1979). Nevertheless, interspecies hydrogen transfer between methanogens and rumen eubacteria is still significant. In co-cultures of the rumen eubacterium *Selenomonas ruminantium* with a methanogen, acetate is the most abundant fermentation product of glucose (Chen and Wolin, 1977). In the absence of the methanogen, lactate is the most abundant product. Thus, the presence of the methanogen causes a shift to more oxidized fermentation products.

Interspecies hydrogen transfer may also be the basis for a physical association between many rumen ciliates and sapropelic protozoa and methanogens (Vogels *et al.*, 1982; Stumm *et al.*, 1982; Bruggen *et al.*, 1983; Krumholz *et al.*, 1983). Methanogenic bacteria are fluorescent due to high levels of coenzyme F_{420} (Edwards and McBride, 1975; Mink and Dugan, 1977; Doddema and Vogels, 1978). Therefore, they can be tentatively identified directly by fluorescent microscopy as shown in Fig. 1 (Vogels *et al.*, 1980; Stumm *et al.*, 1982). The physical association of methanogens with eukaryotes demonstrates the intimate interrelationships of these archaebacteria with other life forms.

III. Taxonomy

In 1979, the taxonomy of the methanogenic bacteria was reorganized to reflect the phylogenetic relationships determined by 16 S rRNA sequence analysis (Balch *et al.*, 1979). Thus, the methanogens are the first major bacterial group with a largely phylogenetic or natural taxonomy. Hopefully. the methanogens will set a precedent for classification of other bacteria.

Based on the analysis of 16 strains, the methanogens were divided into three orders, four families, and seven genera (Table II). Taxa were determined by the following criteria: groups of organisms with an average S_{AB} of 0.22–0.28 (an index of 16 S rRNA homology) were placed in separate orders; S_{AB} of 0.34–0.37 in separate families; S_{AB} of 0.46–0.51 in separate genera; and S_{AB} of 0.55–0.65 in separate species (Fig. 2). When the S_{AB} was higher than 0.84, organisms were considered strains of the same species. A similar depth of divisions is encountered among the eubacteria. The central question that arises is to what extent do differences in 16 S rRNA sequences (S_{AB}) reflect phenotypic variations? Conversely, can the phenotypic properties predict the S_{AB}? At present there is no definitive answer. Although the taxonomy is consistent with a considerable amount of descriptive data, including cell-wall structure, morphology, lipid structure, mol % G + C, and nutrition (Balch *et al.*, 1979), much of the data are incomplete or the number of methanogens tested is too small to draw many conclusions. In addition, a number of new isolates have been described that appear to represent new genera. Thus, the phylogeny is known to be incomplete.

TABLE II

Taxonomy of Methanogenic Bacteria[a]

	Mol % G + C
Order I. Methanobacteriales	
Family I. Methanobacteriaceae	
Genus I. *Methanobacterium*	33–50
Genus II. *Methanobrevibacter*	27–32
Family II. Methanothermaceae	
Genus I. *Methanothermus*	33
Order II. Methanococcales	
Family I. Methanococcaceae	
Genus I. *Methanococcus*	31–40
Order III. Methanomicrobiales	
Family I. Methanomicrobiaceae	
Genus I. *Methanomicrobium*	45–49
Genus II. *Methanogenium*	51–61
Genus III. *Methanospirillum*	46–50
Family II. Methanosarcinaceae	
Genus I. *Methanosarcina*	41–43
Genus II. *Methanothrix*	52

[a] Based upon the classification of Balch *et al.* (1979) as amended by Stackebrandt *et al.* (1982) and Stetter *et al.* (1981).

With these qualifications in mind, a few generalizations can be made. All members of the same genus have a similar morphology. However, all genera do not have a different morphology. The mol % G + C of all strains of a species are within 2–4%. Similarly, the range within all genera (excluding thermophilic members) is within 10%. Pseudomurein is found only in the order Methanobacteriales. Protein cell walls are typical of all other orders. With one exception, the methanogenic substrates acetate, methanol, and methylamines are utilized only by members of the family Methanosarcinaceae. Hydrogen, formate, CO, and CO_2 are the only substrates of the remaining families. The distribution of polyamines also conforms to the classification of families (Scherer and Kneifel, 1983).

In addition, comparison of the immunology of formalin-fixed whole cells is a useful if indirect measure of the S_{AB} (Conway de Macario *et al.*, 1981, 1982a,b,c). All strains of a species either react with antisera prepared against another member of the species or produce antisera that react with other members of the species. The sole exception is strains of *Methanobrevibacter arboriphilus*. In addition, weak reactions are frequently observed between members of the same genus and occasionally between members of the same family. In no cases

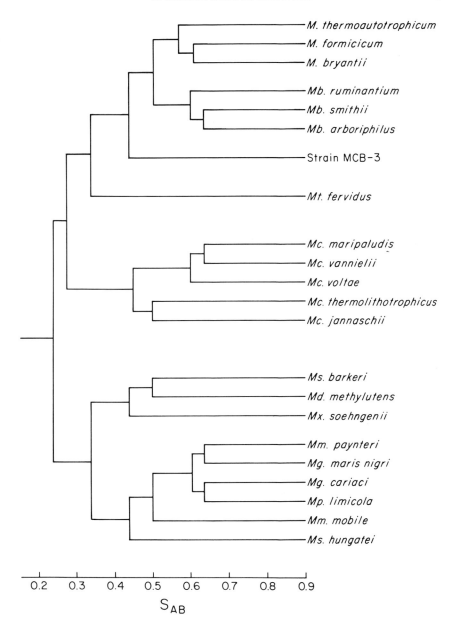

Fig. 2. Dendrogram of the relationships of methanogenic bacteria determined by sequence analysis of 16 S rRNA. *M.* is *Methanobacterium, Mb.* is *Methanobrevibacter, Mc.* is *Methanococcus, Md.* is *Methanococcoides, Mg.* is *Methanogenium, Mm.* is *Methanomicrobium, Mp.* is *Methanoplanus, Msp.* is *Methanospirillum, Ms.* is *Methanosarcina, Mt.* is *Methanothermus,* and *Mx.* is *Methanothrix.* Modified from Balch *et al.* (1979).

are cross-reactions found between members of different families. Thus, the immunology provides a rapid, if not conclusive, estimate of taxonomic position.

Other estimates have also been used. Hybridization of 16 S rRNA with DNA should be a rapid, independent measurement of rRNA homology (Tu *et al.*, 1982). Antisera prepared to the DNA-dependent RNA polymerase from *Methanobacterium thermoautotrophicum* only cross-reacts with extracts prepared from members of the Methanobacteriales (Stetter *et al.*, 1981). Note, however, that antisera prepared against the methyl reductase cross-reacts with extracts from members of several families but not with all members of any family (Ellefson and Wolfe, 1981). Therefore, this general technique must be used with caution. Variations in the quaternary structure of the ribosome and the acidity of ribosomal proteins are also consistent, the classes being defined by S_{AB}'s (Matheson and Yaguchi, 1982; Schmid and Böck, 1981, 1982).

At present, it is very difficult to assign the phylogenetic status of a new isolate without S_{AB} or immunological data. Immunological data are most useful only when the new isolate is closely related to a well-described organism. Thus, creation of new genera or higher taxa in the absence of S_{AB} data can only be considered tentative. No other taxonomic technique has been demonstrated to be as reliable. For the purposes of this chapter, new isolates are discussed with their apparent close relatives (Fig. 2). Without prejudice, the presumptive taxonomic status is noted while awaiting definitive S_{AB} characterization.

A. METHANOBACTERIALES

The Methanobacteriales contains all the methanogens with pseudomurein cell walls. C_{20} and C_{40} isopranyl glycerol ethers are abundant in all members tested. It contains two families. The first family, Methanobacteriaceae, contains two genera, *Methanobacterium* and *Methanobrevibacter*. The second family includes an extremely thermophilic isolate, *Methanothermus fervidus* (Stetter *et al.*, 1981). Creation of a second family was based largely on the failure of antisera prepared against *Methanobacterium thermoautotrophicum* DNA-dependent RNA polymerase to react with extracts of *Methanothermus fervidus* and DNA–RNA hybridization experiments (Tu *et al.*, 1982), but it has been confirmed by S_{AB} data (C. R. Woese, personal communication).

Members of the genus *Methanobacterium* are long, rod-shaped organisms that are nonmotile. All three species are amenable to mass culture and have been the subject of numerous biochemical and physiological studies.

Methanobacterium bryantii strain M.O.H. was isolated from the co-culture "*Methanobacillus omelianskii*" (Bryant *et al.*, 1967). Additional strains have also been obtained from deep water aquifers and freshwater sediments (Godsy, 1980; Ward and Frea, 1980). Morphologically, it is indistinguishable from *Methanobacterium formicicum* (Bryant, 1974), but it is unable to use formate as an electron donor. *Methanobacterium bryantii* is a mesophile, which grows at neutral pH. The mol % G + C of strain M.O.H. is 32.7 (Balch *et al.*, 1979). It will grow autotrophically, but acetate and B vitamins stimulate growth (Bryant *et al.*, 1971; Jarrell *et al.*, 1982).

Methanobacterium formicicum is common in anaerobic sewage digestors and freshwater sediments. Cells are shaped like slender rods ($0.6 \times 2-15$ μm) with blunt rounded ends and are nonmotile. Chains and filaments are common. It is the only member of the genus that will use both formate and hydrogen as electron donors. It is mesophilic and grows at neutral pH. The mol % G + C is 40.7. Acetate and cysteine stimulate growth (Bryant *et al.*, 1971; Bryant, 1974).

Methanobacterium thermoautotrophicum, a thermophile, can grow autotrophically using H_2 as the electron donor. It is shaped like a slender rod ($0.4 \times 3-7$ μm) and often forms filaments (Zeikus and Wolfe, 1972, 1973). Acetate is readily incorporated into cell carbon although it is not required for growth (Fuchs *et al.*, 1978a). The temperature optimum is 65°–70°C, and the pH optimum is 7.2–7.6. Two strains, ΔH and Marburg, are widely used in biochemical studies. Both were isolated from sewage sludge. Other strains have also been isolated from hot springs (Zeikus *et al.*, 1980). The Marburg and ΔH strains are fairly different, and their DNA shows DNA–DNA hybridization binding of only 46% (Brandis *et al.*, 1981). The major sugar in the cell wall of the ΔH strain is glucosamine; the major sugar in the Marburg strain is galactosamine. The sizes of some of the subunits of their DNA-dependent RNA polymerases are also different (Brandis *et al.*, 1981).

Members of the genus *Methanobrevibacter* are lancet-shaped cocci or short rods. They often form pairs or chains. All members are mesophilic and grow at neutral pH. Prior to the revised taxonomy of Balch *et al.* (1979), they were considered members of the genus *Methanobacterium*.

Methanobrevibacter arboriphilus was originally isolated from the wetwood of living trees (Zeikus and Henning, 1975). Additional strains have also been obtained from freshwater sediments and soil (Zehnder and Wuhrmann, 1977). It will grow autotrophically, but growth is greatly stimulated by 10% rumen fluid. Formate will not serve as an electron donor. The mol % G + C for the various strains is between 27.5 and 31.7.

Methanobrevibacter ruminantium is the predominant methanogen in the bovine rumen (Smith and Hungate, 1958). It is a short rod ($0.7 \times 0.8-1.8$ μm), often forms chains, and is nonmotile. Both formate and H_2 serve as electron donors. Strain M-1 requires coenzyme M (Taylor *et al.*, 1974), 2-methyl butyrate, acetate, and several amino acids for growth (Bryant *et al.*, 1971). Strain M-1 has also been used in a bioassay for coenzyme M (Balch and Wolfe, 1976). The temperature optimum for growth is 37°–43°C, and the pH range is 6–8. The mol % G + C is 30.6.

Methanobrevibacter smithii is morphologically similar to *M. ruminantium*. It is abundant in sewage sludge and is the most common isolate from $H_2 + CO_2$

enrichments of human fecal matter (Miller *et al.*, 1982; Miller and Wolin, 1982). Coenzyme M and 2-methyl butyrate are not required for growth. Acetate appears to be the major carbon source (Bryant *et al.*, 1971). The mol % G + C is 31 (Balch *et al.*, 1979).

Recently a new species has been described, which requires both methanol and H_2 as substrates for methanogenesis (Miller and Wolin, 1983). Analysis of its 16 S rRNA suggests that this isolate, strain MCB-3, is related to the Methanobacteriaceae (C. R. Woese, T. L. Miller, and W. J. Jones, personal communication). Thus, it is the sole isolate to date which uses methanol as a substrate for methanogenesis and is not a member of the Methanosarcinaceae. The assignment of strain MCB-3 to the Methanobacteriales is confirmed by the presence of pseudomurein in its cell wall (T. L. Miller, H. König, and M. J. Wolin, personal communication). Other properties of strain MCB-3, its coccoid morphology and inability to utilize H_2 plus CO_2 as substrates for methanogenesis (Miller and Wolin, 1983), make it unique among the Methanobacteriaceae.

Methanothermus fervidus is the only member of the family Methanothermaceae. It is an extreme thermophile isolated from Icelandic volcanic springs and mud holes (Stetter *et al.*, 1981). A nonmotile rod ($0.3-0.4 \times 1-3 \mu$m), it has a double-layered wall. Isolated cell wall sacculi also contain pseudomurein. Hydrogen and carbon dioxide, but not formate and acetate, are growth substrates. Yeast extract is required for growth. The pH optimum for growth is below 7, the temperature optimum is 83°C (range 65°–97°C), and the mol % G + C is 33.

B. METHANOCOCCALES

The order Methanococcales contains one family, Methanococcaceae, and one genus, *Methanococcus*. The six species are all irregular cocci. Their cell walls are probably proteinaceous, and cells lyse in dilute detergents (Jones *et al.*, 1977). C_{20} isopranyl glycerol ethers are abundant and C_{40} isopranyl ethers are absent in the species that have been tested. All members currently classified in this group were isolated from marine or coastal environments. However, their requirement for NaCl varies greatly. Most members use both H_2 and formate as electron donors and can grow autotrophically. In addition, most members have a nutritional requirement for selenium. *Methanococcus vannielii* has been mass cultured and biochemically characterized.

Methanococcus vannielii was isolated from a formate enrichment of black mud from the shore of San Francisco Bay (Stadtman and Barker, 1951). The original culture was subsequently lost, but viable cells were reisolated from a frozen cell paste (Jones and Stadtman, 1977). The average cell diameter is 3–4 μm. Cells are highly motile with polar tufts of flagella (Jones *et al.*, 1977).

M. vannielii grows well on an essentially mineral medium, although breakdown products of cysteine or pyruvate are stimulatory (Jones and Stadtman, 1976). Sodium chloride at 0.5 M completely inhibits growth. Selenium and tungsten are stimulatory to growth. The temperature optimum for growth is 37°–40°C. The pH optimum for growth on formate is 8.0, and the range is 7.4–9.2. Methane formation from $H_2 + CO_2$ is optimal at pH 6.5–7.5. The mol % G + C is 31.1 (Balch *et al.*, 1979).

Methanococcus voltae was isolated from sediment samples taken from Waccasassa estuary in Florida (Ward, 1970). It is a motile, irregular coccus with a diameter of 0.5–3.0 μm. Cells rapidly lyse in distilled water or dilute detergents. The pH optimum for growth with formate or H_2 is 6.0–7.0 (Whitman *et al.*, 1982). Isoleucine, leucine, and acetate are required for growth. Isovalerate and 2-methyl butyrate will substitute for the amino acids. Pantothenate, selenium, and cobalt are stimulatory. Optimal growth occurs at 0.4 M NaCl and 1 mM CaCl$_2$. The temperature optimum is 38°C. The mol % G + C is 30.7 (Balch *et al.*, 1979).

Methanococcus maripaludis is the most abundant H_2-utilizing methanogen in certain salt marsh and estuarine sediments (Jones and Paynter, 1980). It is a weakly motile, irregular coccus with a diameter of 1.2–1.6 μm. Cells rapidly lyse in distilled water or dilute detergents (Jones *et al.*, 1983). The pH optimum for growth is 6.8–7.2, and the temperature optimum is 38°C. Cells grow autotrophically, and acetate, amino acids, and vitamins are not stimulatory. NaCl is not required, but it is slightly stimulatory. Magnesium, 15 mM, is required for optimum growth. The mol % G + C is 33.

Methanococcus jannaschii was isolated from a hydrothermal vent on the East Pacific rise (Jones *et al.*, 1984). It is an extreme thermophile. A motile, irregular coccus, it has an osmotically sensitive cell wall. $H_2 + CO_2$, but not formate, are substrates for growth. At its temperature optimum of 86°C, the doubling time is 30 min. Thus, it is the fastest growing methanogen described. The pH optimum for growth is 6.5, the optimum concentration of NaCl is 0.34–0.51 M, the mol % G + C is 31. Organic supplements to media do not stimulate growth.

Methanococcus thermolithotrophicus was isolated from a geothermally heated marine sediment (Huber *et al.*, 1982). Its classification is based in part on DNA–RNA hybridization experiments (Tu *et al.*, 1982). The coccoid cells are 1.5 μm in diameter, possess a tuft of flagella, and lyse in 2% sodium dodecyl sulfate. The temperature optimum for growth is 65°C, the pH optimum is about 7.0, and the optimal concentration of NaCl is 0.68 M. The mol % G + C is 31.3. In a mineral medium, the doubling time is 55 min. Organic compounds do not stimulate growth.

Methanococcus deltae was isolated from sediment obtained near the mouth of the Mississippi River delta in the Gulf of Mexico (Corder *et al.*, 1983). It is an irregular-shaped coccus with a diameter of 1.0–1.5 μm. It is nonmotile, and no flagella are observed. The temperature optimum for growth is 37°C, the optimum concentration of NaCl is 0.6–0.7 M, and the mol % G + C is 40.5. Acetate and 20 mM magnesium stimulate growth. The classification of this species has not been confirmed by analysis of the 16 S rRNA.

C. METHANOMICROBIALES

The order Methanomicrobiales contains two families classified by 16 S rRNA sequence homologies. A third family has been proposed within this order on the basis of DNA–RNA hybridization experiments of a recent isolate, *Methanoplanus limicola*. The family Methanomicrobiaceae contains most of the members restricted to H$_2$ and formate as electron donors. This family includes rods, a spirillum, and irregular cocci. The cell walls are proteinaceous. The lipid content has been determined for only one species, where it was found to include both C$_{20}$ and C$_{40}$ isopranyl glycerol ethers. All members use both H$_2$ and formate as electron donors. They have been isolated from a variety of marine and freshwater sediments. The mol % G + C ranges from 45.0 to 61.2.

Methanospirillum hungatei is abundant in mesophilic sewage sludge. Cells are curved rods 0.5 × 7.4 μm and often form filaments several hundred microns in length. Cells have polar, tufted flagella and are sheathed. The cell wall composition contains 70% amino acids, 11% lipid, and 6.6% carbohydrate (Sprott and McKellar, 1980). The cytoplasmic membrane and cell sheath have been isolated and their compositions have been determined (Sprott *et al.*, 1983). At least four strains have been described (Ferry *et al.*, 1974; Patel *et al.*, 1976; Breuil and Patel, 1980a,b). Strain JF1 has a pH optimum of 6.6–7.4, a temperature optimum of 40°–45°C, and a mol % G + C of 49.5. Strain GP1 has a pH optimum of 7.0, a temperature optimum of 35°C, and a mol % G + C of 46.5. Both strains grow in mineral medium. Acetate is stimulatory to strain GP1. Casamino acids and B vitamins are stimulatory to strain JF1 (Ferry and Wolfe, 1977). Spheroplasts have been prepared from strain GP1 by treatment with dithiothreitol at pH 9.0 (Sprott *et al.*, 1979b).

The genus *Methanomicrobium* contains three species. Cells are typically short rods and have complex growth requirements. Species are either mesophilic or thermophilic. The mol % G + C of the mesophilic species readily distinguishes them from *Methanogenium* species. Otherwise these genera have similar growth and nutritional requirements.

Methanomicrobium mobile was isolated from the bovine rumen (Paynter and Hungate, 1968). It is a weakly motile rod (0.7 × 1.5 μm) with a single, polar flagellum. The temperature optimum is 40°C, the pH optimum is between 6.1 and 6.9, and the mol % G + C is 48.8 (Balch *et al.*, 1979). It has complex nutritional requirements, which include acetate, isobutyrate, 2-methyl butyrate, tryptophan, thiamin, pyridoxine, and *p*-aminobenzoate. In addition, an unidentified factor, which is abundant in extracts of methanogens and rumen fluid, is required for growth (Tanner and Wolfe, 1982).

Methanomicrobium paynteri is a nonmotile coccobacillus (0.6×1.5–2.5 μm) isolated from marine sediments obtained in the Cayman Islands (Rivard *et al.*, 1983). Cells lyse in 0.1% sodium dodecyl sulfate (SDS), deoxycholate, or Triton X-100. The temperature optimum for growth is 40°C, the pH optimum is 6.5–7.0, and the NaCl optimum is 0.1–0.2 *M*. The mol % G + C is 44.9. Hydrogen but not formate serves as an electron donor. Acetate is required for growth, but it is not a substrate for methanogenesis.

An unnamed thermophilic species has also been isolated from an anaerobic kelp digestor (Ferguson and Mah, 1983). Cells are very short rods or cocci, are 0.7–1.8 μm in diameter, and are nonmotile. Cells lyse in dilute SDS. The temperature optimum for growth is 60°C, the pH optimum is 7.2, and the mol % G + C is 60.0. Acetate and additional compounds in Trypticase or rumen fluid are required for growth. This organism may be a species of *Methanomicrobium* because cells cross-react weakly with antiserum prepared against *M. mobile*.

The genus *Methanogenium* contains two species in the original classification of Balch *et al.* (1979). Both species were isolated from marine sediments (Romesser *et al.*, 1979). They are irregular cocci, require NaCl for optimal growth, and have protein cell walls composed of regular subunits. A thermophilic marine isolate, *M. thermophilicum*, has been added to this genus based upon the similarity of its phenotypic properties and immunochemical evidence (Rivard and Smith, 1982). A mesophilic species, *M. olentangyi*, has also been assigned to this genus based on its mol % G + C and phenotypic properties, but in contradiction to immunological evidence (Corder *et al.*, 1983). In addition, 16 S rRNA partial sequencing data suggest that *Methanoplanus limicola* should probably be considered a species of *Methanogenium* (C. R. Woese, personal communication) and not a separate family as proposed (Tu *et al.*, 1982). Aside from its unusual morphology, the mol % G + C, nutrition, and growth characteristics of *M. limicola* are very similar to those of *Methanogenium*. Thus, until these contradictions are resolved, the classification of *M. limicola* is in doubt.

Methanogenium cariaci is an irregular coccus (2.6 μm in diameter) and is peritrichously flagellated. Cells readily lyse in dilute SDS. The temperature optimum for growth is 20°C, the pH optimum is 6.8–7.3, and the optimal concentration of NaCl is 0.46 *M*. Acetate and yeast extract are required for growth. The mol % G + C is 51.6.

Methanogenium marisnigri is an irregular coccus (1.3 μm in diameter) and is peritrichously flagellated. The temperature optimum for growth is 20°–25°C, the pH optimum is 6.2–6.6, and the optimal concentration of NaCl is 0.1 *M*. Trypticase is required for growth, but not acetate, yeast extract, or casamino acids. The mol % G + C is 61.2.

Methanogenium thermophilicus is a nonmotile, irregular coccus (1.0–1.3 μm) that was isolated from marine sediments near the effluent of a nuclear power station (Rivard and Smith, 1982). The temperature optimum for growth is 55°C, the pH optimum is 7.0, and the optimal concentration of NaCl is 0.2 *M*. Trypticase is required for growth. Cells cross-react with antisera prepared against *Methanogenium cariaci, Methanogenium marisnigri,* and *Methanomicrobium mobile.* The mol % G + C is 59.

Methanogenium olentangyi is included in this genus based entirely on its phenotypic properties (Corder *et al.,* 1983). Whole cells cross-react strongly with antiserum to *Methanobrevibacter smithii* strain ALI and not to antiserum from other members of the proposed genus. However, its mol % G + C of 54.4, sensitivity to detergents, and morphology clearly delineate this isolate from *Methanobrevibacter* and argue for classification within the genus *Methanogenium.* It is a nonmotile, irregular coccus (1.0–1.5 μm in diameter) and lyses in 0.01 SDS. The temperature optimum for growth is 37°C, and the optimal concentration of NaCl is about 0.2 *M*. Hydrogen but not formate is a substrate for methanogenesis. Acetate is required for growth. *Methanogenium olentangyi* was isolated from sediments of the Olentangyi River in Ohio.

The proposed family Methanoplanoceae contains only one species. Creation of this family is based on DNA–RNA hybridization experiments (Tu *et al.,* 1982), but S_{AB} data suggest that it may in fact be a species of *Methanogenium* or a closely related genus (C. R. Woese, personal communication). *Methanoplanus limicola* was isolated from a swamp fed by waste water from a drilling site (Wildgruber *et al.,* 1982). A plane-shaped bacterium, it has dimensions 1–3 × 1–2 × 0.2–0.3 μm. A polar tuft of flagella is present. Cells lyse in dilute SDS, and the cell wall contains at least one major glycoprotein. $H_2 + CO_2$ and formate are the only substrates for methanogenesis. The temperature optimum for growth is 40°C, the pH optimum is 6.5–7.5, and the optimal concentration of NaCl is 0.1 *M*. Acetate is probably a required carbon source. Yeast extract and peptone are stimulatory. The mol % G + C is 47.5.

The second family of the *Methanomicrobiales, Methanosarcinaceae,* contains one genus and one species in the original classification of Balch *et al.* (1979). At that time, all methanogens that used acetate, methanol, or methylamines belonged to this species. Most of the members of this species also used H_2 as an electron donor, and no member used formate. They also had a characteristic pseudosarcina morphology. Recently, several isolates have been described which use acetate or methylamines as substrates for methanogenesis and have a coccoid morphology. An intermediate form, *Methanosarcina mazei* (''*Methanococcus mazei*''), has a morphology similar to both the pseudosarcina and the cocci during different phases of its life cycle (Mah, 1980). The 16 S rRNA sequence analysis suggests that *M. mazei* should be considered a species of *Methanosarcina* (Mah and Kuhn, 1984). The 16 S rRNA sequence data of one of the coccoid isolates (*Methanococcoides methylutens*) also suggests that they are closely related to *Methanosarcina* (C. R. Woese, personal communication). Thus, these coccoid isolates will be discussed after *Methanosarcina barkeri*. A second genus, *Methanothrix*, has also been added to include an acetotrophic rod-shaped organism. This assignment has been confirmed by 16 S rRNA sequence analysis (Stackebrandt *et al.*, 1982).

Methanosarcina barkeri is cosmopolitan, and considerable diversity exists among the reported strains. The type strain, strain MS, was isolated from sewage sludge. Other strains have been isolated from freshwater mud and a thermophilic sludge digestor. The pseudosarcina form is an irregular coccus (1–2 μm in diameter), which forms large clumps of cells frequently visible to the unaided eye. Cells are surrounded by an amorphous outer layer composed of acid hetero-polysaccharide that contains galactosamine, neutral sugars, and uronic acids (Kandler and Hippe, 1977). The strain examined contained only C_{20} isopranyl glycerol ethers. The pH optimum for growth is about 7, and the temperature optimum for mesophilic strains is 37°–40°C. Contrary to earlier reports, the mol % G + C for a number of strains is 41–43 (R. A. Mah, personal communication). Nutritional requirements vary. Strains MS and 227 grow on essentially mineral media (Weimer and Zeikus, 1978a; Mah *et al.*, 1978; Hutten *et al.*, 1980). Growth of strain Fusaro is greatly stimulated by riboflavin (Scherer and Sahm, 1981a). Growth of the thermophilic strain TM-1, which has a temperature optimum near 50°C, is dependent on an unidentified factor in sewage sludge (Zinder and Mah, 1979). All strains except TM-1 use H_2 + CO_2 as substrates for methanogenesis in addition to acetate, methanol, and methylamines. Strain TM-1 uses H_2 + CO_2 only after a long period of adaptation. In addition, strains with gas vacuoles have been reported (Zhilina, 1971; Mah *et al.*, 1977a). *Methanosarcina mazei* forms cell clusters very similar to some strains of *M. barkeri*. Older cultures form cysts that rupture, releasing highly irregular cocci (Mah, 1980). The coccoid cells are 1–3 μm in diameter and nonmotile. Cells do not lyse in dilute SDS. Substrates for methanogenesis include acetate, methanol, and methylamines. H_2 + CO_2 is a poor substrate. The pH optimum for growth is 6–7, and the temperature optimum is 30°–40°C.

A recent isolate, *Methanococcoides methylutens*, is immunologically related to some strains of *Methanosarcina barkeri* and *Methanosarcina mazei*. It is a nonmotile, irregular coccus (1 μm in diameter) isolated from sediment below a kelp bed (Sowers and Ferry, 1983). Cells lyse readily in SDS. Trimethylamine, methylamine, and methanol are substrates for methanogenesis. H_2 + CO_2, formate, and acetate are not substrates. The temperature optimum for growth is 30°–35°C, the pH optimum is 7.0–7.5, and the optimal concentration of NaCl is 0.1–0.8 M. High concentrations of magnesium are also required, and 50 mM is optimal. In addition, yeast extract, Trypticase, rumen fluid, or B vitamins stimulate. The mol % G + C is 42. Analysis of the 16 S rRNA suggests that it is specifically related to *Methanosarcina barkeri* (C. R. Woese, personal communication).

Two similar isolates have been reported. First, a coccus isolated from the Great Salt Lake uses only methylamines, methanol, and methionine as substrates (Paterek and Smith, 1983). The optimal concentration of NaCl for growth is 1.0 *M*. Secondly, *Methanolobus tindarius* is an irregular coccus (0.8–1.25 μm in diameter), which was isolated from coastal sediments (König and Stetter, 1982). A single flagellum is present. Methanol and methylamines are substrates. The temperature optimum for growth is 25°C, the pH optimum is 6.5, and the optimal concentration of NaCl is 0.49 *M*. Vitamins stimulate growth. The mol % G + C is 45.9. Whether or not these isolates and *Methanococcoides methylutens* represent one or more novel species has not been determined. Nevertheless, the similar morphology and range of substrates suggest that they may be related. Considering the importance of acetate and methylamines as methanogenic substrates, the relationship of these isolates with the acetotrophic methanogens is of great interest.

Methanothrix soehngenii is common in sewage digestors of human and animal wastes. It is the only species of its genus. The rod-shaped cells, 0.8 × 2 μm, normally form long filaments and are sheathed (Zehnder *et al.*, 1980; Huser *et al.*, 1982). The outer layer of the cell wall consists of proteins. C_{20} isopranyl glycerol ethers are present. C_{40} isopranyl glycerol ethers are absent. Acetate is the sole substrate for methanogenesis and growth. Methane is not formed from H_2 + CO_2, formate, methanol, and methylamines. However, formate is oxidized. The temperature optimum for growth is 37°C, the pH optimum is 7.4–7.8, and the mol % G + C is 51.9. Growth is obtained on defined medium containing only acetate and vitamins, and other organic supplements have no effect.

IV. Ultrastructure

The complex and diverse ultrastructure of methanogens belies the idea that these are primitive bacteria. Most morphological features associated with eubacteria have also been found in methanogens. Cell-wall types vary from regular protein layers (Romesser *et al.*, 1979) to complex lamination (Zhilina, 1979) and are discussed in detail in a subsequent chapter. Additional ultrastructural features include flagella, fimbriae, intracytoplasmic membranes, cytoplasmic inclusions, cell spacers, and gas vacuoles.

Flagella have been observed in most methanogens. The exceptions include only the genera *Methanobacterium* and *Methanosarcina* (Doddema *et al.*, 1979a; Zhilina, 1971). Polar tufts of flagella are present in *Methanococcus* and *Meth-*

anoplanus (Jones *et al.*, 1977; Huber *et al.*, 1982; Wildgruber *et al.*, 1982). *Methanogenium* is peritrichously flagellated (Romesser *et al.*, 1979). *Methanobrevibacter, Methanolobus,* and *Methanomicrobium* contain a single polar flagellum (Doddema *et al.*, 1979a; König and Stetter, 1982; Paynter and Hungate, 1968). The dimensions of methanogen flagella vary in thickness from 8 to 20 nm in *Methanomicrobium mobile* and *Methanococcus thermolithotrophicus* and in length from 3 to 32 μm in *M. thermolithotrophicus* and *Methanoplanus limicola*. Although some flagella are unusually thin, they otherwise appear to be similar to eubacterial flagella.

Methanobacterium and *Methanogenium* have fimbriae (Doddema *et al.*, 1979a; Romesser *et al.*, 1979). Their diameter is 4–5 nm and typical of fimbriae in eubacteria. The fimbriae of *Methanobacterium thermoautotrophicum* are resistant to treatment with proteases, lipase, and nucleases. Fimbriae have been observed occasionally in *Methanococcus vannielii* (Jones *et al.*, 1977).

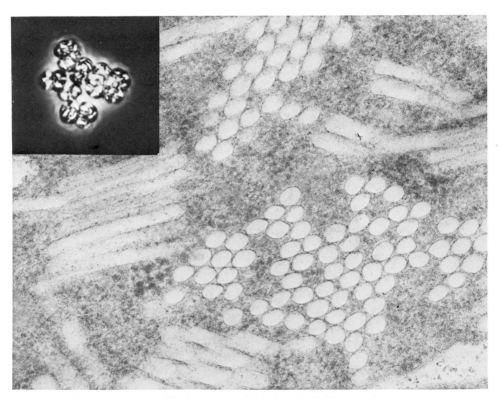

FIG. 3A.   Gas vacuoles in *Methanosarcina barkeri* strain W. Electron micrograph of a cross-section showing honeycomb and tubular appearance of gas vacuoles. Electron micrograph by J. Pangborn. (Inset) Phase contrast micrograph of whole cells by R. Mah.

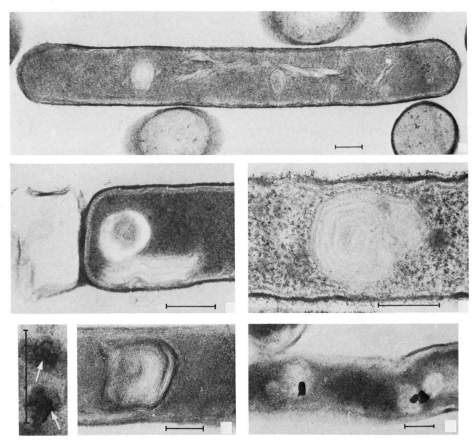

FIG. 3B. Intracytoplasmic membranes of *Methanobacterium thermoautotrophicum*. Electron micrographs of thin sections of cells fixed with glutaraldehyde and OsO_4-$K_2Cr_2O_7$. Markers represent 0.2 μm. (Upper) Thin section of whole cells fixed with glutaraldehyde at room temperature prior to postfixation. (Middle) Cells were fixed with OsO_4 (left) or prefixed with glutaraldehyde at 65°C (right) to show different intracytoplasmic membranes. (Bottom) Thin sections of cells after staining with BSPT[2-(2'-benzothiazolyl)-styryl-3-(4'-phthalhydrazidyl)tetrazolium chloride] to demonstrate hydrogenase association with intracytoplasmic membranes (left and middle). Thin sections after testing for ATPase activity (right). The reaction product is associated with intracytoplasmic membranes. From Doddema *et al.* (1979b).

Intracytoplasmic membranes have been frequently noted in the genus *Methanobacterium* (Langenberg *et al.*, 1968; Zeikus and Wolfe, 1973; Zeikus and Bowen, 1975a). They appear as invaginations of the cytoplasmic membrane and are stacked in concentric circles. In *Methanobacterium thermoautotrophicum*, the enzymes hydrogenase and ATPase appear to be localized on the intracytoplasmic membranes (Fig. 3; Doddema *et al.*, 1979b). Intracytoplasmic membranes are observed less frequently in other methanogens. In *Methanobrevibacter*, mesosome-like structures were occasionally seen associated with cell division (Zeikus and Bowen, 1975a) or cell-wall synthesis (Langenberg *et al.*, 1968). *Methanothermus fervidus* frequently contains concentric stacked membranes similar to *Methanobacterium* (Stetter *et al.*, 1981). Internal membranes have also been observed in *Methanosarcina barkeri*, *Methanogenium marisnigri*, and *Methanospirillum hungatei* (Zeikus and Bowen, 1975a,b; Romesser *et al.*, 1979).

Cytoplasmic inclusions are frequently observed in electron micrographs of methanogens. In most cases, their identity has not been determined. In the Methanosarcinaceae, inclusions were shown to be largely mineral and contained P, Ca, and Fe (Scherer and Bochem, 1983).

Complex spacers have been observed between cells in filaments of *Methanospirillum hungatei* (Zeikus and Bowen, 1975a,b; Sprott *et al.*, 1979b). Spacers are multilayered and about 0.2 μm in length. An amorphous, adhesive material appears to anchor to the wall of the filament. Gas vacuoles have been observed in some strains of *Methanosarcina barkeri* (Zhilina, 1971; Mah *et al.*, 1977a). These tubular structures have a honeycomb appearance in cross section when closely packed in the cytoplasm (Fig. 3). In liquid medium, they allow the dense cell clumps to float in the medium. In addition, a structure of hexagonally packed tubes of unknown function has also been observed in *Methanosarcina* (Archer and King, 1983).

V. Growth Conditions and Nutrition

With the description of greater than 20 species of methanogens, several generalities have emerged concerning their nutrition and growth requirements. First, all methanogens are obligatedly methanogenic. Although many organisms, particularly among the Methanosarcinaceae, may use more than one substrate, methane synthesis is required for growth. Contrary to early work with methanogenic enrichments, organic acids (other than formate and acetate) and alcohols (other than methanol) are not methanogenic substrates in pure cultures. The only substrates presently identified are carbon dioxide, carbon monoxide, hydrogen, formate, acetate, methanol, methylamine, dimethylamine, trimethylamine, and ethyldimethylamine.

Secondly, all methanogens are obligate anaerobes, and exposure to oxygen is lethal. However, the reported oxygen toxicity varies greatly (Paynter and Hungate, 1968; König and Stetter, 1982). For a species of *Methanosarcina*, the half-time for survival in air-equilibrated suspensions of growth medium was 4 min as determined by the decrease in colony-forming units (Zhilina, 1972). In studies with washed cells in nonreduced medium, cultures of *Methanobacterium thermoautotrophicum, Methanobrevibacter arboriphilus,* and *Methanosarcina barkeri* retained 100% viability for 30 hr upon exposure to oxygen (Kiener and Leisinger, 1983). In contrast, cultures of *Methanococcus voltae* and *Methanococcus vannielii* retained only 1% viability after 10 hr under similar conditions. Thus, the reported differences in oxygen toxicity appear to be due to both species differences and differences in culture conditions. The presence of superoxide dismutase in *Methanobacterium bryantii* (Kirby *et al.,* 1981) suggests that in nature at least some methanogens encounter and survive exposures to oxygen.

All methanogens isolated grow best near neutral pH. In pure cultures, the pH optima are generally between 6 and 8. However, methanogenesis in nature is not similarly restricted. A methanol enrichment from the Great Soda Lake in Nevada has a pH optimum of 9.7 (Oremland *et al.,* 1982). Methanogens are also abundant in bogs where the pH is close to 4.0 (King *et al.,* 1981; Risatti, 1978). Thus, alkali- and acid-tolerant methanogens may exist, but they have not been isolated.

Methanogens are abundant in mesophilic and thermophilic environments. The most extreme thermophilic species, *Methanococcus jannaschii,* has a growth optimum at 86°C (Jones *et al.,* 1984). The lowest temperature optimum reported is 20°C for *Methanogenium cariaci* (Romesser *et al.,* 1979). No psychrophilic methanogens have been reported.

A. CARBON SOURCES

The majority of species of methanogens in pure culture can grow autotrophically (Table III). Organic carbon sources do not stimulate growth of a few species, namely *Methanococcus thermolithotrophicus, Methanococcus maripaludis,* and *Methanobacterium thermoautotrophicum.* However, growth of

TABLE III

CARBON SOURCES AND VITAMIN REQUIREMENTS OF METHANOGENS

Organism	Capable of autotrophic growth[a]	Stimulatory or required carbon sources and vitamins
Methanobacterium bryantii	+	Acetate, B vitamins
Methanobacterium formicicum	+	Acetate, cysteine
Methanobacterium thermoautotrophicum	+	None
Methanobrevibacter arboriphilus	+	Rumen fluid
Methanobrevibacter ruminantium	−	Acetate, 2-methyl butyrate, amino acids, coenzyme M

TABLE III (*Continued*)

Organism	Capable of autotrophic growth[a]	Stimulatory or required carbon sources
Methanobrevibacter smithii	−	Acetate
Methanococcus vannielii	+	Breakdown products of pyruvate and cysteine
Methanococcus voltae	−	Acetate, 2-methyl butyrate (or isoleucine), isovalerate (or leucine), pantothenate
Methanococcus maripaludis	+	None
Methanococcus thermolithotrophicus	+	None
Methanococcus jannaschii	+	None
Methanococcus deltae	+	Acetate
Methanospirillum hungatei	+	Strain GP1:acetate; strain JF1:casamino acids, B vitamins
Methanomicrobium mobile	−	Acetate, 2-methyl butyrate, isobutyrate, tryptophan, thiamin, B-6, PABA, and unknown factor
Methanomicrobium paynteri	−	Acetate (yeast extract is stimulatory)
Methanogenium cariaci	−	Acetate, yeast extract
Methanogenium marisnigri	−	Trypticase
Methanogenium olentangyi	−	Acetate
Methanogenium thermophilicum	−	Trypticase
Methanosarcina barkeri	+	Strain 227:B vitamins, strain Fusaro:riboflavin, strain TM-1:unknown factor
Methanothrix soehngenii	Unknown	Acetate, vitamins
Methanolobus tindarius	+	Vitamins
Methanococcoides methylutens	+	Yeast extract, Trypticase, rumen fluid, B vitamins
Methanoplanus limicola	−	Acetate (yeast extract, peptones are stimulatory)
Methanothermus fervidus	−	Yeast extract

[a] Organisms that are capable of growth in mineral medium were considered autotrophs (+). Vitamin requirements are neglected. In cases where only a requirement for a complex additive has been demonstrated, i.e., Trypticase for *Methanogenium marisnigri*, the organism was considered incapable of autotrophic growth (−). This conclusion must be considered tentative until the nature of the nutrient in the complex additive is known. An absolute requirement for acetate was considered indicative of heterotrophic growth.

most is stimulated by acetate, amino acids, or vitamins. Acetate is also incorporated into cell carbon by two autotrophs, *M. thermoautotrophicum* and *Methanosarcina barkeri* (Fuchs *et al.*, 1978a; Weimer and Zeikus, 1978b). Except for the acetotrophic methanogens *M. barkeri* and *Methanothrix soehngenii*, very little carbon from acetate is found in methane (Taylor *et al.*, 1976).

Some methanogens are heterotrophic and require organic carbon sources for growth (Table III). Acetate is required by all these heterotrophs except *Methanogenium cariaci*, which requires Trypticase, and *Methanothermus fervidus*, which requires yeast extract. The nature of the nutrient in Trypticase and yeast extract or its role is not known (Romesser *et al.*, 1979). *Methanobrevibacter ruminantium* and *Methanococcus voltae* obtain 60 and 30%, respectively, of their cell carbon from acetate (Bryant *et al.*, 1971; Whitman *et al.*, 1982). Acetate is incorporated into protein, nucleic acid, and lipid.

In addition to acetate, branched-chain fatty acids are occasionally required. *Methanobrevibacter ruminantium*, *Methanococcus voltae*, and *Methanomicrobium mobile* require 2-methyl butyrate for growth (Bryant *et al.*, 1971; Tanner and Wolfe, 1982; Whitman *et al.*, 1982). Isovalerate and isobutyrate are required by *M. voltae* and *M. mobile*, respectively. *Methanococcus voltae* obtains 40% of its cell carbon from isoleucine and leucine, which substitute for 2-methyl butyrate and isovalerate. The carbon is used almost entirely for protein synthesis. Less than 30% of the cell carbon in *M. voltae* is derived from CO_2. Thus, it is likely that *M. voltae* and probably other methanogens are incapable of autotrophic CO_2 fixation. However, this point cannot be proven until the pathway of CO_2 fixation is known.

B. VITAMIN REQUIREMENTS

Vitamins are stimulatory or required for the growth of many methanogens. In most cases, the vitamin requirement is fulfilled by a trace vitamin solution containing 10 water-soluble vitamins (Balch *et al.*, 1979). The specific nature of the vitamin requirement has been determined in a few cases. Growth of *Methanosarcina barkeri* strain Fusaro is stimulated by riboflavin (Scherer and Sahm, 1981a). Growth of *Methanococcus voltae* is stimulated by pantothenate (Whitman *et al.*, 1982). *Methanomicrobium mobile* requires thiamin, pyridoxine, and *p*-aminobenzoate (Tanner and Wolfe, 1982).

A number of species require factors that are apparently unique to methanogens. *Methanobrevibacter ruminantium* strain M-1 requires coenzyme M for growth (Taylor *et al.*, 1974). *Methanomicrobium mobile* requires an unidentified substance abundant in rumen fluid or hot water extracts of methanogens (Paynter and Hungate, 1968; Tanner and Wolfe, 1982). *Methanosarcina barkeri* strain TM-1 requires an unidentified factor (or factors) extracted from digested sewage sludge (Zinder and Mah, 1979).

C. Nitrogen and Sulfur Sources

All methanogens use ammonia and sulfide as nitrogen and sulfur sources. Even in the presence of amino acids and peptides, ammonia is required for growth (Bryant et al., 1971; Zehnder and Wuhrmann, 1977). *Methanococcus voltae*, which requires amino acids as a carbon source, also requires ammonia for growth (Whitman et al., 1982). Thus, there is little ability to use organic nitrogen sources among methanogens.

Sulfide is the sulfur source for most methanogens (Bryant et al., 1971). However, some species can use cysteine (Bryant et al., 1971; Wellinger and Wuhrmann, 1977; Kenealy et al., 1982; Rivard et al., 1983) or methionine (Scherer and Sahm, 1981b). In addition, an unnamed thermophilic species from a kelp digestor grows in the virtual absence of sulfide in complex medium (Ferguson and Mah, 1983). Optimum total sulfide concentrations vary greatly from 0.3 to 3 mM (Rönnow and Gunnarsson, 1981; Scherer and Sahm, 1981b). In part, this must reflect differences in pH of the media and the distribution of sulfide between the aqueous forms (S^{2-} and HS^-) and volatile H_2S near neutral pH. Sulfide is also toxic to many methanogens (Cappenburg, 1975; Mountfort and Asher, 1979; Scherer and Sahm, 1981b). Addition of a redox resin, Serdoxit, or iron sulfide, may stimulate growth by poising sulfide at its optimal concentration (Scherer and Sahm, 1981b). Sulfate has been generally found not to be a sulfur source of methanogens (Bryant et al., 1971; Rönnow and Gunnerson, 1981; Whitman et al., 1982). In one study, low concentrations of sulfate, 0.2–0.4 mM, were found to be required or stimulatory for growth of some methanogens (Patel et al., 1978). Elemental sulfur is reduced to sulfide by many methanogens (Stetter and Gaag, 1983). In a survey of 12 diverse species, all had this capacity to some extent. However, its biological significance is not well understood.

D. Mineral Requirements

The metals nickel, cobalt, and iron are required by all methanogens that have been tested. These include at least one representative of each of the three orders of methanogens (Schönheit et al., 1979; Taylor and Pirt, 1977; Whitman et al., 1982; Scherer and Sahm, 1981a). Media concentrations of 1–5 μM generally satisfy the nickel requirement. Uptake is in the range of 17–180 μg/g cell dry weight (Diekert et al., 1981; Scherer et al., 1983). Nickel is a component of factor F_{430} (Whitman and Wolfe, 1980; Diekert et al., 1980a) as well as hydrogenase (Lancaster, 1982b; Jacobson et al., 1982; Graf and Thauer, 1981). Similar concentrations of cobalt are required and uptake is in the range of 10–120 μg/g cell dry weight (Whitman et al., 1982; Scherer et al., 1983). Cobalt is probably used largely for corrin biosynthesis because the level of corrinoids, 0.65–0.91 nmol/mg of cell dry weight, are equivalent to 38–54 μg Co/g cell dry weight (Krzycki and Zeikus, 1980; Scherer and Sahm, 1981a; Shapiro, 1982a; Pol *et al.*, 1982). Iron is required in higher levels, about 0.3–0.8 mM total iron in media (Patel *et al.*, 1978). However, the level of iron available for growth is probably lower because of the formation of insoluble FeS in most media (Taylor and Pirt, 1977; Patel *et al.*, 1978). Uptake is also high, 1–3 mg/g cell dry weight for *Methanococcus voltae* (Whitman *et al.*, 1982; Scherer *et al.*, 1983). These levels are within the range found for many eubacteria (G. E. Jones *et al.*, 1979). Presumably, these metal requirements reflect the importance of Ni, Co, and Fe in the biochemistry of methanogens.

Additional metal requirements have been found for some methanogens. The generality of these requirements is difficult to assess because they are often difficult to demonstrate in routine experiments. Molybdenum is stimulatory to growth of both *Methanobacterium thermoautotrophicum* and *Methanosarcina barkeri* (Schöheit *et al.*, 1979; Scherer and Sahm, 1981a) and accumulated by whole cells (Scherer *et al.*, 1983). Selenium is stimulatory to the members of the genus *Methanococcus* that have been tested (Jones and Stadtman, 1977; Whitman *et al.*, 1982; Jones *et al.*, 1983, 1984) and to *Methanosarcina barkeri* (Scherer and Sahm, 1981a). Tungsten is stimulatory to *Methanococcus vannielii* (Jones and Stadtman, 1977). Calcium is required for growth of *Methanococcus voltae* (Whitman *et al.*, 1982) and accumulated by whole cells (Scherer *et al.*, 1983). Especially high concentrations of magnesium are required by *Methanococcus maripaludis* (Jones *et al.*, 1983) and *Methanococcoides methylutens* strain TM-10 (Sowers and Ferry, 1983). Low concentrations of sodium are required for whole cell methanogenesis by *Methanobacterium thermoautotrophicum* (Perski *et al.*, 1981), and higher concentrations are inhibitory (Patel and Roth, 1977).

VI. Bioenergetics

With the isolation of essentially autotrophic organisms capable of growth with H_2 and CO_2, formate, or methanol, it became obvious that methanogens can obtain all their energy for growth from methanogenesis (Stadtman, 1967; Wolfe, 1971). In fact, there is a sufficient change in the standard free energy during methanogenesis from H_2 + CO_2 to synthesize ATP (reaction 1). In methanogenic habitats where the H_2 concentration is about 1 μM, the

$$4H_2 + CO_2 \rightarrow CH_4 + 2H_2O \qquad \Delta G^{\circ\prime} = -31.3 \text{ kcal/mol} \qquad (1)$$

free energy released is reduced to -15 kcal/mol (Thauer *et al.*, 1977). Thus, no more than one ATP could be synthesized per mole of methane.

The reduction of CO_2 to CH_4 can be considered a series of stepwise reductions (reactions 2–5). In the stepwise reduction, most of the free energy released is at the methyl level (reaction 5; Stadtman, 1967). Thus, methanol

$$HCO_3^- + H_2 \rightarrow HCOO^- + H_2O \qquad \Delta G^{\circ\prime} = +0.3 \text{ kcal/mol} \qquad (2)$$

$$HCOO^- + H^+ + H_2 \rightarrow CH_2O + H_2O \qquad \Delta G^{\circ\prime} = +4.0 \text{ kcal/mol} \qquad (3)$$

$$CH_2O + H_2 \rightarrow CH_3OH \qquad \Delta G^{\circ\prime} = -10.7 \text{ kcal/mol} \qquad (4)$$

$$CH_3OH + H_2 \rightarrow CH_4 + H_2O \qquad \Delta G^{\circ\prime} = -26.9 \text{ kcal/mol} \qquad (5)$$

and methylamines are readily reduced by some methanogens. Similarly, formate and hydrogen are nearly equivalent, and formate is also an electron donor (reaction 2).

In the case of the acetotrophic methanogens, the free energy released in the acetoclastic reaction (6) is fairly small (Thauer et al., 1977). The mechanism of coupling to ATP synthesis is not known. Because the magnitude

$$CH_3COO^- + H_2O \rightarrow HCO_3^- + CH_4 \qquad \Delta G^{\circ\prime} = -7.4 \text{ kcal/mol} \qquad (6)$$

of the free energy released seems to preclude substrate level phosphorylation, several authors have proposed likely chemiosmotic mechanisms (Anthony, 1982; Wolfe, 1979; Wolfe and Higgins, 1979; Zehnder and Brock, 1979b).

A. Growth Yields

In eubacteria, cellular growth yields are used to predict the bioenergetics of bacterial growth and the pathways of substrate utilization (Stouthamer and Bettenhaussen, 1973; Forrest and Walker, 1971). In methanogens, the pathways of ATP generation and biosynthesis of cellular components are not well understood. Thus, strict comparisons of growth yields with eubacterial work may be misleading. In general, the growth yields of methanogens have been less than expected (Stadtman, 1967). For growth on $H_2 + CO_2$ in batch cultures, growth yields are generally between 2 and 4 g cell dry weight/mol of methane (Taylor and Pirt, 1977; Roberton and Wolfe, 1970; Zehnder and Wuhrmann, 1977). However, for *Methanosarcina barkeri* a yield of 8.7 g dry weight/mol of methane has been reported (Smith and Mah, 1978). In one study with *Methanobacterium thermoautotrophicum* in mineral medium, the growth yield varied between 1.6 and 3.0 g dry weight/mol of methane depending on the concentration of H_2 and CO_2 (Schönheit et al., 1980). At high hydrogen concentrations, growth was uncoupled from methanogenesis. This effect may in part explain the low growth yields often reported. In addition, growth rates were one-half of maximal at 20% hydrogen and 11% carbon dioxide. These values probably overestimate the true K_S because they were not determined at infinite concentration of each substrate and the gases were not in equilibrium with the aqueous phase (Schönheit *et al.*, 1980). In contrast, growth and methane formation by *Methanobacterium formicicum* are tightly coupled throughout growth (Schauer and Ferry, 1980; Schauer *et al.*, 1982; Chau and Robinson, 1983). The cell yields with $H_2 + CO_2$ and formate are 3.5 and 4.7–6.9 g dry weight/mol of methane, respectively.

Growth yields on methanol and methylamines tend to be higher than growth on $H_2 + CO_2$. Yields of 4–6 g dry weight/mol of methane have been reported (Hippe *et al.*, 1979; Smith and Mah, 1978, 1980; Stadtman, 1967; Weimer and Zeikus, 1978a; Zinder and Mah, 1979). Yields during growth with acetate are reduced. For *Methanosarcina barkeri,* the yields are 2–3 g dry weight/mol of methane (Smith and Mah, 1978, 1980; Zinder and Mah, 1979). In addition, the growth rate is one-half of maximal at 5 mM acetate (Smith and Mah, 1978). For *Methanothrix soehngenii,* yields of 1.1–1.4 g dry weight/mol of methane were found (Huser *et al.*, 1982). The lower yields may reflect the smaller $-\Delta G^{o\prime}$ of the acetoclastic reaction (Smith and Mah, 1978) or the slower growth rate and consequently higher maintenance energy during acetotrophic growth.

Growth yields would be expected to depend somewhat on the source of cell carbon. In eubacteria, the theoretically maximum growth yields during heterotrophic and autotrophic growth (via the Calvin cycle) are about 28 and 5 g dry weight/mol of ATP (Forrest and Walker, 1971). Detailed studies on the effect of heterotrophy on the growth yields of methanogens have not been performed. Comparison of the growth yields of heterotrophic and autotrophic species would give considerable insight into the bioenergetics of methane synthesis and the pathway of CO_2 fixation.

B. ATP SYNTHESIS

In whole cells of *Methanobacterium bryantii,* the rate of methanogenesis is proportional to the intracellular concentration of ATP and inversely proportional to the concentration of AMP (Roberton and Wolfe, 1970). The uncouplers carbonyl cyanide *m*-chlorophenylhydrazone (CCCP), pentachlorophenol (PCP), and dinitrophenol (DNP), and the inhibitors air and chloroform cause an immediate drop in both the rate of methanogenesis and the intracellular concentration of ATP. Thus, methane and ATP synthesis appear to be closely coupled.

Three likely interpretations have been proposed. One, uncouplers inhibit methanogenesis directly. The effect on intracellular levels of ATP is either a consequence of the inhibition of methanogenesis or an additional direct effect of the uncoupler. In support of this hypothesis, methanogenesis in extracts is also inhibited by CCCP, PCP, and DNP even in the presence of exogenous ATP (Roberton and Wolfe, 1970). A variety of other chlorinated compounds are also potent inhibitors including: methane analogs like chloroform, chloral hydrate, dichlorodifluoromethane (Bauchop, 1967; Prins *et al.*, 1972; Wood and Wolfe, 1966b); chloramphenicol (Gunsalus and Wolfe, 1978a); and 1,1,1-trichloro-2,2-bis-*p*-chlorophenylethane or DDT (McBride and Wolfe, 1971a). In addition, the fluorinated uncoupler, carbonyl cyanide *p*-trifluoromethoxyphenylhydrazone (FCCP) is inhibitory in extracts only at extremely high concentrations, 0.5 mM (Whitman and Wolfe, 1983). Therefore, the inhibition by CCCP in extracts is probably due to direct, chemical inhibition of the methyl reductase. The second possibility is that uncouplers inhibit ATP synthesis. Because ATP is required for methanogenesis *in vitro* (M. J. Wolin *et al.*, 1963), methane synthesis is also inhibited. Furthermore, AMP, which accumulates as the ATP concentration declines, may also inhibit the methyl reductase (Mountfort, 1980). The third possibility is that a proton motive force (PMF) is required for both methane and ATP synthesis. Thus, uncouplers inhibit PMF formation and independently inhibit both ATP and methane synthesis (Sauer *et al.*, 1977, 1979). This last hypothesis was proposed to describe methanogenesis in a membrane fraction prepared from *Methanobrevibacter ruminantium*. In this preparation, methanogenesis was inhibited by CCCP and DNP as well as membrane permeable ions and detergents. Although net ATP synthesis was shown not to occur in the absence of uncouplers in these membranes, simultaneous ATP synthesis and degradation could have taken place. Thus, ATP would be available to activate methanogenesis. The failure of low concentrations of FCCP, membrane permeable ions, and detergents to inhibit methanogenesis in extracts of *Methanobacterium bryantii* indicates that a PMF is not required in the soluble methyl reductase system (Whitman and Wolfe, 1983).

Attempts to demonstrate ATP synthesis coupled to methanogenesis *in vitro* have been unsuccessful. However, indirect evidence suggests that ATP synthesis is coupled to a proton motive force generated during methanogenesis. Whole cells of *Methanobacterium thermoautotrophicum* synthesize ATP in response to a rapid drop in pH or the addition of valinomycin (Doddema *et al.*, 1978). In addition, ATP synthesis in the presence of valinomycin and a K$^+$ gradient is stimulated by low concentrations of Na$^+$ (Schönheit and Perski, 1983). However, 0.1 mM *N,N*-dicyclohexylcarbodiimide (DCCD) had no effect on whole cell ATP synthesis (Doddema *et al.*, 1978). Likewise, other methanogens with pseudomurien in their cell walls are also insensitive to DCCD (Sprott and Jarrell, 1982). However, growth of *Methanospirillum hungatei*, which has a protein cell wall, is inhibited by low concentrations of DCCD. Therefore, the insensitivity of *M. thermoautotrophicum* to DCCD may be because DCCD fails to penetrate the cell envelope.

In the presence of valinomycin, the uncouplers CCCP and DNP actually stimulate ATP synthesis in *M. thermoautotrophicum*. These apparently contradictory results may be reconciled if ATP synthesis occurs on intracytoplasmic membranes (Doddema *et al.*, 1979b). Using cytochemical techniques, the

ATPase and hydrogenase have been localized on intracytoplasmic membranes in *M. thermoautotrophicum* (Fig. 3; Doddema *et al.*, 1979b). Whole cells of *Methanosarcina barkeri* also synthesize ATP during a rapid drop in pH (Mountfort, 1978). Uncouplers and DCCD inhibit ATP synthesis at low concentrations. Intracytoplasmic membranes have been observed only rarely in *M. barkeri*. These results suggest that a PMF and an active ATPase are necessary for ATP synthesis.

Several measurements have been made of the PMF in whole cells. At neutral pH, the ΔpH is small in *Methanobacterium thermoautotrophicum, Methanospirillum hungatei,* and *Methanobacterium bryantii* (Sauer *et al.*, 1981; Jarrell and Sprott, 1981, 1983). As the external pH varies from about 5.6 to 7.8, the internal pH varies from 6.3 to 7.2 (Jarrell and Sprott, 1981). During methanogenesis, protons are taken up by *M. thermoautotrophicum,* and a ΔpH of 2.2 at an external pH of 8.5 has been reported (Sauer *et al.*, 1981). This measurement may be an overestimate because methylamine, the radiolabeled tracer used, may be metabolized by these cells (Jarrell and Sprott, 1981). Measurement of the electrical potential has varied greatly. Using thiocyanate as a tracer, no electrical potential was found in *M. thermoautotrophicum* (Sauer *et al.*, 1981). With either rubidium chloride and valinomycin or triphenylmethylphosphonium (TPMP) bromide, an electrical potential of about 130 mV was found (Jarrell and Sprott, 1981; Butsch and Bachofen, 1982). The electrical potential in whole cells of *M. hungatei* as measured with TPMP was about 80 mV (Jarrell and Sprott, 1981). Of special interest, low concentrations of acetylene inhibit growth and maintenance of a pH gradient in methanogens (Sprott *et al.*, 1982). Although the mechanism is unknown, it is specific for methane-producing bacteria in that neither *Halobacterium* nor representative eubacteria are inhibited by acetylene.

A membrane-bound ATPase is present in intracytoplasmic membranes of *M. thermoautotrophicum* (Doddema *et al.*, 1978, 1979b). In a washed membrane fraction, the pH optimum is 7.5–8.0, and the temperature optimum is 65°–70°C. Divalent cations are required for activity. Magnesium^{2+} and Mn^{2+} are the most effective. At an ATP:Mg^{2+} ratio of 1:2, the K_m for ATP is 2 mM. At high concentrations, GTP and UTP are also hydrolyzed well. DCCD, 0.1 mM, and trypsin inactivate the ATPase. Vesicles containing ATPase also contained hydrogenase activity (Doddema *et al.*, 1979b). ATP was synthesized in response to a potassium gradient or in the presence of H$_2$. Synthesis was inhibited by DNP, CCCP, or DCCD. Nigericin also inhibited the H$_2$-dependent ATP synthesis. Unlike whole cells, ATP was not synthesized upon a pH shift. Furthermore, the electron acceptor for the H$_2$-dependent ATP synthesis was not determined. Exogenous electron acceptors were not required, and methane was not formed. Of interest, ATP synthesis, ATP hydrolysis, and ADP uptake were inhibited by inhibitors of ADP/ATP translocase and an atractyloside binding protein was identified in membranes (Doddema *et al.*, 1980). An ADP/ATP translocase had not been previously identified in bacterial systems. Therefore, the ATPase was probably inside the membrane vesicles.

The suggestion that the site of ATP synthesis in *M. thermoautotrophicum* is an intracytoplasmic membrane (Doddema *et al.*, 1979b) is also consistent with the complex effects of uncouplers and inhibitors of methanogenesis on the PMF of whole cells (Sauer *et al.*, 1981). In addition, methane synthesis from a membrane fraction has been described (Sauer *et al.*, 1980a). Whether or not methanogenesis occurs on the same membranes as ATP synthesis has not been demonstrated. Studies on the deuterium isotope effect for whole cell methanogenesis also show that the hydrogenase is intracellular (Spencer *et al.*, 1980). However, no one has yet reported a membrane preparation capable of coupled ATP and methane synthesis. One difficulty may be the inability to hydrolyze the pseudomurein wall of some methanogens and prepare vesicles by gentle lysis. Thus, the definitive evidence on the role of the PMF in ATP synthesis is lacking. Furthermore, the generality of the results obtained with *M. thermoautotrophicum* is questionable, because intracytoplasmic membranes are not common in many methanogens (Sprott *et al.*, 1984).

C. Active Transport

Two active transport systems have been described in methanogens. *Methanobrevibacter ruminantium* strain M-1 takes up coenzyme M with an apparent K_m and V_{max} of 73 nM and 312 pmol/min/mg of dry weight, respectively (Balch and Wolfe, 1979b). Both H_2 and CO_2 are required. Air, bromethanesulfonate (a coenzyme M analog), azide, DNP, iodoacetate, and valinomycin inhibit. CCCP and DCCD have no effect. *Methanobacterium bryantii* accumulates Ni^{2+} against a concentration gradient of free Ni^{2+} of at least tenfold (Jarrell and Sprott, 1982). The K_m is 3.1 μM, and other cations except Co^{2+} are not inhibitory. The pH optimum is 4.9. H_2 and CO_2 stimulated uptake. Uptake could also be driven by an artificial pH gradient. Gramicidin D, nigericin, and monesin were potent inhibitors. Thus, nickel transport may be coupled to a proton gradient.

Methanogens accumulate significant amounts of K^+ and Mg^{2+} intracellularly (Sprott and Jarrell, 1981). In *Methanospirillum hungatei,* the intracellular concentrations are 140 mM for K^+ and 10–30 mM for Mg^{2+}. For *Methanobacterium thermoautotrophicum*, the concentration of K^+ is 600–800 mM, and the concentration of Mg^{2+} is 1–10 mM. *Methanobacterium thermoautotrophicum* also accumulates Na^+, although the intracellular concentration is much lower (4–30 mM). Although it is of interest, the transport of these metals has not been studied.

D. Cyclic 2,3-diphosphoglycerate

Cyclic 2,3-diphosphoglycerate was independently discovered in whole cells of *M. thermoautotrophicum* by two laboratories (Kanodia and Roberts, 1983; Seely and Fahrney, 1983a). The structure of this unique pyrophosphate compound is shown below (Kanodia and Roberts, 1983). The glycerate is the D-enantiomer (Seely and Fahrney, 1983b). Although this compound contains a high en-

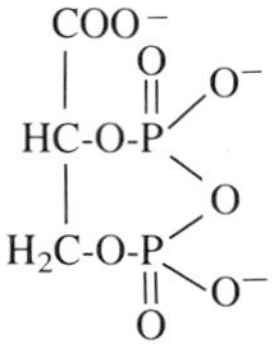

$$
\begin{array}{c}
\text{COO}^- \\
| \quad\quad \text{O} \\
| \quad\quad \| \quad /\text{O}^- \\
\text{HC-O-P} \\
| \quad\quad\quad \backslash \\
| \quad\quad\quad\quad \text{O} \\
\text{H}_2\text{C-O-P} \\
\quad\quad \|\backslash\text{O}^- \\
\quad\quad \text{O}
\end{array}
$$

ergy phosphate diester, its role in energy metabolism is not established. It is the most abundant form of intracellular phosphate during growth on phosphate sufficient medium. During phosphate starvation, the levels are at least ten-fold lower (Seely *et al.*, 1983). Other possible functions of the cyclic diphosphoglycerate may include phosphate storage, an allosteric effector, or metal chelation.

VII. Carbon Metabolism

The carbon metabolism of methanogens has only been characterized in a few organisms. Because of the diversity of heterotrophic and autotrophic species, only a single generalization can be made. It is that acetate is a major carbon source or major intermediate in most methanogens. Among the heterotrophic species, acetate is the most common carbon source. In the two autotrophs studied, acetyl-CoA or acetate is the first product of CO_2 fixation. Frequently, autotrophic species also use acetate as an alternative carbon source. Furthermore, acetate is abundant in many methanogenic environments. Therefore it is a readily available carbon source.

A. Autotrophy

The majority of methanogens isolated can grow autotrophically. Nevertheless, the pathway of autotrophic CO_2 fixation is not known. In short-term labeling studies, $^{14}CO_2$ rapidly labels amino acids and some coenzymes of methanogenesis (Daniels and Zeikus, 1978; Stupperich and Fuchs, 1981). Intermediates of the reductive pentose cycle, serine pathway, or hexulose phosphate pathways are not labeled. Ribulose-1, 5-bisphosphate carboxylase, the key enzyme in the reductive pentose cycle, hydroxypyruvate reductase, a key enzyme in the serine pathway, and hexulose phosphate synthetase, a key enzyme in the hexulose phosphate pathway, are undetectable in extracts of *M. thermoautotrophicum* (Daniels and Zeikus, 1978; Taylor *et al.*, 1976). In addition, intermediates of the acetogenic pathway found in some clostridia are not labeled by $^{14}CO_2$, and the levels of formyltetrahydrofolate synthetase and methylenetetrahydrofolate dehydrogenase, key enzymes of the acetogenic pathway, are too low to be involved in a major pathway of CO_2 reduction (Daniels and Zeikus, 1978; Ferry *et al.*, 1976). Thus, the acetogenic pathway is also absent. Because isocitrate dehydrogenase cannot be detected in *M. thermoautotrophicum*, the reductive tricarboxylic acid pathway is probably absent.

Instead, labeling and enzyme studies support the hypothesis that acetyl-CoA (or acetate) is formed from the condensation of two molecules of CO_2 by a novel pathway as shown in Fig. 4 (Fuchs and Stupperich, 1980). Evidence for this hypothesis is as follows. Alanine is rapidly labeled by $^{14}CO_2$ or [^{14}C]acetate. After labeling with $^{14}CO_2$ for 5 sec, about 25% of the radiolabel in the ethanol-soluble fraction of whole cells is found in alanine (Stupperich and Fuchs, 1981). This is a large amount considering that 90% of the CO_2 fixed eventually forms methane, and the intermediates of methanogenesis must also be labeled. Furthermore, CO_2-fixation by phosphoenolpyruvate carboxylase and α-ketoglutarate synthase tend to dilute radiolabel from the early intermediates of the pathway (Zeikus *et al.*, 1977; Fuchs and Stupperich, 1978, 1982). Degradation of the alanine formed within 2 sec showed that 61, 23, and 16% of the radioactivity was in the C-1, C-2, and C-3 positions, respectively (Stupperich and Fuchs, 1981). Incorporation of radiolabel into the C-2 and C-3 of alanine is consistent with formation of acetyl-CoA by condensation of two molecules of CO_2. The abundance of label in the C-1 position is due to the carboxylation of acetyl-CoA by pyruvate synthase (Fig. 4). Although *M. thermoautotrophicum* contains acetate thiokinase (acetyl-CoA synthetase), it probably functions during acetate assimilation and not CO_2 fixation (Oberlies *et al.*, 1980). Because acetate kinase was not detected, acetyl-CoA and not acetate may be the early product of CO_2 fixation. In contrast, *Methanosarcina barkeri* contains high levels of acetate kinase during autotrophic growth (Kenealy and Zeikus, 1982). Thus, in this organism, acetate has been proposed as an early product in CO_2 fixation.

Little is known about the mechanism of the initial step in CO_2 fixation. However, the labeling data are reminiscent of the initial results found with the acetogenic bacteria (Schulman *et al.*, 1973). In the acetogenic bacteria, the methyl group of acetate is derived from CO_2 via the folate pathway and a corrinoid protein. The carboxyl group is obtained from the C-1 of pyruvate or one carbon compounds. Methanogens lack high levels of the folate enzymes

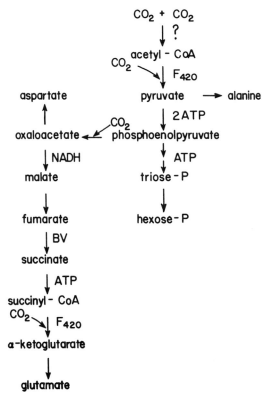

FIG. 4. Pathway of autotrophic CO_2 assimilation in *Methanobacterium thermoautotrophicum*. BV is benzyl viologen. F_{420} is coenzyme F_{420} (Fuchs and Stupperich, 1980).

necessary for synthesis of the methyl group of acetate, so an alternative mechanism must take place. An elegant solution may be that the methyl group of acetate is derived from an intermediate in methanogenesis. Because CO_2 activation appears to be coupled to the terminal step in methanogenesis, the removal of intermediates of methanogenesis for acetate synthesis requires that intermediates be restored by anaplerotic reactions involving heterotrophic carbon fixation.

Unfortunately, the evidence in support of this solution is not as elegant. During growth on $H_2 + CO_2$ and methanol, *Methanosarcina barkeri* preferentially incorporates radiolabel from methanol into the C-3 of alanine (Kenealy and Zeikus, 1982). The specific activity of the C-3 of alanine is nearly equal to the specific activity of the methane formed and is much higher than the specific activity of CO_2. Thus, the C-3 of alanine (methyl of acetate) appears to be derived from an intermediate of methanogenesis. However, when the CO_2 is labeled instead of methanol, the specific activity of the C-3 of alanine is 4–10 times higher than that of methane. The reason for this discrepancy cannot be readily explained. In short-term labeling experiments in the presence of H_2 and methanol, $^{14}CH_3OH$ was rapidly incorporated into acetate and coenzyme M (CoM) derivatives. Little methanol was oxidized to CO_2 (Kenealy and Zeikus, 1982). Thus, methanol is incorporated into acetate without prior conversion to CO_2. These results suggest that under some conditions methanol is a precursor of the methyl group of acetate. However, methanol may have been only partially oxidized. Therefore, the incorporation of radiolabel into acetate does not necessarily mean that the label must have arisen during the reduction of methanol, i.e., methanogenesis.

In acetogenic bacteria, a corrinoid protein and a methyltransferase are required in the terminal step of acetyl-CoA synthesis (Drake *et al.*, 1981; Hu *et al.*, 1982). In methanogens, a methyltransferase activity has been identified in *Methanobacterium bryantii* and *Methanosarcina barkeri* (Taylor and Wolfe, 1974b; Shapiro, 1982a; Meijden *et al.*, 1983a,b,c). This activity transfers the methyl group of methylcobalamin to HS-CoM. Its physiological role is not known, although it has been suggested to play a role in methanogenesis (Stadtman, 1967). In addition, extracts of *M. barkeri* methylate cobalamin(I) with methanol (Blaylock, 1968). This activity requires a low molecular weight compound that is probably identical to coenzyme M. Thus, the methyltransferase may be reversible. During growth on methanol, *M. barkeri* also contains an abundant corrinoid protein (Wood *et al.*, 1982). The corrinoid protein may be similar to the corrinoid protein of acetogens. However, this activity has not been determined. When methylated with methyl iodide, it is a substrate for methanogenesis. While this evidence suggests that CH_3—S—CoM can transfer methyl groups from methanogenesis to a corrinoid or corrinoid protein likely to function in acetate synthesis, there has been no direct demonstration that this in fact occurs.

Methanogens grow on carbon monoxide as the sole carbon and energy source, although the growth rate is 1% of that on $H_2 + CO_2$ (Daniels *et al.*, 1977; Kluyver and Schnellen, 1947). Carbon monoxide is a substrate of acetate synthesis by acetogenic bacteria (Hu *et al.*, 1982; Genther *et al.*, 1981). Although methanogens are the only bacteria to form methane from CO, they may have other pathways in common with CO-utilizing bacteria. In the presence of ^{14}CO, extracts of *M. barkeri* incorporate radiolabel into acetate either by net synthesis or an exchange reaction (Kenealy and Zeikus, 1982; Hu *et al.*, 1982; Stupperich *et al.*, 1983). In addition, extracts catalyze an exchange reaction between $[1-^{14}C]$pyruvate and CO_2 or CO (Kenealy and Zeikus, 1982). However, the rates are very low, 2.5 and 0.1 nmol/min/mg of protein, respectively. Furthermore, *Methanobacterium thermoautotrophicum* produces small amounts of CO during growth on $H_2 + CO_2$ (Conrad and Thauer, 1983). This result provides further evidence for the importance of CO in the metabolism of this bacterium.

Recently, acetyl-CoA synthesis has been demonstrated in detergent-treated cells of *M. thermoautotrophicum* (Stupperich and Fuchs, 1983). Although the rate of acetyl-CoA synthesis is very low, 0.067 nmol/mg protein/min, the cell-free system has many properties predicted from whole-cell experiments. Cyanide inhibits cell-free acetyl-CoA synthesis, and the inhibition is reversed by CO. This result provides direct evidence for the involvement of CO dehydrogenase, which behaves similarly (Daniels *et al.*, 1977). CO dehydrogenase is also required for the terminal reactions of acetyl-CoA synthesis in the acetogenic bacteria (Drake *et al.*, 1981). Cell-free acetyl-CoA synthesis is also dependent on methyl coenzyme M. However, the methyl group is apparently not incorporated into acetyl-CoA. This result suggests that an intermediate of methyl coenzyme M-dependent CO_2 reduction (section IX,D) and not methyl coenzyme per se is the methyl donor to acetyl-CoA.

If intermediates of methanogenesis are used for acetate synthesis, they must be replenished by anaplerotic reactions. There are a number of C-1 and C-3 compounds that are substrates for methanogenesis and could be synthesized by anaplerotic reactions. These include formaldehyde, serine, and pyruvate (Blaylock and Stadtman, 1966; Roberton and Wolfe, 1969; Romesser and Wolfe, 1982a). Pyruvate is especially intriguing because it is an early intermediate both in acetate assimilation by methanogens and acetate synthesis by the acetogenic bacteria. However, labeling studies with whole cells of *M. thermoautotrophicum* indicate that acetate (or acetyl-CoA), the presumed product of an anaplerotic reaction involving pyruvate, is not formed from pyruvate *in vivo* (Fuchs and Stupperich, 1980). The possibility that CO_2 is activated by a CH_3—S—CoM-independent as well as a CH_3—S—CoM-dependent mechanism can also not be discounted (Vogels *et al.*, 1982).

In conclusion, the first detectible intermediate in carbon assimilation is acetyl-CoA (or acetate). Some circumstantial evidence suggests that a modified acetogenic pathway similar to that found in other anaerobes may be involved. The major modification would have to be that the methyl group of acetate is derived from intermediates in methanogenesis rather than the folate pathway. However, there has been no definitive demonstration of a modified acetogenic or any other known pathway of autotrophic CO_2 fixation in methanogens.

B. Incomplete Tricarboxylic Acid Cycle

Like many autotrophic eubacteria, *Methanobacterium thermoautotrophicum* and *Methanosarcina barkeri* use incomplete TCA cycles to form intermediates for amino acid and tetrapyrrole biosynthesis. However, the cycle is different in each organism. The pathway in *M. thermoautotrophicum* is included in Fig. 4. Oxaloacetate is formed from acetyl-CoA by a coenzyme F_{420}-dependent pyruvate synthase, phosphoenolpyruvate synthase, and phosphoenolpyruvate carboxylase (Fuchs and Stupperich, 1982). The phosphoenolpyruvate synthase has been

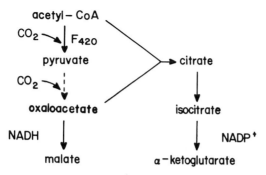

Fig. 5. Incomplete TCA cycle in *Methanosarcina barkeri*. F_{420} is coenzyme F_{420} (Weimer and Zeikus, 1979).

purified 50-fold. *In vivo*, it probably functions only in the direction of phosphoenolpyruvate synthesis because it has a high K_m for phosphoenolpyruvate and is inhibited by AMP, ADP, and α-ketoglutarate (Eyzaguirre *et al.*, 1982). Oxaloacetate is reduced to succinyl-CoA and carboxylated by a coenzyme F_{420}-dependent α-ketoglutarate synthase (Zeikus *et al.*, 1977; Fuchs *et al.*, 1978b). Extracts contain sufficient levels of all these enzymes, and radiolabeling data is consistent with this pathway (Fuchs and Stupperich, 1982).

In *M. barkeri*, α-ketoglutarate is synthesized via isocitrate (Fig. 5). Sufficient levels of pyruvate synthase, aconitase (aconitate hydratase), isocitrate dehydrogenase, and malate dehydrogenase are present in extracts (Weimer and Zeikus, 1979). However, the levels of citrate synthase that have been found are very low, and the enzymes responsible for the conversion of pyruvate to oxaloacetate have not been determined. In addition the following enzymes were not detected in extracts: α-ketoglutarate dehydrogenase, α-ketoglutarate synthase, fumarase (fumarate hydratase), fumarate reductase, phosphoenolpyruvate carboxylase, malate synthase, and citrate lyase.

A malate dehydrogenase has also been purified from *Methanospirillum hungatei* (Sprott *et al.*, 1979a). Because of an unusually low affinity for NAD^+, this enzyme functions only in the direction of oxaloacetate reduction (Storer *et al.*, 1981). This property is consistent with the biosynthetic role of malate dehydrogenase in both *M. barkeri* and *M. thermoautotrophicum*.

C. OTHER PATHWAYS

Radiolabeling studies of whole cells of *M. thermoautotrophicum* with [^{14}C]pyruvate and acetate are consistent with the synthesis of hexoses via pyruvate and the synthesis of pentoses by reactions involving transketolase and transaldolase (Fuchs and Stupperich, 1980; Jansen *et al.*, 1982; Fuchs *et al.*, 1983).

In addition, the following enzymes have been identified in extracts of *M. thermoautotrophicum:* enolase, phosphoglyceromutase, phosphoglycerate kinase, glyceraldehyde-phosphate dehydrogenase, fructose-bisphosphate aldolase, fructose-bisphosphate phosphatase, and triosephosphate isomerase. Glyceraldehyde-phosphate dehydrogenase and phosphoglycerate kinase have also been observed in extracts of *M. barkeri* (Weimer and Zeikus, 1979). Thus, some of the enzymes of gluconeogenesis are present in methanogens.

Similarly, the distribution of ^{13}C in sugars following long-term labeling of *M. hungatei* with [^{13}C]acetate and $^{13}CO_2$ is consistent with the biosynthesis of hexoses from pyruvate (Ekiel *et al.,* 1983). Pentoses arise by the oxidative decarboxylation of hexoses. In the same study, the labeling of nucleosides, lipids, and amino acids were also investigated. The labeling of pyrimidines was consistent with biosynthesis from aspartate. The labeling of the C-2 and C-8 of purines was consistent with biosynthesis from the C-1 donor serine. The labeling of the isoprenoid lipids was consistent with biosynthesis from mevalonate which was formed by a condensation of three acetates. The labeling of most of the amino acids was consistent with expected pathways. Alanine, serine, and aspartate could be biosynthesized from pyruvate. Glycine could be biosynthesized from serine. Threonine and methionine could be biosynthesized from aspartate. Lysine could be biosynthesized from pyruvate and aspartate by the diaminopimelate pathway. Arginine and proline could be biosynthesized from glutamate. Phenylalanine and tyrosine could be biosynthesized from chorismate. Leucine and valine could be biosynthesized from α-acetolactate. Only for isoleucine were the labeling patterns inconsistent with the expected pathway of biosynthesis. Instead, isoleucine appears to be biosynthesized from pyruvate via citramalate and not from threonine. These results have been confirmed in *M. thermoautotrophicum* (Eikmanns *et al.,* 1983a,b). In whole cells, [^{14}C]pyruvate, propionate, and succinate are readily incorporated into isoleucine, and labeled threonine was not incorporated.

Methanococcus voltae obtains 40% of its cell carbon from isoleucine and leucine (Whitman *et al.,* 1982). Radiolabeling studies indicate that carbon from the amino acids is not mixed with carbon from acetate. This is surprising because in eubacteria acetyl-CoA is a product of leucine and isoleucine degradation. Therefore, methanogens may use a novel mechanism for assimilation and degradation of these amino acids.

Incorporation of ^{13}C substrates and [^{13}C]NMR of the tetrapyrrole Factor F_{430} indicate that *M. thermoautotrophicum* has a typical pathway of porphobilinogen biosynthesis (Pfaltz *et al.,* 1982). In whole cells, factor F_{430} is labeled by [^{14}C]succinate and δ-aminolevulinic acid (Diekert *et al.,* 1980c). δ-Aminolevulinic acid is probably synthesized via the C-5 pathway, and δ-aminolevulinic acid synthase is not found in extracts of *M. thermoautotrophicum* (Gilles *et al.,* 1983). Factor F_{430} biosynthesis is inhibited by levulinic acid

(Jaenchen *et al.*, 1981b). Thus, δ-aminolevulinic acid dehydratase (porphobilinogen synthase) is in the pathway. Uroporphyrinogen III accumulates during growth in nickel-free medium, and it is quantitatively converted to factor F_{430} upon the addition of nickel (Gilles and Thauer, 1983). Thus, uroporphyrinogen III is an intermediate. Methionine is also a methyl donor for factor F_{430} biosynthesis (Jaenchen *et al.*, 1981a).

Methanobacterium thermoautotrophicum incorporates ribonucleosides into DNA after reduction of the ribose (Sprengel and Follmann, 1981). However, an adenosylcobalamin-dependent ribonucleotide reductase could not be demonstrated in cell-free extracts. In addition, *M. thermoautotrophicum* excretes alanine into the medium during growth (Schönheit and Thauer, 1980). Thus, cross-bridge formation in pseudomurein may occur by a mechanism similar to that found in peptidoglycan synthesis.

VIII. Unique Coenzymes

A number of coenzymes have been discovered in methanogens that appear to be unique to these organisms. Most of the coenzymes are involved in methanogenesis, the only biochemical pathway investigated in detail in these organisms. The structure and function of many of these coenzymes are not well understood. Their abundance and the abundance of some typical vitamins in methanogens is shown in Table IV. The levels of the typical vitamins were determined in extracts of *Methanobacterium thermoautotrophicum* and *Methanococcus voltae* (Leigh, 1983a). Both methanogens contained levels of folic acid, pantothenate, and thiamin, which were one to two orders of magnitude lower than for nonmethanogens. The folic acid detected may have been a hydrolysis product of methanopterin. In *M. thermoautotrophicum*, the levels of biotin and nicotinic acid were one order of magnitude lower than for nonmethanogens. Only the levels of riboflavin and pyridoxine (and nicotinic acid in *M. voltae*) were as high as found in nonmethanogens. Consistent with this result, FAD is present in extracts of *Methanobacterium bryantii*, *Methanospirillum hungatei*, and *M. thermoautotrophicum* (Lancaster, 1981; McKellar *et al.*, 1981; Jacobson *et al.*, 1982). Not surprising, many of the coenzymes unique to methanogens are present in greater abundance than the typical vitamins. Presumably this fact reflects the importance of the coenzymes in metabolism.

A. COENZYME M

Methane formation from methylcobalamin by extracts of *Methanobacterium bryantii* and *Methanosarcina barkeri* requires a heat-stable co-factor called co-

TABLE IV

CONCENTRATIONS OF COENZYMES AND VITAMINS IN
METHANOGENS[a]

Compound	Level in whole cells (μmol/g dry weight)
Coenzyme M[b]	0.3–16
Coenzyme F_{420}[c]	0.7–2.7
Methanopterin[d]	0.7–1.3
Factor F_{430}	0.2–0.8
Component B	
Methanofuran	>0.3
Corrinoids	0.1–4.0
α-Tocopherolquinone[e]	0.010–0.014
Nicotinic acid	0.07–2.0
Pantothenate	0.02
Riboflavin	0.07–0.09
Thiamin	7×10^{-4}
Folate[f]	$<4 \times 10^{-5}$

[a] Values were obtained as described in the text for novel coenzymes from the stated sources. Levels for the common vitamins
were obtained from Leigh (1983a). The dry weight was assumed to
be equal to 0.20 × wet weight.
[b] Includes reduced, oxidized, methyl, disulfide forms, and unidentified forms.
[c] Based on a formula weight of 740.
[d] Based on a formula weight of about 700.
[e] Includes quinol form.
[f] May be an overestimate due to interference by methanopterin.

enzyme M (McBride and Wolfe, 1971b; Blaylock, 1968). The structure of
coenzyme M is shown in Fig. 6 and has been confirmed by organic synthesis
(Taylor and Wolfe, 1974a). The reduced coenzyme is enzymatically methylated
by methylcobalamin and extracts of *M. bryantii* (Taylor and Wolfe, 1974b) and
by methanol and extracts of *M. barkeri* (Shapiro and Wolfe, 1980; Meijden *et
al.*, 1983a,b,c). Methyl coenzyme M is demethylated by the methyl reductase
system (Taylor and Wolfe, 1974a; Gunsalus and Wolfe, 1980) with the formation of methane. In addition, (S—CoM)$_2$ is enzymatically reduced by NADPH
and extracts of *M. bryantii* (Taylor and Wolfe, 1974b).

A survey of 14 strains of methanogens representing all the major taxonomic
groups showed that coenzyme M was abundant in all methanogens (Balch and
Wolfe, 1979a). Whole cell concentrations range from 0.3–16 nmol/mg dry
weight. In addition, coenzyme M is essential for growth of *Methanobrevibacter
ruminantium* strain M-1, which has been used as a bioassay (Taylor *et al.*, 1974;

$$HS-CH_2-CH_2-SO_3^-$$ HS-CoM

$$^-O_3S-CH_2-CH_2-S-S-CH_2-CH_2-SO_3^-$$ $(S-CoM)_2$

$$CH_3-S-CH_2-CH_2-SO_3^-$$ $CH_3-S-CoM$

$$HOCH_2-S-CH_2-CH_2-SO_3^-$$ $HOCH_2-S-CoM$

FIG. 6. Structures of coenzyme M, HS-CoM; oxidized coenzyme M, $(S-CoM)_2$; methyl coenzyme M, CH_3-S-CoM; and hydroxycoenzyme M, $HOCH_2$-S-CoM (Taylor and Wolfe, 1974a; Romesser ard Wolfe, 1981).

Balch and Wolfe, 1976). Using ^{35}S-labeled coenzyme, the relative abundance of the different forms in whole cells of *M. ruminantium* was determined (Balch and Wolfe, 1979b). When incubated with limiting concentrations of $H^{35}S$—CoM, CH_3—S—CoM, HS—CoM, and $(S$—$CoM)_2$ accounted for 17, 12, and 9%, respectively, of the total intracellular coenzyme. The majority of the intracellular coenzyme M (62%), failed to migrate in the chromatography systems used without treatment by dithiothreitol. After dithiothreitol treatment two unidentified coenzyme M-containing compounds were formed. The biological significance of these compounds is not known. Interestingly, coenzyme M may be a component of factor F_{430}, the chromophore associated with the methyl reductase system (Keltjens *et al.*, 1982). Thus, the actual intracellular nature of coenzyme M is not fully known.

Reduced coenzyme M contains a sulfhydryl group and can participate in a number of chemical reactions. The redox potential of the thiol is pH dependent below pH 9.35 and becomes more positive as the pH decreases (Kell and Morris, 1979). At pH 7 and 18°C, the E_o' is -193 mV. In addition, reduced coenzyme M forms a number of adducts with formaldehyde (Romesser and Wolfe, 1981). The sodium salt forms 2-(hydroxymethylthio)ethane sulfonate or $HOCH_2$—S—CoM. The ammonium salt forms iminobis[2-(methylthio)ethane sulfonate]or NH=$(CH_2$—S—$CoM)_2$. In water, NH=$(CH_2$—S—$CoM)_2$ decomposes to NH_4 and 2 moles of $HOCH_2$—S—CoM. $HOCH_2$—S—CoM is a substrate of methanogenesis in extracts of *M. thermoautotrophicum* (Romesser and Wolfe, 1981; Escalante-Semerena and Wolfe, 1984). Like other thiols, reduced coenzyme M is chemically methylated by methylcobalamin at neutral and alkaline pH (Schrauzer *et al.*, 1978). In this reaction, the methyl acceptor may be the free radical ·S-CoM (Frick *et al.*, 1976). HS-CoM reacts with methylcobalamin and methylcobaloximes to produce methane at significant rates

(Schrauzer *et al.*, 1978). In addition, HS-CoM reacts with methyl iodide, which forms the basis for the chemical synthesis of methyl coenzyme M (Romesser and Balch, 1980). A rapid method for the identification and quantification of minute amounts of coenzyme M by isotachophoresis has been reported (Hermans *et al.*, 1980). Coenzyme M derivatives may also be identified or purified by high-performance liquid chromatography (Apostolides *et al.*, 1982).

B. Coenzyme F_{420}

Coenzyme F_{420} was discovered in extracts of *Methanobacterium bryantii* (Cheeseman *et al.*, 1972) and has been shown to be ubiquitous among methanogens (Eirich *et al.*, 1979). Most methanogens tested contained 0.5–2.0 mg coenzyme F_{420}/g dry weight of cells (Eirich *et al.*, 1979; van Beelen *et al.*, 1983b). A two electron carrier, its physiological function is similar to the role of ferredoxin in many nonmethanogenic anaerobes (Tzeng *et al.*, 1975a). Coenzyme F_{420}-dependent enzymes include hydrogenase (Tzeng *et al.*, 1975a; Jacobson *et al.*, 1982; Yamazaki, 1982), formate dehydrogenase (Tzeng *et al.*, 1975b; Jones and Stadtman, 1980, 1981), NADP$^+$ reductase (Tzeng *et al.*, 1975a,b; Jones and Stadtman, 1980; Yamazaki and Tsai, 1980), pyruvate synthase and α-ketoglutarate synthase (Zeikus *et al.*, 1977), and carbon monoxide dehydrogenase (Daniels *et al.*, 1977). Coenzyme F_{420} also supports methyl coenzyme M reduction in a partially purified enzyme system (Ellefson and Wolfe, 1980).

The structure of coenzyme F_{420} from *M. thermoautotrophicum* is shown in Fig. 7 (Eirich *et al.*, 1978). The structure has been confirmed by synthesis of the chromophore, 7,8-didemethyl-8-hydroxy-5-deazaflavin (Ashton *et al.*, 1979; Ashton and Brown, 1980) and by synthesis of deazaflavin analogs (Grauert, 1980; Pol *et al.*, 1980). The coenzyme from *Methanosarcina barkeri* is similar except the side chain contains four or five glutamates instead of two (van Beelen *et al.*, 1983b). The oxidized coenzyme absorbs light at 420 nm in neutral and alkaline buffers and fluoresces (Eirich *et al.*, 1978). Upon reduction, the coenzyme is colorless and very weakly fluorescent (Cheeseman *et al.*, 1972; Eirich *et al.*, 1979). The redox potential is $E_o' = -0.373$ V, consistent with its biological function (Eirich *et al.*, 1978). Reduction occurs by a direct hydride process, and the coenzyme is readily radiolabeled by reaction with [4-^{3}H]NADPH in the presence of purified NADP$^+$ reductase from *Methanococcus vannielii* (Fig. 8; Yamazaki *et al.*, 1980). Photolysis is rapid in the presence of oxygen (Cheeseman *et al.*, 1972). The decomposition is probably dependent on the presence of the side chain because the acid-stable chromophore [presumably (FO)] is uneffected. The products of the decomposition have not been identified. In addition, a light-independent, oxygen-dependent decomposi-

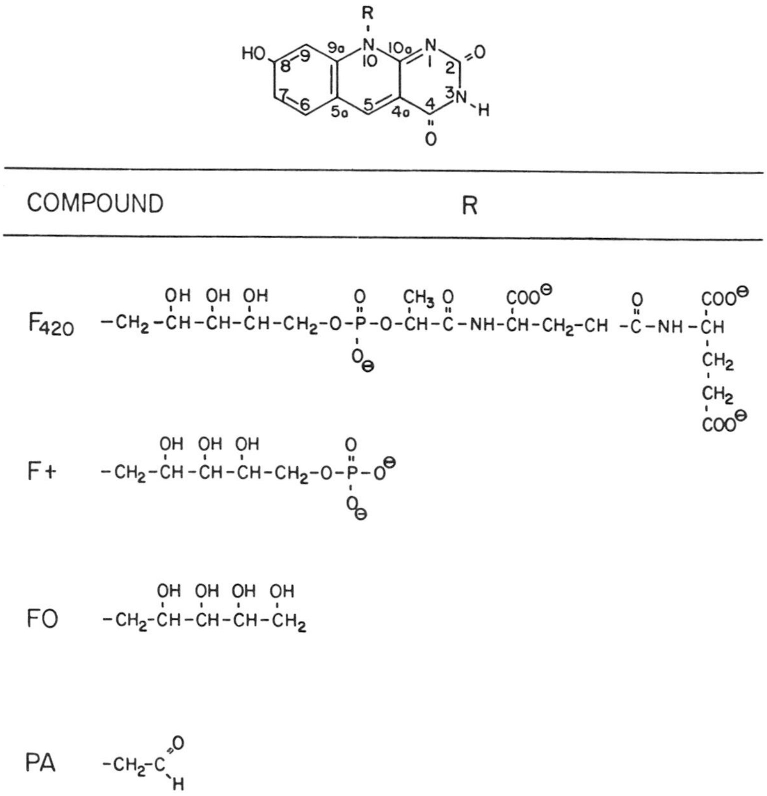

COMPOUND	R

F_{420}

$F+$

FO

PA

FIG. 7. Structure of coenzyme F_{420} and its degradation products formed by acid hydrolysis. From Eirich *et al.* (1979).

tion of coenzyme F_{420} in intact cells has been reported (Schönheit *et al.*, 1981). Coenzyme F_{420} is also susceptible to hydrolysis by 1 *N* HCl at 110°C (Eirich *et al.*, 1978).

A number of fragments of coenzyme F_{420} retain biological activity (Eirich *et al.*, 1979). Fragments F^+ and FO are formed upon acid hydrolysis of the coenzyme (Fig. 7). FO has also been synthesized (Ashton and Brown, 1980).

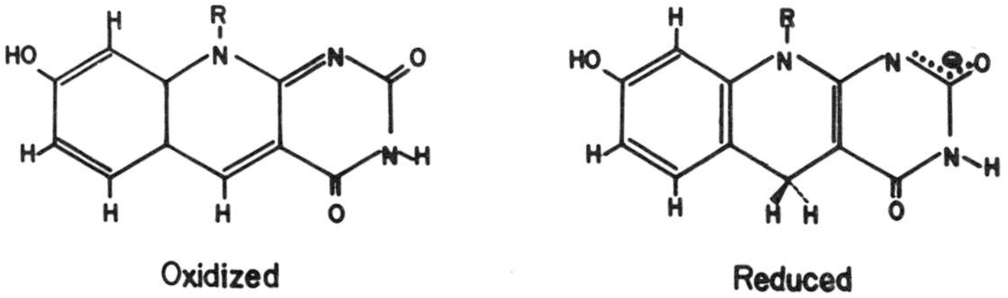

Oxidized

Reduced

FIG. 8. The oxidized and reduced forms of coenzyme F_{420}.

Analogs of flavin mononucleotide, F^+ and FO have nearly identical absorption and fluorescent spectra as coenzyme F_{420}. Both are active in the hydrogenase and NADPH-linked hydrogenase assay with extracts of *Methanobacterium bryantii*. A fragment prepared by periodate oxidation (PA) is inactive in these assays and has different absorption and fluorescence spectra. In contrast, 8-hydroxy-2,4-dioxopyrimido[4,5-*b*]quinoline (R = H, Fig. 7) was active with the $NADP^+$ reductase from *M. vannielii* (Yamazaki *et al.*, 1982). Thus, the importance of the side chain in coenzyme F_{420} depends upon the specific enzyme system studied.

A deazaflavin similar to coenzyme F_{420} is also a cofactor of the photoreactivity enzyme of *Streptomyces griseus* (Eker *et al.*, 1980). This deazaflavin may also be found in *Anacystis nidulans* and other organisms (Eker *et al.*, 1981). It differs from coenzyme F_{420} only in the number of glutamyl residues in the side chain.

The quantity of coenzyme F_{420} in sludge has been measured by high-performance liquid chromatography (van Beelen *et al.*, 1983a,b). This provides a sensitive assay for the number of methanogenic bacteria in natural material.

C. FACTOR F_{430}

An abundant chromophore in extracts of methanogens (Gunsalus and Wolfe, 1978b), the complete structure of factor F_{430} is not known. The structure of the chromophore has been proposed based upon biosynthetic and spectroscopic studies of $F_{430}M$, a fragment produced by methanolysis (Pfaltz *et al.*, 1982; Fig. 9). In addition to the tetrahydro derivative of corphin $F_{430}M$, the factor contains nickel (Whitman and Wolfe, 1980; Diekert *et al.*, 1980a). Other reported constituents include glutamate and aspartate (Anthony, 1982), coenzyme M (Keltjens *et*

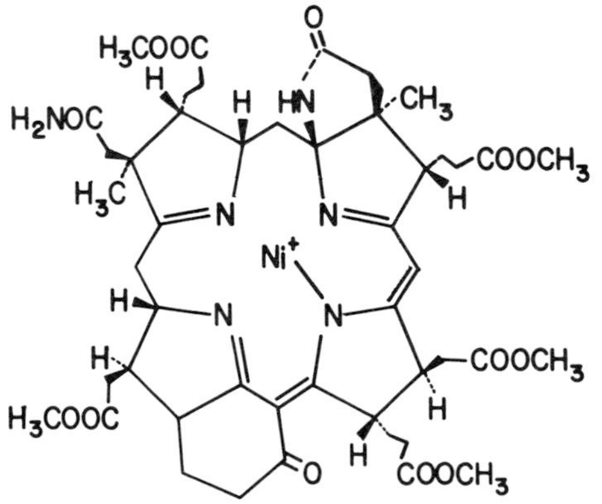

FIG. 9. Structure of the methanolysis degradation product of factor F_{430} from *Methanobacterium thermoautotrophicum* (Pfaltz *et al.*, 1982).

al., 1982), carbohydrate, and a base, probably 6,7-dimethyl-8-ribityl-5,6,7,8-tetra-hydrolumazine (Keltjens *et al.*, 1983c). However, recent reports suggest that the free pentaacid of $F_{430}M$ may be the *in vivo* form (Livingston *et al.*, 1984; Hausinger *et al.*, 1984). Factor F_{430} is the chromophore found in component C of the methyl reductase system from *M. thermoautotrophicum* (Ellefson *et al.*, 1982; Moura *et al.*, 1983). Although its function has not been established, its presence in stoichiometric amounts in the enzyme catalyzing the terminal step in methanogenesis certainly testifies to its importance.

When isolated from whole cells or component C of the methyl reductase, factor F_{430} is yellow and does not fluoresce (Gunsalus and Wolfe, 1978b; Ellefson *et al.*, 1982; Keltjens *et al.*, 1983b). The extinction coefficients of the two major absorption peaks at 430 and 275 nm are about 2×10^5 cm^{-1} M^{-1}. Upon acid hydrolysis, a number of brown, red, and purple derivatives are formed (Keltjens *et al.*, 1983b). Similarly, a number of yellow and red derivatives are usually obtained from whole cells (Whitman and Wolfe, 1980; Diekert *et al.*, 1980c). The derivatives can be distinguished from the parent compound spectroscopically or by thin-layer chromatography (Keltjens *et al.*, 1982, 1983b,c). Presumably, these derivatives are degradation products formed during isolation of the compound. The resonance Raman spectrum of some of these derivatives is relatively sparse, as expected for a highly saturated tetrapyrrole (Shiemke *et al.*, 1983).

In the six species of methanogens tested, the abundance of factor F_{430} varies between 0.23 and 0.80 μmol/g dry weight of cells (Diekert *et al.*, 1980b, 1981). It accounts for most but not all of the nickel required for growth. It has not been detected in acetogenic bacteria or *Escherichia coli*. In *M. thermoautotrophicum*, radiolabeled carbon from succinate, δ-aminolevulinic acid, and the methyl group of methionine are incorporated into the factor (Diekert *et al.*, 1980c; Jaenchen *et al.*, 1981a). Levulinic acid inhibits both growth and F_{430} biosynthesis (Jaenchen *et al.*, 1981b). These results indicate that δ-aminolevulinic acid dehydrase is present in the pathway of factor F_{430} biosynthesis as found in the pathway of tetrapyrrole biosynthesis in eubacteria.

D. Methanopterin and Formaldehyde Activating Factor (FAF)

Upon short-term radiolabeling of whole cells of *Methanobacterium thermoautotrophicum, Methanobrevibacter smithii*, and *Methanosarcina barkeri* with $^{14}CO_2$ or $^{14}CH_3OH$, radiolabel is rapidly incorporated into a yellow, fluorescent compound called YFC (Daniels and Zeikus, 1978). Although YFC is present in whole cells in low amounts, it is readily converted by cell extracts into a much more abundant blue fluorescent compound, methanopterin (Keltjens and

Vogels, 1981; Keltjens *et al.*, 1983d,e; Vogels *et al.*, 1982). The structure of methanopterin is not fully known, but it appears to contain a pterin substituted at the C-6 and C-7 position, glutamate, hexosamine, and phosphate as well as additional unidentified elements. 7-Methylpterin is also abundant in whole cells and is probably identical to a fluorescent compound, F_{342}, previously identified in cell extracts (Keltjens *et al.*, 1983a; Gunsalus and Wolfe, 1978b). Presumably, 7-methylpterin is a degradation product or an intermediate in methanopterin biosynthesis. Methanopterin is reduced in cell extracts by H_2 and converted to unidentified products during methane synthesis (Keltjens and Vogels, 1981; Van Beelen *et al.*, 1983c). It is also required for CH_3—S—CoM-dependent reduction of CO_2 (Leigh and Wolfe, 1983). The levels of methanopterin in whole cells of *M. thermoautotrophicum* are 0.5–1.0 mg/g dry weight (Keltjens and Vogels, 1981; Keltjens *et al.*, 1983c).

When formaldehyde is the carbon donor for methane synthesis by cell free extracts, an oxygen-sensitive co-factor similar to methanopterin is required for activity (Escalante-Semerena *et al.*, 1984a,b,c). This co-factor, formaldehyde activating factor (FAF), also substitutes for methanopterin in the CH_3—S—CoM-dependent CO_2 reduction activity. However, methanopterin substitutes for FAF only after a prolonged incubation. Therefore, FAF may be the physiological or an activated form of methanopterin. Moreover, when FAF is exposed to oxygen, it irreversibly loses the ability to reconstitute formaldehyde-dependent methanogenesis. Present evidence suggests that FAF is tetrahydromethanopterin. It has a molecular weight of 776 compared to 772 for methanopterin. The molecular formula is $C_{30}H_{45}N_6O_{16}P$ compared to $C_{30}H_{41}N_6O_{16}P$ for methanopterin (Escalante-Semerena *et al.*, 1984b).

E. Component B

Fractionation of the methyl reductase system on ion exchange resins resolves two protein fractions from a low molecular weight compound called component B (Gunsalus and Wolfe, 1980). Component B is required for both the reduction of CH_3—S—CoM with either H_2 or NADPH as the electron donor (Ellefson and Wolfe, 1980) and CH_3—S—CoM-dependent CO_2 reduction (Romesser and Wolfe, 1982b). High levels of component B are also required for formaldehyde reduction to methane (Escalante-Semerena and Wolfe, 1984). The structure of component B is not known. Partially purified preparations are oxygen-labile and colorless (Gunsalus and Wolfe, 1980). With a molecular weight of about 1000, component B contains phosphate, a sugar, and nitrogen. Amino acids are absent, and it is acid-labile (Tanner, 1982).

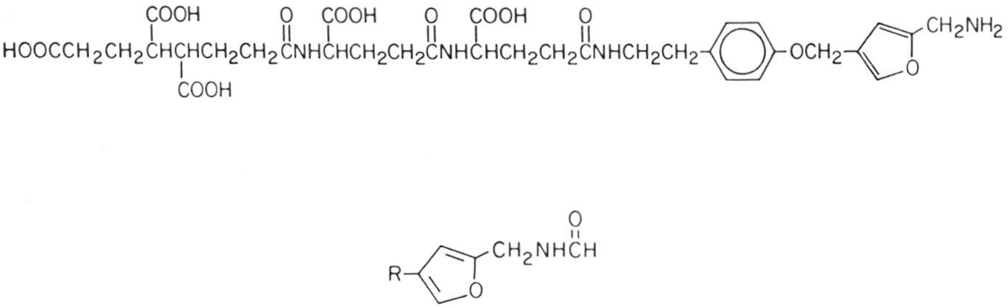

$$\underset{\underset{\text{COOH}}{|}}{\text{HOOCCH}_2\text{CH}_2\text{CHCHCH}_2\text{CH}_2}\overset{\underset{\text{COOH}}{|}}{\overset{\text{O}}{\overset{||}{\text{C}}}\text{NHCHCH}_2\text{CH}_2}\overset{\underset{\text{COOH}}{|}}{\overset{\text{O}}{\overset{||}{\text{C}}}\text{NHCHCH}_2\text{CH}_2}\overset{\text{O}}{\overset{||}{\text{C}}}\text{NHCH}_2\text{CH}_2$$

FIG. 10. Proposed structure of the carbon dioxide reducing factor methanofuran (from Leigh *et al.*, 1984, 1985). (Upper) The complete structure of methanofuran. (Lower) The partial structure of formylmethanofuran showing the addition of the C-1 to the aminomethyl furan moiety.

F. CARBON DIOXIDE REDUCING FACTOR OR METHANOFURAN

After gel filtration on G-25 Sephadex, extracts of *Methanobacterium thermoautotrophicum* require three low molecular weight compounds for CH_3—S—CoM-dependent CO_2 reduction (Leigh and Wolfe, 1983). Two of these, component B and methanopterin, are described above. The third factor is called carbon dioxide reducing factor or CDR (Romesser and Wolfe, 1982b). The proposed structure of CDR is shown in Fig. 10. In addition to the substituted furan, CDR contains a substituted phenolic group, two glutamyl residues, and a 4,5-dicarboxyloctanedioic acid group (Leigh *et al.*, 1984). Because of the presence of a furan moiety, the trivial name methanofuran (MFR) has been proposed. The molecular weight of MFR is 748 and about 50 mg of MFR is purified from 1 kg of whole cells. However, the total amount of MFR present is probably considerably greater than that. When incubated with cell-free extracts under the conditions of methane synthesis, $^{14}CO_2$ is incorporated into MFR at the aminomethyl group of the furan (Leigh *et al.*, 1985). Formyl MFR, the C-1 derivative of MFR formed from CO_2, is also a C-1 donor for methanogenesis in cell-free extracts. Thus, MFR may be the C-1 carrier at the formyl level in methanogenesis.

G. CORRINOIDS

First identified as an abundant corrinoid in sewage sludge (Friedrich and Berhnauer, 1953), 5-hydroxybenzimidazoylcobamide (or Factor III) is the most abundant corrinoid in pure cultures of methanogens (Pol *et al.*, 1982). In extracts of the co-culture "*Methanobacillus omelianskii*," both the coenzyme and monocyanide forms are present (Lezius and Barker, 1965). The common corrinoid,

5,6-dimethylbenzimidazoylcobamide, is absent in extracts of *Methanosarcina barkeri* (Pol *et al.*, 1982). The function of corrinoids in methanogens is not known. Because methylcobalamin and protein bound (5-HOBza)MeCba are substrates for methane synthesis in extracts, it has been suggested that corrinoids are involved in the pathway of methanogenesis (Stadtman, 1967; Wood *et al.*, 1982; Meijden *et al.*, 1983a). The methylation of arsenate by methylcobalamin is also catalyzed by extracts of *Methanobacterium bryantii* (McBride and Wolfe, 1971c). Corrinoids may also play a role in carbon assimilation (Kenealy and Zeikus, 1981) or poising the intracellular redox potential (Shapiro, 1982a). Because of their abundance in whole cells, 0.1–4.0 μmol/g dry weight (Krzycki and Zeikus, 1980; Scherer and Sahm, 1981a; Shapiro, 1982a; Pol *et al.*, 1982), corrinoids probably have a central role in methanogenesis or carbon metabolism.

IX. Biochemistry of Methane Synthesis

In 1956, H. A. Barker recognized that methanogensis was the central unifying factor in the physiology of a diverse group of bacteria (Barker, 1956). At that time, it was also clear that the intermediates of the pathway were probably bound to coenzymes. Nearly 30 years later, the pathway of methane formation is still an area of active research.

Prior to the early 1970s, research concentrated on pathways suggested by C-1 metabolism in eubacteria (Stadtman, 1967). Extracts formed methane from CO_2, formate, formaldehyde, methanol, pyruvate, serine, methylene tetrahydrofolate, methyl tetrahydrofolate, and methylcobalamin (Blaylock and Stadtman, 1963, 1964, 1966; Wolin *et al.*, 1963a, 1964; Wood *et al.*, 1965, 1966; Wood and Wolfe, 1965, 1966a,b,c). Two problems were foremost: the pathway of methane synthesis from methylcobalamin, the most reduced methyl donor then known, and the elucidation of the mechanism of CO_2 activation. Although ATP was required for methanogenesis, much less than one ATP was required per mole of CO_2 reduced (Roberton and Wolfe, 1969). All known autotrophic mechanisms of CO_2 activation were ATP-dependent. Thus, the elucidation of CO_2 activation was a major problem.

The discovery of methyl coenzyme M rapidly changed this perspective (McBride and Wolfe, 1971b; Gunsalus *et al.*, 1976). It became clear that the C-1 carriers in methanogenesis are novel coenzymes and probably unique to that system.

A. METHANOGENIC PATHWAYS

Whole-cell studies have delineated the major pathways of methanogenesis. Essentially two types of reactions have been described. In the first, CO_2 is reduced to methane (reactions 7–9). The reductant is either hydrogen gas, formate (Wood *et al.*, 1965), or carbon monoxide (Kluyver and Schnellen, 1947). In the second class of reactions, methyl-containing compounds are reduced to

$$CO_2 + 4H_2 \rightarrow CH_4 + 2H_2O \tag{7}$$

$$4HCOOH \rightarrow CH_4 + 3CO_2 + 2H_2O \tag{8}$$

$$4CO + 2H_2O \rightarrow CH_4 + 3CO_2 \tag{9}$$

methane without prior oxidation of the methyl group (Pine and Barker, 1956; Pine and Vishniac, 1957; Walther *et al.*, 1981). In reactions 10 and 11, the reductant is methanol or methylamine, although H_2 can also be utilized. In reaction 12, no formal reduction takes place (Zehnder and Brock, 1979b).

$$4CH_3OH \rightarrow 3CH_4 + 2H_2O + CO_2 \tag{10}$$

$$4CH_3NH_3Cl + 2H_2O \rightarrow 3CH_4 + CO_2 + 4NH_4Cl \tag{11}$$

$$CH_3COOH \rightarrow CH_4 + CO_2 \tag{12}$$

Although a small amount of acetate oxidation may occur during this fermentation (Krzycki *et al.*, 1982), the predominant source of methane is the acetoclastic reaction (Baresi *et al.*, 1978; Mah *et al.*, 1978). To date, six new coenzymes have been identified which may play a role in methane biosynthesis (Section VIII). The structure of only some are known with certainty.

B. Methyl Coenzyme M Reductase System

Coenzyme M was discovered because it is an abundant methyl acceptor from [^{14}C]methylcobalamin and $^{14}CO_2$ in extracts (McBride and Wolfe, 1971b). Methyl coenzyme M is also rapidly reduced to methane (McBride and Wolfe, 1971b; Taylor and Wolfe, 1974a). The methyl reductase responsible for this activity in *Methanobacterium thermoautotrophicum* was resolved into three components: A, B, and C (reaction 13; Gunsalus and Wolfe, 1980). Component A contains hydrogenase and has

$$CH_3\text{-S-CoM} + H_2 \xrightarrow[\text{A, B, C}]{\text{ATP, Mg}^{2+}} CH_4 + \text{HS-CoM} \tag{13}$$

been further resolved into three protein fractions (Nagle and Wolfe, 1983). Two of these fractions, A2 and A3, are oxygen-labile. The third fraction, A1, is oxygen stable and contains deazaflavin (F_{420})-dependent hydrogenase activity. Component B is a low molecular weight co-factor whose structure is not known (Section VIII, E). Component C is an oxygen-stable protein, which has been purified to homogeneity (Ellefson and Wolfe, 1981). The enzyme has a native molecular weight of 300,000 and contains three subunits with molecular weights of 68,000, 45,000, and 38,500. The subunit stoichiometry is α_2, β_2, γ_2. Antisera prepared against the *M. thermoautotrophicum* enzyme cross-reacts with proteins with similar electrophoretic properties from *Methanospirillum hungatei, Methanobacterium formicicum,* and *Methanobacterium bryantii.* No cross-reaction was observed with extracts of *Methanobrevibacter ruminantium* and *Methanogenium marisnigri* although enzyme activity was present. Thus, an enzyme similar to the *M. thermoautotrophicum* methyl reductase is widely distributed among methanogens. Methyl reductase activity has also been found in undialyzed extracts of *Methanosarcina barkeri* (Baresi and Wolfe, 1981). Partially purified component C is active when either NADPH or reduced coenzyme F_{420} are substituted for H_2 (reaction 14). However, the enzymes in component A are required for activity with H_2 (Ellefson and Wolfe, 1980). Thus, component C is probably the actual site of methyl coenzyme M reduction. However, no activity from any electron donor is found in the homogeneous preparations of component C without reconstitution with other protein fractions.

$$CH_3\text{-S-CoM} + F_{420} \text{ (red)} \xrightarrow[\text{B, C}]{\text{ATP, Mg}^{2+}} CH_4 + F_{420} \text{ (ox)} + \text{HS-CoM} \qquad (14)$$

Homogeneous component C contains tightly bound factor F_{430} (Section VIII, C: Ellefson *et al.*, 1982; Keltjens *et al.*, 1982). Because F_{430} cannot be resolved from the protein without strong denaturants, its role in enzyme activity has not been proven. Presumably, it is the prosthetic group of component C. Component C also contains tightly bound coenzyme M, possibly in the form of a coenzyme M–F_{430} adduct, CoMF$_{430}$ (Keltjens and Vogels, 1981; Keltjens *et al.*, 1982). Therefore, free or protein-bound coenzyme M may function as a methyl carrier. Whole cells rapidly incorporate $^{14}CO_2$ into unidentified coenzyme M-containing compounds (Daniels and Zeikus, 1978). Whether or not these compounds are CoMF$_{430}$ derivatives is not known.

The methyl reductase system requires ATP and high concentrations of Mg^{2+} for optimal activity (Gunsalus and Wolfe, 1978a). The concentration of Mg^{2+}, 20–40 mM, is several orders of magnitude greater than the concentration of ATP required. Therefore, Mg^{2+} must function as more than a co-factor of ATP activation. In dialyzed extracts of *M. bryantii*, high concentrations of Mg^{2+} increase the velocity of the methyl reductase without effecting the rate of activation by ATP (Whitman and Wolfe, 1983). These results suggest that Mg^{2+} may be an effector of the methyl reductase.

Catalytic amounts of ATP are required for methanogenesis from $H_2 + CO_2$ as well as more reduced substrates (Roberton and Wolfe, 1969). Similarly, ATP is also required for activation of the methyl reductase (Gunsalus and Wolfe, 1978a). In extracts of *M. bryantii,* the half-life of the activated methyl reductase is only 5–15 min (Whitman and Wolfe, 1983). Moreover, CH_3—S—CoM (or catalysis) stabilizes the activated enzyme, although it is not required for the initial activation. ATP analogs, AMP, and corrinoids inhibit the activation (Gunsalus and Wolfe, 1978a; Mountfort, 1980; Whitman and Wolfe, 1982, 1983). Once activated, the methyl reductase retains activity after gel filtration in the cold (Whitman and Wolfe, 1983). These results are consistent with the modification of a protein or protein-bound co-factor by ATP. However, the nature of this modification has not been determined.

In contrast, a membrane fraction from *Methanobrevibacter ruminantium* is reported not to require ATP for methyl reductase activity (Sauer *et al.,* 1979). Because methanogenesis is inhibited by the uncouplers CCCP and DNP, membrane permeable ions, and deoxycholate, methanogenesis is proposed to be activated by the proton motive force directly (Sauer *et al.,* 1979, 1980a,b). Evidence in support of this hypothesis is meager. Although the membrane fraction does not catalyze net ATP synthesis, ATP may be available if ATP is simultaneously synthesized and degraded by a membrane-bound ATPase. In extracts of *Methanobacterium bryantii,* the steady state concentration of ATP is very low, whereas ATP is synthesized at a rate of 0.7 nmol/min/mg protein (Whitman and Wolfe, 1983). In the membrane fraction, uncouplers probably inhibit ATP synthesis and exert an indirect effect on methanogenesis. In extracts of *M. bryantii,* the uncoupler FCCP is not inhibitory below a concentration of 0.5 mM and membrane permeable ions and detergents have no effect at all (Whitman and Wolfe, 1983). Therefore, intact membranes are not a requirement for methanogenesis in cell extracts.

The methyl reductase system is very specific for CH_3—S—CoM. Only one alternative substrate is known, CH_3CH_2—S—CoM (Gunsalus *et al.,* 1978). In this case, ethane rather than methane is the product. Other modifications of CH_3—S—CoM produce inactive or inhibitory compounds. These modifications include (1) increasing the length of the ethylene bridge between the sulfide and sulfonate moieties, (2) addition of a second methyl group to the sulfide moiety [i.e., 2-(dimethylsulfonium)ethane sulfonate], (3) replacement of the sulfide by nitrogen, (4) replacement of the sulfonate by alcohol, esters, amines, or other groups, and (5) replacement of the methyl group by a propyl group. In addition, HS—CoM and some analogs are inhibitors. These analogs include (1) compounds where the methylthiol is substituted by chlorine or bromine, (2) compounds with an increased length of the ethylene bridge between the thiol and sulfonate moieties, and (3) oxidized coenzyme M and its analogs. The great specificity indicates that CH_3—S—CoM reduction is not a fortuitous activity and that it has some biological function.

Low concentrations of corrinoids activate the methyl reductase system three- to five-fold (Whitman and Wolfe, 1982). The activation is nonspecific, and the apparent binding constants vary from 1 μM for cobinamides to 10 μM for cobamides. Adenosylcobalamin is inactive. In the same concentration range, corrinoids also inhibit the activation by ATP and the CH_3—S—CoM-dependent reduction of CO_2. Thus, the effects of cobamides appear to be nonphysiological.

Subsequent to dialysis, extracts of *Methanosarcina barkeri* lose methyl reductase activity (Hutten *et al.*, 1981). However, methane is still formed from H_2 + CO_2 or methanol in the presence of HS—CoM. CH_3—S—CoM inhibits these activities. Therefore, it was proposed that HS—CoM is required for synthesis of a protein-bound co-factor other than CH_3—S—CoM. This proposal seems reasonable in light of the discovery of $CoMF_{430}$ (Vogels *et al.*, 1982). However, it does not explain the consistent observation of methyl reductase activity in purified extracts or protein fractions prepared from other methanogens. Furthermore, it is possible that the failure to demonstrate methyl reductase activity may have been due to the removal of cobamides or other low molecular weight compounds upon dialysis. This important point needs further clarification.

C. METHYLATION OF COENZYME M

Three methyl donors for HS—CoM have been described. Free methyl cobalamin chemically methylates HS—CoM, and a methyltransferase, which catalyzes this reaction, has been purified from *M. bryantii* (Taylor and Wolfe, 1974b). Nevertheless, it is unlikely that this reaction is physiologically important in methanogenesis from CO_2. When intermediates of methanogenesis are radio-labeled with $^{14}CO_2$, methyl cobalamin is never detected (Daniels and Zeikus, 1978). Cobamides are not required to reconstitute CO_2 reduction in extracts resolved from low molecular weight co-factors (Leigh and Wolfe, 1983). In addition, exogenous cobamides actually inhibit CO_2 reduction in such extracts (Whitman and Wolfe, 1982).

Methanol is a substrate for methanogenesis in extracts of *M. barkeri* (Shapiro and Wolfe, 1980). Methyl coenzyme M is formed as a transitory intermediate when HS—CoM is also present. When nitrogen is substituted for the hydrogen atmosphere, the initial rate of methyl coenzyme M synthesis is uneffected and methyl coenzyme M accumulates. Other mercaptans are not methylated. The methylation of HS—CoM requires magnesium and catalytic amounts of ATP. The activation by ATP is time dependent and stable after dialysis (Shapiro, 1982b). Inactivation is caused by oxygen, FAD, FMN, or an active methyl reductase in the absence of hydrogen (Meijden *et al.*, 1983b). Methylamine is a poor substrate for this methyltransferase in extracts of H_2:CO_2 grown cells. Acetate is not a substrate at all.

Two protein fractions are required for the methylation of HS—CoM by methanol in M. *barkeri* (Meijden *et al.*, 1983a). One fraction, MT_1, is oxygen sensitive and contains an oxygen-sensitive cobamide. In the presence of methanol, ATP, and magnesium, the cobamide in MT_1 is methylated (Meijden *et al.*, 1983c). This activity is independent of HS—CoM. The second protein fraction, MT_2, is oxygen stable and has been purified 86-fold (Meijden *et al.*, 1983a). In addition to being required for methylation of HS—CoM by methanol, MT_2 also catalyzes the methylation of HS—CoM by methyl cobamides. Although MT_2 was not purified to homogeneity, highly purified fractions contained two subunits after SDS-polyacrylamide gel electrophoresis with M_r of 38,000 and 43,000. Thus, the sequence of events for the methylation of HS—CoM by methanol appears to be (1) the methylation of the protein-bound cobamide in MT_1 by methanol and (2) the transfer of the methyl group from MT_1 to HS—CoM by MT_2 (Meijden *et al.*, 1983a). The first step requires activation by ATP and H_2. These conclusions are consistent with earlier results of Blaylock (Blaylock and Stadtman, 1966; Blaylock, 1968). Shapiro (1982a) was unable to detect radiolabeled cobamides in extracts of M. *barkeri* incubated with radiolabeled methanol. However, the cobamide in MT_1 is tightly bound (Meijden *et al.*, 1983c), and it may not have been detected in the experiments by Shapiro.

A cobamide protein has been purified aerobically from M. *barkeri* (Wood *et al.*, 1982). It has a molecular weight of about 200,000. It is composed of subunits with a molecular weight of 44,000. After methylation with methyl iodide, the protein is a substrate for methanogenesis in crude extracts. However, the rates are 30-fold lower than those observed in similar extracts with methanol as the substrate (Shapiro and Wolfe, 1980). No specificity for methylation of HS—CoM over other thiols was demonstrated; and only 50% of the protein substrate actually formed methane at the completion of the reaction (Wood *et al.*, 1982). Thus, the identity of this cobamide protein with MT_1 is not certain. Cobamides are abundant in methanogens that do not utilize methanol as a substrate for methanogenesis. In these methanogens, cobamides are probably not intermediates in methanogenesis. Thus, cobamides must have additional roles in the biochemistry of these organisms (Section VII,A). The function of cobamide proteins must therefore be interpreted with caution.

In extracts of M. *thermoautotrophicum*, formaldehyde is a substrate of methanogenesis (Romesser and Wolfe, 1981; Escalante-Semerena and Wolfe, 1984). Methane formation is dependent on HS—CoM. Hydroxymethyl coenzyme M, a formaldehyde adduct of HS—CoM, will substitute for HS—CoM. Although hydroxymethyl coenzyme M has been proposed to be an intermediate in CO_2 reduction (Romesser and Wolfe, 1982a), its formation in cell-free extracts has not been demonstrated (Section IX,D). When the methyl reductase is inhibited by polyphosphate, methyl coenzyme M accumulates from formaldehyde and HS—CoM (J. C. Escalante-Semerena, personal communication).

D. Methyl Coenzyme M-Dependent CO_2 Reduction

In extracts of *M. thermoautotrophicum,* carbon dioxide reduction to methane is stimulated more than ten-fold by methyl coenzyme M (Gunsalus and Wolfe, 1977). In addition, methanogenesis requires H_2, magnesium, and ATP. After dialysis, two cofactors (methanofuran and methanopterin) are required for CO_2 reduction but not methyl reductase activity (Romesser and Wolfe, 1982b; Keltjens and Vogels, 1981; Leigh and Wolfe, 1983). Although the possibility that methyl coenzyme M acts as an allosteric effector in CO_2 reduction cannot be rigorously excluded, it appears likely that the mechanism of methyl coenzyme M reduction is coupled to CO_2 activation (Fig. 11). Evidence in support of this hypothesis is as follows.

The initial steps of CO_2 reduction are thermodynamically unfavorable. By

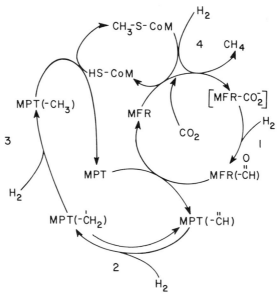

Fig. 11. Sequential reduction of CO_2 to methane (adapted from Escalante-Semerena *et al.,* 1984a). MFR is methanofuran (carbon dioxide reducing factor), MPT is tetrahydromethanopterin (formaldehyde activating factor), HS-CoM is coenzyme M. Coenzyme M may be free or covalently bound to factor F_{430} as $CoMF_{430}$. The reduction of CH_3-S-CoM to CH_4 (step 4) is coupled to CO_2 activation; however, the mechanism is not known. The first intermediate of CO_2 reduction, [MFR-CO_2^-], has not been demonstrated. The formation of methenyltetrahydromethanopterin, MPT ($\equiv$CH), from formylmethanofuran, MFR(-CHO), requires CH_3-S-CoM. The reduction of methenyltetrahydromethanopterin to methylenetetrahydromethanopterin, MPT($=CH_2$), is reversible (step 2). Methylenetetrahydromethanopterin may also be formed by a chemical reaction between formaldehyde and MPT. Methylenetetrahydromethanopterin is reduced to methyltetrahydromethanopterin, MPT (-CH_3), prior to the transfer of the methyl group to HS-CoM.

coupling to the exergonic methyl reductase reaction, the activation of CO_2 becomes feasible. No alternative mechanism of CO_2 activation is known in methanogens, and stoichiometric amounts of ATP are not required. Thus, there is no other obvious energy source for the activation of CO_2. Alternative substrates for methanogenesis as formaldehyde, serine, pyruvate, and CH_3CH_2—S—CoM will substitute for methyl coenzyme M in CO_2 activation (Romesser and Wolfe, 1982a). Formaldehyde, serine, and pyruvate presumably donate C-1 groups to the methyl reductase via either methyl coenzyme M or another mechanism (as $CoMF_{430}$). In any case, high concentrations of methyl coenzyme itself are not required, as might be expected if it were an effector. Mercaptans including HS—CoM inhibit CO_2 reduction. Fifty percent inhibition of CO_2 reduction requires five-fold less HS—CoM than inhibition of the methyl reductase. So some "product" of the methyl reductase but not HS—CoM is involved in CO_2 activation. Mercaptans are potent free radical scavengers, thus a free radical intermediate may be involved (Romesser and Wolfe, 1982a). However, inhibition by thiols does not rule out other mechanisms.

Corrinoids stimulate the methyl reductase and inhibit CO_2 reduction (Whitman and Wolfe, 1982). The specificity and binding constants for both effects are identical. These results are consistent with the action of corrinoids on an intermediate in the methyl reductase reaction involved in CO_2 activation. In this hypothesis, the velocity of the methyl reductase is limited by the rate of decomposition of this intermediate. The intermediate can react with CO_2 or a C-1 carrier to form methane and initiate the reduction of CO_2. Likewise, carbon dioxide stimulates the rate of the methyl reductase even in the presence of saturating concentrations of methyl coenzyme M (Gunsalus and Wolfe, 1977; Romesser and Wolfe, 1982a). This effect of CO_2 would be expected if CO_2 stimulated the rate of decomposition of an intermediate formed in the methyl reductase reaction. If corrinoids catalyze the breakdown of the intermediate as well, they would likewise stimulate the methyl reductase. Because the intermediate is no longer available to initiate CO_2 reduction, methyl coenzyme M-dependent-CO_2 reduction would also be inhibited by corrinoids.

When the methyl coenzyme M-dependent CO_2 reaction is performed with $^{14}CO_2$ and methanofuran in the absence of methanopterin, radiolabel is incorporated into methanofuran (Leigh et al., 1985). NMR studies demonstrate that formyl MFR is formed (Fig. 11). Further incubation of radiolabeled formyl MFR with methanopterin (or FAF), methyl coenzyme M, and other components of the methyl reductase assay leads to formation of radiolabeled CH_4. Thus, methanofuran is the C-1 carrier at the formyl level of oxidation. Presumably, methanofuran is also the CO_2 acceptor. Like methyl coenzyme M-dependent CO_2 reduction, the conversion of $[^{14}CHO]MFR$ to $^{14}CH_4$ requires methyl coenzyme M. Thus, the coupling of methane synthesis and CO_2 reduction may occur at an intermediate step. Whether coupling also occurs in the initial activation of CO_2

per se is not known. Although these results do not contradict earlier work (Romesser and Wolfe, 1982a; Whitman and Wolfe, 1982), they certainly suggest that the coupling of the early and terminal steps of methanogenesis is more complex than originally anticipated.

Incubation of tetrahydromethanopterin (FAF) with formaldehyde under nitrogen leads to formation of methenyl-FAF (Escalante-Semerena et al., 1984c). Under these conditions, electrons from the oxidation of FAF can be coupled to methyl coenzyme M reduction (Escalante-Semerena and Wolfe, 1984). Under hydrogen, methenyl-FAF is reduced to methylene-FAF and methyl-FAF (Fig. 11). These C-1 derivatives of FAF are substrates for methane formation in extracts. Thus, FAF is the C-1 carrier at intermediate levels of CO_2 reduction. Therefore, the sequence of events in the reduction of CO_2 appears to be (1) activation of CO_2 and the initial reduction of the C-1 on methanofuran, (2) transfer of the C-1 to FAF by formyl-MFR, (3) reduction of methenyl-FAF to methyl-FAF in two steps, (4) transfer of the C-1 to methyl coenzyme M or $CoMF_{430}$ by methyl-FAF, and (5) reduction of methyl coenzyme M or $CoMF_{430}$ to methane (Fig. 11).

Conclusive proof of the role of methyl coenzyme M in CO_2 reduction will require further purification of the enzymes and the complete elucidation of the co-factor structures. Methyl coenzyme M-dependent CO_2 reduction has only been observed in a few methanogen extracts (Romesser and Wolfe, 1982a). Whether this is due to a lability of the system or to the presence of an alternative mechanism of CO_2 activation is not known. In dialyzed extracts of M. barkeri, methyl coenzyme M inhibits CO_2 reduction (Hutten et al., 1981). A methyl coenzyme M-independent mechanism of CO_2 activation has been proposed for these extracts (Vogels et al., 1982).

E. ELECTRON DONORS

In the reconstituted methyl reductase system, hydrogen is commonly used as an electron donor. In partially purified extracts, NADPH and coenzyme F_{420} will substitute for hydrogen (Ellefson and Wolfe, 1980). Nevertheless, it has not been established that these coenzymes are intermediates in vivo. In the more highly resolved system, which includes components A1, A2, A3, and C, FAD is required for activity but not F_{420} (Nagle and Wolfe, 1983). FAD is also present in partially purified deazaflavin-dependent hydrogenase (Section X,A), which is in component A1. However, neither $FADH_2$ nor dihydro-F_{420} substituted for components A1, A2, or A3.

It has been established that the hydrogen atoms in methane are derived from water and not hydrogen gas (Daniels et al., 1980). However, little else is known about the path of electrons from hydrogen gas or formate. Isolation of the hydrogenase and formate dehydrogenase from several methanogens will facilitate these investigations (Section X). In addition, virtually nothing is known about methanol and methylamine oxidation. Because methanogenesis is the major electron sink in these organisms, elucidation of the pathway of electrons in the reduction of CO_2 and other C-1 intermediates is of major importance.

F. ISOTOPE FRACTIONATION

Methane formed by pure cultures from carbon dioxide is enriched in ^{12}C over ^{13}C. At 65°C, the $\Delta\delta^{13}C$ for methane and *Methanobacterium thermoautotrophicum* cell carbon is −3.4 and −2.4‰, respectively (Fuchs *et al.*, 1979a). At 40°C, the $\Delta\delta^{13}C$ for methane produced by *Methanosarcina barkeri* and *Methanobacterium bryantii* is −4.5 and −6.1‰, respectively (Games *et al.*, 1978). At 37° and 46°C, the $\Delta\delta^{13}C$ for methane by an unnamed species of *Methanobacterium* is −4.4 and −6.1‰ (Belyaev *et al.*, 1983). These high fractionations suggest that CO_2 and not bicarbonate is the substrate of methanogenesis. In addition, methane is depleted in deuterium relative to water (Spencer *et al.*, 1980). A deuterium isotope effect of 1.20 has been estimated.

G. ACETOCLASTIC REACTION

Methane synthesis from acetate procedes via an acetoclastic reaction where deuterium label in the methyl group of acetate is found in methane (reaction 15; Buswell and Sollo, 1948; Pine and Barker, 1956). Significant oxidation of the methyl group of acetate does not occur (Smith and Mah, 1980). In this

$$CD_3COOH \rightarrow CD_3H + CO_2 \tag{15}$$

fashion, the acetoclastic reaction differs from reductant-dependent methanogenesis from CO_2, methanol, and methylamines. The mechanism of this reaction is unknown. Of special interest is whether or not methyl coenzyme M is an intermediate. Acetate and $H_2 + CO_2$ grown *Methanosarcina barkeri* contain equal levels of coenzyme M and methyl reductase activity (Baresi and Wolfe, 1981). However, in cell-free extracts, the acetoclastic activity is associated with a cell envelope fraction, whereas most but not all of the methyl reductase is found in the supernatant fraction (Baresi, 1984). It cannot be determined whether or not methyl coenzyme M is an intermediate at the present time.

During biphasic growth on trimethylamines and acetate, some methane is derived from acetate prior to the exhaustion of trimethylamine (Blaut and Gottschalk, 1982). When CD_3COOH is the substrate, equal amounts of CD_2H_2 and CD_3H are formed. In the absence of trimethylamine, CD_3H predominates by a factor of 3 to 1. These results suggest that an exchange reaction or oxidation of the methyl group of acetate can occur on the pathway of methane synthesis.

X. Oxidation–Reduction Enzymes

Little is known about electron carriers in methanogens. Menaquinones, ubiquinones, and cytochromes are probably absent in most methanogens (Balch *et al.*, 1979). An exception is *Methanosarcina barkeri*, which contains cytochromes *b* and *c* (Kühn *et al.*, 1979; Kühn and Gottschalk, 1983). The benzoquinones, α-tocopherolquinone and α-tocopherolquinol, are also present at levels comparable to eubacteria in *Methanobacterium* and *Halobacterium* (Hughes and Tove, 1982). Other electron carriers found include coenzyme F_{420}, FAD, NADPH, and NADH. Iron–sulfur proteins are also abundant (Lancaster, 1980, 1982a), and ferredoxins are observed in whole cells of *Methanosarcina* by Mössbauer spectroscopy (Scherer and Sauer, 1982). To date, only a few enzymes involved in oxidation–reduction reactions have been purified. These include hydrogenase, formate dehydrogenase, coenzyme F_{420}-NADPH oxidoreductase, and NADPH diaphorase. A ferredoxin and sulfite reductase have also been purified from *M. barkeri*. A malate dehydrogenase from *Methanospirillum hungatei* has also been described (Section VII, B).

A. HYDROGENASE

Hydrogenase from *Methanobacterium thermoautotrophicum* is membrane-bound in whole cells (Doddema *et al.*, 1979b). A membrane-bound hydrogenase is also found in extracts of *Methanobacterium* strain G2R, a methanogen similar to *Methanobacterium formicicum* (McKellar and Sprott, 1979). The membrane-bound hydrogenase is solubilized by detergents and nearly homogeneous upon electrophoresis. The molecular weight is about 900,000, and it contains subunits with molecular weights of 38,500, 50,700, and 80,000. During electrophoresis, a faster migrating form appears, which presumably has a lower molecular weight. Substrates of the solubilized hydrogenase are benzyl viologen, methyl viologen, methylene blue, 2,6-dichlorophenol-indophenol, FAD, and riboflavin 5′-phosphate. Coenzyme F_{420}, $NADP^+$, NAD^+, and ferricyanide are inactive. All activity is lost upon exposure to air, but the enzyme is reactivated by dithionite or a glucose oxidase system. The pH optimum is 8.5–9.0. Inhibitors include *p*-hydroxymercuribenzoate, cyanide, chloroform, and chloramphenicol.

A hydrogenase with a molecular weight greater than 500,000 is also found in extracts of *M. thermoautotrophicum* (Jacobson *et al.*, 1982). Unlike the enzyme from *Methanobacterium* strain G2R, it is not membrane associated in cell-free extracts. After exposure to air, the enzyme can be reactivated with high salt anaerobically. Both coenzyme F_{420} and methyl viologen are substrates. It contains three subunits with molecular weights of 40,000, 31,000, and 26,000 in a stoichiometry of 2:2:1. Thus, the monomer molecular weight is 170,000. The enzyme contains 33–43 g atom Fe, 24–30 g atom of labile sulfur, and 2.3 molecules of FAD per monomer. In addition, the paramagnetic center associated with nickel that was first described in membrane fractions of *Methanobacterium bryantii* (Lancaster, 1980, 1982b) is also present. Approximately 2.5–3.1 g atoms of nickel per monomer are present in the hydrogenase (Jacobson *et al.*, 1982; Kojima *et al.*, 1983).

A similar hydrogenase is also found in *Methanococcus vannielii* (Yamazaki, 1982). When purified, two interconvertible forms are obtained with molecular weights of 130,000 and 1,300,000. The subunit molecular weights are 42,000, 35,000, and 27,000. The largest subunit contains at least one selenocystienyl residue. Substrates include the deazaflavin FO and tetrazolium dyes. The presence of selenium has not been investigated in the hydrogenase from other sources.

In addition to the high molecular weight hydrogenase, a soluble dye-linked hydrogenase is also found in extracts of *M. thermoautotrophicum* (Fuchs *et al.*, 1979b; Jacobson *et al.*, 1982; Kojima *et al.*, 1983). One group reports an M_r of 60,000 for the monomer, which forms high molecular weight aggregates of 500,000 or greater (Fuchs *et al.*, 1979b). Another group reports that this hydrogenase contains two subunits, M_r 52,000 and 40,000 as well as 14–21 atoms of Fe per Ni (Kojima *et al.*, 1983). It is not inhibited by cyanide, azide, or fluoride. Inhibition by oxygen is rapidly reversible under assay conditions. Carbon monoxide is a reversible inhibitor with a K_I of 2 μM. It contains 0.8 g atoms of Ni per 60,000 g equivalent in partially purified preparations and the Ni-associated paramagnetic center (Graf and Thauer, 1981; Albracht *et al.*, 1982). The low molecular weight hydrogenase differs from the high molecular weight species in its insensitivity to oxygen and cyanide, K_m for methyl viologen, and lack of activity with deazaflavins. Whether it is a fragment of the high molecular weight hydrogenases obtained upon fractionation of extracts or a separate enzyme is not known. *Methanobacterium formicicum* also contains coenzyme F_{420}-dependent and F_{420}-independent hydrogenases (Jin *et al.*, 1983). In this bacterium, the hydrogenases differ greatly in subunit structure and are probably not related. The presence of nickel in methanogen hydrogenases suggests that these enzymes may have a similar mode of action to the nickel-containing hydrogenases from *Alcaligenes eutrophus* (Friedrich *et al.*, 1982), *Desulfovibrio gigas* (LeGall *et al.*,

1982; J. J. G. Moura *et al.*, 1982b), *Desulfovibrio desulfuricans* (Krüger *et al.*, 1982), and *Vibrio succinogenes* (Unden *et al.*, 1982).

B. FORMATE DEHYDROGENASE

Extracts of *Methanococcus vannielii* contain selenium-dependent and selenium-independent formate dehydrogenases (Jones and Stadtman, 1980, 1981). The selenium-dependent formate dehydrogenase contains a selenocysteinyl residue, Fe, Mo, and has a broad absorption maximum at 380 nm (J. B. Jones *et al.*, 1979). The ratio Fe:Mo is 15–30 in highly purified fractions. Tungsten partially substitutes for Mo. The selenium-independent enzyme has been highly purified. It has a native molecular weight of 105,000, although it readily aggregates. One equivalent of Mo and 10 equivalents each of Fe and S are present. However, metals are easily lost during the purification. Both formate dehydrogenases have a pH optimum of 8.5–9.0. Electron acceptors include FAD, FMN, 8-hydroxy-5-deazaflavin, methyl viologen, benzyl viologen, and 2,3,5-triphenyltetrazolium. $NADP^+$ and NAD^+ are not reduced. The addition of electron acceptors in the absence of formate rapidly and irreversibly inactivates both enzymes. In addition, both enzymes are extremely oxygen sensitive and irreversibly lose greater than 50 of their maximal activity after less than 1 min in air. Other inhibitors include cyanide, iodoacetamide, and metal chelators. Azide, 10 mM, does not inhibit.

Only a single formate dehydrogenase is present in extracts of *Methanobacterium formicicum* (Schauer and Ferry, 1982). The highly purified enzyme has a molecular weight of 288,000 and an absorption maximum at 390 nm. Upon treatment with detergent, a fluorescent compound (or compounds) is released with excitation maxima of 385, 302, and 277 nm. The emission maximum is 455 nm. EPR studies demonstrate two paramagnetic centers in the reduced enzyme including a Fe/S cluster and an unusual signal attributed to Mo (Barber *et al.*, 1983). In addition, the enzyme requires FAD for activity (Schauer and Ferry, 1983). The pH optimum for activity is 7.9. Electron acceptors include coenzyme F_{420}, FMN, and FAD. $NADP^+$ and NAD^+ are not utilized (Schauer and Ferry, 1982). Electron acceptors are not inhibitory in the absence of formate. Oxygen irreversibly inactivates the enzyme. In air, greater than 90% inactivation is obtained in 2 min. However, 10 mM azide partially protects against oxygen inactivation. Inhibitors of the enzyme include azide, cyanide, cyanate, thiocyanate, fluoride, nitrite, nitrate, and metal chelators. Thus, the formate dehydrogenase from *Methanobacterium formicicum* resembles the enzymes from *Methanococcus vannielli* in absorption spectrum, presence of Mo and Fe, sensitivity to oxygen, and substrate specificity. Some differences have been observed, especially in their sensitivity to azide.

C. Coenzyme F_{420}:NADPH Oxidoreductase

This enzyme has been purified to homogeneity from *M. vannielli* (Jones and Stadtman, 1980; Yamazaki and Tsai, 1980). The molecular weight of the native protein is 85,000, and it contains two identical subunits with molecular weights of 43,000. Oxygen has no effect on activity. Sulfhydryl reagents rapidly inactivate the enzyme. The pH optimum for deazaflavin reduction is 4.8, and the optimum for oxidation is 7.9. At pH 7.0, the velocity of deazaflavin oxidation is 24 times faster than deazaflavin reduction. Catalysis proceeds by a direct hydride transfer process (Yamazaki *et al.*, 1980). In addition, the enzyme is stereospecific with respect to NADPH and 8-hydroxy-5-deazaflavin. The substrate specificity has also been investigated in detail (Yamazaki and Tsai, 1980; Yamazaki *et al.*, 1982). NAD^+ and NADH do not substitute for $NADP^+$. The basic heterocyclic system of coenzyme F_{420} is the minimum structural requirement for activity. The polyglutamyl side chain is not important in activity. Modification of the C-5, C-7, or C-8 positions effected either catalysis or substrate binding (see Section VIII, B). This specificity is not shared by other deazaflavin-dependent enzymes. For instance, the presence of the polyglutamyl side chain decreases the K_m for deazaflavin of the formate dehydrogenase of *M. formicicum* by tenfold (Schauer and Ferry, 1982).

D. Other Oxidation–Reduction Proteins

A NADH-diaphorase has been purified from *Methanospirillum hungatei* (McKellar *et al.*, 1981). Electron acceptors include oxygen, 2,6-dichlorophenolindophenol, and cytochrome *c*. NADPH is not an electron donor. Activity requires FAD, which is removed during the purification. The pH optimum is 7.0–8.5. Sulfhydryl reagents and metal chelators inhibit. The diaphorase inhibitor dicoumarol inhibits weakly. Although the purified enzyme lacks NAD^+ reductase activity, it may function as a reductase *in vivo*.

Ferredoxin has been isolated and sequenced from two strains of *Methanosarcina barkeri* (I. Moura *et al.*, 1982; Hausinger *et al.*, 1982; Hatchikian *et al.*, 1982). Although very similar in structure, the sequence of amino acids differ slightly. When isolated, the ferredoxin contains exclusively [3Fe-3S] clusters. However, the amino acid sequence shows the most homology with the clostridial ferrodoxin, which has [4Fe-4S] centers. The *M. barkeri* ferredoxin is also unusual because it contains no aromatic amino acid residues, methionine, arginine, or histidine. Ferredoxins have not been found in other methanogens although there is a report of a similar ferredoxin isolated from the mixed culture "*Methanobacillus omelianskii*" (Buchanan and Rabinowitz, 1964). *Methanosarcina barkeri* also contains a protein which catalyzes the reduction of sulfite to sulfide. The purified enzyme is spectrally similar to the assimilatory sulfite reductases and may contain siroheme (J. J. G. Moura *et al.*, 1982a).

Methanosarcina species contain several *b*-type cytochromes and low levels of *c*-type cytochromes (Kühn and Gottschalk, 1983). In cells grown on methanol or methylamines, two membrane-bound *b*-type cytochrome are present. The midpoint potentials are −325 and −183 mV. During growth on acetate, an additional *b*-type cytochrome is formed with a midpoint potential of −250 mV. Although cytochromes are found in H_2 + CO_2 grown cells of *Methanosarcina*, cytochromes are apparently absent from other species of methanogens that are incapable of growth on methylamines or acetate.

XI. Genetics

Because of their extreme sensitivity to oxygen and unique physiology, the study of genetics of methanogens is proceeding slowly. At present, very little is known about the structure of the chromosome or genetic elements. For one species, *M. thermoautotrophicum*, the chromosome is very small, $1.1–1.2 \times 10^9$ daltons, or about 40% of the size of the *Escherichia coli* chromosome (Mitchell *et al.*, 1979; Brandis *et al.*, 1981). Although no satellite DNA was detected, 6% of the total DNA may be highly repeated DNA (Mitchell *et al.*, 1979). Because of the great diversity of methanogens, it is premature to speculate whether or not a reduced genome size is a general feature.

Although extrachromosomal elements are well described in other archaebacteria, none have been described in methanogens until very recently. Two cryptic plasmids have been isolated. A plasmid was found in a coccoid isolate and cloned (Thomm *et al.*, 1983). A second plasmid was also found in high copy number in *M. thermoautotrophicum* strain Marburg (Meile *et al.*, 1983). This second plasmid has a length of about 4500 base pairs (bp). Two phages have also been isolated from rumen fluid (L. Baresi and G. Bertani, personal communication).

To date, two classes of mutants have been isolated in methanogens. Bromoethanesulfonate (BES) resistance mutants have been obtained in *M. barkeri* (Smith and Mah, 1981; Smith, 1983). Bromoethanesulfonate is an analog of coenzyme M and a potent inhibitor of the methyl reductase. In addition, several conditional auxotrophs have been isolated in *Methanococcus voltae* (G. Bertani and L. Baresi, personal communication). The specific nature of these mutants is not known.

Cloned methanogen DNA has been expressed in *E. coli* (Reeve *et al.*, 1982; Bollschweider and Klein, 1982). In addition, cloned methanogen DNA complements *E. coli* auxotrophs at the *argG, purE,* and a *his* locus (Reeve *et al.*, 1982; Hamilton and Reeve, 1984; Wood *et al.*, 1983). The protein products have been identified. Whether these proteins are the normal products of methanogen translation or abnormal proteins produced by fortuitous promoters in *E. coli* is not certain. Only a small fraction of the auxotrophs tested were complemented by methanogen DNA. Thus, the cloned fragments may represent an exceptional class of methanogen genes (or *E. coli* auxotrophs) and may not be typical of the methanogen chromosome.

The structure of the ribosomal RNA genes has been investigated in *Methanoeoccus vannielii* (Jarsch *et al.*, 1983). The three ribosomal RNA genes are closely linked on the chromosome in the order 16 S–23 S–5 S. No intervening sequences were detected in the 16 S and 23 S sequences. The two "operons" containing the 16 S and 23 S genes have spacer regions of 156 and 242 bp. The larger spacer contains the sequence of tRNA^{Ala}, an arrangement similar to the eubacterial pattern (Jarsch and Böck, 1983). A separate 5 S rRNA gene is also found and is not linked to other ribosomal RNA genes, being separated from them by at least 6.6 kb.

In summary, substantial progress is being made in methanogen genetics in spite of a number of severe obstacles. These include the requirement for anaerobic growth conditions, insensitivity to common antibiotics, and ignorance of the physiology of methanogens. Nevertheless, the genetics of methanogenes is likely to be a rewarding area of research and contribute greatly to our understanding of methanogens.

ACKNOWLEDGMENTS

I would like to thank the following people for providing me with manuscripts prior to their publication: G. D. Vogels, G. D. Sprott, R. A. Mah, M. J. Wolin, J. G. Ferry, J. Escalante-Semerena, and J. Leigh. I would also like to thank L. Baresi for stimulating discussions of ongoing research and P. Paradise for typing the manuscript and subsequent revisions. This work was supported in part by research grant PCM-82140681 from the National Science Foundation.

REFERENCES

Abram, J. W., and Nedwell, D. B. (1978a). Inhibition of methanogenesis by sulfate reducing bacteria competing for transferred hydrogen. *Arch. Microbiol.* **117,** 89–92.
Abram, J. W., and Nedwell, D. B. (1978b). Hydrogen as a substrate for methanogenesis and sulphate reduction in anaerobic saltmarsh sediment. *Arch. Microbiol.* **117,** 93–97.
Albracht, S. P. J., Graf, E.-G., and Thauer, R. K. (1982). The EPR properties of nickel in hydrogenase from *Methanobacterium thermoautotrophicum. FEBS Lett.* **140,** 311–313.
Anthony, C. (1982). "The Biochemistry of Methylotrophs." Academic Press, London.

Apostolides, Z., Vermeulen, N. M. J., and Potgieter, P. J. J. (1982). High-performance liquid chromatography of some coenzyme M (2-mercaptoethanesulfonic acid) derivatives by ion pairing on reversed-phase columns. *J. Chromatogr.* **246**, 304–307.

Archer, D. B., and King, N. R. (1983). A novel ultrastructural feature of a gas-vacuolated *Methanosarcina*. *FEMS Microbiol. Lett.* **16**, 217–223.

Ashton, W. T., and Brown, R. D. (1980). Synthesis of 8-demethyl-8-hydroxy-5-deazariboflavins. *J. Heterocycl. Chem.* **17**, 1709–1712.

Ashton, W. T., Brown, R. D., Jacobson, F., and Walsh, C. (1979). Synthesis of 7,8-didemethyl-8-hydroxy-5-deazariboflavin and confirmation of its identity with the deazaisoalloxazine chromophore of *Methanobacterium* redox coenzyme F_{420}. *J. Am. Chem. Soc.* **101**, 4419–4420.

Balch, W. E., and Wolfe, R. S. (1976). New approach to the cultivation of methanogenic bacteria: 2-mercaptoethanesulfonic acid (HS-CoM)-dependent growth of *Methanobacterium ruminantium* in a pressurized atmosphere. *Appl. Environ. Microbiol.* **32**, 781–791.

Balch, W. E., and Wolfe, R. S. (1979a). Specificity and biological distribution of coenzyme M (2-mercaptoethanesulfonic acid). *J. Bacteriol.* **137**, 256–263.

Balch, W. E., and Wolfe, R. S. (1979b). Transport of coenzyme M (2-mercaptoethanesulfonic acid) in *Methanobacterium ruminantium*. *J. Bacteriol.* **137**, 264–273.

Balch, W. E., Fox, G. E., Magrum, L. J., Woese, C. R., and Wolfe, R. S. (1979). Methanogens: Reevaluation of a unique biological group. *Microbiol. Rev.* **43**, 260–296.

Balderston, W. L., and Payne, W. J. (1976). Inhibition of methanogenesis in salt marsh sediments and whole-cell suspensions of methanogenic bacteria by nitrogen oxides. *Appl. Environ. Microbiol.* **32**, 264–269.

Barber, M. J., Siegel, L. M., Schauer, N. L., May, H. D., and Ferry, J. G. (1983). Formate dehydrogenase from *Methanobacterium formicicum*. *J. Biol. Chem.* **258**, 10839–10845.

Baresi, L. (1983). Methanogenic cleavage of acetate by lysates of *Methanosarcina barkeri*. *J. Bacteriol.* **160**, 365–370.

Baresi, L., and Wolfe, R. S. (1981). Levels of coenzyme F_{420}, coenzyme M, hydrogenase, and methylcoenzyme M methylreductase in acetate-grown *Methanosarcina*. *Appl. Environ. Microbiol.* **41**, 388–391.

Baresi, L., Mah, R. A., Ward, D. M., and Kaplan, I. R. (1978). Methanogenesis from acetate: enrichment studies. *Appl. Environ. Microbiol.* **36**, 186–197.

Barker, H. A. (1956). "Bacterial Fermentations." Wiley, New York.

Bauchop, T. (1967). Inhibition of rumen methanogenesis by methane analogues. *J. Bacteriol.* **94**, 171–175.

Belyaev, S. S., Wolkin, R., Kenealy, W. R., Deniro, M. J., Epstein, S., and Zeikus, J. G. (1983). Methanogenic bacteria from the Bondyuzhskoe oil field: General characterization and analysis of stable-carbon isotopic fractionation. *Appl. Environ. Microbiol.* **45**, 691–697.

Blaut, M., and Gottschalk, G. (1982). Effect of trimethylamine on acetate utilization by *Methanosarcina barkeri*. *Arch. Microbiol.* **133**, 230–235.

Blaylock, B. A. (1968). Cobamide-dependent methanol-cyanocob(I)alamin methyl-transferase of *Methanosarcina barkeri*. *Arch. Biochem. Biophys.* **124**, 314–324.

Blaylock, B. A., and Stadtman, T. C. (1963). Biosynthesis of methane from the methyl moiety of methylcobalamin. *Biochem. Biophys. Res. Commun.* **11**, 34–38.

Blaylock, B. A., and Stadtman, T. C. (1964). Enzymic formation of methylcobalamin in *Methanosarcina barkerii* extracts. *Biochem. Biophys. Res. Commun.* **17**, 475–480.

Blaylock, B. A., and Stadtman, T. C. (1966). Methane biosynthesis by *Methanosarcina barkeri*. Properties of the soluble enzyme system. *Arch. Biochem. Biophys.* **116**, 138–152.

Bollschweiler, C., and Klein, A. (1982). Polypeptide synthesis in *Escherichia coli* directed by cloned *Methanobrevibacter arboriphilus* DNA. *Zentralbl. Bakteriol., Mikrobiol. Hyg., Abt. 1, Orig. C* **3**, 101–109.

Boone, D. R., and Bryant, M. P. (1980). Propionate-degrading bacterium, *Syntrophobacter wolinii* sp. nov. gen. nov., from methanogenic ecosystems. *Appl. Environ. Microbiol.* **40,** 626–632.

Brandis, A., Thauer, R. K., and Stetter, K. O. (1981). Relatedness of strains ΔH and Marburg of *Methanobacterium thermoautotrophicum. Zentralbl. Bakteriol., Mikrobiol. Hyg.,* Abt. 1, Orig. C **2,** 311–317.

Breuil, C., and Patel, G. B. (1980a). Composition of *Methanospirillum hungatii* GP1 during growth on different media. *Can. J. Microbiol.* **26,** 577–582.

Breuil, C., and Patel, G. B. (1980b). Viability and depletion of cell constituents of *Methanospirillum hungatii* GP1 during starvation. *Can. J. Microbiol.* **26,** 887–892.

Bruggen, J. J. A., Stumm, C. K., and Vogels, G. D. (1983). Symbiosis of methanogenic bacteria and sapropelic protozoa. *Arch. Microbiol.* **136,** 89–95.

Bryant, M. P. (1974). Part 13. Methane-producing bacteria. *In* "Bergey's Manual of Determinative Bacteriology" (R. E. Buchanan and N. E. Gibbons, eds.), pp. 472–477. Williams & Wilkins, Baltimore, Maryland.

Bryant, M. P. (1979). Microbial methane production—theoretical aspects. *J. Anim. Sci.* **48,** 193–201.

Bryant, M. P., Wolin, E. A., Wolin, M. J., and Wolfe, R. S. (1967). *Methanobacillus omelianskii,* a symbiotic association of two species of bacteria. *Arch. Mikrobiol.* **59,** 20–31.

Bryant, M. P., Tzeng, S. F., Robinson, I. M., and Joyner, A. E., Jr. (1971). Nutrient requirements of methanogenic bacteria. *Adv. Chem. Ser.* **105,** 23–40.

Bryant, M. P., Campbell, L. L., Reddy, C. A., and Crabill, M. R. (1977). Growth of desulfovibrio in lactate or ethanol media low in sulfate in association with H_2-utilizing methanogenic bacteria. *Appl. Environ. Microbiol.* **33,** 1162–1169.

Buchanan, B. B., and Rabinowitz, J. C. (1964). Some properties of *Methanol acterium omelianskii* ferredoxin. *J. Bacteriol.* **88,** 806–807.

Buswell, A. M., and Sollo, F. W. (1948). The mechanism of the methane fermentation. *J. Am. Chem. Soc.* **70,** 1778–1780.

Butsch, B. M., and Bachofen, R. (1982). Measurement of the membrane potential in *Methanobacterium thermoautotrophicum. Experientia* **38,** 1377.

Cappenberg, T. E. (1974a). Interrelations between sulfate-reducing and methane-producing bacteria in bottom deposits of a fresh-water lake. 1. Field observations. *Antonie van Leeuwenhoek* **40,** 285–295.

Cappenberg, T. E. (1974b). Interrelations between sulfate-reducing and methane-producing bacteria in bottom deposits of a fresh-water lake. II. Inhibition experiments. *Antonie van Leeuwenhoek* **40,** 297–306.

Cappenberg, T. E. (1975). Relationships between sulfate reducing and methane producing bacteria. *Plant Soil* **43,** 123–129.

Cappenberg, T. E., and Prins, R. A. (1974). Interrelations between sulfate-reducing and methane-producing bacteria in bottom deposits of a fresh-water lake. III. Experiments with ^{14}C-labeled substrates. *Antonie van Leeuwenhoek* **40,** 457–469.

Carroll, E. J., and Hungate, R. E. (1955). Formate dissimilation and methane production in bovine rumen contents. *Arch. Biochem. Biophys.* **56,** 525–536.

Cheeseman, P., Toms-Wood, A., and Wolfe, R. S. (1972). Isolation and properties of a fluorescent compound, factor$_{420}$, from *Methanobacterium* strain M.o.H. *J. Bacteriol.* **112,** 527–531.

Chen, M., and Wolin, M. J. (1977). Influence of CH_4 production by *Methanobacterium ruminantium* on the fermentation of glucose and lactate by *Selenomonas ruminantium. Appl. Environ. Microbiol.* **34,** 756–759.

Chua, H. B., and Robinson, J. P. (1983). Formate-limited growth of *Methanobacterium formicicum* in steady-state cultures. *Arch. Microbiol.* **135,** 158–160.

Conrad, R., and Thauer, R. K. (1983). Carbon monoxide production by *Methanobacterium thermoautotrophicum. FEMS Microbiol. Lett.* **20,** 229–232.

Conway de Macario, E., Wolin, M. J., and Macario, A. J. L. (1981). Immunology of archaebacteria that produce methane gas. *Science* **214**, 74–75.

Conway de Macario, E., Wolin, M. J., and Macario, A. J. L. (1982a). Antibody analysis of relationships among methanogenic bacteria. *J. Bacteriol.* **149**, 316–319.

Conway de Macario, E., Macario, A. J. L., and Wolin, M. J. (1982b). Specific antisera and immunological procedures for characterization of methanogenic bacteria. *J. Bacteriol.* **149**, 320–328.

Conway de Macario, E., Macario, A. J. L., and Wolin, M. J. (1982c). Antigenic analysis of *Methanomicrobiales* and *Methanobrevibacter arboriphilus. J. Bacteriol.* **152**, 762–764.

Corder, R. E., Hook, L. A., Larkin, J. M., and Frea, J. I. (1983). Isolation and characterization of two new methane-producing cocci: *Methanogenium olentangyi,* sp. nov., and *Methanococcus deltae,* sp. nov. *Arch. Microbiol.* **134**, 28–32.

Daniels, L., and Zeikus, J. G. (1978). One-carbon metabolism in methanogenic bacteria: Analysis of short-term fixation products of $^{14}CO_2$ and $^{14}CH_3OH$ incorporated into whole cells. *J. Bacteriol.* **136**, 75–84.

Daniels, L., Fuchs, G., Thauer, R. K., and Zeikus, J. G. (1977). Carbon monoxide oxidation by methanogenic bacteria. *J. Bacteriol.* **132**, 118–126.

Daniels, L., Fulton, G., Spencer, R. W., and Orme-Johnson, W. H. (1980). Origin of hydrogen in methane produced by *Methanobacterium thermoautotrophicum. J. Bacteriol.* **141**, 694–698.

Diekert, G., Klee, B., and Thauer, R. K. (1980a). Nickel, a component of Factor F_{430} from *Methanobacterium thermoautotrophicum. Arch. Microbiol.* **124**, 103–106.

Diekert, G., Weber, B., and Thauer, R. K. (1980b). Nickel dependence of factor F_{430} content in *Methanobacterium thermoautotrophicum. Arch. Microbiol.* **127**, 273–278.

Diekert, G., Jaenchen, R., and Thauer, R. K. (1980c). Biosynthetic evidence for a nickel tetrapyrrole structure of factor F_{430} from *Methanobacterium thermoautotrophicum. FEBS Lett.* **119**, 118–120.

Diekert, G., Konheiser, U., Piechulla, K., and Thauer, R. K. (1981). Nickel requirement and factor F_{430} content of methanogenic bacteria. *J. Bacteriol.* **148**, 459–464.

Doddema, H. J., and Vogels, G. D. (1978). Improved identification of methanogenic bacteria by fluorescence microscopy. *Appl. Environ. Microbiol.* **36**, 752–754.

Doddema, H. J., Hutten, T. J., Van der Drift, C., and Vogels, G. D. (1978). ATP hydrolysis and synthesis by the membrane-bound ATP synthetase complex of *Methanobacterium thermoautotrophicum. J. Bacteriol.* **136**, 19–23.

Doddema, H. J., Derksen, J. W. M., and Vogels, G. D. (1979a). Fimbrae and flagella of methanogenic bacteria. *FEMS Microbiol. Lett.* **5**, 135–138.

Doddema, H. J., Van der Drift, C., Vogels, G. D., and Veenhuis, M. (1979b). Chemiosmotic coupling in *Methanobacterium thermoautotrophicum:* Hydrogen-dependent adenosine 5'-triphosphate synthesis by subcellular particles. *J. Bacteriol.* **140**, 1081–1089.

Doddema, H. J., Claesen, C. A., Kell, D. B., Van der Drift, C., and Vogels, G. D. (1980). An adenine nucleotide translocase in the procaryote *Methanobacterium thermoautotrophicum. Biochem. Biophys. Res. Commun.* **95**, 1288–1293.

Drake, H. L., Hu, S.-I., and Wood, H. G. (1981). Purification of five components from *Clostridium thermoaceticum* which catalyze synthesis of acetate from pyruvate and methyltetrahydrofolate. *J. Biol. Chem.* **256**, 11137–11144.

Edwards, T., and McBride, B. C. (1975). New method for the isolation and identification of methanogenic bacteria. *Appl. Microbiol.* **29**, 540–545.

Ehhalt, D. H. (1974). The atmospheric cycle of methane. *Tellus* **26**, 58–70.

Ehhalt, D. H., and Schmidt, V. (1978). Sources and sinks of atmospheric methane. *Pure Appl. Geophys.* **116**, 452–463.

Eikmanns, B., Jaenchen, R., and Thauer, R. K. (1983a). Propionate assimilation by methanogenic bacteria. *Arch. Microbiol.* **136**, 106–110.

Eikmanns, B., Linder, D., and Thauer, R. K. (1983b). Unusual pathway of isoleucine biosynthesis in *Methanobacterium thermoautotrophicum*. *Arch. Microbiol.* **136**, 111–113.

Eirich, L. D., Vogels, G. D., and Wolfe, R. S. (1978). Proposed structure for coenzyme F_{420} from *Methanobacterium*. *Biochemistry* **17**, 4583–4593.

Eirich, L. D., Vogels, G. D., and Wolfe, R. S. (1979). Distribution of coenzyme F_{420} and properties of its hydrolytic fragments. *J. Bacteriol.* **140**, 20–27.

Eker, A. P. M., Pol, A., Van der Meyden, P., and Vogels, G. D. (1980). Purification and properties of 8-hydroxy-5-deazaflavin derivatives from *Streptomyces griseus*. *FEMS Microbiol. Lett.* **8**, 161–166.

Eker, A. P. M., Dekker, R. H., and Berends, W. (1981). Photoreactivating enzyme from *Streptomyces griseus*. IV. On the nature of the chromophoric cofactor in *Streptomyces griseus* photoreactivating enzyme. *Photochem. Photobiol.* **33**, 65–72.

Ekiel, I., Smith, I. C. P., and Sprott, G. D. (1983). Biosynthetic pathways in *Methanospirillum hungatei* as determined by ^{13}C nuclear magnetic resonance. *J. Bacteriol.* **156**, 316–326.

Ellefson, W. L., and Wolfe, R. S. (1980). Role of component C in the methyl-reductase system of *Methanobacterium*. *J. Biol. Chem.* **255**, 8388–8389.

Ellefson, W. L., and Wolfe, R. S. (1981). Component C of the methylreductase system of *Methanobacterium*. *J. Biol. Chem.* **256**, 4259–4262.

Ellefson, W. L., Whitman, W. B., and Wolfe, R. S. (1982). Nickel-containing factor F_{430}: Chromophore of the methylreductase of *Methanobacterium*. *Proc. Natl. Acad. Sci. U.S.A.* **79**, 3707–3710.

Escalante-Semerena, J. C., and Wolfe, R. S. (1984). Formaldehyde oxidation and methanogenesis. *J. Bacteriol.* **158**, 721–726.

Escalante-Semerena, J. C., Leigh, J. A., and Wolfe, R. S. (1984a). New insights into the biochemistry of methanogenesis from H_2 and Co_2. *In* "Microbial Growth on C_1 Compounds" (R. L. Crawford and R. S. Hanson, eds.), pp. 191–198. Amer. Soc. Microbiol., Washington, D.C.

Escalante-Semerena, J. C., Leigh, J. A., Reinhart, K. L., and Wolfe, R. S. (1984b). Formaldehyde activation factor, tetrahydromethanopterin, a coenzyme of methanogenesis. *Proc. Natl. Acad. Sci. U.S.A.* **81**, 1976–1980.

Escalante-Semerena, J. C., Rinehart, K. L., and Wolfe, R. S. (1984c). Tetrahydromethanopterin, a carbon carrier in methanogenesis. *J. Biol. Chem,* **259**, 9447–9455.

Eyzaguirre, J., Jansen, K., and Fuchs, G. (1982). Phosphoenolpyruvate synthetase in *Methanobacterium thermoautotrophicum*. *Arch. Microbiol.* **132**, 67–74.

Fathepure, B. Z. (1983). Isolation and characterization of an aceticlastic methanogen from a biogas digester. *FEMS Microbiol. Lett.* **19**, 151–156.

Ferguson, T. J., and Mah, R. A. (1983). Isolation and characterization of an H_2-oxidizing thermophilic methanogen. *Appl. Environ. Microbiol.* **45**, 265–274.

Ferry, J. G., and Wolfe, R. S. (1977). Nutritional and biochemical characterization of *Methanospirillum hungatii*. *Appl. Environ. Microbiol.* **34**, 371–376.

Ferry, J. G., Smith, P. H., and Wolfe, R. S. (1974). *Methanospirillum*, a new genus of methanogenic bacteria, and characterization of *Methanospirillum hungatii*, sp. nov. *Int. J. Syst. Bacteriol.* **24**, 465–469.

Ferry, J. G., Sherod, R. D., Peck, H. D., and Ljungdahl, L. G. (1976). Autotrophic fixation of CO_2 via tetrahydrofolate intermediates by *Methanobacterium thermoautotrophicum*. *In* "Microbial Production and Utilization of Gases (H_2, CH_4, CO)" (H. Schlegel, G. Gottschalk, and N. Pfennig, eds.), pp. 151–155. Akad. Wiss. Göttingen, Goltze, Göttingen.

Forrest, W. W., and Walker, D. J. (1971). The generation and utilization of energy during growth. *Adv. Microb. Physiol.* **5**, 213–274.

Frick, T., Francia, M. D., and Wood, J. M. (1976). Mechanism for the interaction of thiols with methylcobalamin. *Biochim. Biophys. Acta* **428**, 808–818.

Friedrich, C. G., Schneider, K., and Friedrich, B. (1982). Nickel in the catalytically active hydrogenase of *Alcaligenes eutrophus*. *J. Bacteriol.* **152,** 42–48.

Friedrich, W., and Bernhauer, K. (1953). Uber einen neuen vitamin B_{12}-faktor. *Angew. Chem.* **65,** 627–628.

Fuchs, G., and Stupperich, E. (1978). Evidence for an incomplete reductive carboxylic acid cycle in *Methanobacterium thermoautotrophicum*. *Arch. Microbiol.* **118,** 121–125.

Fuchs, G., and Stupperich, E. (1980). Acetyl CoA, a central intermediate of autotrophic CO_2 fixation in *Methanobacterium thermoautotrophicum*. *Arch. Microbiol.* **127,** 267–272.

Fuchs, G., and Stupperich, E. (1982). Autotrophic CO_2 fixation pathway in *Methanobacterium thermoautotrophicum*. *Zentralbl. Baketeriol., Mikrobiol. Hyg. Abt. 1, Orig. C* **3,** 277–288.

Fuchs, G., Stupperich, E., and Thauer, R. K. (1978a). Acetate assimilation and the synthesis of alanine, aspartate, and glutamate in *Methanobacterium thermoautotrophicum*. *Arch. Microbiol.* **117,** 61–66.

Fuchs, G., Stupperich, E., and Thauer, R. K. (1978b). Function of fumarate reductase in methanogenic bacteria (*Methanobacterium*), *Arch. Microbiol.* **119,** 215–218.

Fuchs, G., Thauer, R., Ziegler, H., and Stichler, W. (1979a). Carbon isotope fractionation by *Methanobacterium thermoautotrophicum*. *Arch. Microbiol.* **120,** 135–139.

Fuchs, G., Moll, J., Scherer, P., and Thauer, R. (1979b). Activity, acceptor specificity and function of hydrogenase in *Methanobacterium thermoautotrophicum*. *In* "Hydrogenases: Their Catalytic Activity, Structure and Function" (H. G. Schlegel, ed.), pp. 83–92. Akad. Wiss. Göttingen, Goltze, Göttingen.

Fuchs, G., Winter, H., Steiner, I., and Stupperich, E. (1983). Enzymes of gluconeogenesis in the autotroph *Methanobacterium thermoautotrophicum*. *Arch. Microbiol.* **136,** 160–162.

Games, L. M., Hayes, J. M., and Gunsalus, R. P. (1978). Methane-producing bacteria: Natural fractionations of the stable carbon isotopes. *Geochim. Cosmochim. Acta* **42,** 1295–1297.

Genthner, B. R. S., Davis, C. L., and Bryant, M. P. (1981). Features of rumen and sewage sludge strains of *Eubacterium limosum,* a methanol- and H_2-CO_2-utilizing species. *Appl. Environ. Microbiol.* **42,** 12–19.

Gilles, H., and Thauer, R. K. (1983). Uroporphyrinogen III, an intermediate in the biosynthesis of the nickel-containing factor F_{430} in *Methanobacterium thermoautotrophicum*. *Eur. J. Biochem.* **135,** 109–112.

Gilles, H., Jaenchen, R., and Thauer, R. K. (1983). Biosynthesis of 5-aminolevulinic acid in *Methanobacterium thermoautotrophicum*. *Arch. Microbiol.* **135,** 237–240.

Godsy, E. M. (1980). Isolation of *Methanobacterium bryantii* from a deep aquifer by using a novel broth-antibiotic disk method. *Appl. Environ. Microbiol.* **39,** 1074–1075.

Graf, E.-G., and Thauer, R. K. (1981). Hydrogenase from *Methanobacterium thermoautotrophicum,* a nickel-containing enzyme. *FEBS Lett.* **136,** 165–169.

Grauert, R. W. (1980). Donator substituted 5-deazaflavins. II. Hydroxy and 8-dimethylamino-5-deazaflavins - Model compounds of naturally occurring (deaza)-flavocoenzymes. *Arch. Pharm. (Weinheim Ger.)* **313,** 937–950.

Gunsalus, R. P., and Wolfe, R. S. (1977). Stimulation of CO_2 reduction to methane by methylcoenzyme M in extracts of *Methanobacterium*. *Biochem. Biophys. Res. Commun.* **76,** 790–795.

Gunsalus, R. P., and Wolfe, R. S. (1978a). ATP activation and properties of the methyl coenzyme M reductase system in *Methanobacterium thermoautotrophicum*. *J. Bacteriol.* **135,** 851–857.

Gunsalus, R. P., and Wolfe, R. S. (1978b). Chromophoric factors F_{342} and F_{430} of *Methanobacterium thermoautotrophicum*. *FEMS Microbiol Lett.* **3,** 191–193.

Gunsalus, R. P., and Wolfe, R. S. (1980). Methyl coenzyme M reductase from *Methanobacterium thermoautotrophicum*. *J. Biol. Chem.* **255,** 1891–1895.

Gunsalus, R. P., Eirich, D., Romesser, J., Balch, W., Shapiro, S., and Wolfe, R. S. (1976). Methyl-transfer and methane formation. *In* "Microbial Production and Utilization of Gases (H_2,

CH$_4$, CO)'' (H. G. Schlegel, G. Gottschalk, N. Pfennig, eds.), pp. 191–198. Akad. Wiss. Göttingen, Goltze, Göttingen.

Gunsalus, R. P., Romesser, J. A., and Wolfe, R. S. (1978). Preparation of coenzyme M analogues and their activity in the methyl coenzyme M reductase system of *Methanobacterium thermoautotrophicum*. *Biochemistry* **17**, 2374–2377.

Hamilton, P. T., and Reeve, J. N. (1984). Cloning and expression of archaebacterial DNA from methanogens in *Escherichia coli*. *In* "Microbial Chemoautotrophy" (W. R. Strohl and O. H. Tuovinen, eds.), pp. 291–308. Ohio State Univ. Press, Columbus.

Hatchikian, E. C., Bruschi, M., Forget, N., and Scandellari, M. (1982). Electron transport components from methanogenic bacteria: The ferredoxin from *Methanosarcina barkeri* (strain Fusaro). *Biochem. Biophys. Res. Commun.* **109**, 1316–1323.

Hausinger, R. P., Moura, I., Moura, J. J. G., Xavier, A. V., Santos, M. H., LeGall, J., and Howard, J. B. (1982). Amino acid sequence of a 3Fe:3S ferredoxin from the ''Archaebacterium'' *Methanosarcina barkeri* (DSM 800). *J. Biol. Chem.* **257**, 14192–14197.

Hausinger, R. P., Orme-Johnson, W. H., and Walsh, C. (1984). Nickel tetrapyrrole cofactor F$_{430}$: Comparison of the forms bound to methyl coenzyme M reductase and protein free in cells of *Methanobacterium thermautotrophicum* ΔH. *Biochemistry* **23**, 801–804.

Hermans, J. M. H., Hutten, T. J., Van der Drift, C., and Vogels, G. D. (1980). Analysis of coenzyme M (2-mercaptoethanesulfonic acid) derivatives by isotachophoresis. *Anal. Biochem.* **106**, 363–366.

Hines, M. E., and Buck, J. D. (1982). Distribution of methanogenic and sulfate-reducing bacteria in near-shore marine sediments. *Appl. Environ. Microbiol.* **43**, 447–453.

Hippe, H., Caspari, D., Fiebig, K., and Gottschalk, G. (1979). Utilization of trimethylamine and other N-methyl compounds for growth and methane formation by *Methanosarcina barkeri*. *Proc. Natl. Acad. Sci. U.S.A.* **76**, 494–498.

Hu, S.-I., Drake, H. L., and Wood, H. G. (1982). Synthesis of acetyl coenzyme A from carbon monoxide, methyltetrahydrofolate, and coenzyme A by enzymes from *Clostridium thermoaceticum*. *J. Bacteriol.* **149**, 440–448.

Huber, H., Thomm, M., König, H., Thies, G., and Stetter, K. O. (1982). *Methanococcus thermolithotrophicus*, a novel thermophilic lithotrophic methanogen. *Arch. Microbiol.* **132**, 47–50.

Hughes, P. E., and Tove, S. B. (1982). Occurrence of α-tocopherolquinone and α-tocopherolquinol in microorganisms. *J. Bacteriol.* **151**, 1397–1402.

Hungate, R. E. (1967). Hydrogen as an intermediate in the rumen fermentation. *Arch. Mikrobiol.* **59**, 158–164.

Hungate, R. E., Smith, W., Bauchop, T., Yu, I., and Rabinowitz, J. C. (1970). Formate as an intermediate in the bovine rumen fermentation. *J. Bacteriol.* **102**, 389–397.

Huser, B. A., Wuhrmann, K., and Zehnder, A. J. B. (1982). *Methanothrix soehngenii* gen. nov. sp. nov., a new acetotrophic non-hydrogen-oxidizing methane bacterium. *Arch. Microbiol.* **132**, 1–9.

Hutten, T. J., Bongaerts, H. C. M., Van der Drift, C., and Vogels, G. D. (1980). Acetate, methanol, and carbon dioxide as substrates for growth of *Methanosarcina barkeri*. *Antonie van Leeuwenhoek* **46**, 601–610.

Hutten, T. J., de Jong, M. H., Peeters, B. P. H., Van der Drift, C., and Vogels, G. D. (1981). Coenzyme M derivatives and their effects on methane formation from carbon dioxide and methanol by cell extracts of *Methanosarcina barkeri*. *J. Bacteriol.* **145**, 27–34.

Jacobson, F. S., Daniels, L., Fox, J. A., Walsh, C. T., and Orme-Johnson, W. H. (1982). Purification and properties of an 8-hydroxy-5-deazaflavin-reducing hydrogenase from *Methanobacterium thermoautotrophicum*. *J. Biol. Chem.* **257**, 3385–3388.

Jaenchen, R., Diekert, G., and Thauer, R. K. (1981a). Incorporation of methionine-derived methyl groups into factor F$_{430}$ by *Methanobacterium thermoautotrophicum*. *FEBS Lett.* **130**, 133–136.

Jaenchen, R., Gilles, H. H., and Thauer, R. K. (1981b). Inhibition of factor F_{430} synthesis by levulinic acid in *Methanobacterium thermoautotrophicum*. *FEMS Microbiol. Lett.* **12**, 167–170.

Jansen, K., Stupperich, E., and Fuchs, G. (1982). Carbohydrate synthesis from acetyl CoA in the autotroph *Methanobacterium thermoautotrophicum*. *Arch. Microbiol.* **132**, 355–364.

Jarrell, K. F., and Sprott, G. D. (1981). The transmembrane electrical potential and intracellular pH in methanogenic bacteria. *Can. J. Microbiol.* **27**, 720–728.

Jarrell, K. F., and Sprott, G. D. (1982). Nickel transport in *Methanobacterium bryantii*. *J. Bacteriol.* **151**, 1195–1203.

Jarrell, K. F., and Sprott, G. D. (1983). The effects of ionophores and metabolic inhibitors on methanogenesis and energy-related properties of *Methanobacterium bryantii*. *Arch. Biochem. Biophys.* **225**, 33–41.

Jarrell, K. F., Colvin, J. R., and Sprott, G. D. (1982). Spontaneous protoplast formation in *Methanobacterium bryantii*. *J. Bacteriol.* **149**, 346–353.

Jarsch, M., and Böck, A. (1983). DNA sequence of the 16S rRNA/23S rRNA intercistronic spacer of two rDNA operons of the archaebacterium *Methanococcus vannielii*. *Nucleic Acids Res.* **11**, 7537–7544.

Jarsch, M., Altenbuchner, J., and Böck, A. (1983). Physical organization of the genes for ribosomal RNA in *Methanococcus vannielii*. *Mol. Gen. Genet.* **189**, 41–47.

Jeris, J. S., and McCarty, P. L. (1965). The biochemistry of methane fermentation using C^{14} tracers. *J. Water Pollut. Control Fed.* **37**, 178–192.

Jin, S.-L.C., Blanchard, D. K., and Chen, J.-S. (1983). Two hydrogenases with distinct electron-carrier specificity and subunit composition in *Methanobacterium formicicum*. *Biochim. Biophys. Acta* **748**, 8–20.

Jones, G. E., Royle, L. G., and Murray, L. (1979). Cationic composition of 22 species of bacteria grown in seawater medium. *Appl. Environ. Microbiol.* **38**, 800–805.

Jones, J. B., and Stadtman, T. C. (1976). *Methanococcus vannielii*: Growth and metabolism of formate. *In* "Microbial Production and Utilization of Gases (H_2, CH_4, CO)" (H. G. Schlegel, G. Gottschalk, and N. Pfennig, eds.), pp. 199–205. Akad. Wiss. Göttingen, Goltze, Göttingen.

Jones, J. B., and Stadtman, T. C. (1977). *Methanococcus vannielii*: Culture and effects of selenium and tungsten on growth. *J. Bacteriol.* **130**, 1404–1406.

Jones, J. B., and Stadtman, T. C. (1980). Reconstitution of a formate-NADP⁺ oxidoreductase from formate dehydrogenase and a 5-deazaflavin-linked NADP⁺ reductase isolated from *Methanococcus vannielii*. *J. Biol. Chem.* **255**, 1049–1053.

Jones, J. B., and Stadtman, T. C. (1981). Selenium-dependent and selenium-independent formate dehydrogenases of *Methanococcus vannielli*. Separation of the two forms and characterization of the purified selenium-independent form. *J. Biol. Chem.* **256**, 656–663.

Jones, J. B., Bowers, B., and Stadtman, T. C. (1977). *Methanococcus vannielii*: ultrastructure and sensitivity to detergents and antibiotics. *J. Bacteriol.* **130**, 1357–1363.

Jones, J. B., Dilworth, G. L., and Stadtman, T. C. (1979). Occurrence of selenocysteine in the selenium-dependent formate dehydrogenase of *Methanococcus vannielii*. *Arch. Biochem. Biophys.* **195**, 225–260.

Jones, W. J., and Paynter, M. J. B. (1980). Populations of methane-producing bacteria and in vitro methanogenesis in salt marsh and estuarine sediments. *Appl. Environ. Microbiol.* **39**, 864–871.

Jones, W. J., Paynter, M. J. B., and Gupta, R. (1983). Characterization of *Methanococcus maripaludis* sp. nov., a new methanogen isolated from salt marsh sediment. *Arch. Microbiol.* **135**, 91–97.

Jones, W. J., Leigh, J. A., Mayer, F., Woese, C. R., and Wolfe, R. S. (1984). *Methanococcus jannaschii* sp. nov., an extremely thermophilic methanogen from a submarine hydrothermal vent. *Arch. Microbiol.* **136**, 254–261.

Kandler, O., and Hippe, H. (1977). Lack of peptidoglycan in the cell walls of *Methanosarcina barkeri*. *Arch. Microbiol.* **113**, 57–60.

Kanodia, S., and Roberts, M. F. (1983). Methanophosphagen: Unique cyclic pyrophosphate isolated from *Methanobacterium thermoautotrophicum*. *Proc. Natl. Acad. Sci. U.S.A.* **80**, 5217–5221.

Kell, D. B., and Morris, J. G. (1979). Oxidation-reduction properties of coenzyme M (2-mercaptoethanesulfonate) at the mercury electrode. *FEBS Lett.* **108**, 481–484.

Keltjens, J. T., and Vogels, G. D. (1981). Novel coenzymes of methanogens. *In* "Microbial Growth on C_1 Compounds" (H. Dalton, ed.), pp. 152–158. Heyden, London.

Keltjens, J. T., Whitman, W. B., Caerteling, C. G., van Kooten, A. M., Wolfe, R. S., and Vogels, G. D. (1982). Presence of coenzyme M derivatives in the prosthetic group (coenzyme MF_{430}) of methylcoenzyme M reductase from *Methanobacterium thermoautotrophicum*. *Biochem. Biophys. Res. Commun.* **108**, 495–503.

Keltjens, J. T., van Beelen, P., Stassen, A. M., and Vogels, G. D. (1983a). 7-Methylpterin in methanogenic bacteria. *FEMS Microbiol. Lett.* **20**, 259–262.

Keltjens, J. T., Caerteling, C. G., van Kooten, A. M., van Dijk, H. F., and Vogels, G. D. (1983b). Chromophoric derivatives of coenzyme MF_{430}, a proposed coenzyme of methanogenesis in *Methanobacterium thermoautotrophicum*. *Arch. Biochem. Biophys.* **223**, 235–253.

Keltjens, J. T., Caerteling, C. G., van Hooten, A. M., van Dijk, H. F., and Vogels, G. D. (1983c). 6,7-Dimethyl-8-ribityl-5,6,7,8-tetrahydrolumazine, a proposed constituent of coenzyme MF_{430} from methanogenic bacteria. *Biochim. Biophys. Acta* **743**, 351–358.

Keltjens, J. T., Huberts, M. J., Laarhoven, W. H., and Vogels, G. D. (1983d). Structural elements of methanopterin, a novel pterin present in *Methanobacterium thermoautotrophicum*. *Eur. J. Biochem.* **130**, 537–544.

Keltjens, J. T., Daniels, L., Jannsen, H. G., Borm, P. J., and Vogels, G. D. (1983e). A novel one-carbon carrier (carboxy-5,6,7,8-tetrahydromethanopterin) isolated from *Methanobacterium thermoautotrophicum* and derived from methanopterin. *Eur. J. Biochem.* **130**, 545–552.

Kenealy, W., and Zeikus, J. G. (1981). Influence of corrinoid antagonists on methanogen metabolism. *J. Bacteriol.* **146**, 133–140.

Kenealy, W. R., and Zeikus, J. G. (1982). One-carbon metabolism in methanogens: Evidence for synthesis of a two-carbon cellular intermediate and unification of catabolism and anabolism in *Methanosarcina barkeri*. *J. Bacteriol* **151**, 932–941.

Kenealy, W. R., Thompson, T. E., Schubert, K. R., and Zeikus, J. G. (1982). Ammonia assimilation and synthesis of alanine, aspartate, and glutamate in *Methanosarcina barkeri* and *Methanobacterium thermoautotrophicum*. *J. Bacteriol.* **150**, 1357–1365.

Kiener, A., and Leisinger, T. (1983). Oxygen sensitivity of methanogenic bacteria. *Syst. Appl. Microbiol.* **4**, 305–312.

King, G. M., and Klug, M. J. (1982). Glucose metabolism in sediments of an eutrophic lake: Tracer analysis of uptake and product formation. *Appl. Environ. Microbiol.* **44**, 1308–1317.

King, G. M., Berman, T., and Wiebe, W. J. (1981). Methane formation in the acidic peats of Okefenokee Swamp, Georgia. *Am. Midl. Nat.* **105**, 386–389.

Kirby, T. W., Lancaster, J. R., Jr., and Fridovich, I. (1981). Isolation and characterization of the iron-containing superoxide dismutase of *Methanobacterium bryantii*. *Arch. Biochem. Biophys.* **210**, 140–148.

Kluyver, A. J., and Schnellen, C. G. J. P. (1947). On the fermentation of carbon monoxide by pure cultures of methane bacteria. *Arch. Biochem.* **14**, 57–70.

Kojima, N., Fox, J. A., Hausinger, R. P., Daniels, L., Orme-Johnson, W. H., and Walsh, C. (1983). Paramagnetic centers in the nickel-containing, deazaflavin-reducing hydrogenase from *Methanobacterium thermoautotrophicum*. *Proc. Natl. Acad. Sci. U.S.A.* **80**, 378–382.

König, H., and Stetter, K. O. (1982). Isolation and characterization of *Methanolobus tindarius*, sp.

nov., a coccoid methanogen growing only on methanol and methylamines. *Zentralbl. Bakteriol., Mikrobiol. Hyg., Abt. 1, Orig. C* **3**, 478–482.

Koyama, T. (1963). Gaseous metabolism in lake sediments and paddy soils and the production of atmospheric methane and hydrogen. *J. Geophys. Res.* **68**, 3971–3973.

Kristjansson, J. K., Schönheit, P., and Thauer, R. K. (1982). Different K_S values for hydrogen of methanogenic bacteria and sulfate reducing bacteria: an explanation for the apparent inhibition of methanogenesis by sulfate. *Arch. Microbiol.* **131**, 278–282.

Krüger, H.-J., Huynh, B. H., Ljungdahl, P. O., Xavier, A. V., DerVartanian, D. V., Moura, I., Peck, H. D., Teixeira, M., Moura, J. J. G., and LeGall, J. (1982). Evidence for nickel and a three-iron center in the hydrogenase of *Desulfovibrio desulfuricans*. *J. Biol. Chem.* **257**, 14620–14623.

Krumholz, L. R., Forsberg, C. W., and Veira, D. M. (1983). Association of methanogenic bacteria with rumen protozoa. *Can. J. Microbiol.* **29**, 676–680.

Krzycki, J. A., and Zeikus, J. G. (1980). Quantification of corrionids in methanogenic bacteria. *Curr. Microbiol.* **3**, 243–245.

Krzycki, J. A., Wolkin, R. H., and Zeikus, J. G. (1982). Comparison of unitrophic and mixotrophic substrate metabolism by an acetate-adapted strain of *Methanosarcina barkeri*. *J. Bacteriol.* **149**, 247–254.

Kühn, W., and Gottschalk, G. (1983). Characterization of the cytochromes occurring in *Methanosarcina* species. *Eur. J. Biochem.* **135**, 89–94.

Kühn, W., Fiebig, K., Walther, R., and Gottschalk, G. (1979). Presence of a cytochrome b_{559} in *Methanosarcina barkeri*. *FEBS Lett.* **105**, 271–274.

Lancaster, J. R., Jr. (1980). Soluble and membrane-bound paramagnetic centers in *Methanobacterium bryantii*. *FEBS Lett.* **115**, 285–288.

Lancaster, J. R., Jr. (1981). Membrane-bound flavin adenine dinucleotide in *Methanobacterium bryantii*. *Biochem. Biophys. Res. Commun.* **100**, 240–246.

Lancaster, J. R., Jr. (1982a). Identification and detection of electron transfer components in methanogens. *In* Methods in Enzymology'' (L. Packer, ed.), Vol. 88, pp. 412–416. Academic Press. New York.

Lancaster, J. R., Jr. (1982b). New biological paramagnetic center: octahedrally coordinated nickel (III) in the methanogenic bacteria. *Science* **216**, 1324–1325.

Langenberg, K. F., Bryant, M. P., and Wolfe, R. S. (1968). Hydrogen-oxidizing methane bacteria. II. Electron microscopy. *J. Bacteriol.* **95**, 1124–1129.

LeGall, J., Ljungdahl, P. O., Moura, I., Peck, H. D., Jr., Xavier, A. V., Moura, J. J. G., Teixera, M., Huynh, B. H., and DerVartanian, D. V. (1982). The presence of redox-sensitive nickel in the periplasmic hydrogenase from *Desulfovibrio gigas*. *Biochem. Biophys. Res. Commun.* **106**, 610–616.

Leigh, J. A. (1983a). Levels of water-soluble vitamins in methanogenic and non-methanogenic bacteria. *Appl. Environ. Microbiol.* **45**, 800–803.

Leigh, J. A. (1983b). The structure of the carbon dioxide reduction factor, a novel carbon carrier in *Methanobacterium thermoautotrophicum*. Ph.D. Thesis, University of Illinois, Urbana.

Leigh, J. A., and Wolfe, R. S. (1983). Carbon dioxide reduction factor and methanopterin, two coenzymes required for CO_2 reduction to methane by extracts of *Methanobacterium*. *J. Biol. Chem.* **258**, 7536–7540.

Leigh, J. A., Rinehart, K. L., and Wolfe, R. S. (1984). Structure of methanofuran, the carbon dioxide reduction factor of *Methanobacterium thermoautotrophicum*. *J. Am. Chem. Soc.* **106**, 3636–3640.

Leigh, J. A., Rinehart, K. L., and Wolfe, R. S. (1985). Methanofuran (carbon dioxide reduction factor), a formyl carrier in methane production from carbon dioxide in *Methanobacterium*. *Biochemistry* (in press).

Lezius, A. G., and Barker, H. A. (1965). Corrinoid compounds of *Methanobacillus omelianskii*. I. Fractionation of the corrinoid compounds and identification of factor III and factor III coenzyme. *Biochemistry* **4**, 510–518.

Livingston, D. A., Pfaltz, A., Schreiber, J., Eschenmoser, A., Ankel-Fuchs, D., Moll, J., Jaenchen, R., and Thauer, R. K. (1984). Zur kenntnis des faktors F430 aus methanogenen bakterien: Struktur des proteinfreien faktors. *Helv. Chim. Acta* **67**, 334–351.

Lovely, D. R., Dwyer, D. F., and Klug, M. J. (1982). Kinetic analysis of competition between sulfate reducers and methanogens for hydrogen in sediments. *Appl. Environ. Microbiol.* **43**, 1373–1379.

McBride, B. C., and Wolfe, R. S. (1971a). Inhibition of methanogenesis by DDT. *Nature (London)* **234**, 551–552.

McBride, B. C., and Wolfe, R. S. (1971b). A new coenzyme of methyl transfer, coenzyme M. *Biochemistry* **10**, 2317–2324.

McBride, B. C., and Wolfe, R. S. (1971c). Biosynthesis of dimethylarsine by *Methanobacterium*. *Biochemistry* **10**, 4312–4317.

McCarty, P. L. (1972). Energetics of organic matter degradation. *In* "Water Pollution Microbiology" (R. Mitchell, ed.), pp. 91–113. Wiley (Interscience), New York.

McInerney, M. J., and Bryant, M. P. (1981). Anaerobic degradation of lactate by syntrophic associations of *Methanosarcina barkeri* and *Desulfovibrio* species and effect of H_2 on acetate degradation. *Appl. Environ. Microbiol.* **41**, 346–354.

McInerney, M. J., Bryant, M. P., and Pfennig, N. (1979). Anaerobic bacterium that degrades fatty acids in syntrophic association with methanogens. *Arch. Microbiol.* **122**, 129–135.

McInerney, M. J., Mackie, R. I., and Bryant, M. P. (1981a). Syntrophic association of a butyrate-degrading bacterium and *Methanosarcina* enriched from bovine rumen fluid. *Appl. Environ. Microbiol.* **41**, 826–828.

McInerney, M. J., Bryant, M. P., Hespell, R. B., and Costerton, J. W. (1981b). *Syntrophomonas wolfei* gen. nov. sp. nov., an anaerobic syntrophic fatty-acid oxidizing bacterium. *Appl. Environ. Microbiol.* **41**, 1029–1039.

McKellar, R. C., and Sprott, G. D. (1979). Solubilization and properties of a particulate hydrogenase from *Methanobacterium* strain G2R. *J. Bacteriol.* **139**, 231–238.

McKellar, R. C., Shaw, K. M., and Sprott, G. D. (1981). Isolation and characterization of a FAD-dependent NADH diaphorase from *Methanospirillum hungatei* strain GP1. *Can. J. Biochem.* **59**, 83–91.

Mackie, R. I., and Bryant, M. P. (1981). Metabolic activity of fatty acid-oxidizing bacteria and the contribution of acetate, propionate, butyrate, and CO_2 to methanogenesis in cattle waste at 40 and 60°C. *Appl. Environ. Microbiol.* **41**, 1363–1373.

Mah, R. A. (1980). Isolation and characterization of *Methanococcus mazei*. *Curr. Microbiol.* **3**, 321–326.

Mah, R. A. (1982). Methanogenesis and methanogenic partnerships. *Philos. Trans. R. Soc. London, Ser. B* **297**, 599–616.

Mah, R. A., and Kuhn, D. A. (1984). Transfer of the type species of the genus *Methanococcus* to the genus *Methanosarcina*, naming it *Methanosarcina mazei* (Barker 1936) comb. nov. et emend. and conservation of the genus *Methanococcus* (Approved Lists 1980) with *Methanococcus vannielii* (Approved Lists 1980) as the type species. *Int. J. Syst. Bact.* **34**, 263–265.

Mah, R. A., Smith, M. R., and Baresi, L. (1977a). Isolation and characterization of a gas-vacuolated Methanosarcina. *Am. Soc. Microbiol., Abstr. Annu. meet.* p. 160.

Mah, R. A., Ward, D. M., Baresi, L., and Glass, T. L. (1977b). Biogenesis of methane. *Annu. Rev. Microbiol.* **31**, 309–341.

Mah, R. A., Smith, M. R., and Baresi, L. (1978). Studies on an acetate-fermenting strain of *Methanosarcina*. *Appl. Environ. Microbiol.* **35**, 1174–1184.

Martens, C. S., and Berner, R. A. (1974). Methane production in the interstitial waters of sulfate-depleted marine sediments. *Science* **185**, 1167–1169.

Martens, C. S., and Berner, R. A. (1977). Interstitial water chemistry of anoxic Long Island Sound sediments. 1. Dissolved gases. *Limnol. Oceanogr.* **22**, 10–25.

Matheson, A. T., and Yaguchi, M. (1982). The evolution of the archaebacterial ribosome. *Zentralbl. Bakteriol., Mikrobiol. Hyg., Abt. 1, Orig. C* **3**, 192–199.

Meijden, P., Heythuysen, H. J., Pouwels, A., Houwen, F., van der Drift, C., and Vogels, G. D. (1983a). Methyltransferases involved in methanol conversion by *Methanosarcina barkeri*. *Arch. Microbiol.* **134**, 238–242.

Meijden, P., Heythuysen, H. J., Sliepenbeek, H. T., Houwen, F. P., van der Drift, C., and Vogels, G. D. (1983b). Activation and inactivation of methanol: 2-mercaptoethanesulfonic acid methyltransferase from *Methanosarcina barkeri*. *J. Bacteriol.* **153**, 6–11.

Meijden, P., Jansen, L. P. J. M., van der Drift, C., and Vogels, G. D. (1983c). Involvement of corrinoids in the methylation of coenzyme M (2-mercaptoethanesulfonic acid) by methanol and enzymes from *Methanosarcina barkeri*. *FEMS Microbiol. Lett.* **19**, 247–251.

Meile, L., Kiener, A., and Leisinger, T. (1983). A plasmid in the archaebacterium *Methanobacterium thermoautotrophicum*. *Mol. Gen. Genet.* **191**, 480–484.

Miller, T. L., and Wolin, M. J. (1982). Enumeration of *Methanobrevibacter smithii* in human feces. *Arch. Microbiol.* **131**, 14–18.

Miller, T. L., and Wolin, M. J. (1983). Oxidation of hydrogen and reduction of methanol to methane is the sole energy source for a methanogen isolated from human feces. *J. Bacteriol.* (in press).

Miller, T. L., Wolin, M. J., Conway de Macario, E., and Macario, A. J. L. (1982). Isolation of *Methanobrevibacter smithii* from human feces. *Appl. Environ. Microbiol.* **43**, 227–232.

Mink, R. W., and Dugan, P. R. (1977). Tentative identification of methanogenic bacteria by fluorescence microscopy. *Appl. Environ. Microbiol.* **33**, 713–717.

Mitchell, R. M., Loeblich, L. A., Klotz, L. C., and Loeblich, A. R. (1979). DNA organization of *Methanobacterium thermoautotrophicum*. *Science* **204**, 1082–1084.

Mountfort, D. O. (1978). Evidence for ATP synthesis driven by a proton gradient in *Methanosarcina barkeri*. *Biochem. Biophys. Res. Commun.* **85**, 1346–1351.

Mountfort, D. O. (1980). Effect of adenosine 5'-monophosphate on adenosine 5'-triphosphate activation of methyl coenzyme M methylreductase in cell extracts of *Methanosarcina barkeri*. *J. Bacteriol.* **143**, 1039–1041.

Mountfort, D. O., and Asher, R. A. (1978). Changes in proportions of acetate and carbon dioxide used as methane precursors during the anaerobic digestion of bovine waste. *Appl. Environ. Microbiol.* **35**, 648–654.

Mountfort, D. O., and Asher, R. A. (1979). Effect of inorganic sulfide on the growth and metabolism of *Methanosarcina barkeri* strain DM. *Appl. Environ. Microbiol.* **37**, 670–675.

Mountfort, D. O., and Asher, R. A. (1981). Role of sulfate reduction versus methanogenesis in terminal carbon flow in polluted intertidal sediment of Waimea Inlet, Nelson, New Zealand. *Appl. Environ. Microbiol.* **42**, 252–258.

Mountfort, D. O., Asher, R. A., Mays, E. L., and Tiedje, J. M. (1980). Carbon and electron flow in mud and sandflat intertidal sediments at Delaware Inlet, Nelson, New Zealand. *Appl. Environ. Microbiol.* **39**, 686–694.

Moura, I., Moura, J. J. G., Huynh, B.-H., Santos, H., LeGall, J., and Xavier, A. V. (1982). Ferredoxin from *Methanosarcina barkeri*: Evidence for the presence of a three-iron center. *Eur. J. Biochem.* **126**, 95–98.

Moura, I., Moura, J. J. G., Santos, H., Xavier, A. V., Burch, G., Peck, H. D., and LeGall, J. (1983). Proteins containing the factor F_{430} from *Methanosarcina barkeri* and *Methanobacterium thermoautotrophicum*. *Biochim. Biophys. Acta* **742**, 84–90.

Moura, J. J. G., Moura, I., Santos, H., Xavier, A. V., Scandellari, M., and LeGall, J. (1982a).

Isolation of P_{590} from *Methanosarcina barkeri:* Evidence for the presence of sulfite reductase activity. *Biochem. Biophys. Res. Commun.* **108**, 1002–1009.

Moura, J. J. G., Moura, I., Huynh, B. H., Krüger, H.-J., Teixeira, M., DuVarney, R. C., DerVartanian, D. V., Xavier, A. V., Peck, H. D., and LeGall, J. (1982b). Unambiguous identification of the nickel EPR signal in ⁶¹Ni-enriched *Desulfovibrio gigas* hydrogenase. *Biochem. Biophys. Res. Commun.* **108**, 1388–1393.

Nagle, D. P., Jr., and Wolfe, R. S. (1983). Component A of the methyl coenzyme M reductase system of *Methanobacterium:* Resolution into four components. *Proc. Natl. Acad. Sci. U.S.A.* **80**, 2151–2155.

Oberlies, G., Fuchs, G., and Thauer, R. K. (1980). Acetate thiokinase and the assimilation of acetate in *Methanobacterium thermoautotrophicum*. *Arch. Microbiol.* **128**, 248–252.

Oppermann, R. A., Nelson, W. O., and Brown, R. E. (1961). *In vivo* studies of methanogenesis in the bovine rumen: Dissimilation of acetate. *J. Gen. Microbiol.* **25**, 103–111.

Oremland, R. S., and Polcin, S. (1982). Methanogenesis and sulfate reduction: Competitive and noncompetitive substrates in estuarine sediments. *Appl. Environ. Microbiol.* **44**, 1270–1276.

Oremland, R. S., and Taylor, B. F. (1978). Sulfate reduction and methanogenesis in marine sediments. *Geochim. Cosmochim. Acta* **42**, 209–214.

Oremland, R. S., Marsh, L., and DesMarais, D. J. (1982). Methanogenesis in Big Soda Lake, Nevada: An alkaline, moderately hypersaline desert lake. *Appl. Environ. Microbiol.* **43**, 462–468.

Panganiban, A. T., Jr., Patt, T. E., Hart, W., and Hanson, R. S. (1979). Oxidation of methane in the absence of oxygen in lake water samples. *Appl. Environ. Microbiol.* **37**, 303–309.

Patel, G. B., and Roth, L. A. (1977). Effect of sodium chloride on growth and methane production of methanogens. *Can. J. Microbiol.* **23**, 893–897.

Patel, G. B., Roth, L. A., van den Berg, L., and Clark, D. S. (1976). Characterization of a strain of *Methanospirillum hungatii*. *Can. J. Microbiol.* **22**, 1404–1410.

Patel, G. B., Khan, A. W., and Roth, L. A. (1978). Optimum levels of sulphate and iron for the cultivation of pure cultures of methanogens in synthetic media. *J. Appl. Bacteriol.* **45**, 347–356.

Paterek, J. R., and Smith, P. H. (1983). Isolation of a halophilic methanogenic bacterium from the sediments of Great Salt Lake and a San Francisco Bay saltern. *Am. Soc. Microbiol., Abstr. Annu. Meet.* p. 140.

Patterson, J. A., and Hespell, R. B. (1979). Trimethylamine and methylamine as growth substrates for rumen bacteria and *Methanosarcina barkeri*. *Curr. Microbiol.* **3**, 79–83.

Paynter, M. J. B., and Hungate, R. E. (1968). Characterization of *Methanobacterium mobilis,* sp. n., isolated from the bovine rumen. *J. Bacteriol.* **95**, 1943–1951.

Perski, H.-J., Moll, J., and Thauer, R. K. (1981). Sodium dependence of growth and methane formation in *Methanobacterium thermoautotrophicum*. *Arch. Microbiol.* **130**, 319–321.

Pfaltz, A., Jaun, B., Fässler, A., Eschenmoser, A., Jaenchen, R., Gilles, H. H., Diekert, G., and Thauer, R. K. (1982). Zur kenntnis des faktors F430 aus methanogenen bakterien: Struktur des porphinoiden ligandsystems. *Helv. Chim. Acta* **65**, 828–865.

Pine, M. J., and Barker, H. A. (1956). Studies on the methane fermentation. XII. The pathway of hydrogen in the acetate fermentation. *J. Bacteriol.* **71**, 644–648.

Pine, M. J., and Vishniac, W. (1957). The methane fermentations of acetate and methanol. *J. Bacteriol.* **73**, 736–742.

Pol, A., Van der Drift, C., Vogels, G. D., Cuppen, J. H. M., and Laarhoven, W. H. (1980). Comparison of coenzyme F_{420} from *Methanobacterium bryantii* with 7- and 8-hydroxy-10-methyl-5-deazaisoalloxazine. *Biochem. Biophys. Res. Commun.* **92**, 255–260.

Pol, A., Van der Drift, C., and Vogels, G. D. (1982). Corrinoids from *Methanosarcina barkeri:* Structure of the α-ligand. *Biochem. Biophys. Res. Commun.* **108**, 731–737.

Prins, R. A., Van Nevel, C. J., and Demeyer, D. I. (1972). Pure culture studies of inhibitors for methane formation. *Antonie van Leeuwenhoek* **38**, 28–287.

Rasmussen, R. A., and Khalil, M. A. K. (1981). Atmospheric methane (CH_4): Trends and seasonal cycles. *JGR, J. Geophys. Res.* **86**, 9826–9832.

Reeburgh, W. S. (1976). Methane consumption in Cariaco Trench waters and sediments. *Earth Planet. Sci. Lett.* **28**, 337–344.

Reeburgh, W. S., and Heggie, D. T. (1977). Microbial methane consumption reactions and their effect on methane distributions in freshwater and marine environments. *Limnol. Oceanogr.* **22**, 1–9.

Reeve, J. N., Trun, N. J., and Hamilton, P. T. (1982). Beginning genetics with methanogens. *In* "Genetic Engineering of Microorganisms for Chemicals" (A. Hollaender, R. D. DeMoss, S. Kaplan, J. Konisky, D. Savage, and R. S. Wolfe, eds.), pp. 233–244. Plenum, New York.

Risatti, J. B. (1978). Geochemical and microbial aspects of Bolo Bog, Lake County, Illinois. Ph.D. Thesis, University of Illinois, Urbana.

Rivard, C. J., and Smith, P. H. (1982). Isolation and characterization of a thermophilic marine methanogenic bacterium, *Methanogenium thermophilicum* sp. nov. *Int. J. Syst. Bacteriol.* **32**, 430–436.

Rivard, C. J., Henson, J. M., Thomas, M. V., and Smith, P. H. (1983). Isolation and characterization of *Methanomicrobium paynteri* sp. nov., a mesophilic methanogen isolated from marine sediments. *Appl. Environ. Microbiol.* **46**, 484–490.

Roberton, A. M., and Wolfe, R. S. (1969). ATP requirement for methanogenesis in cell extracts of *Methanobacterium* strain M.o.H. *Biochim. Biophys. Acta* **192**, 420–429.

Roberton, A. M., and Wolfe, R. S. (1970). Adenosine triphosphate pools in *Methanobacterium*. *J. Bacteriol.* **102**, 43–51.

Romesser, J. A., and Balch, W. E. (1980). Coenzyme M: Preparation and assay. *In* "Methods in Enzymology" (D. B. McCormick and L. D. Wright, eds.), Vol. 67, pp. 545–552. Academic Press, New York.

Romesser, J. A., and Wolfe, R. S. (1981). Interaction of coenzyme M and formaldehyde in methanogenesis. *Biochem. J.* **197**, 565–572.

Romesser, J. A., and Wolfe, R. S. (1982a). Coupling of methyl coenzyme M reduction with carbon dioxide activation in extracts of *Methanobacterium thermoautotrophicum*. *J. Bacteriol.* **152**, 840–847.

Romesser, J. A., and Wolfe, R. S. (1982b). CDR Factor, a new coenzyme required for carbon dioxide reduction to methane by extracts of *Methanobacterium thermoautotrophicum*. *Zentralbl. Bakteriol., Mikrobiol. Hyg., Abt. 1, Orig. C***3**, 271–276.

Romesser, J. A., Wolfe, R. S., Mayer, F., Spiess, E., and Walther-Mauruschat, A. (1979). *Methanogenium*, a new genus of marine methanogenic bacteria, and characterization of *Methanogenium cariaci* sp. nov. and *Methanogenium marisnigri* sp. nov. *Arch. Microbiol.* **121**, 147–153.

Rönnow, P. H., and Gunnarsson, L. A. H. (1981). Sulfide-dependent methane production and growth of a thermophilic methanogenic bacterium. *Appl. Environ. Microbiol.* **42**, 580–584.

Sansone, F. J., and Martens, C. S. (1981). Methane production from acetate and associated methane fluxes from anoxic coastal sediments. *Science* **211**, 707–709.

Sauer, F. D., Bush, R. S., Mahadevan, S., and Erfle, J. D. (1977). Methane production by cell-free particulate fraction of rumen bacteria. *Biochem. Biophys. Res. Commun.* **79**, 124–132.

Sauer, F. D., Erfle, J. D., and Mahadevan, S. (1979). Methane synthesis without the addition of adenosine triphosphate by cell membranes isolated from *Methanobacterium ruminantium*. *Biochem. J.* **178**, 165–172.

Sauer, F. D., Erfle, J. D., and Mahadevan, S. (1980a). Methane production by the membranous fraction of *Methanobacterium thermoautotrophicum*. *Biochem. J.* **190**, 177–182.

Sauer, F. D., Mahadevan, S., and Erfle, J. D. (1980b). Valinomycin inhibited methane synthesis in *Methanobacterium thermoautotrophicum*. *Biochem. Biophys. Res. Commun.* **95,** 715–721.

Sauer, F. D., Erfle, J. D., and Mahadevan, S. (1981). Evidence for an internal electrochemical proton gradient in *Methanobacterium thermoautotrophicum*. *J. Biol. Chem.* **256,** 9843–9848.

Schauer, N. L., and Ferry, J. G. (1980). Metabolism of formate in *Methanobacterium formicicum*. *J. Bacteriol.* **142,** 800–807.

Schauer, N. L., and Ferry, J. G. (1982). Properties of formate dehydrogenase in *Methanobacterium formicicum*. *J. Bacteriol.* **150,** 1–7.

Schauer, N. L., and Ferry, J. G. (1983). FAD requirement for the reduction of coenzyme F_{420} by formate dehydrogenase from *Methanobacterium formicicum*. *J. Bacteriol.* **155,** 467–472.

Schauer, N. L., Brown, D. P., and Ferry, J. G. (1982). Kinetics of formate metabolism in *Methanobacterium formicicum* and *Methanospirillum hungatei*. *Appl. Environ. Microbiol.* **44,** 549–554.

Scherer, P. A., and Bochem, H. P. (1983). Ultrastructural investigation of 12 *Methanosarcinae* and related species grown on methanol for occurrence of polyphosphatelike inculsions. *Can. J. Microbiol.* **29,** 1190–1199.

Scherer, P. A., and Kneifel, H. (1983). Distribution of polyamines in methanogenic bacteria. *J. Bacteriol.* **154,** 1315–1322.

Scherer, P. A., and Sahm, H. (1981a). Effect of trace elements and vitamins on the growth of *Methanosarcina barkeri*. *Acta Biotechnol.* **1,** 57–65.

Scherer, P. A., and Sahm, H. (1981b). Influence of sulphur-containing compounds on the growth of *Methanosarcina barkeri* in a defined medium. *Eur. J. Appl. Microbiol. Biotechnol.* **12,** 28–35.

Scherer, P. A., and Sauer, C. (1982). State of iron in the archaebacterium *Methanobacterium barkeri* grown on different carbon sources as studied by Mössbauer spectroscopy. *Z. Naturforsch., C: Biosci.* **37C,** 877–880.

Scherer, P. A., Lippert, H., and Wolff, G. (1983). Composition of the major elements and trace elements of 10 methanogenic bacteria determined by inductively coupled plasma emission spectrometry. *Biol. Trace Elem. Res.* **5,** 149–163.

Schmid, G., and Böck, A. (1981). Immunological comparison of ribosomal proteins from archaebacteria. *J. Bacteriol.* **147,** 282–288.

Schmid, G., and Böck, A. (1982). The ribosomal protein composition of five methanogenic bacteria. *Zentralbl. Bakteriol., Mikrobiol. Hyg., Abt. 1, Orig. C* **3,** 347–353.

Schönheit, P., and Perski, H. J. (1983). ATP synthesis driven by a potassium diffusion potential in *Methanobacterium thermoautotrophicum* is stimulated by sodium. *FEMS Microbiol. Lett.* **20,** 263–267.

Schönheit, P., and Thauer, R. K. (1980). L-Alanine, a product of cell wall synthesis in *Methanobacterium thermoautotrophicum*. *FEMS Microbiol Lett.* **9,** 77–80.

Schönheit, P., Moll, J., and Thauer, R. K. (1979). Nickel, cobalt, and molybdenum requirement for growth of *Methanobacterium thermoautotrophicum*. *Arch. Microbiol.* **123,** 105–107.

Schönheit, P., Moll, J., and Thauer, R. K. (1980). Growth parameters (K_s, μ_{max}, Y_s) of *Methanobacterium thermoautotrophicum*. *Arch. Microbiol.* **127,** 59–65.

Schönheit, P., Keweloh, H., and Thauer, R. K. (1981). Factor F_{420} degradation in *Methanobacterium thermoautotrophicum* during exposure to oxygen. *FEMS Microbiol. Lett.* **12,** 347–349.

Schönheit, P., Kristjansson, J. K., and Thauer, R. K. (1982). Kinetic mechanism for the ability of sulfate reducers to out-compete methanogens for acetate. *Arch. Microbiol.* **132,** 285–288.

Schrauzer, G. N., Grate, J. H., and Katz, R. N. (1978). Coenzyme M and methylcobalamin in methane biosynthesis: Results of model studies. *Bioinorg. Chem.* **8,** 1–10.

Schulman, M., Ghambeer, R. K., Ljungdahl, L. G., and Wood, H. G. (1973). Total synthesis of acetate from CO_2. VII. Evidence with *Clostridium thermoaceticum* that the carboxyl of acetate is derived from the carboxyl of pyruvate by transcarboxylation and not by fixation of CO_2. *J. Biol. Chem.* **248,** 6255–6261.

Seely, R. J., and Fahrney, D. E. (1983a). A novel diphospho-P,P'-diester from *Methanobacterium thermoautotrophicum*. *J. Biol. Chem.* **258**, 10835–10838.

Seely, R. J., and Fahrney, D. E. (1983b). The cyclic-2,3-diphosphoglycerate from *Methanobacterium thermoautotrophicum* is the *D* enantiomer. *Curr. Microbiol.* **10**, 85–88.

Seely, R. J., Krueger, R. D., and Fahrney, D. E. (1983). Cyclic-2,3-diphosphoglycerate levels in *Methanobacterium thermoautotrophicum* reflect inorganic phosphate availability. *Biochem. Biophys. Res. Commun.* **16**, 1125–1128.

Senior, E., Lindström, E. B., Banat, I. M., and Nedwell, D. B. (1982). Sulfate reduction and methanogenesis in the sediment of a saltmarsh on the east coast of the United Kingdom. *Appl. Environ. Microbiol.* **43**, 987–996.

Shapiro, S. (1982a). Do corrinoids function in the methanogenic dissimilation of methanol by *Methanosarcina barkeri*? *Can. J. Microbiol.* **28**, 629–635.

Shapiro, S. (1982b). Hysteretic activation of methanol-coenzyme M methyltransferase by ATP. *Can. J. Microbiol.* **28**, 1409–1411.

Shapiro, S., and Wolfe, R. S. (1980). Methyl-coenzyme M, an intermediate in methanogenic dissimilation of C_1 compounds by *Methanosarcina barkeri*. *J. Bacteriol.* **141**, 728–734.

Sheppard, J. C., Westberg, H., Hopper, J. F., Ganesan, K., and Zimmerman, P. (1982). Inventory of global methane sources and their production rates. *JGR, J. Geophys. Res.* **87**, 1305–1312.

Shiemke, A. K., Eirich, L. D., and Loehr, T. M. (1983). Resonance raman spectroscopic characterization of the nickel cofactor, F_{430}, from methanogenic bacteria. *Biochim. Biophys. Acta* **748**, 143–147.

Smith, M. R. (1983). Reversal of 2-bromoethanesulfonate inhibition of methanogenesis in *Methanosarcina* sp. *J. Bacteriol.* **156**, 516–523.

Smith, M. R., and Mah, R. A. (1978). Growth and methanogenesis by *Methanosarcina* strain 227 on acetate and methanol. *Appl. Environ. Microbiol.* **36**, 870–879.

Smith, M. R., and Mah, R. A. (1980). Acetate as sole carbon and energy source for growth of *Methanosarcina* strain 227. *Appl. Environ. Microbiol.* **39**, 993–999.

Smith, M. R., and Mah, R. A. (1981). 2-Bromoethanesulfonate: A selective agent for isolating resistant *Methanosarcina* mutants. *Curr. Microbiol.* **6**, 321–326.

Smith, P. H., and Hungate, R. E. (1958). Isolation and characterization of *Methanobacterium ruminantium* n. sp. *J. Bacteriol.* **75**, 713–718.

Smith, P. H., and Mah, R. A. (1966). Kinetics of acetate metabolism during sludge digestion. *Appl. Microbiol.* **14**, 368–371.

Sowers, K. R., and Ferry, J. G. (1983). Isolation and characterization of a methylotrophic marine methanogen, *Methanococcoides methylutens* gen. nov., sp. nov. *Appl. Environ. Microbiol.* **43**, 684–690.

Spencer, R. W., Daniels, L., Fulton, G., and Orme-Johnson, W. H. (1980). Product isotope effects on in vivo methanogenesis by *Methanobacterium thermoautotrophicum*. *Biochemistry* **19**, 3678–3683.

Sprengel, G., and Follmann, H. (1981). Evidence for the reductive pathway of deoxyribonucleotide synthesis in an archaebacterium. *FEBS Lett.* **132**, 207–209.

Sprott, G. D., and Jarrell, K. F. (1981). K^+, Na^+, and Mg^{2+} content and permeability of *Methanospirillum hungatei* and *Methanobacterium thermoautotrophicum*. *Can. J. Microbiol.* **27**, 444–451.

Sprott, G. D., and Jarrell, K. F. (1982). Sensitivity of methanogenic bacteria to dicyclohexylcarbodiimide. *Can. J. Microbiol.* **28**, 982–986.

Sprott, G. D., and McKellar, R. C. (1980). Composition and properties of the cell wall of *Methanospirillum hungatei*. *Can. J. Microbiol.* **26**, 115–120.

Sprott, G. D., McKellar, R. C., Shaw, K. M., Giroux, J., and Martin, W. G. (1979a). Properties of malate dehydrogenase isolated from *Methanospirillum hungatii*. *Can. J. Microbiol.* **25**, 192–200.

Sprott, G. D., Colvin, J. R., and McKellar, R. C. (1979b). Spheroplasts of *Methanospirillum hungatii* formed upon treatment with dithiothreitol. *Can. J. Microbiol.* **25**, 730–738.

Sprott, G. D., Jarrell, K. F., Shaw, K. M., and Knowles, R. (1982). Acetylene as an inhibitor of methanogenic bacteria. *J. Gen. Microbiol.* **128**, 2453–2462.

Sprott, G. D., Shaw, K. M., and Jarrell, K. F. (1983). Isolation and chemical composition of the cytoplasmic membrane of the archaebacterium *Methanospirillum hungatei. J. Biol. Chem.* **258**, 4026–4031.

Sprott, G. D., Sowden, L. C., Colvin, J. R., and Jarrell, K. F. (1984). Methanogenesis in the absence of intracellular membranes. *Can. J. Microbiol.* **30**, 594–604.

Stackebrandt, E., Seewaldt, E., Ludwig, W., Schleifer, K.-H., and Huser, B. A. (1982). The phylogenetic position of *Methanothrix soehngenii.* Elucidated by a modified technique of sequencing oligonucleotides from 16S rRNA. *Zentralbl. Bakteriol. Mikrobiol. Hyg., Abt. 1, Orig. C* **3**, 90–100.

Stadtman, T. C. (1967). Methane fermentation. *Annu. Rev. Microbiol.* **21**, 121–142.

Stadtman, T. C., and Barker, H. A. (1951). Studies on the methane fermentation. X. A new formate-decomposing bacterium, *Methanococcus vannielii. J. Bacteriol.* **62**, 269–280.

Stetter, K. O., and Gaag, G. (1983). Reduction of molecular sulphur by methanogenic bacteria. *Nature (London)* **305**, 309–311.

Stetter, K. O., Thomm, M., Winter, J., Wildgruber, G., Huber, H., Zillig, W., Janecovic, D., König, H., Palm, P., and Wunderl, S. (1981). *Methanothermus fervidus,* sp. nov., a novel extremely thermophilic methanogen isolated from an Icelandic hot spring. *Zentralbl. Bakteriol., Mikrobiol. Hyg., Abt. 1, Orig. C* **2**, 166–178.

Storer, A. C., Sprott, G. D., and Martin, W. G. (1981). Kinetic and physical properties of the L-malate-NAD$^+$ oxidoreductase from *Methanospirillum hungatii* and comparison with the enzyme from other sources. *Biochem. J.* **193**, 235–244.

Stouthamer, A. H., and Bettenhaussen, L. (1973). Utilization of energy for growth and maintenance in continuous and batch cultures of microorganisms. *Biochim. Biophys. Acta* **301**, 53–70.

Stumm, C. K., Gijzen, H. J., and Vogels, G. D. (1982). Association of methanogenic bacteria with ovine rumen ciliates. *Br. J. Nutr.* **47**, 95–99.

Stupperich, E., and Fuchs, G. (1981). Products of CO_2 fixation and ^{14}C labelling pattern of alanine in *Methanobacterium thermoautotrophicum* pulse-labelled with $^{14}CO_2$. *Arch. Microbiol.* **130**, 294–300.

Stupperich, E., and Fuchs, G. (1983). Autotrophic acetyl coenzyme A synthesis in vitro from two CO_2 in *Methanobacterium. FEBS Lett.* **156**, 345–348.

Stupperich, E., Hammel, K. E., Fuchs, G., and Thauer, R. K. (1983). Carbon monoxide fixation into the carboxyl group of acetyl coenzyme A during autotrophic growth of *Methanobacterium. FEBS Lett.* **152**, 21–23.

Tanner, R. S. (1982). Novel compounds from methanogens: Characterization of component B of the methylreductase system and Mobile factor. Ph.D. Thesis, University of Illinois, Urbana.

Tanner, R. S., and Wolfe, R. S. (1982). Nutrient requirements of *Methanomicrobium mobile. Am. Soc. Microbiol. Abstr. Annu. Meet.* p. 109.

Taylor, C. D., and Wolfe, R. S. (1974a). Structure and methylation of coenzyme M ($HSCH_2CH_2SO_3$). *J. Biol. Chem.* **249**, 4879–4885.

Taylor, C. D., and Wolfe, R. S. (1974b). A simplified assay for coenzyme M ($HSCH_2CH_2SO_3$). *J. Biol. Chem.* **249**, 4886–4890.

Taylor, C. D., McBride, B. C., Wolfe, R. S., and Bryant, M. P. (1974). Coenzyme M, essential for growth of a rumen strain of *Methanobacterium ruminantium. J. Bacteriol.* **120**, 974–975.

Taylor, G. T., and Pirt, S. J. (1977). Nutrition and factors limiting the growth of a methanogenic bacterium (*Methanobacterium thermoautotrophicum*). *Arch. Micribiol.* **113**, 17–22.

Taylor, G. T., Kelley, D. P., and Pirt, S. J. (1976). Intermediary metabolism in methanogenic bacteria (*Methanobacterium*). *In* "Microbial Production and Utilization of Gases (H_2, CH_4,

CO)'' (H. G. Schlegel, G. Gottschalk, and N. Pfennig, eds.), pp. 173–180. Akad. Wiss. Göttingen, Goltze, Göttingen.

Thauer, R. K., Jungermann, K., and Decker, K. (1977). Energy conservation in chemotrophic anaerobic bacteria. *Bacteriol. Rev.* **41**, 100–180.

Thomm, M., Altenbuchner, J., and Stetter, K. O. (1983). Evidence for a plasmid in a methanogenic bacterium. *J. Bacteriol.* **153**, 1060–1063.

Tu, J., Prangishvilli, D., Huber, H., Wildgruber, G., Zillig, W., and Stetter, K. O. (1982). Taxonomic relations between archaebacteria including 6 novel genera examined by cross hybridization of DNAs and 16S rRNAs. *J. Mol. Evol.* **18**, 109–114.

Tzeng, S. F., Wolfe, R. S., and Bryant, M. P. (1975a). Factor 420-dependent pyridine nucleotide-linked hydrogenase system of *Methanobacterium ruminantium*. *J. Bacteriol.* **121**, 184–191.

Tzeng, S. F., Bryant, M. P., and Wolfe, R. S. (1975b). Factor 420-dependent pyridine nucleotide-linked formate metabolism of *Methanobacterium ruminantium*. *J. Bacteriol.* **121**, 192–196.

Unden, G., Böcher, R., Knecht, J., and Kröger, A. (1982). Hydrogenase from *Vibrio succinogenes*, a nickel protein. *FEBS Lett.* **145**, 230–234.

van Beelen, P., Dijkstra, A. C., and Vogels, G. D. (1983a). Quantitation of coenzyme F_{420} in methanogenic sludge by the use of reversed-phase high-performance liquid chromatography and a fluorescence detector. *Eur. J. Appl. Microbiol. Biotechnol.* **18**, 67–69.

van Beelen, P., Geerts, W. J., Pol, A., and Vogels, G. D. (1983b). Quantification of coenzymes and related compounds from methanogenic bacteria by high-performance liquid chromatography. *Anal. Biochem.* **131**, 285–290.

van Beelen, P., Thiemessen, H. L., DeCock, R. M., and Vogels, G. D. (1983c). Methanogenesis and methanopterin conversion by cell-free extracts of *Methanobacterium thermoautotrophicum*. *FEMS Microbiol. Lett.* **18**, 135–138.

Vogels, G. D., Hoppe, W. F., and Stumm, C. K. (1980). Association of methanogenic bacteria with rumen ciliates. *Appl. Environ. Microbiol.* **40**, 608–612.

Vogels, G. D., Keltjens, J. T., Hutten, T. J., and Van der Drift, C. (1982). Coenzymes of methanogenic bacteria. *Zentralbl. Bakteriol., Mikrobiol. Hyg., Abt. 1, Orig. C* **3**, 258–264.

Walther, R., Fahlbusch, K., Sievert, R., and Gottschalk, G. (1981). Formation of trideuteromethane from deuterated trimethylamine or methylamine by *Methanosarcina barkeri*. *J. Bacteriol.* **148**, 371–373.

Ward, J. M. (1970). The microbial ecology of estuarine methanogenesis. Master's Thesis, University of Florida, Gainesville.

Ward, T. E., and Frea, J. I. (1980). Sediment distribution of methanogenic bacteria in Lake Erie and Cleveland Harbor. *Appl. Environ. Microbiol.* **39**, 597–603.

Weimer, P. J., and Zeikus, J. G. (1978a). One carbon metabolism in methanogenic bacteria. Cellular characterization and growth of *Methanosarcina barkeri*. *Arch. Microbiol.* **119**, 49–57.

Weimer, P. J., and Zeikus, J. G. (1978b). Acetate metabolism in *Methanosarcina barkeri*. *Arch. Microbiol.* **119**, 175–182.

Weimer, P. J., and Zeikus, J. G. (1979). Acetate assimilation pathway of *Methanosarcina barkeri*. *J. Bacteriol.* **137**, 332–339.

Wellinger, A., and Wuhrmann, K. (1977). Influence of sulfide compounds on the metabolism of *Methanobacterium* strain AZ. *Arch. Microbiol.* **115**, 13–17.

Whitman, W. B., and Wolfe, R. S. (1980). Presence of nickel in factor F_{430} from *Methanobacterium bryantii*. *Biochem. Biophys. Res. Commun.* **92**, 1196–1201.

Whitman, W. B., and Wolfe, R. S. (1982). Activation of the methylreductase system from *Methanobacterium bryantii* by cobalamins. *Fed. Proc., Fed. Am. Soc. Exp. Biol.* **41**, 1152.

Whitman, W. B., and Wolfe, R. S. (1983). Activation of the methylreductase system from *Methanobacterium bryantii* by ATP. *J. Bacteriol.* **154**, 640–649.

Whitman, W. B., Ankwanda, E., and Wolfe, R. S. (1982). Nutrition and carbon metabolism of *Methanococcus voltae*. *J. Bacteriol.* **149**, 852–863.

Wildgruber, G., Thomm, M,, König, H., Ober, K., Ricchiuto, T., and Stetter, K. O. (1982).

Methanoplanus limicola, a plate-shaped methanogen representing a novel family, the Methanoplanaceae. *Arch. Microbiol.* **132,** 31–36.

Winfrey, M. R., and Zeikus, J. G. (1977). Effect of sulfate on carbon and electron flow during microbial methanogenesis in freshwater sediments. *Appl. Environ. Microbiol.* **33,** 275–281.

Winfery, M. R., and Zeikus, J. G. (1979a). Microbial methanogenesis and acetate metabolism in a meromictic lake. *Appl. Environ. Microbiol.* **37,** 213–221.

Winfrey, M. R., and Zeikus, J. G. (1979b). Anaerobic metabolism of immediate methane precursors in Lake Mendota. *Appl. Environ. Microbiol.* **37,** 244–253.

Wolfe, R. S. (1971). Microbial formation of methane. *Adv. Microb. Physiol.* **6,** 107–146.

Wolfe, R. S. (1979). Methanogens: A surprising microbial group. *Antonie van Leeuwenhoek* **45,** 353–364.

Wolfe, R. S., and Higgins, I. J. (1979). Microbial biochemistry of methane - a study in contrast. *Int. Rev. Biochem.* **21,** 267–350.

Wolin, E. A., Wolin, M. J., and Wolfe, R. S. (1963). Formation of methane by bacterial extracts. *J. Biol. Chem.* **238,** 2882–2886.

Wolin, M. J. (1976). Interaction between H_2-producing and methane-producing species. *In* "Microbial Formation and Utilization of Gases (H_2, CH_4, CO)" (H. G. Schlegel, G. Gottschalk, and N. Pfennig, eds.), pp. 141–146. Akad. Wiss. Göttingen, Goltze, Göttingen.

Wolin, M. J. (1981). Fermentation in the rumen and human large intestine. *Science* **213,** 1463–1468.

Wolin, M. J., and Miller, T. L. (1982). Interspecies hydrogen transfer: 15 years later. *ASM News* **48,** 561–565.

Wolin, M. J., Wolin, E. A., and Wolfe, R. S. (1963). ATP-dependent formation of methane from methylcobalamin by extracts of *Methanobacillus omelianskii. Biochem. Biophys. Res. Commun.* **12,** 464–468.

Wolin, M. J., Wolin, E. A., and Wolfe, R. S. (1964). The cobalamin product of the conversion of methylcobalamin to CH_4 by extracts of *Methanobacillus omelianskii. Biochem. Biophys. Res. Commun.* **15,** 420–423.

Wood, A. G., Redborg, A. H., Cue, D. R., Whitman, W. B., and Konisky, J. (1983). Complementation of *argG* and *hisA* mutations of *Escherichia coli* by DNA cloned from the archaebacterium *Methanococcus voltae. J. Bacteriol.* **156,** 19–29.

Wood, J. M., and Wolfe, R. S. (1965). The formation of CH_4 from N^5-methyltetrahydrofolate monoglutamate by cell-free extracts of *Methanobacillus omelianskii. Biochem. Biophys. Res. Commun.* **19,** 306–311.

Wood, J. M., and Wolfe, R. S. (1966a). Propylation and purification of a B_{12} enzyme involved in methane formation. *Biochemistry* **5,** 3598–3603.

Wood, J. M., and Wolfe, R. S. (1966b). Alkylation of an enzyme in the methane-forming system of *Methanobacillus omelianskii. Biochem. Biophys. Res. Commun.* **22,** 119–123.

Wood, J. M., and Wolfe, R. S. (1966c). Components required for the formation of CH_4 from methylcobalamin by extracts of *Methanobacillus omelianskii. J. Bacteriol.* **92,** 696–700.

Wood, J. M., Allam, A. M., Brill, W. J., and Wolfe, R. S. (1965). Formation of methane from serine by cell-free extracts of *Methanobacillus omelianskii. J. Biol. Chem.* **240,** 4564–4569.

Wood, J. M., Wolin, M. J., and Wolfe, R. S. (1966). Formation of methane from methyl factor B and methyl factor III by cell-free extracts of *Methanobacillus omelianskii. Biochemistry* **5,** 2381–2384.

Wood, J. M., Moura, I., Moura, J. J. G., Santos, M. H., Xavier, A. V., LeGall, J., and Scandellari, M. (1982). Role of vitamin B_{12} in methyl transfer for methane biosynthesis by *Methanosarcina barkeri. Science* **216,** 303–305.

Yamazaki, S. (1982). A selenium-containing hydrogenase from *Methanococcus vannielii. J. Biol. Chem.* **257,** 7926–7929.

Yamazaki, S., and Tsai, L. (1980). Purification and properties of 8-hydroxy-5-deazaflavin-dependent $NADP^+$ reductase from *Methanococcus vannielii. J. Biol. Chem.* **255,** 6462–6465.

Yamazaki, S., Tsai, L., Stadtman, T. C., Jacobson, F. S., and Walsh, C. (1980). Stereochemical studies of 8-hydroxy-5-deazaflavin-dependent NADP+ reductase from *Methanococcus van-nielii. J. Biol. Chem.* **255**, 9025–9027.

Yamazaki, S., Tsai, L., and Stadtman, T. C. (1982). Analogues of 8-hydroxy-5-deazaflavin cofactor: Relative activity as substrates for 8-hydroxy-5-deazaflavin-dependent NADP+ reductase from *Methanococcus vannielii. Biochemistry* **21**, 934–939.

Zehnder, A. J. B., and Brock, T. D. (1979a). Methane formation and methane oxidation by methanogenic bacteria. *J. Bacteriol.* **137**, 420–432.

Zehnder, A. J. B., and Brock, T. D. (1979b). Biological energy production in the apparent absence of electron transport and substrate level phosphorylation. *FEBS Lett.* **107**, 1–3.

Zehnder, A. J. B., and Brock, T. D. (1980). Anaerobic methane oxidation: Occurrence and ecology. *Appl. Environ. Microbiol.* **39**, 194–204.

Zehnder, A. J. B., and Wuhrmann, K. (1977). Physiology of *Methanobacterium* strain AZ. *Arch. Microbiol.* **111**, 199–205.

Zehnder, A. J. B., Huser, B. A., Brock, T. D., and Wuhrmann, K. (1980). Characterization of an acetate-decarboxylating non-hydrogen-oxidizing methane bacterium. *Arch. Microbiol.* **124**, 1–11.

Zehnder, A. J. B., Ingvorsen, K., and Marti, T. (1982). Microbiology of methane bacteria. *In* "Anaerobic Digestion 1981" (D. E. Hughes, D. A. Stafford, B. I. Wheatley, W. Baader, G. Lettinga, E. J. Nyns, W. Verstraete, and R. L. Wentworth, eds.), pp. 45–68. Elsevier Biomedical Press, New York.

Zeikus, J. G. (1977). The biology of methanogenic bacteria. *Bacteriol. Rev.* **41**, 514–541.

Zeikus, J. G., and Bowen, V. G. (1975a). Comparative ultrastructure of methanogenic bacteria. *Can. J. Microbiol.* **21**, 121–129.

Zeikus, J. G., and Bowen, V. G. (1975b). Fine structure of *Methanospirillum hungatii. J. Bacteriol.* **121**, 373–380.

Zeikus, J. G., and Henning, D. L. (1975). *Methanobacterium arbophilicum* sp. nov. an obligate anaerobe isolated from wetwood of living trees. *Antonie van Leeuwenhoek* **41**, 543–552.

Zeikus, J. G., and Ward, J. C. (1974). Methane formation in living trees: a microbial origin. *Science* **184**, 1181–1183.

Zeikus, J. G., and Wolfe, R. S. (1972). *Methanobacterium thermoautotrophicus* sp. n., an anaerobic, autotrophic, extreme thermophile. *J. Bacteriol.* **109**, 707–713.

Zeikus, J. G., and Wolfe, R. S. (1973). The fine structure of *Methanobacterium thermoautotrophicum.* Effect of growth temperature on morphology and ultrastructure. *J. Bacteriol.* **113**, 461–467.

Zeikus, J. G., Fuchs, G., Kenealy, W., and Thauer, R. K. (1977). Oxidoreductases involved in cell carbon synthesis of *Methanobacterium thermoautotrophicum. J. Bacteriol.* **132**, 604–613.

Zeikus, J. G., Ben-Bassat, A., and Hegge, P. W. (1980). Microbiology of methanogenesis in thermal, volcanic environments. *J. Bacteriol.* **143**, 432–440.

Zhilina, T. N. (1971). The fine structure of *Methanosarcina. Microbiology (Engl. Transl.)* **40**, 587–591.

Zhilina, T. N. (1972). Death of *Methanosarcina* in the air. *Microbiology (Engl. Transl.)* **41**, 980–981.

Zhilina, T. N. (1979). Growth of a pure *Methanosarcina* culture, Biotype 2, on acetate. *Microbiology (Engl. Transl.)* **47**, 321–323.

Zimmerman, P. R., Greenberg, J. P., Wandiga, S. O., and Crutzen, P. J. (1982). Termites: A potentially large source of atmospheric methane, carbon dioxide, and molecular hydrogen. *Science* **218**, 563–565.

Zinder, S. H., and Mah, R. A. (1979). Isolation and characterization of a thermophilic strain of *Methanosarcina* unable to use H_2-CO_2 for methanogenesis. *Appl. Environ. Microbiol.* **38**, 996–1008.

CHAPTER 2

Thermoplasma and the Thermophilic Sulfur-Dependent Archaebacteria

K. O. STETTER AND W. ZILLIG

Copyright © 1985 by Academic Press, Inc.
All rights of reproduction in any form reserved.
ISBN 0-12-307208-5

I. Introduction and Definitions

When the concept of a distinct third urkingdom of life, the archaebacteria, was established, it became evident that three phyla of methanogens and the extreme halophiles were relatively closely related. Two genera, however, *Thermoplasma* and *Sulfolobus*, though also clearly belonging to the novel urkingdom, appeared relatively isolated, both from the rest of the archaebacteria and from one another (Fox *et al.*, 1980). Because the latter two grow at low pH values, they have often been distinguished from the methanogens and halophiles and have been grouped together as the thermoacidophilic archaebacteria, in spite of their large phylogenetic distance from each other. The finding that the component patterns of the DNA-dependent RNA polymerases of *Thermoplasma* and *Sulfolobus* were of the same type and obviously different from those of *Halobacterium* and *Methanobacterium* seemed to justify this view (Zillig *et al.*, 1982a). Several sulfur-respiring or even chemolithoautotrophic organisms, placed into the novel order Thermoproteales (Zillig *et al.*, 1981; 1982b), are closer to *Sulfolobus* than to the methanogens and halophiles based on both their RNA polymerase component patterns (Zillig *et al.*, 1982a,c) and on their 16S rRNA sequence (Woese *et al.*, 1984) and cross-hybridization data (Tu *et al.*, 1982). Thus, together with *Sulfolobus*, they constitute a distinct second major branch of the archaebacteria, from both of which *Thermoplasma* remains isolated.

Because some of the new organisms can grow at neutral pH values, but all generate energy by metabolizing sulfur, though in different ways, we propose to call this second division the sulfur-dependent rather than the thermoacidophilic branch and to deal with *Thermoplasma* as a genus of uncertain affiliation.

II. Phylogenetic Considerations

The S_{AB} (a similarity coefficient calculated from partial sequences of 16 S rRNAs) of *Thermoplasma acidophilum* with the methanogenic and the extremely halophilic archaebacteria is about 0.2, and with *Sulfolobus* as low as 0.17 (Fox *et al.*, 1980). Accordingly, an analysis of the phylogenetic distances between archaebacteria on the basis of 16 S rRNA–DNA cross-hybridization data places *Thermoplasma* slightly, though insignificantly, closer to the methanogens and extreme halophiles than to *Sulfolobus* and its relatives. On the other hand, the component pattern of its DNA-dependent RNA polymerase is of the same BAC type as those of *Sulfolobus* and the Thermoproteales (Huet *et al.*, 1983; Schnabel *et al.*, 1983; and this volume, Chapter 11). *Thermoplasma* thus appears so far either as the sole representative of an isolated group of archaebacteria, possibly a third branch of the kingdom, or as a phenotypic link between the known major branches.

As shown by results of 16 S rRNA–DNA cross-hybridization, several novel anaerobic sulfur-respiring or chemolithoautotrophic, extremely thermophilic archaebacteria from solfataric hot springs, water and mud holes, and sea floors, including the genera *Thermoproteus* (Zillig *et al.*, 1981), *Thermofilum* (Zillig *et al.*, 1983a), *Desulfurococcus* (Zillig *et al.*, 1982c), *Thermococcus* (Zillig *et al.*, 1983b), and two recently described marine isolates, *Thermodiscus* (Stetter *et al.*, in preparation), and *Pyrodictium* (Stetter, 1982; Stetter *et al.*, 1983) are not only related to each other but also are phylogenetically closer to *Sulfolobus* than to the methanogens and extreme halophiles (Tu *et al.*, 1982, and unpublished data). Accordingly, the component patterns of their DNA-dependent RNA polymerases, like those of *Sulfolobus* and *Thermoplasma*, are of the BAC type (Huet *et al.*, 1983; Schnabel *et al.*, 1983; and this volume, Chapter 11). The homology of the components of these enzymes has been established by immunochemical cross-reaction, employing antibodies directed against single components of the RNA polymerases of yeast, *Sulfolobus*, and *Methanobacterium thermoautotrophicum;* as well as iodinated *Staphylococcus* protein A in the "Western blotting" technique (Huet *et al.*, 1983; Schnabel *et al.*, 1983). In contrast, no cross-reactions between the RNA polymerases of *Thermoproteus, Thermofilum, Desulfurococcus,* and *Thermococcus* have been observed employing the Ouchterlony immunodiffusion technique, which depends on immunoprecipitation. In the case of eubacteria, this characterizes interfamily or larger distances. We therefore consider these genera representatives of different families of a novel order previously represented only by several species of the genus *Sulfolobus* (Brock *et al.*, 1972; Brierley and Brierley, 1973; Zillig *et al.*, 1980), with which they form the second major division of the archaebacteria, here proposed to be termed the sulfur-dependent branch.

As indicated by this term, sulfur is involved in energy metabolism in both orders of this branch, although *Sulfolobus* oxidizes, whereas the Thermoproteales reduce it. A phylogenetic relation of these two apparently opposite types of metabolism has not been established but appears certainly possible.

As indicated by cross-hybridization data (H. P. Klenk, F. Fischer, K. O. Stetter, and W. Zillig, unpublished) and their metabolism, the novel marine isolates *Thermodiscus* (Stetter *et al.*, in preparation) and *Pyrodictium* belong in the same branch. Their exact taxonomic positions, however, remain to be determined.*

---

*Recently, we were able to isolate novel members of the *Sulfolobaceae*, which can grow the same aerobically by sulfur oxidation as anaerobically by sulfur reduction (Seegerer *et al.*, 1985; Zillig *et al.*, 1985).

III. *Thermoplasma:* Genus of Uncertain Affiliation

The wall-less aerobic heterotroph *Thermoplasma acidophilum,* isolated by Darland *et al.* (1970) from smoldering coal refuse piles, was first considered a "thermophilic mycoplasma." The recognition of several features distinguishing it from the eubacterial mollicutes, e.g., its osmotic stability (Belly and Brock, 1972), the nature of its membrane lipids (Langworthy *et al.,* 1972), and finally the 16 S rRNA sequence data (Woese *et al.,* 1980; Fox *et al.,* 1980) led to the conclusion that it was an archaebacterium. An extensive review of the properties of the organism and a thorough discussion of questions concerning this organism has been written by Brock (1978a). In this chapter we will concentrate on later results, especially on the molecular level and on the changing views of the organism.

A. ECOLOGY

Until recently, *Thermoplasma* had been isolated only from burning coal refuse piles (Brock, 1978a). An earlier statement, that it had also been found in hot springs (Castenholz, 1979), is obviously due to a misunderstanding. In his book on life at extreme temperatures, Brock (1978a) discussed thoroughly the question of the natural habitat of this moderately thermophilic, extremely acidophilic, but (in contrast to *Sulfolobus*) non-acid-producing heterotroph. The frequency of isolation from mineral coal refuse samples taken within the proper temperature and pH range, as recently as 2 years after the pile was established, supports the idea that the organism easily spreads and multiplies under appropriate conditions, but it does not necessarily contradict the so far unchecked possibility that the organism was present already in the coal itself. We have failed to isolate it from either smoldering brown coal refuse or from warm solfataric springs (K. O. Stetter and W. Zillig, unpublished).

The previous discussion about the natural habitat of *Thermoplasma* took a surprising turn with the isolation of a *Thermoplasma* from a Japanese hot spring (Ohba and Oshima, 1982). The composition of its lipids (T. A. Langworthy, personal communication) and the component pattern of its RNA polymerase, which immunochemically cross-reacts with that of *Thermoplasma acidophilum* (R. Schnabel, personal communication), prove it to be closely related to, though significantly different from, that species. In light of the finding that solfataric hot springs are populated by a number of primary producers, such as *Sulfolobus,* the chemolithoautotrophic *Thermoproteales,* and thermophilic methanogens (Stetter *et al.,* 1981), the heterotrophic existence of *Thermoplasma* in this habitat appears unproblematic.

B. Morphology and Chemistry

Because of the absence of an envelope, *Thermoplasma* can formally be considered a mycoplasma, though phylogenetically it is unrelated to the eubacterial mollicutes. In phase contrast microscopy, freshly prepared specimens of growing cells often appear somewhat elongated. After a while, they become spherical, as nongrowing cells normally are. The ostensible presence of an actin-like protein in *T. acidophilum*, which could be involved in such morphological changes, has been reported (Searcy *et al.*, 1978). However, attempts to confirm this claim have so far failed (K. Zechel, personal communication). *Thermoplasma acidophilum* has been reported to be motile and flagellated (Black *et al.*, 1979), a unique feature for an organism lacking an envelope. Confirmation of this observation would be reassuring. The membrane contains a lipopolysaccharide of unique structure (Mayberry-Carson *et al.*, 1974; Smith, 1980) and at least one glycoprotein (Young and Hong, 1979). The basic ("histone-like") protein associated with the DNA (Searcy and Stein, 1980) has been partially sequenced (Searcy and De Lange, 1980). Such proteins, sometimes more than one in a species, have since been shown to occur in other archaebacteria (Thomm *et al.*, 1982; Green *et al.*, 1983). The structures of archaebacterial "chromatins" are however still unknown.

C. Growth and Metabolism

Thermoplasma acidophilum appears to be an aerobic heterotroph (Searcy and Whatley, 1982) requiring yeast extract for growth. Multiplication in the absence of oxygen has been reported to be inhibited by CO_2 (Smith *et al.*, 1973). We have found normal growth in the presence of CO_2 (R. Schnabel, unpublished). Yeast extract can be replaced by a peptide fraction purified therefrom (Smith *et al.*, 1975) and also by a decoction of coal refuse (Bohlool and Brock, 1974). A fully defined medium has, however, not yet been developed, nor are the growth requirements, mode of energy metabolism, and carbon utilization pathways fully understood.

Although the organism grows well in shaken cultures, optimal growth in strongly aerated batches requires addition of CO_2 (about 1%) to the air (unpublished observation of the authors). The mode of cell division remains unknown.

D. Molecular Biology

The numerous families of middle repetitive DNA sequences so characteristic for the genome of *Halobacterium* (Sapienza and Doolittle, 1982a,b) have not been found in the rather small genome of *Thermoplasma* (Searcy and Doyle, 1975; Christiansen *et al.,* 1975). The genes for 16 S, 23 S, and 5 S rRNA occur once per genome and appear unlinked (Tu and Zillig, 1982), quite in contrast to the situation in eubacteria and in other archaebacteria (Hofman *et al.,* 1979; Jarsch *et al.,* 1982; Neumann *et al.,* 1983), again underlining the special position of *Thermoplasma* among the archaebacteria. As already mentioned the DNA is associated in an unknown manner with a strongly basic, histone-like protein present in high quantity (Searcy and Stein, 1980).

The first restriction endonuclease found in an archaebacterium has been isolated from *Thermoplasma* by McConnell *et al.* (1978). It attacks the sequence CGCG, as do several enzymes isolated from eubacteria.

The DNA-dependent RNA polymerase (Sturm *et al.,* 1980) is completely insensitive to the antibiotics rifampicin and streptolydigin, which typically inhibit DNA-dependent RNA polymerases of wild-type eubacteria. The isolated enzyme transcribes native DNA strikingly less efficiently than poly[d(A-T)]·poly[d(A-T)]. The component pattern corresponds to those of the RNA polymerases of *Sulfolobus* (Zillig *et al.,* 1979, 1980) and the Thermoproteales (Zillig *et al.,* 1982a; Prangishvilli *et al.,* 1982), although the sizes of the components are somewhat larger than those of this latter group. By the Western blotting technique, homology of the RNA polymerases of *Sulfolobus* and *Thermoplasma* was demonstrated for components A to E (Schnabel *et al.,* 1983), indicating that *Thermoplasma* is similar in this feature to the sulfur-dependent archaebacteria. This is in contrast to the 16 S rRNA comparisons, which show no specific relationship between the two groups (Fox *et al.,* 1980).

The binding site for ribosomal protein L1 of *Escherichia coli* in the 23 S rRNA of *Thermoplasma* is strikingly different from that in the 23 S rRNAs of *Methanobacterium, Halobacterium,* and *Sulfolobus.* These show significant homology with that of *E. coli* itself, again pointing to a unique position of *Thermoplasma* among the archaebacteria.

In a phylogenetic tree constructed by Fox *et al.* (1982) on the basis of 5 S rRNA primary structure, *Thermoplasma* is placed closest to what the authors call "group IIb," although it is not particularly close to the group. In its secondary structure, *Thermoplasma* 5 S RNA resembles most its counterpart from *Sulfolobus.* Both are unique among archaebacteria having *no* looped-out residue in helix II, having the smallest bulge loops of all between helices IV and V, and possibly being able to stack helix V directly on helix I (see Chapter 5).

The sequence of the initiator tRNA of *Thermoplasma* (Kuchino *et al.,* 1982) shows slightly less homology than that of *Sulfolobus* to the sequences of eukaryotic initiators. With the other known archaebacterial initiator sequences, it shares several features, which thus could be typically archaebacterial.

The methionine tRNA used in elongation has less than 70% homology with any other tRNA sequence so far reported, but almost 90% homology with the so-called "master copy of all tRNAs representing the ancestral quasi-species" deduced by Eigen and Winkler-Oswatitsch (1981; Kilpatrick and Walker, 1982). The modification pattern of the tRNAs of *Thermoplasma* is typically archaebacterial.

The elongation factor EFII of *Thermoplasma* is ADP-ribosylated by the ADP ribosyltransferase of diphtheria toxin (Kessel and Klink, 1980), as are all other archaebacterial EFIIs tested so far and, typically, the EFIIs from eukaryotes (the homologous EFGs of eubacteria remain unaltered).

The respiratory chain appears to consist of only a *b*-type cytochrome (Belly *et al.*, 1973; Holländer, 1978; Searcy *et al.*, 1978; Searcy and Whatley, 1982) and a quinone (Holländer *et al.*, 1977). A membrane-bound ATPase stimulated by magnesium sulfate has been proposed to act as a sulfate-exporting translocase (Searcy and Whatley, 1982). No proton-translocating ATPase has been found. It thus appears possible that protons are pumped out by respiration only and furthermore that the export of sulfate creates a positive surplus charge inside the cell counteracting proton flow along the strong gradient resulting from the low pH of the medium (Searcy and Whatley, 1982). Energy could then be produced only via substrate phosphorylation. This assumption is in accord with the claim that *Thermoplasma* requires oxygen for growth (Searcy and Whatley, 1982).

E. CONCLUSIONS

The occurrence of a histone-like DNA-binding protein and possibly an actin-like protein, the probable absence of a *c*-type, but the presence of a *b*-type cytochrome, the nature of the superoxide dismutase (Searcy and Searcy, 1981), and other features have lead Searcy to the idea that *Thermoplasma* among the archaebacteria might be specifically related to eukaryotes (Searcy *et al.*, 1978). The evidence discussed above shows however that other archaebacteria, particularly *Sulfolobus*, appear as close or even closer to the eukaryotic cytoplasm.

On the other hand, *Thermoplasma* clearly holds an exceptional, possibly "bridging" position among archaebacteria. Further comparative biochemical work is required for a full understanding of its status.

IV. *Sulfolobus*

Extensive reviews covering the work on *Sulfolobus* have been written by Brock (1978b, 1981), one of the discoverers of this extremely thermoacidophilic aerobic organism which lives in solfataric waters. In this section, we will deal mainly with later results, their implications, and some general aspects.

A. History of Discovery

The first representative of the genus *Sulfolobus* was isolated from an acidic hot spring in Yellowstone park by J. A. Brierley and was preliminarily described in his thesis (1966). Independently, T. D. Brock and his collaborators discovered the organism in 1970 (Brock, 1978b) and the first description was published in 1972 (Brock *et al.*, 1972). The isolate of Brierley was fully described in 1973 (Brierley and Brierley, 1973). Two isolates of an organism termed *Caldariella*, which was thought to belong to a highly convergent, but taxonomically distant genus, were obtained from Italian solfataric hot springs by De Rosa *et al.* (1974). The component patterns of the DNA-dependent RNA polymerases of independent isolates from this source, assumed to be closely related to *Caldariella*, and of the enzyme from Brierley's isolate, proved, however, to be very similar to that of *Sulfolobus acidocaldarius* 98-3 (DSM 639) (Zillig *et al.*, 1979, 1980). Furthermore these enzymes showed immunochemical cross-reaction with the latter polymerase in the Ouchterlony test. On this basis, they were named as new species of the genus, namely *Sulfolobus solfataricus* and *Sulfolobus brierleyi* (Zillig *et al.*, 1980).

B. Taxonomy

Until the discovery of the Thermoproteales (see below), *Sulfolobus* appeared even more isolated from the rest of the archaebacteria than did *Thermoplasma*. The S_{AB} of only 0.17 with the others is lower than any known S_{AB} between two members of the same urkingdom (Fox *et al.*, 1980). The nature of the envelope (Weiss, 1974; Michel *et al.*, 1980), the membrane (Langworthy *et al.*, 1974, 1982; De Rosa *et al.*, 1977), the ribosomes (Fox *et al.*, 1980), the elongation factor EFII (Kessel and Klink, 1980), the RNA polymerase (Zillig *et al.*, 1979, 1980), and other features (see below), however, proved to be typically archaebacterial.

The S_{AB} between *Sulfolobus* and *Thermoproteus*, one representative of the new order Thermoproteales, is 0.33 (C. Woese, E. Stackebrandt, and W. Zillig, unpublished). Extensive 16 S rRNA–DNA cross-hybridization studies (see below) showed that *Sulfolobus* and the Thermoproteales form a distinct second branch of the archaebacterial urkingdom (Tu *et al.*, 1982). On this basis, and considering the different (opposite) types of sulfur metabolism, we propose that *Sulfolobus* represents a separate order, the Sulfolobales, which, together with the Thermoproteales, form a separate sulfur-dependent branch of the archaebacteria, distinct from the methanogens and extreme halophiles. This view is supported by the finding that the component patterns of the RNA polymerases of *Sulfolobus* and the Thermoproteales constitute a distinct type significantly differing from the type represented by the enzymes of the methanogens and extreme halophiles (Huet *et al.*, 1983; Schnabel *et al.*, 1983). In this respect, *Thermoplasma* resembles the new branch; in others, however, it appears either intermediate or isolated.

C. Ecology

Sulfolobus has been isolated from many acidic hot waters and soils throughout the world, including Yellowstone National Park (Brierley, 1966; Brock *et al.*, 1972) and Mount Lassen Volcanic National Park (W. Zillig and J. Weber, unpublished) in the United States, Italy (De Rosa *et al.*, 1974), New Zealand (Bohlool, 1975), Japan (Furuya *et al.*, 1977; Yeats *et al.*, 1982), and Iceland (W. Zillig and K. O. Stetter, unpublished). Both *Sulfolobus acidocaldarius* and *S. solfataricus* have been found in North America as well as in Italian and Japanese solfataras, indicating that geographic barriers for their propagation do not exist.

The temperature of the sources ranges from lower than 60° to 100°C, the pH from 1 to more than 5 (Brock, 1978b; W. Zillig and K. O. Stetter, unpublished). Most isolates have been obtained from waters of 80°–90°C and pH 2–3. Many springs yielding *Sulfolobus* isolates contain finely distributed or even colloidal elemental sulfur. The ionic strength of the sources is usually low. The cell density can be as high as 10^6/ml. An iridescent oily layer on the surface of some sources contained particularly high numbers of *Sulfolobus* cells (unpublished observation of the authors).

D. Growth Requirements

The optimal growth temperatures of various isolates differ considerably and can be as high as 87°C (Zillig *et al.*, 1980). Measurements in the field indicate that strains of *Sulfolobus* tolerate and possibly even grow at temperatures up to 100°C. Most isolates, however, show sharp upper limits to their growth temperature. Brock (1978b) has reported pH 4 as the uppermost pH value for growth. We have found that different isolates have different pH optima of growth, usually between pH 2 and 3.5, and some can still survive at pH 5.5.

Many *Sulfolobus* isolates are facultatively autotrophic, growing on yeast extract and various other carbon sources (often particularly well on sucrose) just as well as, in a chemolithoautotrophic manner, on CO_2 as the sole carbon source (and oxidizing elemental sulfur to sulfuric acid as an energy source). The CO_2 assimilation occurs via a reductive carboxylic acid pathway and not via the Calvin–Benson cycle as is usual in eubacteria (Kandler and Stetter, 1981). Others grow only heterotrophically (as indicated by alkalinization of the growth medium in the presence of both yeast extract and sulfur), whereas the isolate B6/2 from Japan is an obligatory chemolithoautotroph (W. Zillig, unpublished). The utilization of organic components is thought to occur via a fermentative mechanism (Brierley, 1978). The oxidation of sulfur and several other electron donors, e.g., Fe^{2+} ions occurs by aerobic respiration. Sulfur can be oxidized by O_2, as well as Fe^{3+} and molybdate (Brierley and Brierley, 1982). Under strong aeration, mixing CO_2 (about 1%) with the air appreciably enhances growth (W. Zillig, unpublished observation). The generation times are about 4 hr under optimal conditions. The adaptation of *Sulfolobus* strains growing on one carbon source, e.g., sucrose, to a different one, e.g., lactose, is often a slow process requiring days (W. Zillig, unpublished). *Sulfolobus* species form colonies or lawns on water glass (polysilicate) plates equlibrated with medium (S. Yeats and W. Zillig, unpublished) or on 10 to 12% starch gels, which can be poured like agar plates (R. Skorko and H. Neumann, unpublished).

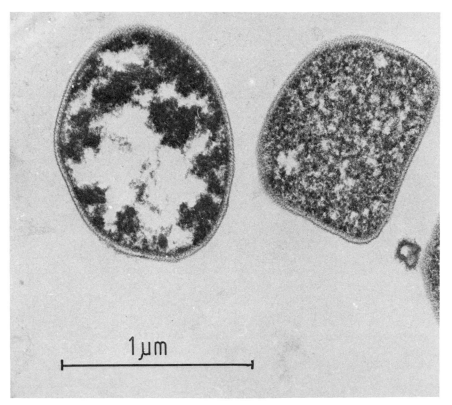

FIG. 1.   Thin sections of *Sulfolobus acidocaldarius* strain B6 "normal," midexponential phase.

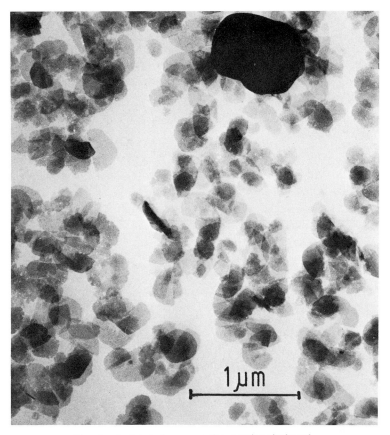

FIG. 2. Isolated glycogen scales, rotation shadowed.

E. MORPHOLOGY

In exponentially growing cultures, *Sulfolobus* cells appear rather spherical with rounded edges and more or less planar surfaces (Fig. 1), but usually unlobed. However, cells from later stages of growth and from colonies grown on plates are often strongly lobed (McClure and Wyckoff, 1982) (see Fig. 3).

In sections, but also in shadowed specimens, the subunit nature of the cell envelope is apparent. Pili are often seen. They have been implicated in sulfur metabolism (Brock, 1978b). Another type of "holdfast," which is involved in the attachment to carbon substrates, resembles pseudopodia (McClure and Wyckoff, 1982). In sections, the cytoplasm appears more or less homogeneous, except for bodies of low electron density, which consist of glycogen formed under favorable conditions (König *et al.*, 1982) (Fig. 2).

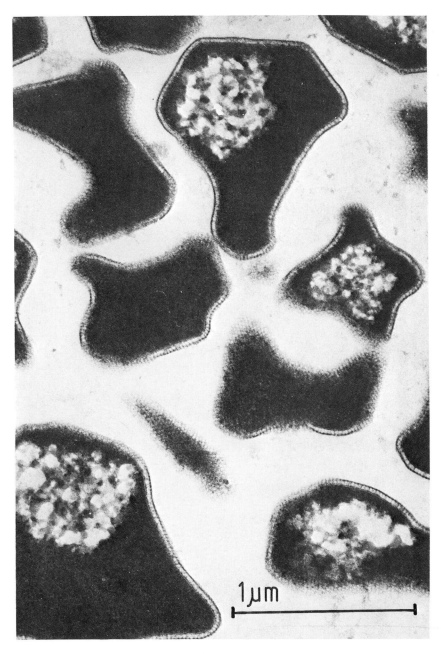

FIG. 3. Thin section of lobed *Sulfolobus acidocaldarius*, strain B6, with glycogen granules, showing subunits in envelope.

F. Biochemistry

The cell envelope is a regular array of subunits in two-dimensional hexagonal packing, consisting of a glycoprotein of a distinct amino acid composition, which is insoluble even in 20 g/liter of sodium dodecyl sulfate (SDS) at 60°C but can be solubilized at pH 9 (Michel *et al.*, 1980; Taylor *et al.*, 1982) (see Fig. 3). The membrane contains the isopranyl glycerol ether lipids, characteristic of archaebacteria, in this case consisting of 95% tetraethers and 5% diethers. [*Thermoplasma* has 90% tetraethers and 10% diethers (Langworthy *et al.*, 1982).]

As in other archaebacteria, the DNA is associated with basic proteins that differ in size and charge density between the different species and that do not cross-react with an antibody directed against the DNA-binding protein of *Thermoplasma* (Thomm *et al.*, 1982; Green *et al.*, 1983). This variance is in sharp contrast to the conservativity of eukaryotic histones and eubacterial DNA-binding proteins, possibly indicating an especial phylogenetic depth to this group or a relatively low selective pressure on these molecular sequences.

Like *Thermoplasma*, *Sulfolobus acidocaldarius* and *S. solfataricus* contain a restriction endonuclease (P. McWilliam, unpublished), which has the same specificity as the *Hae*III enzyme, i.e., CCGG.

The shape of the ribosomal 50 S subunits of *Sulfolobus* (and *Thermoproteus*, see below) is distinct from that of members of the other major branch, the latter resembling more the eubacterial 50 S subunits (Henderson *et al.*, 1984). The 30 S subunits are again similar to those of *Thermoproteus* and *slightly* different from those representing the other major branch (Lake *et al.*, 1982).

The number of proteins in the 30 S subunits of *Sulfolobus* is 27; in *Methanobacterium thermoautotrophicum*, 23; in *E. coli*, 21. The number of proteins in the 50 S subunits of *Sulfolobus* is 34; in *M. thermoautotrophicum*, 32; in *E. coli*, 32. In eukaryotes, the total number of proteins in both subunits together is about 70–84 (Schmid *et al.*, 1982).

The structure of the 5 S rRNA of *Sulfolobus* resembles that of the 5 S rRNA of the cytoplasmic ribosomes of eukaryotes in several respects (Stahl *et al.*, 1981; Fox *et al.*, 1982).

The sequence of the initiator tRNA of *Sulfolobus* shows a closer resemblance than does that of the *Halococcus* initiator tRNA to the initiator from the eukaryotic cytoplasm (Kuchino *et al.*, 1982). [Again, the position of *Thermoplasma* is intermediate.] In contrast, the reverse is true with regard to the eubacterial initiators. Here, the sequence of *Sulfolobus* shows a slightly lower similarity than do those of *Halococcus* and *Thermoplasma* (see Fig. 4). These three archaebacterial initiators and that from *Halobacterium* (Gupta, 1984) (which is practically identical with that from *Halococcus*) contain consensus

	Aminoacyl stem									D stem				D loop									
	1	2	3	4	5	6	7	8	9	10	11	12	13	14	15	16	17	18	19	20	21	22	23
	xxx						xxx					x		xxxxxxxxxxxx									
Yeast	A	G	C	C	G	C	G	U	G	G	C	G	C	A	G	U	G	G	A				A
S.a.	A	G	C	G	G	C	G	U	G	G	G	G	A	A	C	U	G	G	G	A	G	U	A
Th. ac.	A	G	C	G	G	G	G	U	G	G	G	G	U	A	G	U	C	A	G	G	A		A
H.m.	A	G	C	G	G	G	A	U	G	G	G	A	U	A	G	C	C	A	G	G	A	G	A
E.c.	C	G	C	G	G	G	G	U	G	G	A	G	C	A	G	C	C	U	G	G	U		A
				xxx		xxx								xxxxxx					xxxxxx				

	Extra arm					TF stem					TF loop							TF stem			
	46	47	48	49	50	51	52	53	54	55	56	57	58	59	60	61	62	63	64	65	66	67
							xxxxxxxxxxxxxx															
			xxx						xxx										xxx		xxxxxx	
Yeast	A	U	G	D	C	C	U	C	G	G	A	U	C	G	A	A	A	C	C	G	G	G
S.a.	A	G	G	U	C	C	C	U	G	G	U	U	C	G	A	A	U	C	C	A	G	G
Th. ac.	A	G	A	U	C	G	A	U	G	G	U	U	C	G	A	A	U	C	C	A	U	C
H.m.	A	G	A	U	C	A	G	U	A	G	U	U	C	G	A	A	U	C	U	A	C	U
E.c.	A	G	G	U	C	G	U	C	G	G	U	U	C	A	A	A	U	C	C	G	G	C
		xxx		xxx					xxx		xxx						xxx		xxx			

Fig. 4. Sequences of initiator tRNAs of *Sulfolobus, Thermoplasma,* and *Halococcus* as compared to yeast and *E. coli,* modifications not considered. Abbreviations: Yeast, *Saccharomyces cerevisiae; S.a., Sulfolobus acidocaldarius; Th. ac., Thermoplasma acidophilum; H.m., Halococcus morrhuae; E.c., Escherichia coli.*

nucleotides at positions 11 and 24, which form a base pair in the D stem, and at positions 52 and 62, which form a base pair in the TψC stem, that appear specific to the archaebacteria. These data are in perfect accord with the notion that *Sulfolobus* represents a distinct major branch of the archaebacteria. However, it is also less distant from *Thermoplasma* than from *Halococcus* by this measure.

Like the elongation factors EFII of other archaebacteria, that from *Sulfolobus* is ADP ribosylated by the diphtheria toxin ADP ribosyltransferase (Kessel and Klink, 1980).

The DNA-dependent RNA polymerase component pattern conforms with those of the Thermoproteales and, in principle, also with that of *Thermoplasma* (Zillig *et al.*, 1982a). It exhibits a striking resemblance with patterns of eukaryotic RNA polymerases, especially of type I (= A). That this is the result of true homology has been established by the demonstration of immunochemical cross-reaction of components with components of yeast RNA polymerases (Huet *et al.*, 1983).

As typical for archaebacterial RNA polymerases, the enzyme from *Sulfolobus*

D stem		Anti C. stem	Anti C. loop	Anti C. stem
24 25 26 27	28	29 30 31 32 33	34 35 36 37 38 39 40	41 42 43 44 45
G C G C	G	C A G G G	C U C A U A A	C C C U G
U C C C	G	C A G G G	C U C A U A A	C C C U G
A U C C	G	A U G G C	C U C A U A A	C C C G U
U U C C	G	G C G G C	C U C A U A A	C C C G C
G C U C	G	U C G G G	C U C A U A A	C C C G A

Aminoacyl stem
69 70 71 72 73 74 75 76 77 78
G C G G C U A C C A
G C C G C U A C C A
C C C G C U A C C A
C C C G C U A C C A
C C C G C A A C C A

is not inhibited by rifampicin, streptolydigin, and α-amanitin (Zillig *et al.*, 1979). As with those of other archaebacteria and polymerase I of eukaryotes, the enzyme is specifically stimulated in the elongation phase by the flavolignane derivative silybin (Schnabel *et al.*, 1982). Even at elevated temperature, the enzyme does not specifically bind to restriction fragments carrying strong bacterial promoters (Zillig *et al.*, 1980).

Like other archaebacteria of this branch, *Sulfolobus* is able to synthesize glycogen of very small average chain length (König *et al.*, 1982), especially when the cells have been grown on sucrose. The glucosyltransferase involved in its synthesis has been purified with a glycogen protein complex by CsCl density gradient centrifrugation. Its substrate specifity is remarkably low, utilizing both ADP-glucose and UDP-glucose with similarly high K_m values and low turnover numbers.

G. Extrachromosomal Genetic Elements

In an isolate from Beppu, Japan, termed B6, electron micrographs of sections showed the presence of virus-like particles crystallized in hexagonal dense packing (W. Zillig and I. Scholz, unpublished) (Fig. 5). After isolation by PEG, precipitation, and CsCl gradient centrifugation, they were found to contain dou-

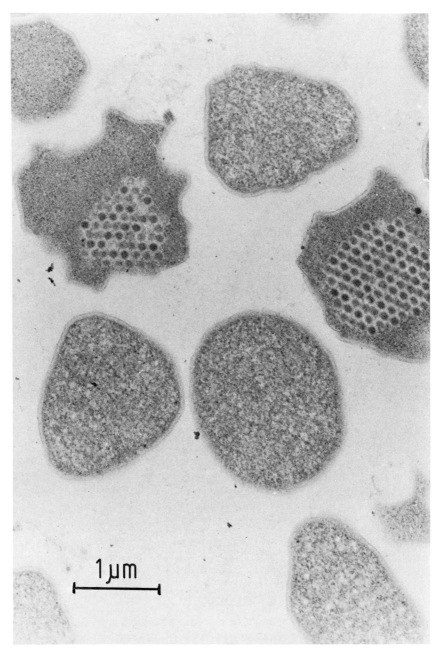

FIG. 5. Thin section of *Sulfolobus acidocaldarius*, strain B6, containing virus-like particles crystallized in hexagonal dense packing.

ble-stranded DNA. Conditions for their reproducible production have not been worked out.

A fraction of the cell clones obtained from this isolate contain a large plasmid, pBG of about 35,000 base pairs (bp), which shows a remarkable genetic instability: the restriction fragment pattern exhibits substoichiometric side bands (P. McWilliam and S. Yeats, unpublished) (Fig. 6). Clones have been isolated in which these are the major bands, and at least one of the changes has been found to have resulted from an inversion.

Another isolate from Beppu, B12, contains a plasmid of 13,000 bp, which exists in both a free circular, supercoiled form and integrated into the host chromosome at a specific site (Yeats *et al.*, 1982) (Fig. 7a–c); it was recently shown to be a provirus. The free plasmid is strongly amplified on irradiation with a low UV dose. Its multiplication is accompanied by destruction of the host chromosome. Six hours after UV induction, spikes protruding from the envelope

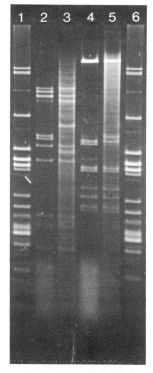

FIG. 6. Two forms of a plasmid from *S. acidocaldarius*, strains B6 and P1, pB6 and pB1, distinguished, among other features, by an inversion. Lane 1, size markers; lane 2, pB6; lane 3, pB1, both restricted with *Eco*R1; lane 4, pB6; lane 5, pB1, both restricted with *Bam*H1; lane 6, size markers.

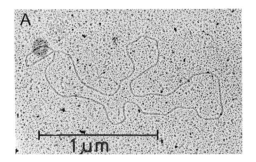

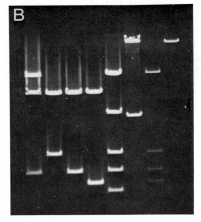

a b c d e f g h

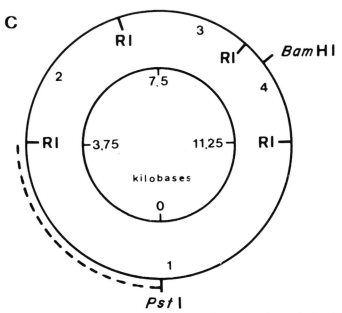

FIG. 7. Plasmid pB12. (A) Electron micrograph of spread plasmid (rotation shadowed). (B) Agarose gel electropherogram of restriction fragments of cloned and native plasmid DNA. Lanes a–d: pBR 325 *Eco*RI clones, digested with *Eco*RI, containing B12 *Eco*RI fragments 1+2, 2, 3, and 4, respectively. Lanes e and f: a PBR322-B12 *Bam*HI clone digested with *Eco*RI and *Bam*HI, respectively. Lanes g and h: B12 plasmid digested with *Eco*RI and *Bam*HI, respectively. (C) Restriction map of *Bam*HI, *Eco*RI, and *Pst*I cleavage sites on plasmid DNA–*Eco*RI sites labeled RI. The region covering the site of integration is shown by the dotted line.

are formed; 18 hr after induction, lemon-shaped short-tailed virus particles containing closed circular DNA are liberated (Martin *et al.,* 1984).

Sulfolobus cells contain protein kinases, which link phosphate to both serine and threonine (R. Skorko, unpublished). No cAMP has been found in starved *Sulfolobus* and *Thermoplasma* cultures (M. Schweiger, unpublished).

H. Conclusions

The structures of the EFII, the DNA-dependent RNA polymerase, and (to some extent) the 5 S rRNA of *Sulfolobus* show a pronounced similarity to their eukaryotic counterparts. The sequence of the initiator tRNA of *Sulfolobus* shows relationships to those of eubacterial as well as to eukaryotic initiators. Among the archaebacterial initiator tRNAs, it is the most "eukaryotic." However, the phylogenetic position of *Sulfolobus* "between" the rest of the archaebacteria and the eukaryotic cytoplasm is more pronounced with some phenotypic features than with others.

V. Thermoproteales

A. Geographic Distribution

Members of four genera of sulfur-dependent bacteria, *Thermoproteus* (Zillig *et al.,* 1981), *Thermofilum* (Zillig *et al.,* 1983a), *Desulfurococcus* (Zillig *et al.,* 1982), and *Thermococcus* (Zillig *et al.,* 1983b), have been isolated from springs, water holes, mud holes, and soils of solfataras. *Thermoproteus* was found at Mount Hengill, the Krafla, Theistareikir, Geysir, Hveradalir in Kerlingarfjoll, Hveravellir, Krisuvik, Reykjanes, Askja, Vonaskard, Landmannalaugar and Kaldaklofsjokull, all in Iceland, and at the Sulfur Works and Bumpas Hell, both on Mount Lassen, in California. *Thermofilum* and *Desulfurococcus* so far have been isolated only from sources in Iceland; *Thermococcus* was obtained from a marine water hole on the beach of Vulcano, Italy. Similar organisms have been reported to exist in hot springs of New Zealand (Morgan and Daniel, 1982). Their exact taxonomic positions in relation to the above genera have not yet been determined.

B. Ecology

Members of the novel order Thermoproteales have so far been isolated only from solfataric hot springs of pH 1.7–6.8 (Table I). Direct microscopy of the

TABLE I

TEMPERATURES AND pH VALUES OF SOURCES OF THERMOPROTEALES

Samples	Number	Temperature (°C)		pH range
		Maximum	Average	
Total taken	267 (86)[a]	100	87	1.0–8.3
Yield				
Thermoproteus total	67	100	92	1.7–6.8
Thermoproteus autotrophic	14 (of 86)[a]	98	92	3.0–6.5
Thermofilum	39	100	92	2.8–6.8
Desulfurococcus	35	98	92	2.2–6.8
Thermococcus	2	96	96	5.5

[a] Independent sampling tour.

samples has, however, revealed the presence of organisms of similar morphology in so-called alkaline hot springs of pH 7–9 (Brock, 1978b). The organisms known so far appear to tolerate the limited sulfide ion concentrations occurring below pH 7. Evidence presented by Brock (1978b) indicates that higher sulfide ion concentrations might be toxic (Brock, 1978b). This could explain the failure of attempts to obtain isolates from alkaline hot waters.

Members of all genera of the Thermoproteales described so far have been isolated from sources with temperatures of more than 90°C, often up to 100°C (Table I). Because marine archaebacteria have been obtained from sources above 100°C, it appears possible that *Thermoproteus, Thermofilum, Desulfurococcus,* and *Thermococcus* species could also exist at higher temperatures provided that liquid water is available. Several isolates have indeed been obtained from 0.5 to 1.5 m depths.

All known Thermoproteales exist by sulfur respiration of organic matter or in a purely chemolithotrophic manner using H_2 plus elemental sulfur as an energy source and CO_2 as a carbon source (Table II). *Desulfurococcus* and *Thermococcus* are able to survive alternatively by relatively inefficient unknown modes of fermentation. Hydrogen sulfide is not only tolerated by all these organisms, sometimes up to 2.10^4 kPa, but appears to be required in some cases. All these requirements are satisfied in solfataric fields, which abound in sulfur and are gassed by emanating superheated steam carrying CO_2, H_2, H_2S, and sometimes CO and CH_4.

Most isolates are from waters of low ionic strength. Though the Italian *Thermococcus* is well adapted to marine life, cultures could not be isolated from a hot saltwater spring at Cape Reykjanes in Iceland.

The anaerobic Thermoproteales share their habitat with the aerobic *Sulfolobus*. It appears that *Sulfolobus* exists at or close to the surface of springs and water holes, whereas most isolates of Thermoproteales have been obtained from the depths of such waters or from mud.

About 40% of all samples from Icelandic solfataras yielded isolates, often more than one, sometimes three or four (in the latter case including *Sulfolobus*) (see Table I). Thirty-seven of the 39 samples that yielded *Thermofilum* also gave rise to *Thermoproteus*. In laboratory culture, most isolates of *Thermofilum* require a polar lipid from *Thermoproteus* for growth. *Desulfurococcus* has never been isolated alone. In enrichment cultures, it is often found attached to other organisms, especially *Thermofilum*. Both *Thermofilum* and *Desulfurococcus* feed on peptides and protein.

In view of the high density of these organisms in their habitat, which can exceed 10^6 cells per ml, it thus appears that the primary producers, *Thermoproteus* and *Sulfolobus,* form the basis of the food chain, whereas *Ther-*

TABLE II

PROPERTIES OF THERMOPHILIC, CHEMOLITHOAUTOTROPHIC, SULFUR-DEPENDENT ARCHAEBACTERIA[a]

Organism	Source	Source		Optimal growth		
		Temperature (°C)	pH	Temperature (°C)	pH	Autotrophy
Thermoproteus tenax, Kra 1, DSM 2078	Krafla, Iceland, mud hole	93	6.0	88	5.5	Facultative
Thermoproteus neutrophilus, Hvv 24 (V24), DSM 2338	Kerlingarfjöll, Iceland, hot spring	85	6.5	85	6.8	Obligate
Thermoproteus sp., H3	Hengill, Iceland, mud from 0.5 m depth	98	5.5	90	5.5	Facultative
Thermoproteus sp., Bu5	Mt. Lassen, California, hot spring	90	4.5	88	5.2	Facultative
Pyrodictium occultum, PL 19	Vulcano, Italy, hot sea floor	103	6.5	105	6.5	Obligate

[a] *Thermoproteus* species were grown under strong agitation in Allen's medium (Allen, 1959) containing 5 g/liter elemental sulfur and gassed with 80% H_2 and 20% CO_2 at pH 5.5 or, in the case of *Thermoproteus neutrophilus,* pH 6.5. *Pyrodictium occultum* was grown without agitation in sea water, with 5 g/liter sulfur and gassed with 80% H_2.

mofilum and *Desulfurococcus* live off organic matter produced by the former organisms.

SAMPLING, STORAGE, AND ENRICHMENT

Samples were usually taken from the depth of the water holes and springs, often after stirring up mud layers. Samples above pH 4 were immediately transferred into syringes and injected into anaerobic tubes containing elemental sulfur, 20 μg of resazurin as redox indicator, and a gas phase of 10^5 kPa of a mixture of 95% N_2 and 5% H_2, plus 10^4 kPa of H_2S. Samples with a pH of less than 4 were rapidly neutralized by the addition of solid $CaCO_3$ before transferring them into the anaerobic tubes.

In this form, the samples could be transported at room temperature and stored at 4°C for at least a year without appreciable loss of viability, provided they remained anaerobic.

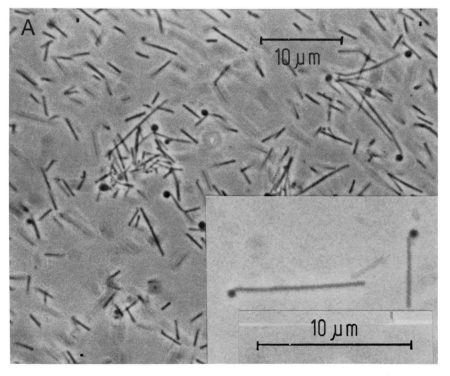

FIG. 8. Phase contrast micrographs of (A) *Thermoproteus tenax,* showing normal cells and "golf clubs." (B) *Thermoproteus tenax,* showing sharp bends and branching and budding cells. (C) *Thermofilum pendens,* "golf clubs". (D) *Desulfurococcus mucosus.*

Enrichment for *Thermoproteus* was performed by shaking in anaerobic tubes at 85°C in 5–15 ml of the following media: (1) a heterotrophic medium containing 0.02% yeast extract, 0.2% sucrose, and 1% elemental sulfur in Allen's salt base (Allen, 1959) with 10^5 kPa N_2 and 10^4 kPa H_2S in the gas phase and (2) a medium for chemolithoautotrophic growth containing Allen's salts plus 1% ele-

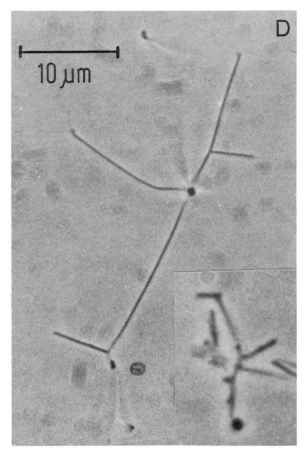

FIG. 8-B

mental sulfur and 2×10^5 kPa of a mixture of 80% H_2 and 20% CO_2 plus 1.10^4 kPa H_2S in the gas phase. The other organisms were enriched in Allen's salt base containing 0.2% yeast extract or bactotryptone $+0.2\%$ sucrose, 1% sulfur, and 10^5 kPa N_2 + 10^4 kPa H_2S in the gas phase, *Thermococcus* with the addition of 3.5–4% (w/v) of NaCl. Isolation was achieved by serial dilution or by plating, either on polyacrylamide or starch plates.

C. Morphology

1. Light Microscopy

As seen in phase contrast microscopy, *Thermoproteus* (Fig. 8a) and *Thermofilum* (Fig. 8b) appear as rods, with diameters of about 0.5 and 0.2 μm,

Fig. 8-C

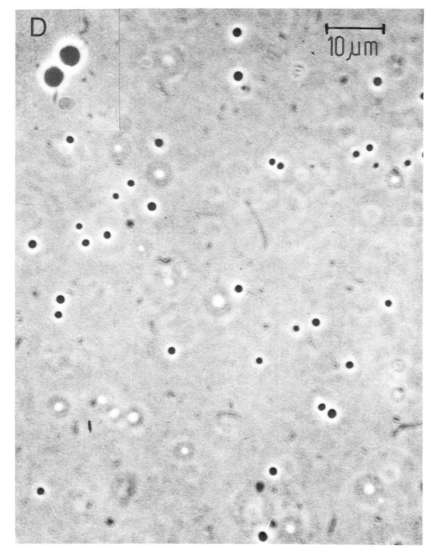

FIG. 8-D

respectively, and a striking variance of length, 1–80 μm for *Thermoproteus tenax* and 5–100 μm for *Thermofilum* (Table III). Both organisms are sometimes branched, *Thermoproteus* much more frequently than *Thermofilum*, and show eventually sharp bends with an angle of about 135°. Both can terminally and laterally expel spheric bodies. Occasionally, *Thermoproteus* forms buds. The

TABLE III

Morphology and G + C Contents of Thermoproteales

Species	DSM number	Shape	Diameter (μm)	Length (μm)	Pili	Flagella	Slime coat	G + C content (moles %)
Thermoproteus tenax	2078	Rods	0.5	1–80	+	–	–	55.5
Thermoproteus neutrophilus	2338	Rods	0.5	1–40	?	?	–	56.2
Thermofilum pendens	2475	Rods	0.2	5–100	+	–	–	57.4
Desulfurococcus mucosus	2162	Cocci	1	n.d.[a]	–	–	+	50.8
Desulfurococcus mobilis	2161	Cocci	1	n.d.	–	+	–	50.8
Thermococcus celer	2476	Cocci	1	n.d.	–	+	–	50.8

[a] n.d., no data given.

average length and the degree of branching increase with decreasing agitation, being strongest in colonies on plates.

Desulfurococcus (Fig. 8c) and *Thermococcus* (Fig. 8d) are spherical organisms of about 1 μm diameter (Table III). In growing cultures, *Thermococcus* usually appears in diploforms, but these constitute only a minor fraction of a growing *Desulfurococcus* culture. At high nutrient concentration, *Desulfurococcus* often forms giant cells up to 10 μm in diameter.

2. ELECTRON MICROSCOPY

Cells of the species *Desulfurococcus mucosus* are surrounded by a slime layer (Fig. 9). An isolate from Hveravellir in central Iceland, Hvv 3, shows long slime threads attached to its envelope (Fig. 10). It appears that these threads are involved in the attachment to other cells, especially of the *Thermofilum* type, as often observed in enrichment cultures (Fig. 11).

Desulfurococcus mobilis and *Thermococcus celer* show monopolar polytrichous flagellation (Fig. 12 and Table III). In negative staining, the flagella show different diameters, 7 nm in the case of *Desulfurococcus*, 10 nm in the case

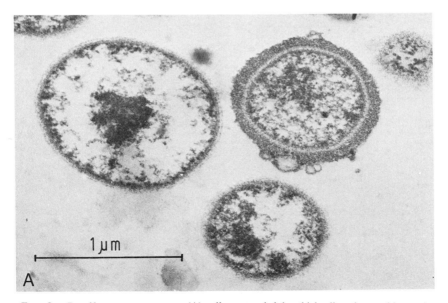

FIG. 9. *Desulfurococcus mucosus* (A) cell surrounded by thick slime layer, thin section. (B) Cells connected by solid slime layers, thin section. (C) Giant form with slime layer, surrounded by normal cells, thin section. (D) Budding cell surrounded by slime layer, rotation shadowed.

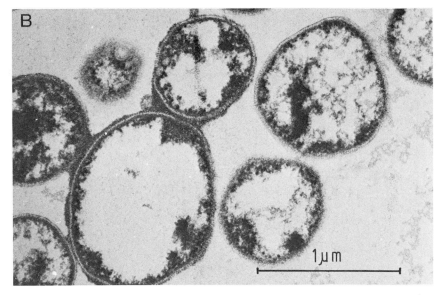

FIG. 9-B

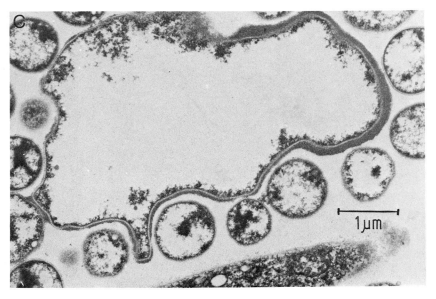

FIG. 9-C

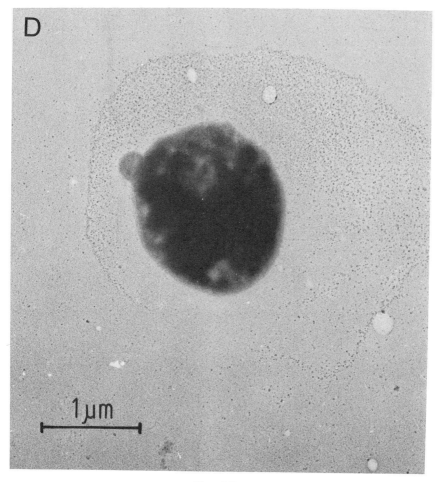

FIG. 9-D

of *Thermococcus*, both lower than that of eubacteria (13 nm for *Pseudomonas fluorescens*). In sections of *Desulfurococcus mobilis*, a peculiar apparatus forming the basis of the flagella has been recognized. Cross sections of *Desulfurococcus* cells appear rather empty, with the electron-dense fraction of the cytoplasm either attached to the membrane or forming bridges through the cell body. In contrast, sections of *Thermococcus* (Fig. 13) have a dense and evenly distributed cytoplasm with enclosed central regions of lower electron density. The envelopes of *Desulfurococcus* and *Thermococcus* appear to consist of a "pavement" of subunits attached to the membrane.

Dividing cells of *Thermococcus* show constrictions of varying extent or are even connected only by narrow strings of cytoplasm surrounded by membrane

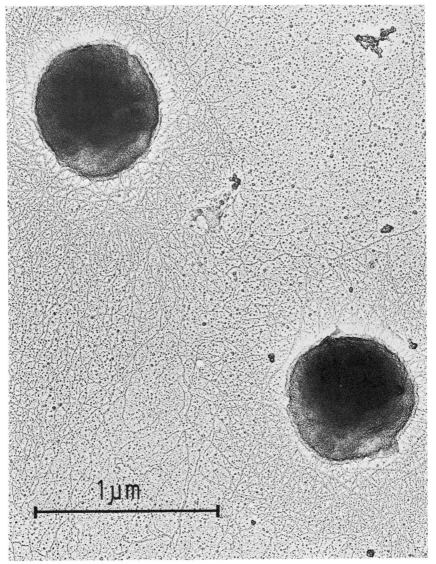

FIG. 10. Electron micrograph of two cells of *Desulfurococcus* sp. Hvv 3 surrounded by slime filaments. Rotation shadowed.

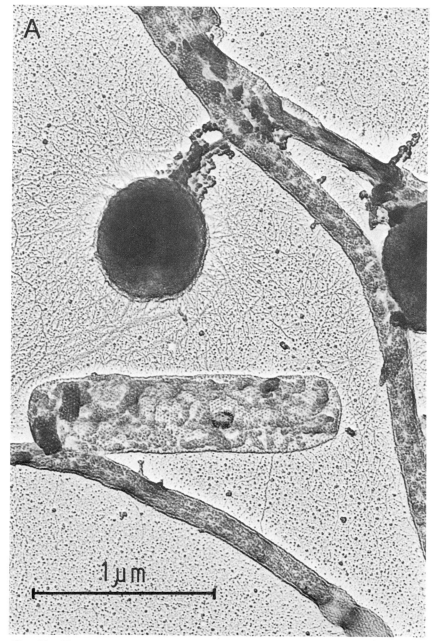

FIG. 11. Electron micrograph of enrichment cultures, showing (A) a *Thermoproteus* cell, segments of three *Thermofilum* cells, and a cell of *Desulfurococcus* sp. with slime filaments. (B) A *Desulfurococcus* cell attached to both an empty and an intact cell of *Thermofilum*.

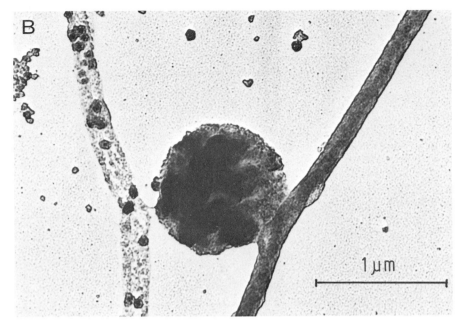

FIG. 11-B

and envelope, implying that cell division does not proceed via septum formation, which has never been observed, but by constriction (Fig. 13A–E). In closed vessels *Thermococcus,* cultures lyse after reaching a cell density of about 2×10^8 per ml, possibly induced by accumulation of H_2S. Hexagonal bodies resembling viruses have been found in such lysates (Fig. 13f).

Both *Thermoproteus* and *Thermofilum* show a regular hexagonally packed array of subunits in their envelopes, most clearly seen in shadowed specimens (Figs. 14 and 15) and cell ghosts (Fig. 16). In thin sections, the subunits appear like paving stones or teeth (Figs. 17 and 18). These subunits appear more distinct than those in the envelopes of *Desulfurococcus* and *Thermococcus.*

The cytoplasm of *Thermoproteus* is normally electron dense and evenly distributed. Inclusions of low density are seen occasionally. In the isolate H3 from Mount Hengill in Iceland, sharp sagittal lines possibly indicating longitudinal septation are occasionally seen (Fig. 19).

The branching or budding regions and the sharp bends of *Thermoproteus* and *Thermofilum* are devoid of septa (Fig. 20). "Golf clubs", i.e., cells with terminal lateral spheric protrusions, of *Thermofilum* and of the isolate H3 of *Thermoproteus* show envelopes around the spheric bodies (Figs. 21 and 22). Occasionally free spheres have been found.

Both *Thermofilum* and *Thermoproteus* show pili attached terminally or later-

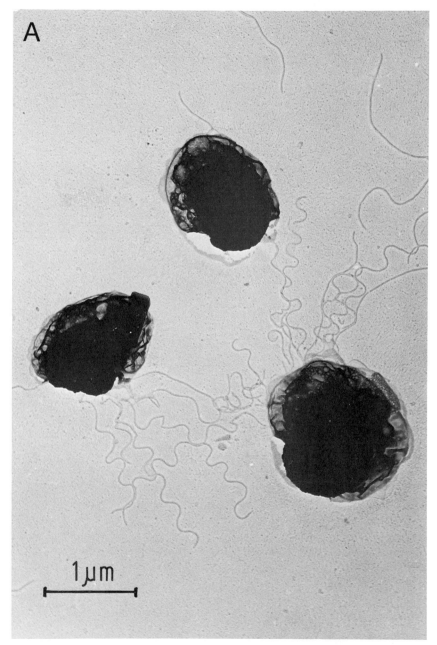

Fig. 12. Flagellation in (A) Thermoproteales and (B) flagellated cells of *Desulfurococcus mobilis*. (A) rotation shadowed and (B) thin section. (C) and (D) flagellated cells of *Thermococcus celer*. (C) Rotation shadowed and (D) negatively stained. (E) Sections of flagella of (E-1) *Pseudomonas fluorescens*, (E-2) *Thermococcus celer*, and (E-3) *Desulfurococcus mobilis* negatively stained.

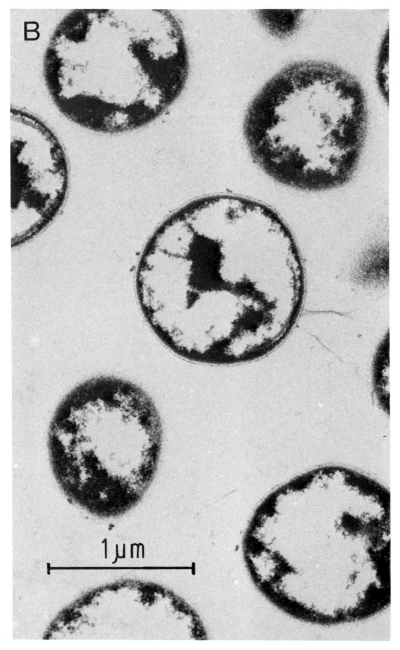

FIG. 12-B

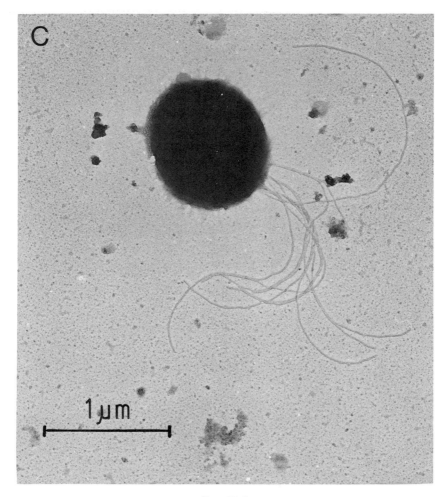

FIG. 12-C

ally (Fig. 23). A single flagellum protrudes from the tip of cells of the isolate V24 Sta of *Thermoproteus neutrophilus*.

Peculiar local swellings along cells of *Thermofilum* appear to be involved in the formation of new cell ends and thus in cell division (Fig. 24).

3. PECULIAR FORMS

Sections of the isolate Hvd 5 of *Thermoproteus* exhibit intracellular filaments of unknown function (Fig. 25). During the adaptation of heterotrophic cultures of

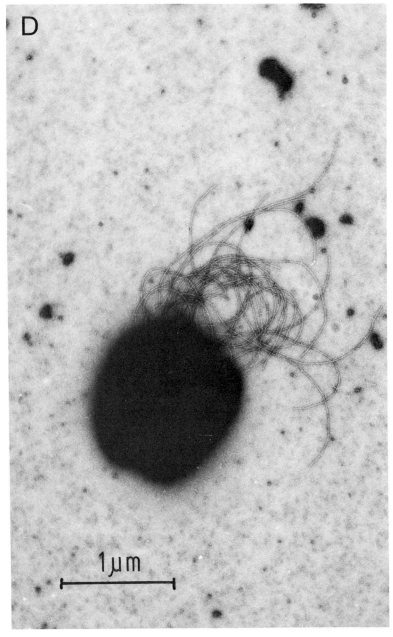

Fig. 12-D

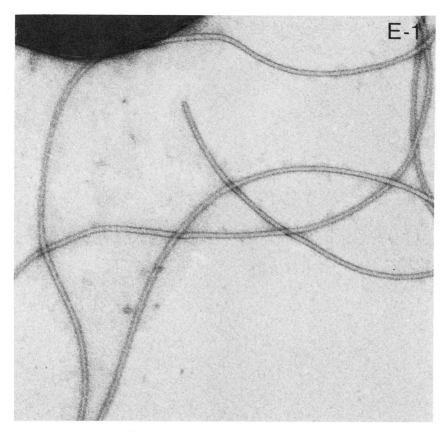

Fig. 12-E-1. (For scale see Fig. 12-E-3.)

Thermoproteus tenax to chemolithoautotrophic growth with limiting amounts of elemental sulfur, similar filaments of three types, distinguished by diameter and length, were liberated from the cells, of which a large fraction was lysed (Fig. 26A and B, see also Fig. 16). These were shown to be three different viruses, TTV 1, 2, and 3, each containing linear, double-stranded DNA (Janekovic *et al.*, 1983).

In electron micrographs of *Thermofilum,* ultrathin rods with a diameter of about 0.1 μm and the same surface structure have been observed occasionally (Fig. 27).

Electron micrographs of *Desulfurococcus* sometimes exhibit long rod-shaped or filamentous protrusions (Fig. 28).

In stationary cultures of *Thermococcus,* the cells often show long protrusions, sometimes many per cell, sometimes branched and interrupted by swellings (Fig. 29).

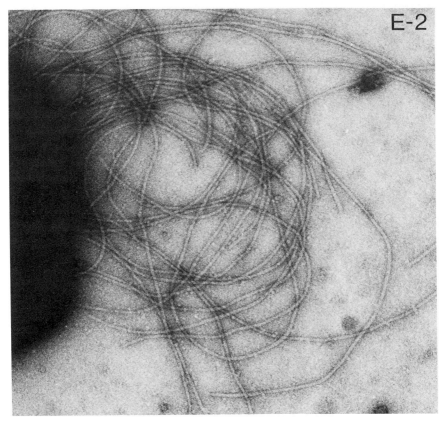

Fig. 12-E-2.   (For scale see Fig. 12-E-3.)

All these forms might be artifacts of decay, but they could be functional as well under certain unknown circumstances.

D. Growth and Metabolism

The growth and metabolism of Thermoproteales is outlined in Tables IV and V. *Thermoproteus* can grow in strictly chemolithoautotrophic manner with H_2 plus elemental sulfur as energy source and CO_2 as sole carbon source, or, alternatively, by sulfur respiration of different substrates like yeast extract, glucose, starch, ethanol, methanol, formate, malate, fumarate, casamino acids and even carbon monoxide. The products of heterotrophic growth are CO_2 and

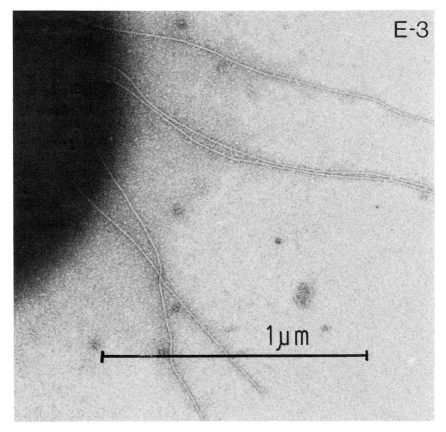

FIG. 12-E-3

H_2S. Instead of sulfur, cystine, or malate can serve as terminal electron acceptors.

Chemolithoautotrophic growth has not been observed with *Thermofilum, Desulfurococcus,* and *Thermococcus,* all of which live by sulfur respiration of substrates like yeast extract, peptides (tryptone), or protein (casein) but not of casamino acids. One isolate of *Desulfurococcus* from an almost neutral hot spring was able to feed on glucose. In addition to the carbon source, *Thermofilum pendens* requires a polar lipid fraction from *Thermoproteus* for multiplication. At least in the cases of *Thermoproteus* and *Thermofilum,* molecular hydrogen has been found as a by-product of growth.

These organisms, especially *Thermoproteus,* are able to emulsify the sulfur used as an electron acceptor. In growing cultures of *Thermofilum pendens,* the

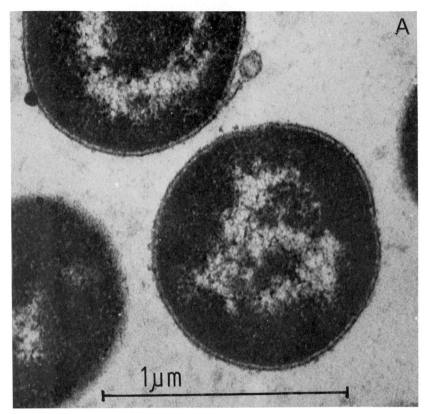

FIG. 13. Thin section of *Thermococcus* celer. (A) single cells. (B) Diploforms, beginning constriction. (C) Diploform, strong constriction. (D) Diploform shortly before division. (E) Single cells still connected by bridge of cytoplasm surrounded by membrane and envelope. (F) Single cells, one empty, with "virus-like" hexagonal particles.

usually evenly distributed sulfur changes its structure drastically, forming either needles with rounded edges or even semisolid droplets at temperatures, where normally the rhombic forms of sulfur are stable.

Sulfite, thiosulfate, and sulfate did not serve as terminal electron acceptors in the growth of Thermoproteales.

The temperature optima of the growth of the Thermoproteales are about 88°C. The temperature curves, however, differ significantly. The growth rate of *Thermoproteus* declines strongly in both directions, e.g., to about one-tenth of the optimal rate at 78°C and at 96°C. A culture previously grown at 96°C immediately resumes fast growth at 88°C, indicating that the lower growth rate at the higher temperature is not due to the decay of cells.

The temperature dependence of the growth of *Desulfurococcus* and, similarly,

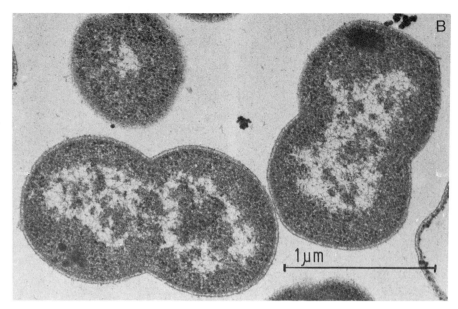

FIG. 13-B

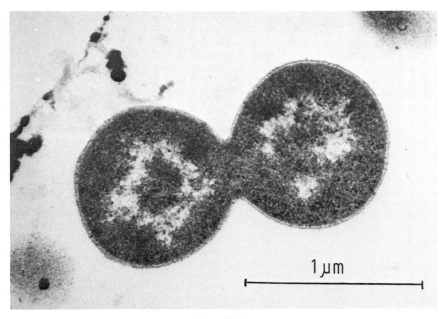

FIG. 13-C

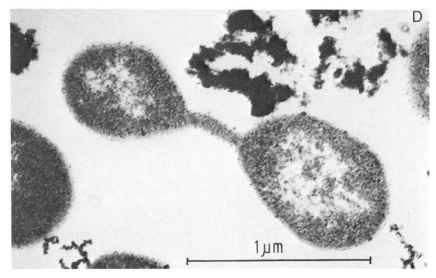

FIG. 13-D

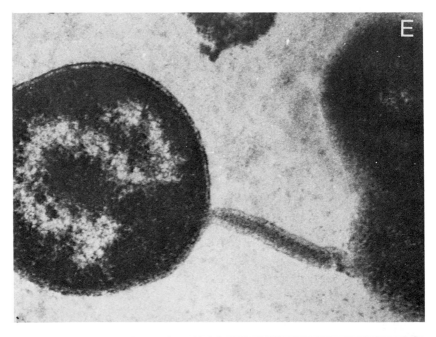

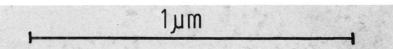

FIG. 13-E

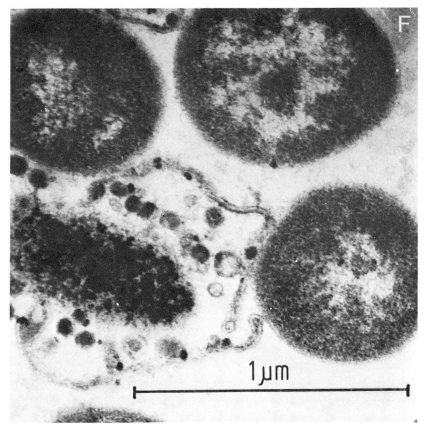

FIG. 13-F

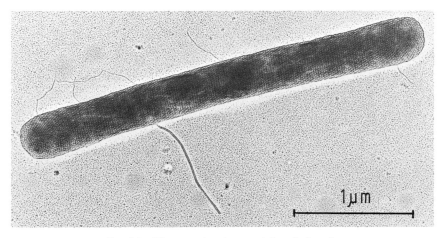

FIG. 14. *Thermoproteus tenax*, rotation shadowed.

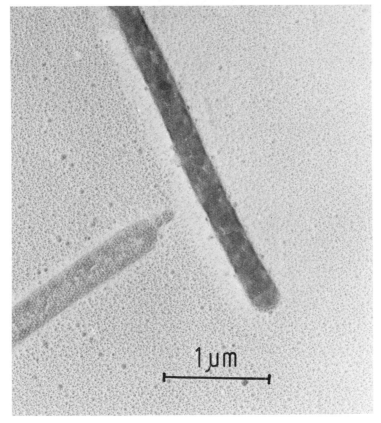

FIG. 15.   *Thermofilum pendens*, rotation shadowed.

Thermococcus is much less marked. Details on the influence of temperature on the multiplication of *Thermofilum* have not been published.

Thermoproteus tenax grows optimally at pH 5.2–5.8. It dies around pH 6. Another species, *Thermoproteus neutrophilus*, still multiplies at pH 7.0. *Thermoproteus tenax* can tolerate a pH as low as 2.5 provided that the growth conditions are otherwise optimal. *Desulfurococcus* grows optimally at pH 6, but still significantly at pH 7 and 4.5. At pH 4, it can grow only after adaptation. The growth optima of *Thermococcus* and *Thermofilum* are both close to pH 6.

Thus, in contrast to *Thermoplasma* and *Sulfolobus* the Thermoproteales cannot be considered strict acidophiles. Several species grow even above pH 7, and all tolerate pH values below pH 4.5 only under conditions allowing high metabolic activity.

During chemolithoautotrophic growth of *Thermoproteus*, about 1.5 moles H_2S were found to be produced per gram cell mass (dry weight) formed. The corresponding figure for the growth of *Thermoproteus* by sulfur respiration was 0.4 moles H_2S per gram cell mass, and, for the growth of *Desulfurococcus* by sulfur respiration, 0.044 moles H_2S per gram cell mass (dry weight). In the latter instance, this was accompanied by the formation of 0.034 moles of CO_2 from the carbon source (tryptone). These results are in accord with the expectation that a reaction yielding as little energy as H_2S formation during chemolithoautotrophic growth should result in a growth yield lower than that of sulfur respiration of organic substrates.

For heterotrophic growth of *Thermoproteus*, the growth yield relative to H_2S

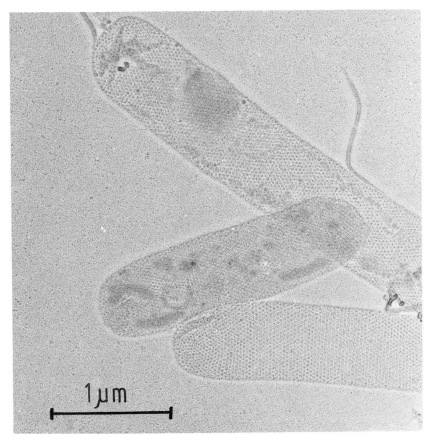

FIG. 16. Ghosts of *Thermoproteus tenax*, probably produced by virus lysis.

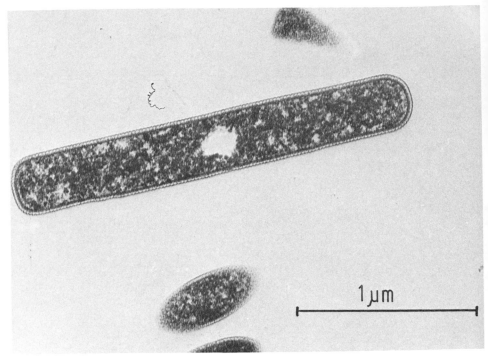

FIG. 17.   *Thermoproteus tenax*, thin section.

formation decreased steeply at extreme temperature, e.g., at 96°C. In the stationary state, uncoupled H_2S production was observed, possibly for maintenance metabolism.

Desulfurococcus and probably also *Thermococcus* can grow in the absence of sulfur, though with considerably lower rate and yield. Little CO_2 is produced under these conditions, implying that energy is conserved via some unknown mode of fermentation.

When growing by sulfur respiration (only), both *Desulfurococcus* and *Thermococcus* produce a characteristically fetid compound, possibly a mercaptan.

In closed vessels *Thermococcus*, *Desulfurococcus*, and *Thermoproteus* cease growing at a density of about 2×10^8 cells/ml. High amounts of H_2S (more than 0.2 atm) also prohibit multiplication. Gassing of the culture, particularly with CO_2, improves the final yield for example to more than 2×10^9 cells/ml for *Thermococcus*.

As *Sulfolobus*, *Thermoproteus*, and at least one species of *Desulfurococcus* can utilize various sugars, e.g., glucose as carbon sources. The growth of *Thermoproteus*, *Thermofilum*, and *Thermococcus* is stimulated by the addition of

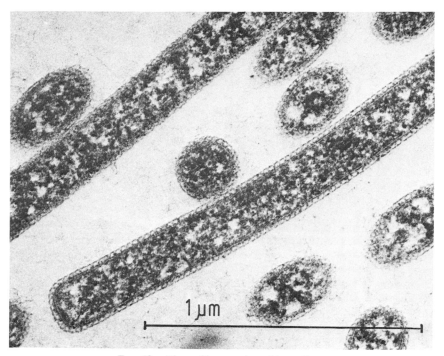

FIG. 18. *Thermofilum pendens*, thin section.

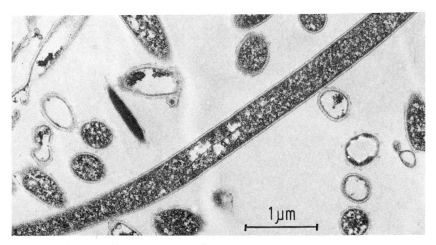

FIG. 19. *Thermoproteus* sp. H3, thin section, exhibiting sagittal line.

sucrose. All Thermoproteales are able to produce glycogen, particularly when sucrose is available (König *et al.*, 1982).

MULTIPLICATION

The complete absence of septa in the rods, even at sharp bends and in branching and budding regions, as well as in dividing cocci of the genera *Thermococcus* and *Desulfurococcus* indicates that multiplication might not proceed by equal division, as is usual in eubacteria.

For *Thermoproteus,* one mode of multiplication seems evident: The terminally laterally protruding spheroids frequent in growing cultures (''golf clubs'') are released and grow out into daughter rods, either with envelopes as appears to be the case with the isolate H3, or possibly after the formation of juvenile rods within the uncoated cytoplasm of the spheroid, as seems to be the case in *Thermoproteus tenax* (Fig. 21). As an alternative way of multiplication of *Thermoproteus,* ''legitimate fragmentation,'' for example at sharp bends or branch-

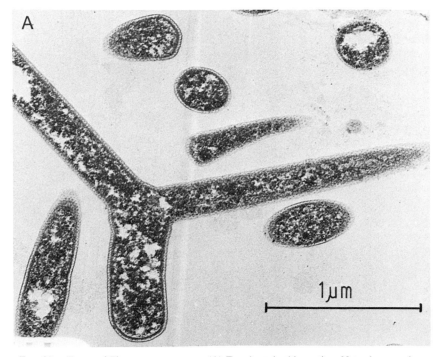

FIG. 20. Forms of *Thermoproteus tenax*. (A) True branch, thin section. Note absence of septa. (B) True branch, rotation shadowed. (C) Budding cell, rotation shadowed.

ing and budding regions, has been discussed. This assumption is based on the observation that the average length of the rods decreases with increasing agitation. The frequency of "golf club forms" in growing cultures of *Thermofilum* suggests that it also multiplies by the first of the two modes discussed for *Thermoproteus*. An alternative mode of division is evident. New cell tips can form within the swellings which are often observed along the rods (see Fig. 24). Similar to eukaryotic cells *Thermococcus* divides by constriction (Fig. 13). The

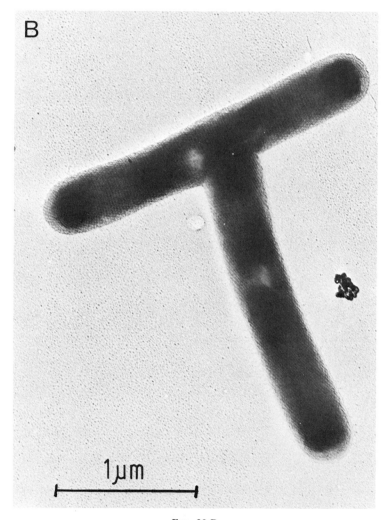

FIG. 20-B

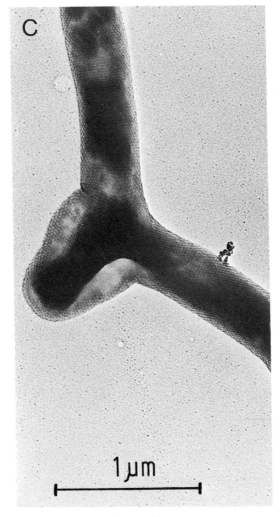

Fig. 20-C

same mechanism appears possible for *Desulfurococcus,* where doublets of cells linked by a thin thread of unknown nature have often been observed in growing cultures. The electron microscopic evidence, however, is by far not as clear as in the case of *Thermococcus.* On the other hand, budding has been observed (Fig. 30).

E. Chemistry

The slime of *Desulfurococcus mucosus* consists mainly of neutral sugars but contains a small fraction of amino sugars. The nature of the outer layer attached to the envelope of the *Thermoproteus* isolate H3 (Fig. 31) is unknown. The envelopes of the Thermoproteales resemble that of *Sulfolobus*, except for the rod shape in the cases of *Thermoproteus* and *Thermofilum*, and a lower rigidity of the two-dimensional crystal structure in *Desulfurococcus* and *Thermococcus* thus allowing the cells to be spherical instead of showing planar regions in their surfaces. Like the envelope of *Sulfolobus*, that of *Thermoproteus* is insoluble in 2% SDS, even at 100°C, and in 6 M urea + 20% (w/v) formamide. All attempts

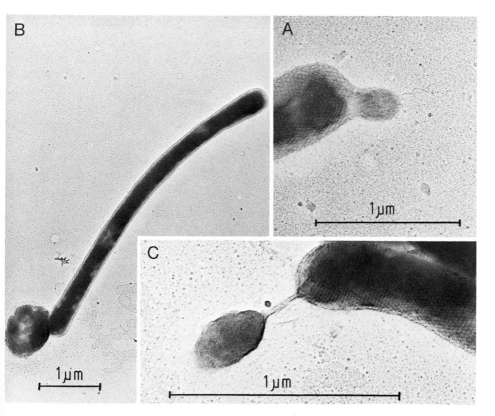

Fig. 21. "Golf club" forms of *Thermoproteus*. (A) Young protrusion, isolate H3. (B) Protrusion with stem, *T. tenax*. (C) Mature "golf club," *T. tenax*. (D) One "golf club," one normal cell, and one released protrusion of isolate H3.

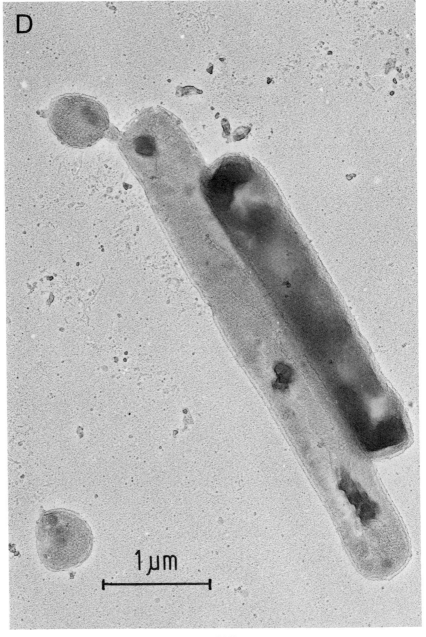

FIG. 21-D

to dissociate it into subunits have so far failed, implying the possibility of chemical cross-links between the subunits. Isolated envelopes contain all 20 amino acids, and, in addition, sugars. Thus, the subunit is probably a glycoprotein.

The membrane contains the typical archaebacterial lipids. Tetra- and diethers occur in a ratio similar to that found in *Sulfolobus*. At least seven different glycoproteins have been demonstrated in the membrane of *Thermoproteus* by periodate Schiff staining of SDS-polyacrylamide gels.

The polar lipid fraction from *Thermoproteus* acting as growth factor for *Thermofilum* does not contain phospholipids. The fraction is not involved in the change of the structure of sulfur observed in growing *Thermofilum* cultures.

Hydrophobic compounds of the compositions $(CH_2)_2S_3$ and $(CH_2)_2S_5$ (W. Schäfer, unpublished) have been isolated from the residual sulfur of chemolithoautotrophic *Thermoproteus* cultures. Soon after the start of such cultures, the previously light yellow sulfur turns grayish green and becomes evenly distributed in an almost colloidal manner. Concomitantly, the previously smooth surface of the sulfur grains turns rough.

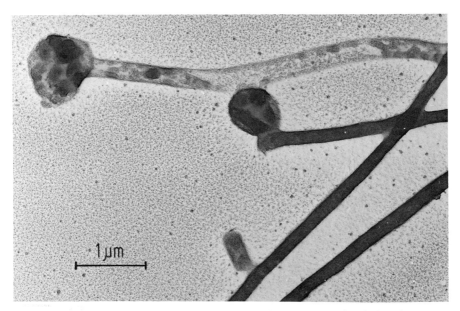

FIG. 22.   Golf club forms of *Thermofilum*, various stages, rotation shadowed.

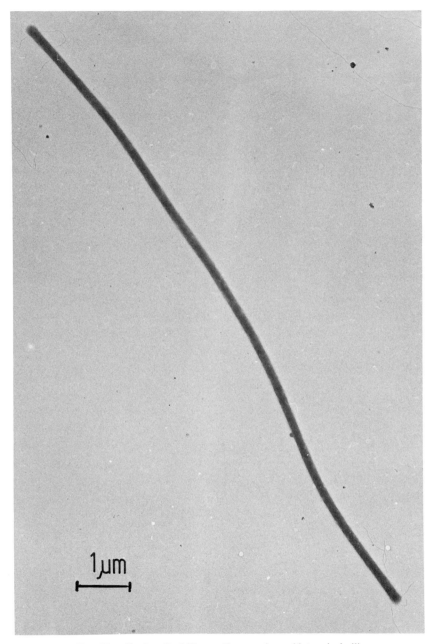

FIG. 23. Single cell of *Thermofilum pendens* with terminal pili.

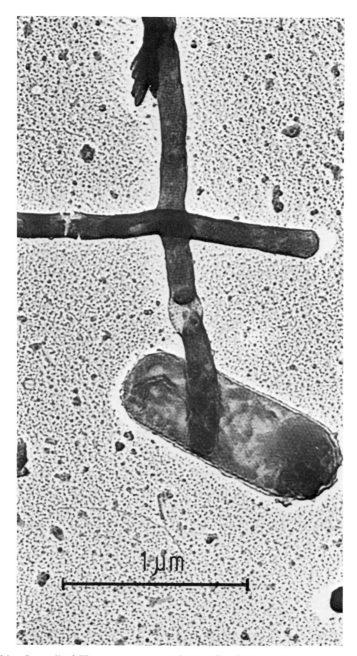

FIG. 24. One cell of *Thermoproteus* sp. and two cells of *Thermofilum* sp. from enrichment culture, one of the latter exhibiting swelling with new tips formed.

F. Molecular Biology

The G + C contents of the deoxyribonucleic acids of the Thermoproteales range from 50 to 57% and thus do not explain the temperature stability required for their genetic material.

Little is known about the organization of the genome of the Thermoproteales. The structural genes for rRNA in the *Thermoproteus* chromosome are arranged in the order 16 S, 23 S, 5 S, as in eubacteria, but with an unusually large linker of 11 kbp between the 23 S and the 5 S rRNA gene (Neumann *et al.*, 1983). Whether this cluster is a single transcription unit is not known.

Ribosomal subunits of the Thermoproteales have been purified by conventional methods. It appears, that these are the stable dissociation products of the 70 S particles which at 25 mM Mg^{2+} do not exist as such. The shape of the 30 S subunits of *Thermoproteus* shows the archaebacterial "bill" (Lake *et al.*, 1982) and thus resembles that of other archaebacteria. The shape of the 50 S subunit of *Thermoproteus*, like that of *Sulfolobus*, is somewhat reminiscent of the eukaryotic cytoplasmic 60 S subunit (Henderson *et al.*, 1984).

The DNA-dependent RNA polymerases of the Thermoproteales are all of the BAC-type (Schnabel *et al.*, 1983, this volume, Chapter 11). The homology of corresponding components of the different enzymes has been established by

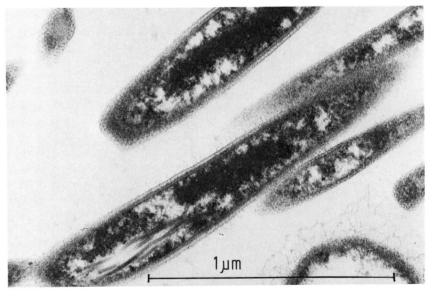

Fig. 25. *Thermoproteus* sp. Hvv 5. Thin sections showing intracellular filament structures.

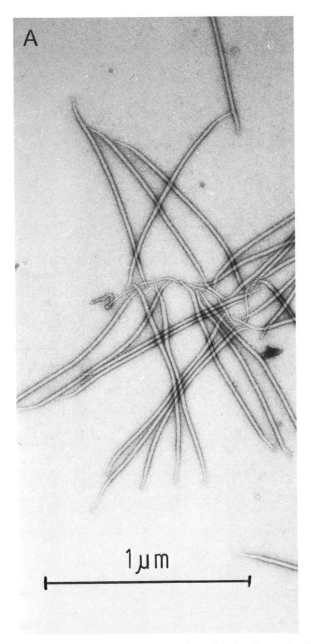

FIG. 26.   Products of lysis of *Thermoproteus tenax* formed during change from heterotrophic to autotrophic culture upon consumption of sulfur. (A) Thin filaments, negatively stained. (B) Few thin filaments, many rods and residues of cell envelopes.

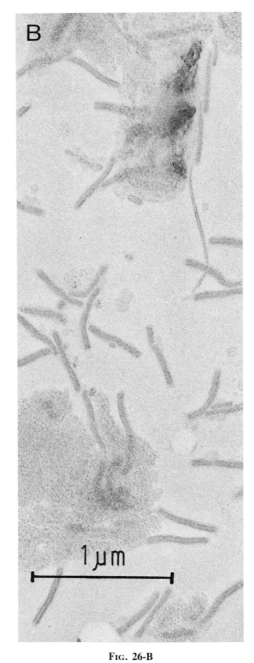

Fig. 26-B

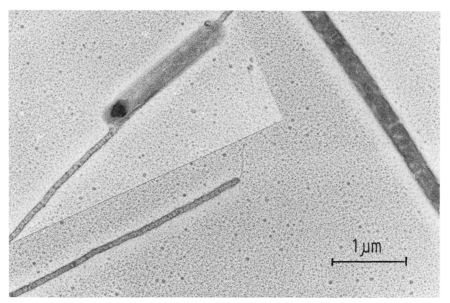

FIG. 27. Ultrathin and normal cells of *Thermofilum pendens*, rotation shadowed.

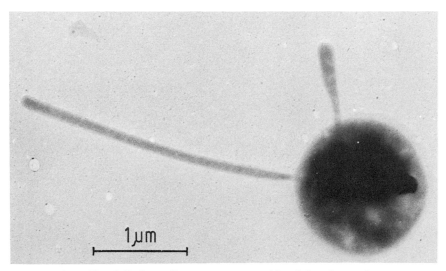

FIG. 28. Cell of *Desulfurococcus mucosus* with rod-shaped protrusions.

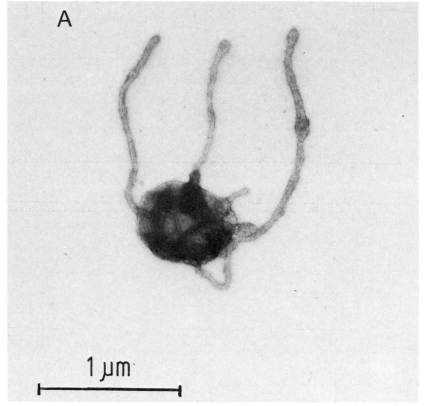

FIG. 29.   (A) and (B) Peculiar forms of *Thermococcus celer*.

challenging "Western blots" of SDS-polyacrylamide gel patterns with anti-bodies against single components of *Sulfolobus* RNA polymerase.

G. THE THERMOPROTEALES ARE ARCHAEBACTERIA

Like the envelope of *Sulfolobus,* the walls of members of all four genera of Thermoproteales known so far are devoid of murein and consist of protein or glycoproteid subunits. The membrane lipids contain isopranyl glycerol di- and tetraethers in a ratio similar to that in *Sulfolobus.*

The growth of these organisms is not inhibited by 100 μg/ml each of the antibiotics vancomycin, chloramphenicol, streptomycin, and rifampicin. The DNA-dependent RNA polymerases, all of the archaebacterial BAC-type, are completely resistant to 100 μg/ml of the antibiotics rifampicin and strep-

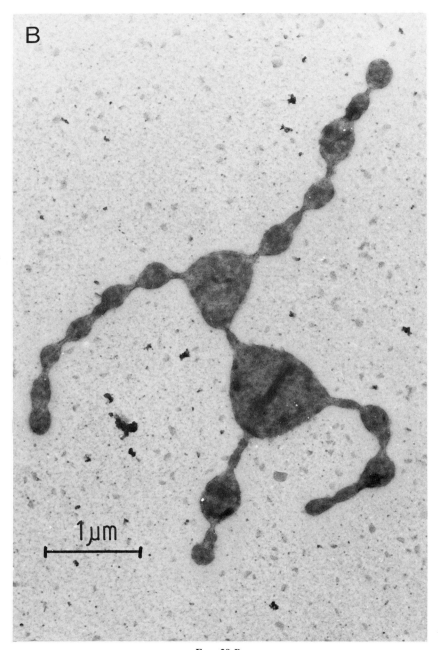

FIG. 29-B

TABLE IV

Requirements and Modes of Growth of Thermoproteales

Species	Temperature (°C)		Optimal pH	pH range allowing growth	Minimal generation time (hr)	Heterotrophic growth (by sulfur respiration)	Chemolithoautotrophic growth (by H_2S formation)	Growth in absence of S^0
	Optimal growth	Least growth						
Thermoproteus tenax	88	98	5.5	2.5–5.9	1.7	+	+	−
Thermoproteus neutrophilus	85	95	6.5	5.0–7.5	9	−	+	−
Thermofilum pendens	88	95	5.5	4.0–6.5	10	+	−	−
Desulfurococcus mucosus	85	93	6.0	4.0–7.5	3	+	−	+
Thermococcus celer	92	93	5.8	4.0–7.0	0.8	+	−	?

TABLE V

Carbon Sources and Electron Acceptors of Thermoproteales

Factor	Thermoproteus	Thermofilum	Desulfurococcus	Thermococcus
Substrates	Glucose, sucrose, glycogen, amylopectin, amylose, ethanol, methanol, malate, fumarate, formate, (propionate) (casamino acids), CO, CO_2 (with $H_2 + S^0$)	Yeast extract peptides: Bacto-tryptone, gelatin	Yeast extract peptides: Bacto-tryptone protein: casein starch, glycogen (D. saccharovorans)	Yeast extract peptides: Bacto-tryptone protein: casein
"Growth factor" required	—	Polar lipid from Thermoproteus	Yeast extract (for D. saccharovorans)	—
Stimulatory factors	Yeast extract	Sucrose, starch	Yeast extract	Sucrose
Electron acceptor	S^0, cystine, malate	S^0	S^0	S^0
Alternative fermentation	—	—	+	

147

tolydigin, and also to α-amamitin. Moreover, they are stimulated by silybin (Schnabel *et al.*, 1982). The EFII of *Thermoproteus* is ADP ribosylated by diphtheria toxin (Kessel and Klink, 1982). The sum of these features establishes the archaebacterial nature of the Thermoproteales beyond reasonable doubt.

H. Phylogeny

The DNA-dependent RNA polymerases of *Thermoproteus, Thermofilum, Desulfurococcus,* and *Thermococcus* are all of the BAC-type. In this respect, the Thermoproteales, *Sulfolobus* and *Thermoplasma* appear specifically related to one another. Though homologous, the enzymes of each of these genera do not form precipitation lines with antibodies against the polymerases of other Thermoproteales, or of *Sulfolobus* and *Thermoplasma* in the Ouchterlony in-gel diffu-

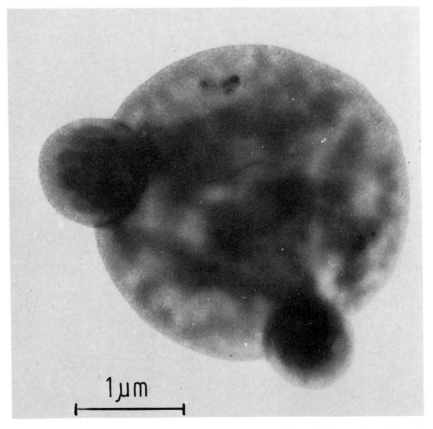

Fig. 30. Budding of *Desulfurococcus mucosus,* giant cell with two buds, rotation shadowed.

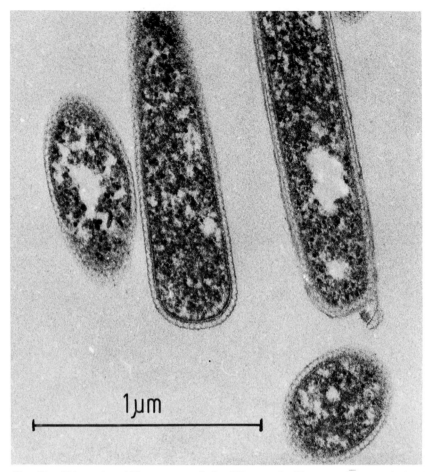

FIG. 31. Thin section of *Thermoproteus* isolate H3, one cell with an outer layer attached to the envelope.

sion test. Comparable results with eubacterial RNA polymerases suggest that all these phyla have at least interfamily distance from each other.

The S_{AB} between *Sulfolobus* and *Thermoproteus* is 0.33 (Woese *et al.*, 1984), indicating a less phylogenetic distance than between *Sulfolobus* on the one hand and *Thermoplasma* and the rest of the archaebacteria on the other. This is verified by 16 S rRNA–DNA cross-hybridization, which reveals that the genera of Thermoproteales known so far might be considered representatives of different families of a novel order (Tu *et al.*, 1982). Measured by this method, the phylogenetic distance of the Thermoproteales and *Sulfolobus* is again considerably less than between the newly established branch and the methanogens and extreme halophiles. By this measure, *Thermoplasma* appears closer to the latter than to the sulfur-dependent archaebacteria (as already discussed).

Cross-hybridization of 16 S rRNA–DNA shows that *Thermococcus* is phylogenetically more distant from the rest of the Thermoproteales, than these are from each other (Zillig *et al.,* 1983b). Its distance from *Sulfolobus* appears as large as that of *Halobacterium* (unpublished). So far it appears to be the only archaebacterium showing a closer relation with Thermoplasma, thus possibly forming a link between the latter and the Thermoproteales among which it belongs by its metabolism and the structure of its envelope and membrane. In spite of the rather small phylogenetic distance, the marked differences between the modes of existence (aerobic in the case of *Sulfolobus,* anaerobic in the case of the Thermoproteales) and in sulfur metabolism (oxidative in *Sulfolobus* and reductive in the Thermoproteales) are strong reasons for considering the Sulfolobales and the Thermoproteales as different orders within the same branch. The aerobic *Sulfolobus* strains could be descended from the Thermoproteales, branching off after oxygen became available on earth. This relation resembles that of the extreme halophiles and the methanogens. The similarity of many features of both phyla, e.g., envelope, membrane, ribosome, RNA polymerase, and glycogen synthesis, is in line with this view.

The Thermoproteales utilize a mode of energy production, which appears possible under early earth conditions. Other ostensibly primitive features of the Thermoproteales, e.g., the apparent absence of a mechanism of equal division (in *Thermoproteus* and *Thermofilum*) and the low specifity and efficiency of glucosyltransferases (in *Sulfolobus* and *Thermoproteus*) strengthen this argument.

VI. Organisms Occurring in Submarine Hot Volcanic Areas

To check for the possible existence of life at temperatures above 100°C, samples were taken from geothermally heated sea floors around Vulcano, Baia (Naples), and Ischia, in Italy. As expected, at elevated hydrostatic pressure, areas with water temperatures exceeding 100°C were found. Samples taken around Vulcano yielded extremely thermophilic organisms of unusual shape, some of which grow optimally at temperatures above 100°C.

A. THE HABITAT

These isolates originate from a submarine solfataric field off the beach of Vulcano characterized by strong emanation of gases including H_2S and of hot water. In depths from 2 to 10 m, the sea floor consisted of sandy sediments with many small craters and of rock formations with gas spewing forth from sulfur encrusted cracks and holes. From some of the cracks, hot water flowed. The temperatures in the sandy sediments and in the cracks were up to 103°C. The pH of all samples was about 6, and the conductivity was that of sea water (Stetter, 1982).

B. SAMPLING AND ISOLATION OF THE ORGANISMS

Using scuba diving equipment, samples of the hot sediments were taken with syringes fitted with enlarged inlets. Loosened sulfur crusts were also sampled by means of syringes. The hot samples were taken to the surface and transferred into storage bottles immediately sealed by plugs. Oxygen was removed by adding sodium dithionite until resazurin became colorless. The samples were transported to the laboratory without temperature control and stored at 4°C.

For enrichment, sea water sterilized by ultrafiltration was inoculated with 5% of its volume of the samples and incubated at 85°C with strict exclusion of oxygen in the presence of elemental sulfur (1% w/v), yeast extract (0.02% w/v), and an H_2/CO_2 atmosphere (80 : 20 v/v; 200 kpa). After 1–3 days, disk-shaped organisms had grown in 20 of 26 samples, with the simultaneous formation of H_2S. The organisms were transferred into synthetic sea water supplemented with trace minerals (Balch et al., 1979), 0.05% (w/v) each of KH_2PO_4 and NH_4Cl, 0.02% (w/v) of yeast extract, and 1% (w/v) flowers of sulfur, pH 5.5. Oxygen was removed by the addition of sodium dithionite until resazurin became colorless. Isolation was performed by serial dilutions in the same medium. Two main types of organisms distinct in morphology and physological properties have been isolated and classified as representatives of two novel genera, *Thermodiscus* and *Pyrodictium* (Stetter et al., 1983).

C. MORPHOLOGY

1. LIGHT MICROSCOPY

Under the phase contrast microscope, *Thermodiscus* appears as highly irregular disks, sometimes dish-shaped and sometimes resembling flat crystals. The diameter of the plates varies from about 0.3 to 3 μm. The disks are only about 0.2 μm thick. In older cultures, the cells lyse, leaving behind empty envelopes containing dark grana, possibly composed of polyphosphate. In the isolate *Thermodiscus maritimus,* cells are usually single. In other isolates, e.g., PL-5, the cells occur in aggregates, mainly short chains of up to five individuals. *Pyrodictium,* which grows as a cobweb-like layer on the sulfur is visible in the phase contrast microscope as disks, highly variable in diameter between 0.3 and 2.5 μm. The disks are arranged at varying distances from each other, which, however, remain fixed. Under dark field illumination, the disks appear to be fixed within a huge network of ultrathin fibers that are invisible under phase contrast. However, in the late exponential phase, under phase contrast the fibers can be traced by sulfur granules, which are sticking to them and appear to be linearly assorted.

2. ELECTRON MICROSCOPY

Thermodiscus appears as a flat organism, surrounded by a subunit envelope (Fig. 32). Sometimes pili-like structures of 0.01 μm in diameter and up to 15 μm

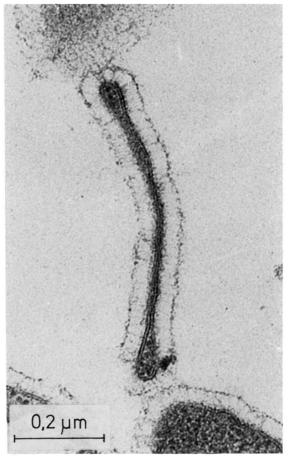

0,2 μm

FIG. 32. *Thermodiscus maritimus,* thin section.

FIG. 33. *Thermodiscus maritimus,* platinum-shadowed. Two cells connected by a pilus-like structure.

in length connect the surfaces of two individuals (Fig. 33). Often, extremely small (less than 0.2 μm diameter) disks are seen, surrounded by branched root-like appendices (Fig. 34), the latter possibly composed of protein subunits. This observation could explain the phenomenon that the titer as determined by serial dilution is always at least 10 times higher than that determined by direct counting

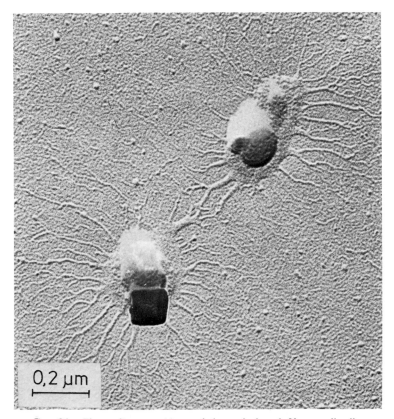

FIG. 34. *Thermodiscus maritimus,* platinum shadowed. Very small cells.

in the light microscope. Thin sections exhibit the extreme flatness of these cells. Their protein subunit envelope is about 33 nm thick. Within the envelope, tubes enclosed by unit membranes are inserted, possibly connecting cells or being involved in cell propagation (Fig. 35).

Pyrodictium consists of huge loose networks composed of thin filaments (diameter 0.04–0.08 μm), often in bundles and strongly reminiscent of fungal plectenchyms (Fig. 36). They can cover areas of more than 10,000 μm². Disks appear to be fixed laterally and terminally on these networks. Sometimes the disks seem to be embraced by several fibers (Fig. 37). The disks are surrounded by an envelope composed of protein subunits around 30 nm in diameter. They often show strongly indented areas, possibly in order to increase their surface-to-volume ratio (Fig. 38). If the culture is grown without stirring, almost all disks appear attached to fibers. Gentle stirring (200 rpm) leads to the occurrence of free disks with short fibers. The disks as well as the filaments contain pili-like

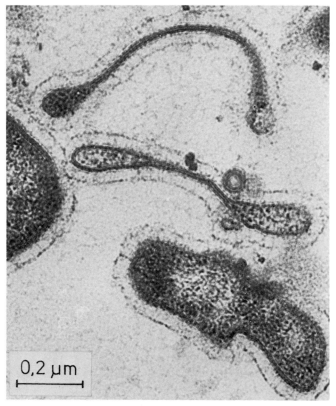

0,2 μm

FIG. 35. *Thermodiscus maritimus,* thin sections.

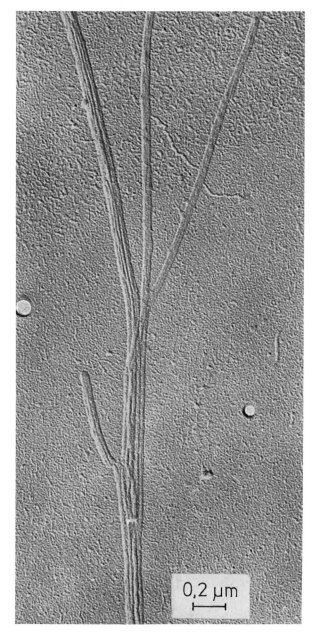

Fig. 36. *Pyrodictium occultum*, platinum-shadowed. Plectenchyme-like bundles of thin filaments.

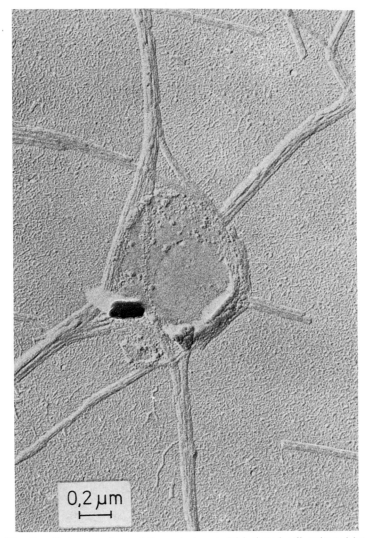

Fig. 37. *Pyrodictium occultum*, platinum-shadowed. Disk-shaped cell embraced by several filaments.

appendices, sometimes branched and often cross-linking two individuals (Fig. 39).

D. Physiology and Growth Requirements

Thermodiscus grows heterotrophically on yeast extract (table V) and sulfur by sulfur respiration. In some isolates, growth is stimulated by hydrogen. All isolates show good growth on yeast extract without sulfur and hydrogen. Sometimes, even when growing optimally, they do not form H_2S even in the presence of sulfur and hydrogen.

Pyrodictium grows by a sulfur–hydrogen autotrophy similar to *Thermoproteus*. Although growth may be stimulated by the addition of yeast extract, e.g., as in strain S1, it is strictly dependent on hydrogen and sulfur. The salt requirement of both genera is quite different: *Thermodiscus* is a moderate halophile growing

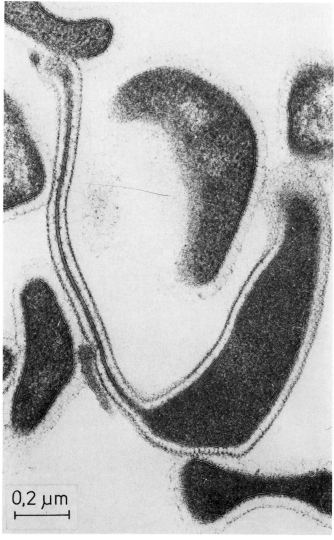

Fig. 38. *Pyrodictium occultum*, thin sections.

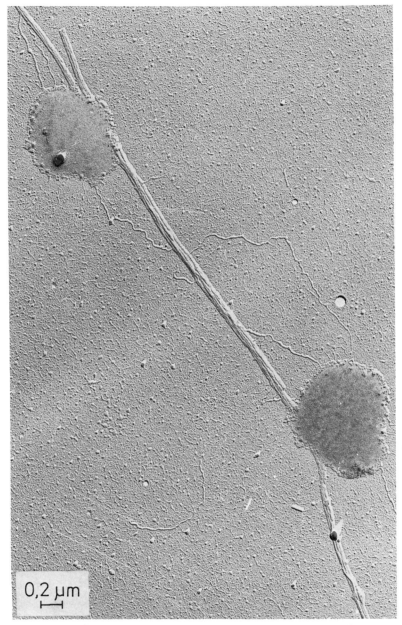

0,2 µm

FIG. 39. *Pyrodictium occultum*, platinum-shadowed pili-like appendices.

between 1 and 4% NaCl, whereas *Pyrodictium* is extremely halotolerant, growing between 0.1 and 12% NaCl. Both genera grow between pH 5 and pH 7 with an optimum around pH 5.5.

The temperature optimum for *Thermodiscus* is around 85°C, the maximum about 95°C. *Pyrodictium* grows optimally at 105°C (under low pressure) (Stetter, 1982), showing a doubling time of only 110 min as determined by ATP measurements. No growth occurs at 80°C or below. At 85°C, the doubling time is 550 min; at 100°C, it is 200 min. Even at 110°C, a doubling time of around only 2 hr was observed if the cultures remained unshaken. No growth could be obtained at 115°C or above. The organisms could be reisolated from the original samples, which had been stored in the cold room or at −20°C, for at least 12 months. *Pyrodictium* is the most thermophilic organism identified to date. Owing to its stability at low temperatures, it may be widely distributed in submarine volcanic areas.

E. MOLECULAR BIOLOGY

The DNA of *Thermodiscus* has a G + C content of 53 mol%. Isolate F1 is different with a G + C content of 33 mol%. *Pyrodictium* species have a G + C content of 62 mol% (Stetter *et al.*, 1983). Both genera contain phytanyl ether lipids in their membranes (T. Langworthy, personal communication). Furthermore, *Thermodiscus* and *Pyrodictium* contain proteins that can be ADP ribosylated by diphtheria toxin (F. Klink, personal communication). The cell envelopes contain neither muramic acid nor mesodiaminopimelic acid. Preparations of envelopes of both genera contain one dominating glycoprotein with apparent molecular weights of about 84,000 in *Thermodiscus* and 172,000 in *Pyrodictium* (H. König, personal communication).

Cataloging of the 16 S rRNAs of *Thermodiscus* and *Pyrodictium* species revealed a low but significant relationship to the 16 S rRNAs of *Sulfolobus* and *Thermoproteus* and a much closer relationship to each other (E. Stackebrandt, personal communication). As is the case for other sulfur-dependent archaebacteria (Woese, *et al.*, 1984), the 16 S rRNA of *Pyrodictium* and *Thermodiscus* contains a substantial member of posttranscriptionally modified nucelotides. These latter catalogs are, however, incomplete, so no S_{AB} values can be given for them.

F. TAXONOMY

The nature of the lipids, the ADP ribosylation of proteins by diphtheria toxin, and the lack of muramic acid show that both genera belong to the archaebacteria.

This is further substantiated by the relatively close relationship between the 16 S rRNAs of *Thermodiscus* and *Sulfolobus,* by cross-hybridization of the DNA of *Thermodiscus* and *Pyrodictium* with 16 S rRNAs from members of the sulfur-dependent archaebacteria, and by the sulfur metabolism of the novel organisms, e.g., the same type of chemolithoautotrophy for *Thermoproteus* and *Pyrodictium* species. Therefore, both genera are proposed to be members of the same branch of the archaebacteria.

The unusual properties of *Pyrodictium,* particularly its growth at extremely high temperatures, suggest that its phenotype may be an extremely primitive one.

VII. Requirements and Upper Boundary of Thermophilic Life

The molecular basis of thermophily was previously studied with eubacteria, e.g., *Bacillus stearothermophilus* and *Thermus thermophilus* (Zuber, 1981; Schär and Zuber, 1979; Jaenicke, 1981; Fujita *et al.,* 1976), growing optimally around 70°C (Table VI). The main structures in cells that are subject to destruction at high temperatures are those of proteins, nucleic acids, membranes, and energy-rich small molecules.

Comparative studies of proteins from thermophilic and closely related mesophilic species e.g., L-lactate dehydrogenases (Schär and Zuber, 1979) and glyceraldehyde-3-phosphate dehydrogenases (Biesecher *et al.,* 1977) of bacilli yielded only small differences in enzyme structure, which could not be definitely related to thermostability, e.g., the presence or absence of clusters of basic amino acids like lysine and arginine allowing the formation of stabilizing bonds. At present, no common concept for the basis of thermostability of protein structure has been worked out. In the course of secondary adaptation to heat (Zuber, 1979), thermostability might indeed be achieved in different ways. It appears that in extremely thermophilic archaebacteria, proteins are generally stable and function optimally at extremely high temperatures, as known for DNA-dependent RNA polymerases (Zillig *et al.,* 1982b) and glucosyltransferases of sulfur-dependent archaebacteria (König *et al.,* 1982) and for a restriction enzyme from *Sulfolobus* (P. McWilliams, unpublished). The glucosyltransferase of *Sulfolobus* involved in glycogen synthesis has a low affinity and specifity, and a low turnover number for substrates (König *et al.,* 1982), possibly because the enzyme structure is largely required for thermostability and could thus not be optimally adapted to function. It thus appears of interest to compare structures and functions of enzymes of extremely thermophilic archaebacteria with those of their mesophilic counterparts.

The secondary structures of DNA and the different RNA species, especially their paired regions, are sensitive to heat. In DNA, the heat stability increases

TABLE VI

Properties of the Submarine Thermophilic Isolates

Designation of isolate	Morphology	Proposed name	Growth temperature (°C)			Salt (NaCl) requirement (%, w/v)	Nutrition	G + C content of DNA (mol %)
			minimum	optimal	maximum			
S2	Disks, 0.3–3 μm diameter, 0.2 μm thick, occurring singly, with pili-like appendices (up to 10 μm long; 0.01 μm φ)	*Thermodiscus maritimus*	70	87	95	2–4	Mixotrophic	53
PL-5	Disks, occurring in aggregates, very often *Streptococcus*-like. Pili-like structures very rare	*Thermodiscus* species	70	85	95	2–4	Heterotrophic	Not done
PL-19	Network of thin hyphae (up to 40 μm long; 0.04–0.06 μm φ). In the late growth phase, disks protrude laterally and terminally	*Pyrodicium occultum*	82	105	110	0.1–12	Autotrophic	62

with the G + C content. Although the latter varies between about 20 and 80% in different organisms, no relation between thermophily and the amount of G + C is obvious. Surprisingly, the DNAs of some extreme thermophiles have low G + C contents, e.g., *Sulfolobus* 38 mol% (Zillig *et al.*, 1980) and *Methanothermus* only 33 mol% (Stetter *et al.*, 1981). Isolated DNA of *Methanothermus* is denatured at 83°C in 1 × SSC buffer, but the organism grows at temperatures as high as 97°C. The mechanism of protection is unclear. In *Methanothermus,* evidence suggests a relatively high internal ionic strength (0.6 *M* KCl, M. Schupfner, personal communication), which could stabilize the DNA structure at temperatures approaching 100°C. In addition, basic DNA binding proteins have been found in large amounts (Searcy, 1975; Thomm *et al.*, 1982). These could stabilize DNA structure.

Nothing is yet known about the stabilization of RNA structure in thermophilic archaebacteria. In the eubacterium *Thermus thermophilus,* tRNA structure seems stabilized by modification (Watanabe *et al.*, 1978) and by the fact that the stems contain more G ± C pairs than do those of comparable tRNA from mesophilic bacteria. Thermoadaptation can be achieved by increased thiolation of thymidine at elevated temperatures (Oshima, 1978). Such modifications can increase the melting temperatures of tRNAs by 10°C. In *Thermus thermophilus,* polyamines appear to stabilize the mRNA–ribosome translational complex (Oshima, 1978).

The cell membranes show a high degree of order, which is destroyed at high temperature. In eubacteria, the lipid structure is normally based on fatty acid esters of glycerol (Langworthy *et al.*, 1982). Thermophilic eubacteria contain more unsaturated fatty acids in their membranes than do mesophilic bacteria (Esser, 1979). The lipids of all archaebacteria contain ether linkages between glycerol and isopranyl alcohols instead (Langworthy *et al.*, 1982, this volume, Chapter 10). The lipids of extremely thermophilic archaebacteria are composed mainly of tetraethers with C_{40}-isopranediols (containing up to four 5-membered rings) covalently linking the two glycerols. This forms a very stable symmetric "monomeric bilayer," instead of the conventional, less thermostable bilayer of normal membranes. The ether linkages are stable even in acidic or alkaline environments. In *Sulfolobus* at least, a large fraction of the glycerol residues is replaced by calditol, an aliphatic nonylpolyalcohol, apparently present only in the outer membrane layer (De Rosa *et al.*, 1980).

UPPER TEMPERATURE LIMITS OF LIFE

In view of the fact that extreme thermophiles grow even in boiling and super-heated water two questions arise: (1) What is the upper temperature limit for the existence of life? (2) Where are the possible habitats for such extreme thermophiles? An absolute prerequisite for life, liquid water, is available at high temperatures under increased pressure, e.g., at the sea floor. For submarine hydrothermal systems on the deep sea floor, water temperatures of more than 300°C have been reported (Spiess *et al.*, 1980). Furthermore, temperatures in the same range have been measured in aquifers of geothermal power plants (Rau, 1978). At increased pressure (critical pressure 225.5 bar), water can be liquid at temperatures up to the critical point at 374.2°C. But above 300°C, the specific volume already increases significantly in the fluid state.

A further upper bound for life may be defined by the chemical stability of biomolecules (Bernhardt *et al.*, 1984), e.g., energy-rich small organic molecules like ATP, the latter having its chemical decomposition point at 208°C. Could organisms existing at such temperatures have alternative energy-rich compounds, e.g., pyrophosphate or polyphosphate? If life exists above 208°C, its principle energy-yielding mechanisms and its biomolecules should be very different from those known to us.

VIII. Key to *Thermoplasma* and the Genera of the Thermophilic Sulfur-Dependent Archaebacteria

  I. Obligately heterotrophic, moderately thermophilic, aerobic, envelope absent:
  Thermoplasmales
  Regular or elongated cocci, growing between 50° and 64°C and pH 0.8 to 3.0 on yeast extract:
  Thermoplasma
  II. Chemolithoautotrophic or heterotrophic. Obligately, facultatively or normally sulfur-dependent synthesizing glycogen, extremely thermophilic. Envelope hexagonal array of protein or glycoprotein subunits:
  Sulfur-dependent branch of archaebacteria
II.A. Aerobic, oxidizing S^0 to sulfate:
  Irregular cocci, heterotrophic and/or chemolithoautotrophic growth between 50° and 90°C (at least) and pH from 1 to pH 5.5:
  Sulfolobales, only one genus known: *Sulfolobus*
II.B. Anaerobic, reducing S^0 to H_2S:
  Thermoproteales
   1. Rigid rods of variable length up to 100 μm, frequently with terminal lateral spheric protrusions. Sometimes budding and/or branching. Optimal growth temperature around 88°C. Growth occurs between 70° and 96°C (at least) and between pH 4.0 and (some species) 7.
   1a. About 0.5 μm diameter, facultatively or obligately chemolithoauto-

trophic. Heterotrophic growth on many carbon sources, including carbon monoxide by sulfur respiration. Autotrophic growth with $H_2 + S^0$ as energy source and CO_2 as sole carbon source:

Thermoproteaceae, only one genus known: *Thermoproteus*

1b. About 0.2-μm diameter heterotrophic growth on yeast extract or bactotryptone by sulfur respiration. H_2S required. A polar lipid fraction from *Thermoproteus* acts as growth factor:

Thermofilaceae, only one genus known: *Thermofilum*

2. Regular cocci, about 1 μm diameter

2a. During growth, two cells often connected by thin thread, frequently budding. At high nutrient concentration, there are giant forms with up to 10 μm diameter. Cells occasionally with rod-shaped protrusions. In sections, dense parts of cytoplasm attached to membrane. Heterotrophic growth on protein (casein), bactotryptone, or yeast extract; one species on glucose, by sulfur respiration. Alternatively, inefficient growth by unknown type of fermentation. Optimal growth temperature 87°C, growth occurs between pH 4.0 and 7. Some species with slime coats or threads attached to envelope, others with monopolar polytrichous flagellation:

Desulfurococceae, only one genus known: *Desulfurococcus*

2b. Cells divide by progressive constriction. In growing cultures, they are normally in constricted diplo stages. Cytoplasm appears evenly distributed and dense in thin sections. Heterotrophic growth on protein (casein), bactotryptone, or yeast extract by sulfur respiration. Growth between 70° and 95°C (at least) and pH 4.5 to 6.5. Marine, moderately halophilic: 3 to 4% NaCl required.

Thermococceae, only one genus known: *Thermococcus*

3. Irregular disks, 0.3 to 3 μm diameter, about 0.2-μm thick. Appendices sometimes connecting two cells. Growth on yeast extract, most likely by fermentation. In the presence of S^0, H_2S and CO_2 may be formed by sulfur respiration. A few isolates grow alternatively mixotrophically on H_2 and S^0. Moderately halophilic: between 1 and 4% NaCl required. Growth between 70° and 95°C (at least):

Thermodiscus

4. Network of thin fibers, about 0.04- to 0.08-μm thick, often in bundles, laterally or terminally associated with disks of 0.3- to 2.5-μm diameter and about 0.2-μm thick. Growth as a cobweb-like layer above the sulfur. Chemolithoautotrophic growth by H_2S formation from S^0 and H_2. Growth may be stimulated by acetate or yeast extract. Marine. Halotolerant: growth between 0.1 and 12% NaCl. Optimal growth temperature 105°C, grow between 83° and 110°C:

Pyrodictium

ACKNOWLEDGMENTS

The electron micrographs in Figs. 1, 2, 3, 6, 7, and 9–31 were supplied by Davorin Janekovic; that in Fig. 5, by Ingrid Scholz; and those in Figs. 32–39, by Helmut König and Elisabeth Kleemeier. We thank Peter Palm and Alfons Gierl for critical reading of the manuscript.

REFERENCES

Allen, M. B. (1959). Studies with *Cyanidium caldarium,* an anomalously pigmented chlorophyte. *Arch. Microbiol.* **32,** 270–277.

Balch, W. E., Fox, C. E., Magrum, L. J., Woese, C. R., and Wolfe, R. S. (1979). *Methanogens:* Reevaluation of a unique biological group. *Microbiol. Rev.* **43,** 260–296.

Belly, R. T., and Brock, T. D. (1972). Cellular stability of a thermophilic, acidophilic mycoplasma. *J. Gen. Microbiol.* **73,** 465–469.

Belly, R. T., Bohlool, B. B., and Brock, T. D. (1973). The genus *Thermoplasma. Ann. N.Y. Acad. Sci.* **225,** 94–107.

Bernhardt, G., Lindemann, H. D., Jaenicke, R., König, H., and Stetter, K. O. (1984). Biomolecules are unstable under "Black Sm ker" conditions. *Naturwissenschaften* **71,** 583–585.

Biesecher, G., Harris, J. I., Thiery, J. C., Walker, J. E., and Wonacott, A. J. (1977). Sequence and structure of D-glyceraldehyde-3-phosphate dehydrogenase from *Bacillus stearothermophilus. Nature (London)* **266,** 328.

Black, F. T., Freundt, E. A., Vinther, O., and Christiansen, C. (1979). Flagellation and swimming motility of *Thermoplasma acidophilum. J. Bacteriol.* **137,** 456–460.

Bohlool, D. B. (1975). Occurence of *Sulfolobus acidocaldarius,* an extremely thermophilic acidophilic bacterium, in New Zealand hot springs. Isolation and immunofluorescence characterization. *Arch. Microbiol.* **106,** 177–194.

Bohlool, D. B., and Brock, T. D. (1974). Immunofluorescence approach to the study of the ecology of *Thermoplasma acidophilum* in coal refuse material. *Appl. Microbiol.* **28,** 11–16.

Brierley, C. L. (1978). Bacterial leaching. *CRC Crit. Rev. Microbiol.* **6,** 207–262.

Brierley, C. L., and Brierley, J. A. (1973). A chemolithoautotropic and thermophilic microorganism isolated from an acid hot spring. *Can. J. Microbiol.* **19,** 183–188.

Brierley, C. L., and Brierley, J. A. (1982). Anaerobic reduction of molybdenum by *Sulfolobus* species. *Zentralbl. Bakteriol., Mikrobiol. Hyg. Abt., 1, Orig. O* **3,** 289–294.

Brierley, J. A. (1966). Contribution of chemolithoautotrophic bacteria to the acid thermal waters of the Geysir springs group in Yellowstone National Park. Ph.D. Thesis, pp. 58–60. Montana State University, Bozeman.

Brock, T. D. (1978a). "Thermophilic Microorganisms and Life at High Temperatures," pp. 92–116. Springer-Verlag, Berlin and New York.

Brock, T. D. (1978b). "Thermophilic Microorganisms and Life at High Temperatures," pp. 117–179; 303–336; 386–418. Springer-Verlag, Berlin and New York.

Brock, T. D. (1981). Extreme thermophiles of the genera *Thermus* and *Sulfolobus. In* "The Prokaryotes" (M. P. Starr, H. Stolp, H. G. Truper, A. Balows, and H. G. Schlegel, eds.), pp. 978–984. Springer-Verlag, Berlin and New York.

Brock, T. D., Brock, K. M., Belley, R. T., and Weiss, R. L. (1972). *Sulfolobus:* A new genus of sulfur oxidizing bacteria living at low pH and high temperature. *Arch. Mikrobiol.* **84,** 54–68.

Castenholz, R. W. (1979). *In* "Strategies of Microbial Life in Extreme Environments" (M. Shilo, ed.), Dahlem Konf. Life Sci. Res. Rep., Vol. 13, p. 380. Verlag Chemie, Weinheim.

Christiansen, C., Freundt, E. A., and Black, F. T. (1975). Genome Size and Deoxyribonucleic acid base composition of *Thermoplasma acidophilum. Int. J. Syst. Bacteriol.* **25,** 99–101.

Darland, G., Brock, T. D., Samsonoff, W., and Conti, S. F. (1970). A thermophilic acidophilic mycoplasm isolated from a coal refuse pile. *Science* **170,** 1416–1418.

De Rosa, M., Gambocorta, A., Millonig, G., and Bullock, J. D. (1974). Convergent characters of extremely thermophilic acidophilic bacteria. *Experientia* **30,** 866.

De Rosa, M., De Rosa, S., Gambacorta, A., Minale, L., and Bullock, J. D. (1977). Chemical structure of the ether lipids of thermophilic acidophilic archaebacteria of the *Caldariella* group. *Phytochemistry* **16,** 1961–1965.

De Rosa, M., De Rosa, S., Gambacorta, A., and Bullock, J. D. (1980). Structure of calditol, a new branched-chain nonitol and of the derived tetraether lipids in thermoacidophilic archaebacteria of the *Caldariella* group. *Phytochemistry* **19,** 249–254.

Esser, A. F. (1979). Physical chemistry of thermostable membranes. *In* "Strategies of Microbial Life in Extreme Environments" (M. Shilo, ed.), Dahlem Konf. Life Sci. Res. Rep., Vol. 13, pp. 433–454. Verlag Chemie, Weinheim.

Fox, G. E., Stackebrandt, E., Hespell, R. B., Gibson, J., Maniloff, H., Dyer, T. A., Wolfe, R. S., Balch, W. E., Tanner, R. S., Magrum, L. J., Zablen, L. B., Blakemore, R., Gupta, R., Bonen, L., Lewis, B. J., Stahl, D. A., Luehrsen, K. R., Chen, K. N., and Woese, C. R. (1980). The phylogeny of prokaryotes. *Science* **209,** 457–463.

Fox, G. E., Luehrsen, K. R., and Woese, C. R. (1982). Archaebacterial 5S ribosomal RNA. *Zentralbl. Bakterial., Mikrobiol. Hyg.,* Abt. 1, *Orig. C* **3,** 330–345.

Fujita, S. C., Oshima, T., and Imahori, K. (1976). Purification and properties of D-glyceraldehyde-3-phosphate dehydrogenase from an extreme thermophile, *Thermus thermophilus* strain HB8. *Eur. J. Biochem.* **64,** 57–68.

Furuya, T., Nagumo, T., Itoh, T., and Kaneko, H. (1977). A thermophilic acidophilic bacterium from hot springs. *Agric. Biol. Chem.* **41,** 607–1612.

Green, G. R., Searcy, D. G., and DeLange, R. J. (1983). Histone-like protein in the archaebacterium *Sulfolobus acidocaldarius. Biochim. Biophys. Acta* **741,** 251–257.

Gupta, R. (1984). *Halobacterium volcanii* t-RNAs: Identification of 41t-RNAs covering all aminoacids and the sequences of 33 class I t-RNAs. *J. Biol. Chem.* **259,** 9461–9471.

Henderson, E., Oakes, M., Clark M. W., Lake, J. A., Matheson, A. T., and Zillag, W. (1984). A new ribosome structure. *Science* **225,**510–512.

Hofman, H., Lan, R. H., and Doolittle, W. F. (1979). The number, physical organization and transcription of ribosomal RNA cistrons in an archaebacterium: *Halobacterium halobium. Nucleid Acids Res.* **7,** 1321–1333.

Holländer, R. (1978). The cytochromes of *Thermoplasma acidophilum. J. Gen. Microbiol.* **108,** 165–167.

Holländer, R., Wolf, G., and Mannheim, W. (1977). Lipoquinones of some bacteria and mycoplasmas, with consideration of their function and significance. *Antonie van Leeuwenhoek* **43,** 177–185.

Huet, J., Schnabel, R., Sentenac, A., and Zillig, W. (1983). Archaebacteria and eukaryotes possess DNA-dependent RNA polymerases of a common type. *EMBO J.* **2,** 1291–i294.

Jaenicke, R. (1981). Enzymes under extremes of physical conditions. *Annu. Rev. Biophys. Bioeng.* **10,** 1–67.

Janekovic, D., Wunderl, S., Holz, I., Zillig, W., Gierl, A., and Neumann, H. (1983). TTV1, TTV2 and TTV3, a family of viruses of the extremely thermophilic, anaerobic, sulfur reducing archaebacterium *Thermoproteus tenax. Mol. Gen. Genet.* **192,** 39–45.

Jarsch, M., Altenbuchner, J., and Böck, A. (1982). Physical organization of the genes for ribosomal RNA in *Methanococcus vannielii. Mol. Gen. Genet.* **189,** 41–47.

Kandler, O., and Stetter, K. O. (1981). Evidence for autotrophic CO2 assimilation in *Sulfolobus brierleyi* via a reductive carboxylic acid pathway. *Zentralbl. Bakteriol., Mikrobiol. Hyg.,* Abt. *1. Orig. C* **2,** 111–121.

Kessel, M., and Klink, F. (1980). Archaebacterial elongation factor is ADP-ribosylated by diphtheria toxin. *Nature (London)* **287**, 250.

Kessel, M., and Klink, F. (1982). Identification and comparison of eighteen Archaebacteria by means of the diphtheria toxin reaction. *Zentralbl. Bakteriol. Mikrobiol. Hyg., Abt. 1, Orig. C* **3**, 140–148.

Kilpatrick, M. W., and Walker, R. T. (1982). The nucleotide sequence of the tRNA from the archaebacterium *Thermoplasma acidophilum. Zentralbl. Bakteriol. Mikrobiol. Hyg., Abt. 1, Orig. C* **3**, 79–89.

König, H., Skorko, R., and Zillig, W. (1982). Glycogen in thermoacidophilic archaebacteria of the genera *Sulfolobus, Thermoproteus, Desulfurococcus* and *Thermococcus. Arch. Microbiol.* **132**, 297–303.

Kuchino, Y., Hiara, M., Yabusaki, Y., and Nishimura, S. (1982). Initiator tRNAs from archaebacteria show common unique sequence characteristics. *Nature (London)* **298**, 684–685.

Lake, J. A., Henderson, E., Clark, M. W., and Matheson, A. T. (1982). Mapping evolution with ribosome structure: Intralineage constancy and interlineage variation. *Proc. Natl. Acad. Sci. U.S.A.* **79**, 5948–5952.

Langworthy, T. A., Smith, P. F., and Mayberry, W. R. (1972). Lipids of *Thermoplasma acidophilum. J. Bacteriol.* **112**, 1193–1200.

Langworthy, T. A., Mayberry, W. R., and Smith, P. F. (1974). Long chain glycerol diether and polyol dialkyl glycerol triether lipids of *Sulfolobus acidocaldarius. J. Bacteriol.* **119**, 106–116.

Langworthy, T. A., Tornabene, T. G., and Holzer, G. (1982). Lipids of archaebacteria. *Zentralbl. Bakteriol., Mikrobiol. Hyg., Abt. 1, Orig. C* **3**, 228–244.

McClure, M., and Wyckoff, W. G. (1982). Ultrastructural characteristics of *Sulfolobus acidocaldarius. J. Gen. Microbiol.* **128**, 433–437.

McConnell, D. J., Searcy, D. G., and Sutclifle, J. G. (1978). A restriction enzyme Tha I from the thermophilic mycoplasma *Thermoplasma acidophilum. Nucleic Acids Res.* **5**, 1729.

Martin, A., Yeats, S., Janekovic, D., Reiter, W. D., Aicher, W., and Zillig, W. (1984). SAV 1, a temperate u.v.-inducible DNA virus-like particle from the archaebacterium *Sulfolobus acidocaldarius* isolate B12. *EMBO J.* **3**, 2165–2168.

Mayberry-Carson, K. J., Langsworthy, T. A., Mayberry, W. R., and Smith, P. F. (1974). A new class of lipopolysaccharide from *Thermoplasma acidophilum. Biochim. Biophys. Acta* **360**, 217–229.

Michel, H., Neugebauer, D. C., and Oesterhelt, D. (1980). The 2-D crystalline cell wall of *Sulfolobus acidocaldarius:* Structure, solubilization and reassembly. *In* "Electron Microscopy at Molecular Dimension" (W. Baumeister and W. Vogell, eds.), pp. 27–35, Springer-Verlag, Berlin and New York.

Morgan, H. W., and Daniel, R. M. (1982). Isolation of a new species of sulphur reducing extreme thermophile. *Proc. Int. Congr. Microbiol., 13th, 1982* Abstracts, p. 86.

Neumann, H., Gierl, A., Tu, J., Leibrock, J., Staiger, D., and Zillig, W. (1983). Organization of the genes for ribosomal RNA in archaebacteria. *Mol. Gen. Genet.* **192**, 66–72.

Ohba, M., and Oshima, T. (1982). Some biochemical properties of the protein synthesizing machinery of acidothermophilic archaebacteria isolated from Japanese hot springs. *In* "Archaebacteria" (O. Kandler, ed.), p. 353. Fischer, Stuttgart.

Oshima, T. (1978). Novel polyamines of extremely thermophilic bacteria. *In* "Biochemistry of Thermophily" (S. D. Friedman, ed.), pp. 211–220. Academic Press, New York.

Oshima, T. (1979). Molecular basis for unusual thermostabilities of cell conditions from an extreme thermophile, *Thermus thermophilus. In* "Strategies of Microbial Life in Extreme Environments" (M. Shilo, ed.), Dahlem Konf. Life Sci. Res. Rep., Vol. 13, pp. 455–469. Verlag Chemie, Weinheim.

Prangishvilli, D., Zillig, W., Gierl, A., Biesert, L., and Holz, I. (1982). DNA-dependent RNA polymerases of thermoacidophilic archaebacteria. *Eur. J. Biochem.* **122,** 471–477.

Rau, H. (1978). "Geothermische Energie." Udo Pfriemer Verlag, München.

Sapienza, C., and Doolittle, W. F. (1982a). Repeated sequence in the genomes of *Halobacteria, Zentralbl. Bakteriol., Mikrobiol. Hyg., Abt. 1, Orig. C* **3,** 120–127.

Sapienza, C., and Doolittle, W. F. (1982b). Unusual physical organization of the halobacterial genome. *Nature (London)* **295,** 384–389.

Schär, H. P., and Zubar, H. (1979). Structure and function of L-lactate dehydrogenases from thermophilic and mesophilic bacteria. *Hoppe-Seyler's Z. Physiol. Chem.* **360,** 795–807.

Schmid, G., Pecher, T., and Böck, A. (1982). Properties of the translational apparatus of archaebacteria. *Zentralbl. Bakteriol., Mikrobiol. Hyg., Abt. 1, Orig. c* **3,** 209–217.

Schnabel, R., Sonnenbichler, J., and Zillig, W. (1982). Stimulation by silybin, a eukaryotic feature of archaebacterial RNA polymerases. *FEBS Lett.* **150,** 400–402.

Schnabel, R., Thomm, M., Gerardy-Schahn, R., Zillig, W., Stetter, K. O., and Huet, J. (1983). Structural homology between different archaebacterial DNA-dependent RNA polymerases analyzed by immunological comparison of their components. *EMBO J.* **2,** 751–755.

Searcy, D. G. (1975). Histone-like protein in the prokaryote *Thermoplasma acidophilum*. *Biochim. Biophys. Acta* **395,** 535–547.

Searcy, D. G., and De Lange, R. J. (1980). *Thermoplasma acidophilum* histone-like protein: Partial aminoacid sequence suggestive of homology to eukaryotic histones. *Biochim. Biophys. Acta* **609,** 197–200.

Searcy, D. G., and Doyle, E. K. (1975). Characterization of *Thermoplasma acidophilum* deoxyribonucleic acid. *Int. J. Syst. Bacteriol.* **25,** 286–289.

Searcy, K. B., and Searcy, D. G. (1981). Superoxiddisumtase from the archaebacterium *Thermoplasma acidophilum*. *Biochim. Biophys. Acta* **670,** 39–46.

Searcy, D. G., and Stein, D. B. (1980). Nucleoprotein subunit structure in an unusual prokaryotic organism: *Thermoplasma acidophilum*. *Biochim. Biophys. Acta* **609,** 180–195.

Searcy, D. G., and Whatley, F. R. (1982). *Thermoplasma acidophilum* cell membrane: Cytochrome *b* and sulfate-stimulated ATPase. *Zentralbl. Bakteriol. Mikrobiol. Hyg., Abt. 1, Orig. c* **3,** 245–257.

Searcy, D. G., Stein, D. B., and Green, G. R. (1978). Phylogenetic affinities between eukaryotic cells and a thermophilic mycoplasma. *BioSystems* **10,** 19–28.

Seegerer, A., Stetter, K. O., and Klink, F. (1985). Two contrary modes of lithotrophy in the same archaebacterium. *Nature (London)* (in press).

Smith, P. F. (1980). Sequence and glycosidic bond arrangement of sugars in lipopolysaccharide from *Thermoplasma acidophilum*. *Biochim. Biophys. Acta* **619,** 367–373.

Smith, P. F., Langworthy, T. A., Mayberry, W. R., and Hoagland, A. E. (1973). Characterization of the membranes of *Thermoplasma acidophilum*. *J. Bacteriol.* **116,** 1019–1028.

Smith, P. F., Langworthy, T. A., and Smith, M. R. (1975). Polypeptide nature of growth requirement of yeast extract for *Thermoplasma acidophilum*. *J. Bacteriol.* **124,** 884–892.

Spiess, F. N., Macdonald, K. C., Atwater, T., Ballard, R., Carranza, A., Cordoba, D., Cox, C., Diaz Garcia, V. M., Francheteau, J., Guerrero, J., Hawkins, H., Haymon, R., Hessler, R., Juteu, T., Kastner, M., Larson, R., Luyendyk, B., Macdongall, J. D., Miller, S., Normark, W., Orcutt, J., and Rangin, C. (1980). East Pacific Rise: Hot springs and geophysical experiments. *Science* **207,** 1421–1433.

Stahl, D. A., Luehrsen, K. R., Woese, C. R., and Pace, N. R. (1981). An unusual 5S rRNA, from *Sulfolobus acidocaldarius*, and its implications for a general 5S rRNA structure. *Nucleic Acids Res.* **9,** 6129–6137.

Stetter, K. O. (1982). Ultrathin mycelia forming organisms from submarine volcanic areas having an optimum growth temperature of 105°C. *Nature (London)* **300,** 258–260.

Stetter, K. O., Thomm, M., Winter, J., Wildgruber, G., Huber, H., Zillig, W., Janekovic, D., König, H., Palm, P., and Wunderl, S. (1981). *Methanothermus fervidus*, sp. nov., a novel extremely thermophilic methanogen from an Icelandic hot spring. *Zentralbl. Bakteviol. Mikrobiol. Hyg., Abt. 1, Orig. C* **2**, 166–178.

Stetter, K. O., König, H., and Stackebrandt, E. (1983). *Pyrodictium* gen. nov., a new genus of submarine disc-shaped sulfur-reducing archaebacteria growing optimally at 105°C. *Syst. Appl. Microbiol.* **4**, 535–551.

Sturm, S., Schönefeld, V., Zillig, W., Janekovic, D., and Stetter, K. O. (1980). Structure and function of the DNA-dependent RNA polymerase of the archaebacterium *Thermoplasma acidophilum. Zentralbl. Bakteriol., Mikrobiol. Hyg., Abt. 1, Orig. C* **1**, 12–25.

Taylor, K. A., Deatherage, J. F., and Amos, L. A. (1982). Structure of the S-layer of *Sulfolobus acidocaldarius. Nature (London)* **299**, 840–842.

Thomm, M., Stetter, K. O., and Zillig, W. (1982). Histone-like proteins in Eu- and Archaebacteria. *Zentralbl. Bakteriol., Mikrobiol. Hyg., Abt. 1, Orig. C* **3**, 128–139.

Tu, J., and Zillig, W. (1982). Organization of rRNA structural genes in the archaebacterium *Thermoplasma acidophilum. Nucleic Acids. Res.* **10**, 7231–7245.

Tu, J., Prangishvilli, P., Huber, H., Wildgruber, G., Zillig, W., and Stetter, K. O. (1982). Taxonomic relations between archaebacteria including 6 novel genera examined by cross hybridization of DNAs and 16S rRNAs. *J. Mol. Evol.* **18**, 109–114.

Watanabe, K., Kuchino, Y., Yamaizumi, Z., Kato, M., Oshima, T., and Nishimura, S. (1978). Nucleotide sequence of formylmethionine tRNA from an extreme thermophile, *Thermus thermophilus* HB8. *Nucleic Acids Res., Spec. Publ.* **5**, 473–476.

Weiss, R. L. (1974). Subunit cell wall of *Sulfolobus acidocaldarius. J. Bacteriol.* **118**, 275–284.

Woese, C. R., Maniloff, J., and Zablen, L. B. (1980). Phylogenetic analysis of the mycoplasmas. *Proc. Natl. Acad. Sci. U.S.A.* **77**, 494–498.

Woese, C. R., Gupta, R., Hahn, C. M., Zillig, W., and Tu, J. (1984). The phylogenetic relationships of three sulfur-dependent archaebacteria. *Syst. Appl. Microbiol.* **5**, 97–105.

Yeats, S., McWilliam, P., and Zillig, W. (1982). A plasmid in the archaebacterium *Sulfolobus acidocaldarius. EMBO J.* **1**, 1035–1038.

Young, L. L., and Hong, A. (1979). Purification and partial characterization of a procaryotic glygcoprotein from the plasma membrane of *Thermoplasma acidophilum. Biochim. Biophys. Acta* **556**, 265–277.

Zillig, W., Stetter, K. O., and Janekovic, D. (1979). DNA-dependent RNA polymerase from the archaebacterium *Sulfolobus acidocaldarius. Eur. J. Biochem.* **96**, 597–604.

Zillig, W., Stetter, K. O., Wunderl, S., Schulz, W., Priess, H., and Scholz, I. (1980). The *Sulfolobus-* "*Caldariella*" Group: Taxonomy on the basis of the structure of DNA-dependent polymerases. *Arch. Microbiol.* **125**, 259–269.

Zillig, W., Stetter, K. O., Schäfer, W., Janekovic, D., Wunderl, S., Holz, I., and Palm, P. (1981). *Thermoproteales:* A novel type of extremely thermoacidophilic anaerobic archaebacteria isolated from Icelandic solfataras. *Zentralbl. Bakteriol. Mikrobiol. Hyg., Abt. 1, Orig. C* **2**, 200–227.

Zillig, W., Stetter, K. O., Schnabel, R., Madon, J. and Gierl, A. (1982a). Transcription in Archaebacteria. *Zentralbl. Bakteriol., Mikrobiol. Hyg. Abt. 1, Orig. C* **3**, 218–227.

Zillig, W., Schnabel, R., Tu, J., and Stetter, K. O. (1982b). The Phylogeny of Archaebacteria including novel anaerobic thermoacidophiles in the light of RNA polymerase structure. *Naturwissenschaften* **69**, 197–204.

Zillig, W., Stetter, K. O., Prangishvilli, D., Schäfer, H., Wunderl, S., Janekovic, D., Holz, I., and Palm, P. (1982c). *Desulfurococceae*, the second family of the extremely thermophilic, anaerobic, sulfur respiring *Thermoproteales. Zentralbl. Bakteriol., Mikrobiol. Hyg., Abt. 1, Orig. C* **3**, 304–317.

Zillig, W., Gierl, A., Schreiber, S., Wunderl, S., Janekovic, P., Stetter, K. O., and Klenk, H. P. (1983a). The archaebacterium *Thermofilum pendens* represents a novel genus of the Thermophilic, anaerobic, sulfur respiring *Thermoproteales*. *Syst. Appl. Microbiol.* **4,** 79–87.

Zillig, W., Holz, I., Janekovic, D., Schäfer, W., and Reiter, W. D. (1983b). The archaebacterium *Thermococcus* celer represents a novel genus within the thermophilic branch of the archaebacteria. *Syst. Appl. Microbiol.* **4,** 88–94.

Zillig, W., Yeats, S., Holz, I., Böck, A., Gropp, F., Rettenberger, M., and Lutz, S. Plasmid-related anaerobic autotrophy of the novel archaebacterium *Sulfolobus ambivalens*. *Nature* (submitted).

Zuber, H. (1979). Structure and function of enzymes from thermophilic microorganisms. *In* "Strategies of Microbial Life in Extreme Environments" (M. Shilo, ed.), Dahlem Konf. Life. Sci. Res. Rep., Vol. 13, pp. 393–415. Verlag Chemie, Weinheim.

Zuber, H. (1979). Structure and function of enzymes from thermophilic microorganisms. *In* "Strategies of Microbial Life in Extreme Environments" (M. Shilo, ed.), Dahlem Konf. Life. Sci. Res. Rep., Vol. 13, pp. 393–415. Verlag Chemie, Weinheim.

Zuber, H. (1981). Structure and function of thermophilic enzymes. *In* "Structural and Functional Aspects of Enzyme Catalysis" (H. Eggerer and R. Huber, eds.), pp. 114–127. Springer-Verlag, Berlin and New York.

CHAPTER 3

The Halobacteriaceae

D. J. KUSHNER

I. Introduction: What They Are and Why We Study Them

Halobacteria, halococci, and related bacteria were first studied systematically because they attacked salted foods and hides. On long sea voyages, salted fish spoiled and turned red; salted hides developed red patches and disintegrated. These changes were observed for centuries before their cause could be deter-

Copyright © 1985 by Academic Press, Inc.
All rights of reproduction in any form reserved.
ISBN 0-12-307208-5

mined. Accounts of the discovery of the organisms responsible, the so-called red halophiles, bacterial rods (mainly), and cocci, are given in several review articles (Ingram, 1957; Kushner, 1968, 1978; Larsen, 1962, 1967, 1973, 1980; Scott, 1957).

In those countries best able to carry out microbiological research, meat and fish are preserved by refrigeration rather than salt. Few studies are still carried out on red halophilic bacteria as food spoilage organisms. However, interest in the fundamental properties of these organisms has persisted and, in recent years, has grown greatly. Their very existence poses some obvious questions: What kind of adaptations are necessary for the life of any bacteria in strong salt solutions? Indeed, what kind of adaptations made bacteria *need* such high salt concentrations?

As physiological studies were carried out on the red halophilic bacteria, the biochemical peculiarities of these organisms became increasingly evident. Some of these peculiarities can be used to group the Halobacteriaceae with other archaebacteria.

Indeed, students of the archaebacteria owe a special debt to the halobacteria and related bacteria. The latter organisms were the first archaebacteria to be studied in any detail, and, assuming that the report of water reddening during the manufacture of salt from sea water, found in a 5000-year-old Chinese treatise (cited in Baas-Becking, 1931), reflects the presence of halobacteria, they were by far the first to be cited in any kind of literature. One possession of the halobacteria, the "purple membrane," with its bacteriorhodopsin system, has revealed new ways of transducing the energy of light through biological membranes. Indeed, the bacteriorhodopsin system appears to be the simplest and most elegant energy-transducing system in the biological world, one that we are very close to understanding in molecular terms. The possibility that such systems might lead to new, practical sources of energy has attracted much research support. It is not surprising that more than half the papers that have appeared in the past few years deal with bacteriorhodopsin and related systems.

The halobacteria have also been studied as examples of the kind of microorganisms that might grow on other planets, more specifically on Mars, where the level of available water is very low. When, following the Viking Lander missions, Mars apparently proved sterile (see Kushner, 1980), much of the support (from the National Aeronautic and Space Administration in the United States) for investigations of such microorganisms vanished. The halobacteria remain quite interesting enough, however, as examples of microorganisms on our own earth.

The increasing interest in these organisms can be judged by a survey of listings under "Halobacteria" in Biological Abstracts since 1930. (A few classical papers seem to have escaped.) From 1930–1940, one title; 1941–1950, none;

1951–1960, 12; 1961–1970, 89; 1971–1980, 435. Since 1980, there is no evidence of a declining interest.

The present chapter, which draws heavily on recent reviews, gives a general but by no means exhaustive survey of the distribution, ecology, physiology, and genetics of the halobacteria and related organisms and their place in the biological world.

II. Salty Places and Their Inhabitants

Red halophiles have made their presence known for centuries, through the red color they impart to salterns, the ponds used to prepare solar salt from sea water (Baas-Becking, 1931). Such colors are still evident in salterns such as those around San Francisco Bay and in certain salt lakes, e.g., the Great Salt Lake. (Post, 1975, 1977).

The most-studied red halophiles come from salt lakes, especially the Dead Sea, and from salterns, either isolated directly or from spoiled salted foods, sausage casings, or hides preserved with solar salt. Red halophiles have a world-wide distribution. They appear quickly whenever a solar salt facility is set up by the sea or in an inland area and give a reddish-purple tinge to the water. Probably, they can spread from one such saltern to another in dried salt crystals on the wind or on the legs and feathers of birds that feed in the less concentrated ponds of a saltern or around the shores of more concentrated parts.

The ionic composition of water in two salterns and in several salt lakes are shown in Table I. The lakes are defined as *terminal desertic lakes,* with no natural outflows, into which water flows and then evaporates. Thus, the Dead Sea, which is about 400 m below sea level, is fed by the River Jordan; the Wadi Natrun lakes in Egypt, which are about 20 m below sea level, are fed by inflow of water from the River Nile. The water evaporates, leaving the very salty composition shown. In recent years, the use of Jordan water for irrigation in Israel has reduced the inflow of water by about 50%, the level of the Dead Sea has fallen about 5 m, and the salt concentration has increased (Nissenbaum, 1975; Beyth, 1980).

Natural salt lakes may differ widely in ionic composition. The Great Salt Lake has Na^+ as its major cation, whereas in the Dead Sea there is more Mg^{2+} than Na^+. In the Wadi Natrun lakes and other salt lakes in Africa where high concentrations of HCO_3^-/CO_3^{2-} maintain pH values greater than 11, no Mg^{2+} can be detected. The response of certain halobacteria to different concentrations of Na^+ and Mg^{2+} can vary according to the particular lake from which they have been isolated.

TABLE I

IONIC AND MICROBIAL COMPOSITION OF HYPERSALINE LAKES AND BRINES[a]

Composition	Ion concentration (g/liter)			
	Dead Sea[b]	Great Salt Lake[c]	Wadi Natrun[d]	Marine Saltern
Na^+	39.2	105.4	142.0	65.4
K^+	7.3	6.7	2.3	5.2
Mg^{2+}	40.7	11.1	0.0	20.1
Ca^{2+}	16.9	0.3	0.0	0.2
Cl^-	212.4	181.0	154.6	144.0
Br^-	5.1	0.2	—	—
SO_4^{2-}	0.5	27.0	22.6	19.0
HCO_3/CO_3^{2-}	0.2	0.72	67.2	—
Total salinity	322.6	332.5	393.9	253.9
pH	5.9–6.3	7.7	11.0	—
Microorganisms/ml				
Halobacteria + halococci	7×10^7	7×10^7	—	up to 10^8
Dunaliella	4×10^4	10^4	—	—

[a] Summarized from review of Larsen (1980). Detailed discussions of these and other salt lakes are found in Nissenbaum (1980), Edgerton and Brimblecombe (1981), and Javor (1983a,b). (—), figures not given.

[b] The composition of the Dead Sea changes with depth and location and has also changed over recent years (Beyth, 1980). Changes in numbers of bacteria (largely halobacteria) and *Dunaliella* at different seasons and at different depths are described by Oren and Shilo (1981, 1982).

[c] North arm of Great Salt Lake. The lake is divided by a causeway. The south arm, into which fresh water flows, is about one-half as concentrated as the north arm (Post, 1977).

[d] This is composed of a series of lakes. The composition of Lake Zugm (shown) is similar to that of another lake, Lake Gaar (not shown).

Table I also shows the microbial populations of some of these salt lakes and salterns. Halobacteria (and to a lesser extent halococci) make up the greater part of the bacterial flora. Populations up to 10^8/ml have been found. Other major organisms include the green algal genus *Dunaliella,* whose pigmentation (green in *D. viridis* or red in *D. salina,* which requires higher NaCl concentrations for growth) may contribute substantially to the lake's color. Salt lakes also seem to have several cycles of microbial activity. They contain primary producers, including *Dunaliella* species and blue-green algae (cyanobacteria), sulfate reducers, and the halophilic photosynthetic sulfur bacteria *Ectothiorhodospira* species, which utilize the H_2S produced. The heterotrophic halobacteria and halococci act as mineralizers (Post, 1977; Imhoff *et al.,* 1979; Larsen, 1980).

Kushner pointed out (1978) that certain food products (especially soy sauce and miso paste) contain enough NaCl to support the growth of red halophiles.

Though many halophilic and salt-tolerant microorganisms grow in such products (and greatly influence their final composition), red halophiles, fortunately, do not. Possibly, red halophiles, which are well endowed to compete with other microorganisms under aerobic conditions in bright sunlight, are less able to do so anaerobically in the dark. Furthermore, because red halophiles are sensitive to acid [killed at pH below 5 (Kushner *et al.*, 1965)], they could not compete well with salt-tolerant fermenting microorganisms.

III. Halophilic and Halotolerant Microorganisms

The terms *extremely halophilic* or *halophilic* bacteria are sometimes used as if they were synonymous with the halobacteria. They are not. A very wide spec-

TABLE II

SALT RESPONSE OF DIFFERENT MICROORGANISMS[a]

Category	Reaction	Examples
Nonhalophile	Grows best in media containing less than 0.2 M salt	Most normal eubacteria and most freshwater microorganisms
Slight halophile	Grows best in media containing 0.2–0.5 M salt	Many marine microorganisms
Moderate halophile	Grows best in media containing 0.5–2.5 M salt. Organisms able to grow in less than 0.1 M salt are considered facultative halophiles	*Vibrio costicola, Paracoccus halodenitrificans, Pseudomonas* species
Borderline extreme halophile	Grows best in media containing 1.5–4.0 M salt	*Ectothiorhodospira halophila, Actinopolyspora halophila, Halobacterium volcanii, H. mediterranei*
Extreme halophile	Grows best in media containing 2.5–5.2 M (saturated) salt	*Halobacterium salinarium, Halococcus morrhuae*
Halotolerant	Nonhalophile that can tolerate salt. If the growth range extends above 2.5 M salt, it may be considered extremely halotolerant	*Staphylococcus epidermidis,* Solute-tolerant yeasts, fungi, and algae. *Halomonas elongata*

[a] Adapted from Table 3 in Kushner (1978). For a list of *Ectothiorhodospira* species with different levels of NaCl requirement, see Imhoff et al. (1983). For *Halomonas elongata,* see Vreeland and Martin (1980). Other species are listed in Kushner (1978).

trum of response to different salt concentrations exists in the microbial world. As Table II shows, microorganisms can, on the basis of this response, be grouped into nonhalophilic, slightly, moderately, and extremely halophilic, and haloto-lerant categories. There is, furthermore, considerable overlapping between these categories. The *Halobacterium* and *Halococcus* species are classified, by their response to salt, as extreme halophiles, some more extreme than others. They share this position with several species of the photosynthetic bacteria *Ecto-thiorhodospira* and with the actinomycete *Actinopolyspora halophila*. Thus, though salt response is a very important characteristic of the Halobacteriaceae, it does not serve to define this family. The fact that a number of bacteria can grow in similar ranges of NaCl concentrations to the Halobacteriaceae permits us to distinguish between physiological and biochemical properties that are limited to the Halobacteriaceae (and possibly other archaebacteria) and to those associated with growth in high salt concentrations.

IV. Definition and Classification of *Halobacterium* and *Halococcus* Species (Family Halobacteriaceae)

A. CLASSIFICATION IN BERGEY'S MANUAL

The following description is taken mainly from reviews (Kushner, 1978; Larsen, 1984). The family Halobacteriaceae consists of gram-negative rods or disks (genus *Halobacterium*) or cocci (genus *Halococcus*). Many *Halobacterium* species have a number of involution forms. All the Halobacteriaceae reproduce by binary fission, with constriction (*Halobacterium*) or septation (*Halococcus*), and possess no resting stages.* Cells are nonmotile or motile by lophotrichous flagella. Motility is observed only with some *Halobacterium* species. The colonies formed are red, pink, vermillion, orange-red, or mauve-red (occasionally colorless); colonies are opaque or translucent. Most species are strict aerobes; a few can grow anaerobically, with or without nitrate. All are chemoheterotrophic. Many require amino acids for growth and are stimulated by vitamins. Optimal growth temperature is 40°–50°C. DNA is usually composed of a major and minor component, the latter being made of several extrachromosomal DNA fragments. Moles% G + C of major component, 66–68; of minor component, 57–60. Cells require at least 1.5 M (8.8%) NaCl for growth, and usually 3–4 M (17–23%); most can grow in saturated (5.5 M, 32%) NaCl.

---

*Fujii (1980) has isolated spore-forming, extremely halophilic rods, which he has suggested including in the Halobacteriaceae. However, the fact that these cells are gram-positive should exclude them from this family.

Other characteristics, which justify assigning new isolates with the above properties to the Halobacteriaceae are their lipid contents; all of them have diphytanyl ether-linked phospholipids as their major component (Fig. 1). The presence of such lipids can be easily determined, because they are resistant to normal saponification treatments (Kates *et al.*, 1966). A rapid routine procedure for determining such lipids has been proposed by Ross *et al.* (1981).

M. Kates and co-workers have reported new glycolipids and sulfated glycolipids in halobacteria from different sources (Fig. 1; Kushwaha *et al.*, 1982). Their pattern of distribution suggests that they may be very useful in distinguishing halobacterial species.

A further essential "marker" is the absence of usual bacterial cell-wall components, especially muramic acid. The chemistry of the outer envelopes of halobacteria and halococci are quite different from those of gram-negative or

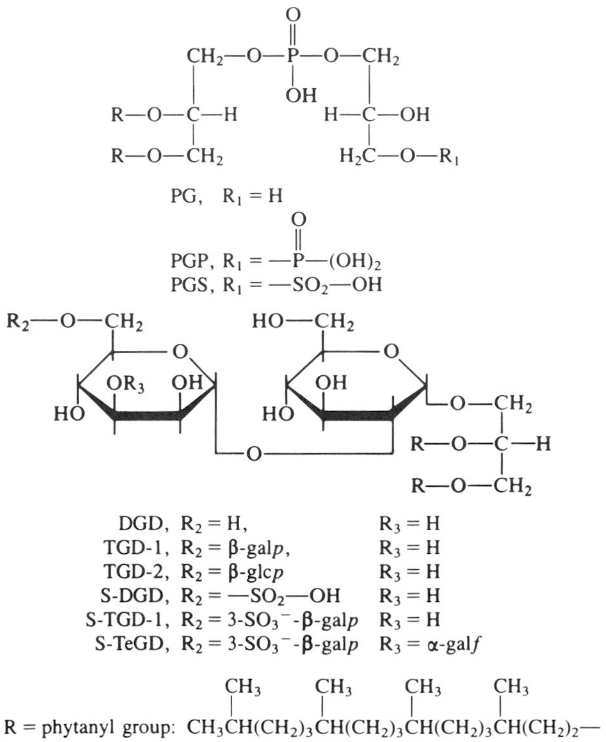

FIG. 1. Structures of phospholipids and glycolipids in extremely halophilic bacteria. PG, phosphatidylglycerol; PGP, phosphatidylglycerophosphate; PGS, phosphatidylglycerosulfate; DGD, diglycosyl diether; TGD-1 and TGD-2, triglycosyl diether; S-DGD, sulfated diglycosyl diether; S-TGD-1, sulfated triglycosyl diether; S-TeGD, sulfated tetraglycosyl diether. From Kushwaha *et al.* (1982).

positive eubacteria. The absence of muramic acid, or of more than a trace amount of hexosamines, characterizes these cells and is convenient in taxonomic studies.

It might be thought difficult to distinguish between disks and spheres, i.e., between pleomorphic halobacteria and halococci. However, halococci divide into pairs, tetrads, sarcina, or irregular clusters. Furthermore, halobacteria, whatever their shape, have very thin outer envelopes that are easy to break mechanically; most of them lyse in water. Halococci have thick cell envelopes that are difficult to break mechanically; they do not lyse in water and, as will be seen, survive exposure to much lower salt concentrations than do halobacteria.

In the 8th edition of Bergey's Manual (Gibbons, 1974) *Halobacterium salinarium* and *H. halobium* are listed as separate species. In the 9th edition, they are considered one species, together with isolates previously called *H. cutirubrum,* under the name *H. salinarium.* (Larsen, 1984). The reclassification is due to the physiological similarities between different isolates and to the observation by Fox *et al.* (1980) that they have similar 16 S rRNA catalogs.

Other species of halobacteria to be listed in the 9th edition of Bergey's Manual include *H. saccharovorum, H. vallismortis, H. volcanii,* and *H. pharaonis.* All named species are distinguished by their ability or lack of ability to use glucose, their ability to grow anaerobically using nitrate as an electron acceptor, their cellular form, whether they are rod-shaped or pleomorphic, and pH and salt ranges permitting growth. Only one species of halococci is listed in the chapter on the Halobacteriaceae, *Halococcus morrhuae* (Larsen, 1984).

B. Other Halobacteria

Colwell *et al.* (1979) described a study of 60 halophilic bacteria from culture collections and 23 fresh isolates from a solar saltern. The latter were chosen by their ability to form red colonies on rich complex media containing 20–25% NaCl. The authors compared a large number of properties of these bacteria by numerical taxonomy. All the extremely halophilic rods were closely related to each other, but there seemed to be enough difference between two groups to designate two species, *Halobacterium salinarium* and *H. cutirubrum;* strains previously called *H. halobium* were included in the *H. cutirubrum* group. The extremely halophilic cocci were clearly different from the extremely halophilic rods, and they displayed more internal variation. Nevertheless, they were all included in the species *Halococcus morrhuae. Halococcus morrhuae* is still the only recognized halococcal species, possibly reflecting the smaller amount of work that has been done with the halococci.

In the same study, Colwell *et al.* (1979) compared 48 strains of salt-tolerant streptococci, staphylococci, micrococci, and a few moderately halophilic bacteria, and they found no apparent relationship between these and the halobacteria and halococci.

The fact that extreme halophiles studied could be grouped into two or three species suggests that these organisms are genetically stable. This seems paradoxical in view of more recent evidence of the great potential that the same organisms may possess for genetic variation. Possible explanations for this seeming contradiction will be discussed later.

Rodriguez-Valera and associates (1980) isolated over 300 extremely halophilic rod forms from solar salterns near Alicante, Spain, by enrichment in growth media of high salt concentration with individual carbohydrates as the main carbon and energy source. Some halobacteria, which did not fit into previously known species, were obtained; one general characteristic of many of these organisms was that pigmentation was highest at the lowest salt concentration that permitted growth, whereas at the maximum possible NaCl concentration, cells grew without producing pigments (Kushwaha et al., 1982). One of these bacteria, strain R4, has been characterized in some detail and was given the name *Halobacterium mediterranei* (Rodriguez-Valera et al., 1983b). Other new species also seem to be present. These species differ from other halobacteria in the thickness of their cell walls (25–35 nm, as opposed to ca 13 nm for other halobacteria). Their cells are more resistant to mechanical disruption by pressure bombs, ultrasonication, freezing, and thawing than are other halobacteria. *Halobacterium mediterranei* grows at about the same range of salt as *H. volcanii*.

Many halobacteria lyse and their cell envelopes dissolve at low salt concentrations. *Halobacterium halobium* and *H. salinarium* need at least 5% salts, of the same composition as found in sea water, to protect them from lysis and death. However, *H. mediterranei* was not lysed in 3% of such salts, and after exposure to 1% of such salt, 10% of the cells remained viable.

C. STILL OTHER HALOBACTERIA

The names of species of *Halobacterium* continue to multiply. Some isolates are named with only a brief description, e.g., *H. tunesiensis, H. capanicum* (Pfeifer et al., 1981a). A brief description of species currently being studied is given by Torsvik and Dundas (1982).

A more recent isolate is *Halobacterium sodomense* from the Dead Sea, which is characterized by its very high magnesium requirement (Oren, 1983). It needs $0.6–1.2\ M$ for optimal growth and can grow well in $0.5\ M$ NaCl if $1.5–2.0\ M$ $MgCl_2$ is also present. In many properties, it is similar to *H. volcanii* (Mullak-hanbhai and Larsen, 1975) but has a somewhat higher Mg^{2+} requirement and tolerance.

The name *Halobacterium maris-mortui* deserves special attention. This was given to one of the first halobacteria isolated from the Dead Sea (Elazari-Volcani, 1940). However, the original isolate was lost, and *H. maris-mortui* was listed in the 8th edition of Bergey's Manual as of *incertae sedis* (Gibbons, 1974). In 1970, Ginzburg *et al.* isolated a halobacterium from the Dead Sea, which has been called, variously, *Halobacterium* of the Dead Sea and *Halobacterium maris-mortui*. This organism has been extensively used by scientists in Israel for physiological studies including the most advanced and sophisticated thermodynamic studies of isolated enzymes yet undertaken. Brief reinvestigations of some of its properties are mentioned in Werber and Mevarech (1978) and Pundak and Eisenberg (1981), with the suggestion that it may well be the same as the original *H. maris-mortui*. The 5 S rRNA fingerprinting study of Nicholson and Fox (1983) suggested a close relationship between this organism and *H. vallismortis*, which was isolated from a salt pond in Death Valley, California. Another study (Juez *et al.*, 1985), in which a number of halobacteria are compared by numerical taxonomy, also strongly suggests the species level of *H. maris-mortui* and *H. vallismortis* is identical. By the strict rules of taxonomy, both should be given the valid name *H. vallismortis*, which seems well established through standard taxonomic criteria. This would seem a pity: *Halobacterium maris-mortui* undoubtedly comes from the Dead Sea, and it has the blessing of the original giver of the name, though it is not the same strain that he named (Pundak and Eisenberg, 1981). The name will probably persist. For that matter, by far the most active lines of research on halobacteria include different *H. halobium* strains (see below). This name is not apt to vanish either, whatever its orthodox taxonomic status.

D. Square and Box-Shaped Bacteria

In 1980, A. E. Walsby described a curious square, flat bacterium containing gas vacuoles isolated from a sabkha, or hypersaline pool, in the coastal plain of the Sinai Peninsula. Such pools vary in salinity throughout the year, being diluted by fresh seawater and flash floods. The square bacteria appeared on salt crusts around the pond when salinity reached about 21% or higher total dissolved salt, a property that groups them among the extreme halophiles. Such square bacteria have also been found in commercial salterns in Israel and Mexico (Baha, California) (Stoeckenius, 1981). These bacteria, which have not yet been grown in pure culture and which may consist of several different species (Kessel and Cohen, 1982), appear in the form of thin square sheets. Up to 7×10^7 organisms per milliliter of brine have been found (Walsby, 1980). In the salt ponds where they are found, they are the dominant species. The squares were only 0.15–0.5

μm thick; many were about 5 μm per side, though cells as large as 11 μm and as small as 1.5 μm per side have been reported. Often, cells remained attached after division; sheets of 64 or more cells have been seen (Kessel and Cohen, 1982).

The surfaces of these cells show the hexagonal (occasionally tetragonal) arrangement of particles characteristic of halobacteria. Some had external sheaths of fibrillar material; others lacked such sheaths (Kessel and Cohen, 1982; Parkes and Walsby, 1981; Stoeckenius, 1981). Such bacteria possessed pigments (bacterioruberins and bacteriorhodopsin) characteristic of many halobacteria (Stoeckenius, 1981). These properties suggest that they may be closely related to the halobacteria, and they should be considered members of the archaebacteria.

Though the "square bacteria" have not yet, to my knowledge, been grown in pure culture, several "box-shaped" halophilic bacteria have been readily isolated in pure culture from similar environments, i.e., brines from the Sinai Peninsula, Lower California (Mexico), and Southern California (U.S.A.) (Javor et al., 1982). These were grown in complex media, showed a strong NaCl requirement (optimal 3–4 M), and also required Mg^{2+}. They were irregular rectangles and squares, 0.3–0.4 μm thick, 1–2 μm on a side; triangular, round, or oval cells of similar dimensions were also seen. All contained bacteriorhodopsin. They also contained diphytanyl glycerol diether (M. Kates, personal communication). The genus *Haloarcula* was tentatively suggested as a name for these species (Javor et al., 1982).

Morphologically, it is difficult to distinguish between disk-shaped, cup-shaped, square, rectangular, and triangular cells; attempts to draw a distinction are rather subjective. Thus, the new box-shaped bacteria do not appear really different from certain extreme halophiles designated as halobacteria, for example, *Halobacterium volcanii* (Mullakhanbhai and Larsen, 1975). More recently, Nicholson and Fox (1983) carried out a study on the 5 S ribosomal RNA of two of the box-shaped bacteria; they found, by ribonuclease finger-printing, that the box-shaped bacteria were essentially identical to *H. vallismortis* and *H. maris-mortui*; they were less closely related to *H. cutirubrum* and *Halococcus morrhuae*. For the present, the box-shaped bacteria should probably be considered members of the halobacteria, closely related to or identical with *H. valismortis*/*H. maris-mortui*.

E. THE NATRONOBACTER AND NATRONOCOCCI

Grant and co-workers (Tindall et al., 1980, 1984) have studied extremely halophilic rods and cocci from an alkali soda lake in Kenya. These organisms are alkaliphilic (pH optimum 9–10) and may have a very low Mg^{2+} requirement (0.1 mM for optimum growth; 10 mM inhibited growth and caused morphological changes in the rods). These organisms have other unusual properties: they contain both $C_{20}:C_{20}$ and $C_{20}:C_{25}$ diether lipids as well as some $C_{25}:C_{25}$ diether lipids (de Rosa *et al.*, 1982, 1983). A very recent study of a number of alkaliphilic and known *Halobacterium* species, using DNA/DNA hybridization and DNA/16 S rRNA hybridization showed that the alkaliphilic bacteria were not closely related to *Halobacterium halobium, H. volcanii,* or *Halococcus morrhuae.* They did seem related to each other and to *H. pharaonis* (Soliman and Trüper, 1982), which has similar physiological properties. The authors proposed new genus names, *Natronobacter* and *Natronococcus,* with *Halobacterium pharaonis* to be included in the former (Tindall *et al.*, 1984).

Considering the wide variety of physiological types previously assigned to the genus *Halobacterium,* it may well be time to rename some members of this group.

V. Physiology of the Halobacteriaceae

A. SALT REQUIREMENTS FOR GROWTH

Although it has been known for some time that K^+ can partly replace Na^+ (Brown and Gibbons, 1955), all species seem to need high NaCl concentrations. Recently, Edgerton and Brimblecombe (1981) studied the growth of species of halobacteria in mixtures of Na^+, K^+, and Mg^{2+}, Cl^- and SO_4^{2-} ions of defined water activity (a_w). They found that a range of ionic conditions (growth areas), which permitted growth of these organisms, was not easily defined in terms of Na^+ or Cl^- activity alone. *Halobacterium salinarium* and *H. halobium* grew rapidly at water activities of about 0.78, whereas *H. volcanii* did so at an a_w of 0.925. The first two bacteria changed from rods to sphere shapes when the Mg^{2+} activity dropped below 0.15 mol kg^{-1}. The authors suggested that the "niches" described reflected the differences in origins of the bacteria studied: solar salt for the first two, the Dead Sea for the last. Now that more extensive studies of the ecology of hypersaline environments are being carried out, the interesting approach of Edgerton and Brimblecombe (1981) may have a wider application (see Oren and Shilo, 1981).

The striking differences between Mg^{2+} requirements (more than 1000-fold) between halobacteria from the Dead Sea, which is rich in Mg^{2+} and alkaline lakes, in which Mg^{2+} cannot be detected (between *Halobacterium volcanii* or *H. sodomense* on one hand and *H. pharaonis* or *Natronobacterium* species on the other) has already been noted. There is also a large difference in Mg^{2+} tolerance in such species: concentrations required for growth of the former would inhibit the latter.

Extreme halophiles grow optimally at 40°–45°C. No growth has been reported below 10°C or above 55°C (Larsen, 1981). Increasing the NaCl concentration increases the growth temperature for *H. salinarium* (Gibbons and Payne, 1961). Many enzymes of halobacteria work well in temperatures of 50°C or higher. Indeed, a number of these enzymes have definitely thermophilic characters. Temperature optima of 70°C for the alanine dehydrogenase and of 60°–65°C for five other enzymes of *H. salinarium* were reported by Keradjopoulos and Wulff (1974). The nitrate reductase of this organism had an optimum of 85°C (Marquez and Brodie, 1973). The temperature optima varied with the salt composition and concentration.

B. STRATEGIES OF ADAPTATION

1. INTRACELLULAR SOLUTES

The Halobacteriaceae are of special interest to students of life in extreme environments because of the striking adaptation they display to high salt concentrations. They are exposed to such concentrations not only outside, but also inside the cytoplasmic membrane. It has, of course, long been realized, and in some cases demonstrated, that most microorganisms that live in high solute concentrations do not exclude high solute concentrations from their interiors. The water activity, a_w, inside the cells is never higher than that outside; in some cases, it may be lower (Brown, 1976; Kushner, 1978). It is well known that microorganisms that live in fresh water or very dilute media have considerably higher solute concentrations inside their cells than outside. Their cell walls protect them from the strong osmotic pressures that would otherwise destroy them.

However, in microorganisms growing in high solute concentrations, the internal solutes differ in composition from those outside, as they do in all living cells. Brown (1976) has used the term "compatible solute" to describe those solutes that adjust the internal a_w of cells but permit vital functions to take place. Examples of such solutes are glycerol in *Dunaliella* and other halophilic and salt-tolerant algae and KCl in species of halobacteria. The former microorganism may grow in NaCl concentrations of 4 M or higher, but its enzymes are inhibited by much lower concentrations of salt. Many or most of the enzymes of the halobacteria, in contrast, function well in high concentrations of NaCl or KCl; in fact, they usually require such high concentrations in order to function at all. Often, they respond equally well to NaCl and KCl; however, as will be seen, allosteric regulation often depends on the presence of KCl rather than NaCl.

Stability of ribosomes of halobacteria and their functioning, as measured by *in vitro* protein synthesis, certainly depend on high KCl, rather than high NaCl concentrations (see below).

There are a number of salt-tolerant and extremely halophilic bacteria, not belonging to the Halobacteriaceae (including those listed in Table II), for which the internal solute composition and the nature of the compatible solute are still unknown. Though Galinsky and Trüper (1982) indicated that betaine may be an important compatible solute in *Ectothiorhodospira* species, we still cannot account for all the internal solutes in these organisms nor in *Vibrio costicola,* which we have been studying.

Analyses of cell-associated KCl of halobacteria have indicated concentrations of up to 5 *M,* which is well beyond the solubility of KCl in water and which can represent a K^+ concentration gradient inside/outside of more than 1000-fold. It has been calculated that K^+ may constitute as much as 40% of the dry weight of halobacterial cells. Because of this, K^+ in the medium can easily become growth limiting. Substantial amounts of K^+, at least 26 m$M,$ may be needed for optimum growth. Otherwise, the cells grows until all K^+ has been assimilated, and then growth stops (Gochnauer and Kushner, 1971; Kushner, 1978).

The state of solutes inside halobacteria has posed an intriguing and still unresolved question: how can these cells maintain such a high differential of K^+ ions between the inside and the outside, especially for long periods and under conditions of low-energy yield, sometimes even in poisoned cells? (Reviewed in Kushner, 1978; Brown and Sturtevant, 1980.) Early nuclear magnetic resonance studies of K^+ in cells suggested that K^+ was somehow "bound" within the cells (Cope and Damadian, 1970), but a more recent study (Shporer and Civan, 1977) showed that most K^+ was free in the cell. These papers should be consulted for details of techniques that lead to such different interpretation.

Lanyi and Silverman (1972) found no evidence for K^+ binding in frozen pastes of *Halobacterium cutirubrum*. More recently, Brown and Sturtevant (1980) and Brown and Duong (1982) measured the state of water in *H. halobium* by differential scanning calorimetry and heat of dilution. Though the results were somewhat difficult to interpret because of the contribution made by intercellular solutes in the thick pastes of bacteria used and because of the changes in proportions of intracellular ions during different phases of growth, the authors concluded that the cell-associated ions were indeed free in the cytoplasm. As we will see, activity of individual enzymes and of *in vitro* protein synthesis is also highest in ionic conditions similar to those shown to be intracellular by direct measurement. This might be taken as indirect evidence that the ions within halobacterial cells are indeed "free." However, recent results with highly purified enzymes (see below) show that these may have very considerable ion-binding ability, so that enzymes may work best when surrounded by a "cloud" of bound cations.

2. THE ADVANTAGE OF HIGH INTERNAL K^+

The experiments on the "state" of intracellular K^+ ions were partly motivated by the idea that cells with such high Na^+/K^+ gradients between the outside and the inside would be under a great physiological strain. More recently, however, it has been realized that these gradients may, instead, confer a physiological advantage. Skulachev and collaborators (Arshavsky *et al.*, 1981) have proposed that in many bacteria the Na^+/K^+ gradient has a proton-motive force (pmf) buffering function. According to this theory, when pmf is low, K^+ will diffuse out of the cells making a change in membrane potential. Sodium will simultaneously enter the cell to maintain electoneutrality. This will lead to the extrusion of H^+ by a Na^+/H^+ antiport and to an increase of the pmf (see also Section V, H and Wagner *et al.*, 1978).

3. ENZYMES AND OTHER PROTEINS

From the beginning, those studying enzymes of halobacteria have realized that these usually require high salt concentrations for activity and always require them for stability. Most enzymes, e.g., the aspartate carbamoyltransferase of *Halobacterium cutirubrum* will become irreversibly denatured within minutes if the salt concentration is reduced below 2 M (Norberg *et al.*, 1973); other enzymes, such as malate dehydrogenase and isocitrate dehydrogenase, lose all activity in the absence of high NaCl or KCl concentrations, but regain this activity if salt is added back slowly (Holmes and Halvorson, 1965; Hubbard and Miller, 1969). Because of their instability in the absence of salts, few enzymes from the halobacteria have been purified: the high salt concentrations needed for stability have made impractical some of the standard chromatographic methods of protein purification, which depend on adsorption and desorption on charged surfaces. Recently, more success has been obtained by exclusion, hydrophobic, hydroxylapatite, and affinity chromatography or salting-out mediated chromatography (de Medicis *et al.*, 1982; Leicht and Pundak, 1981).

The major conclusions drawn from earlier work, using the behavior of impure or partially purified preparations, rather than detailed chemical and physical analyses of their structures (Lanyi, 1974; Dundas, 1977; Kushner, 1978), can be summarized as follows.

A number of halobacterial enzymes need at least 1 M monovalent salt for activity and show maximal activity in 2–4 M salt. These include aspartate carbamoyl-transferase, many amino acid–activating enzymes, and enzymes associated with the cell membrane. A number of enzymes including NADP-specific isocitrate dehydrogenase, show maximal activity at 0.5–1.5 M salt and are inhibited by higher salt concentrations. A few enzymes have the highest activity in the absence of salt and are strongly inhibited by high salt concentrations. Most striking of these is the fatty acid synthetase, which has quite low specific activity in *H. cutirubrum* (about 1/15 the activity of the same enzyme from *Escherichia coli*). *Halobacterium cutirubrum* possesses an acyl carrier protein that can substitute for that of *E. coli*, using the *E. coli* fatty acid synthetase. Addition of NaCl or KCl inhibits the fatty acid synthetase of *H. cutirubrum,* so that in 4 *M* salt only about 30% of the activity is left (Pugh *et al.*, 1971). The inhibitory action of salts and the low specific activity of this enzyme probably account for the fact that the halobacteria make practically no fatty acids. In contrast, the mevalonate pathway for isoprenoid synthesis in halobacteria requires high salt concentrations (Kates and Kushwaha, 1978).

The DNA-dependent RNA polymerase of halobacteria is inhibited by salt concentrations of 0.5 *M* or higher, depending on the substrate. The polymerase of *H. cutirubrum* has been reported to have a molecular weight of only 36,000, being made of two identical subunits (Louis and Fitt, 1972a,b), but Chazan and Bayley (1973) found the molecular weight much higher, between 300,000 and 400,000. A more recent study, which involved purification of the DNA-dependent RNA polymerase of *H. halobium* (Zillig *et al.*, 1978), showed that it is an enzyme of 4–5 different subunits, with a combined molecular weight of 400,000–500,000, similar to the enzymes of nonhalophilic organisms. This enzyme has its maximum activity in 0.15 *M* KCl and has only 10% as much activity in 4 *M* KCl, a similar concentration to that in which it is likely to act in the cell. Still more recently, a polymerase was isolated that transcribes native DNA templates efficiently (Madon and Zillig, 1983).

On the other hand, Fitt (1978) and Louis and Fitt (1972a,b) have reported that a number of halobacterial enzymes that are involved in nucleic acid synthesis are smaller than their nonhalophilic counterparts. However, this does not necessarily hold for other enzymes of these organisms.

Most enzymes studied so far respond about equally to NaCl and KCl, the major salts outside and inside the cell, respectively. Others show substantially more activity in the presence of KCl than of NaCl, as would be appropriate for intracellular enzymes. The malate enzyme of *H. cutirubrum* has its highest activity in 1 *M* NH_4Cl, somewhat less activity in 3 *M* KCl, and no activity in NaCl. However, this enzyme was more susceptible to allosteric inhibitors, such as acetyl-CoA and NADH in 3 *M* KCl than in 1 *M* NH_4Cl (Cazzulo and Vidal, 1972).

Certain enzymes that require high concentrations of monovalent salts for activity can be activated by much lower concentrations of divalent salts, such as those of Mg^{2+}, Mn^{2+}, or Ca^{2+} or of polyvalent cations, such as spermidine or spermine (Lanyi, 1974). Examples include citrate synthase (Cazzulo, 1973) and aspartate carbamoyltransferase (Norberg *et al.*, 1973).

The highly purified alanine dehydrogenase of *Halobacterium salinarium* requires high salt concentrations for activity, but effects of specific ion effects may be quite different in the oxidative and reductive reactions catalyzed by this enzyme. Thus, KCl, NaCl, RbCl, and CsCl have about the same effect on the reductive amination of pyruvate, but KCl is more effective than the other salts with regard to the oxidative deamination of alanine by this enzyme, and no activity is observed in the presence of NaCl (Keradjopolous and Holldorf, 1979, 1980).

The pyruvate kinase of *Halobacterium cutirubrum* is more active in high KCl than in high NaCl concentrations. Moreover, the affinity of the enzyme for phosphoenol pyruvate is considerably higher in KCl than in NaCl (de Medicis *et al.*, 1982).

Though different enzymes of halobacteria will have varying responses to different salts, the overall picture is that normal internal salt concentrations are the ones that work best—either in terms of activity, affinity for substrates that are probably present in low concentrations in the cell, or regulatory properties. As will be seen below, this is also true for the complex of reactions that constitute protein synthesis. KCl seems eminently fitted to be the major "compatible solute" in these organisms.

4. MECHANISMS OF SALT-DEPENDENCE OF PROTEINS OF HALOBACTERIA

Overall, the proteins of halobacteria are more acidic than those of non-halophilic bacteria. This is true of the bulk proteins, the ribosomal proteins, and as will be seen, of several isolated enzymes. Baxter (1959) suggested earlier that salts support the activity of halophilic enzymes by virtue of their cations screening negatively charged groups and by preventing mutual repulsion by these charges distorting protein conformation. We might also expect that cations such as Mg^{2+}, Ca^{2+}, spermidine, and other polyamines, which have a much higher charge density than monovalent cations, would be proportionately more effective in supporting enzyme activity and/or stability. This proves true for a number of enzymes of extreme halophiles (Lanyi, 1974).

Though the charge-shielding action of salts must be very important, Lanyi (1974) pointed out that this action cannot account for the very high concentrations of salts that are required: the maximal charge shielding would be reached in about 0.1 M NaCl and in much lower concentrations of salts of divalent cations. The high salt concentrations that are actually required by enzymes of halobacteria seem due, in part, to their stabilizing action on hydrophobic bonds.

Lanyi (1974) noted that some purified enzymes of halobacteria, as well as bulk cytoplasmic and envelope proteins contained lower amounts of nonpolar amino acids than similar proteins from nonhalophilic bacteria. In some cases, the lower amounts of nonpolar residues was counterbalanced by higher amounts of the "borderline hydrophobic" amino acids (serine and threonine). Rao and Argos (1981) in an examination of the amino acid composition of seven halophilic and nonhalophilic proteins, pointed out that serine and threonine might be considered as completely hydrophobic amino acids when they are buried within the protein's interior. By this criterion, they found that the overall hydrophobicity of the halophilic and nonhalophilic proteins was basically the same. They also found that the bulkiness of amino acid residues in halophilic proteins was greatly reduced.

The work of Rao and Argus (1981) was concentrated on studies of a 2Fe-2S ferredoxin from the cyanobacterium *Spirulina platensis* and of a similar ferredoxin from *Halobacterium halobium* and *H. maris-mortui* (see also Werber *et al.*, 1978 for a discussion of the similarities between halophilic and eukaryotic ferredoxins). The halobacterial ferredoxins, whose sequences were very similar to each other, had a 98 residue sequence homologous to that of *S. platensis*. They also contained an additional 22 amino acid homologs at their N-termini and six more residues at the C-termini. The cyanobacterial ferredoxin's architecture has been determined by X-ray crystallography, and this knowledge was used to assign three-dimensional positions to the different amino acid residues of the halobacterial ferredoxin. A great increase of acidic charges was found at the ferredoxin surface. One of the major differences between nonhalophilic and halophilic ferredoxin was a substitution of glutamate for aspartate in the latter. This should greatly affect water-binding ability of ferredoxin, because glutamate may have the greatest water-binding ability of any amino acid residue (Rao and Angus, 1981). The results suggested that halophilic proteins can compete very effectively with cytoplasmic salts for water, through their external carboxyl groups of dicarboxylic amino acids.

The validity of this concept is also supported by studies on malate dehydrogenase and glutamate dehydrogenase of *Halobacterium maris-mortui* by Pundak and Eisenberg (1981) and Pundak *et al.* (1981). Large amounts of these enzymes were isolated by salting-out mediated chromatography (Leicht and Pundak, 1981), and the purified enzymes were studied by methods of sedimentation, diffusion, and circular dichroism. The enzymes were able to associate with unusually large amounts of water and salts (in this case, NaCl rather than KCl was studied, though the intracellular enzymes would presumably have to deal with the latter salt). The exclusion parameters, which measure "preferentially bound" water and NaCl gave water values of about 1 ml/g protein and NaCl values of about 0.3 g/g protein for both enzymes. In contrast, a nonhalophilic protein, bovine serum albumin, bound about 0.2 ml water and 0.01 g NaCl per gram. The high water- and salt-binding abilities were lost when malate dehydrogenase was denatured and unfolded at low NaCl concentrations. Thus, the high binding properties seem associated with the structure of the intact enzyme, and not only with its high content of dicarboxylic amino acids. That this malate dehydrogenase binds unusually large amounts of NaCl and water was confirmed more recently, through small-angle X-ray scattering studies by Reich *et al.* (1982).

Malate dehydrogenase was stabilized by salts with greater salting-out power, in common with other halophilic enzymes (Pundak *et al.*, 1981). Malate dehydrogenase was also stabilized by low temperatures. In this, it differed from other halophilic enzymes such as citrate synthase and threonine dehydratase, whose cold sensitivity has suggested that hydrophobic bonds are very important in their structure (Kushner, 1978). This enzyme may also be stabilized by immobilization on agarose (Koch-Schmidt *et al.*, 1979).

C. METABOLIC PATHWAYS

The first halobacteria isolated had complex growth requirements and were thought not to use carbohydrates (reviewed in Kushner, 1978). More recently, a number of isolates have been found that grow well on individual sugars as single carbon sources. Hochstein's group (Tomlinson *et al.*, 1974; Hochstein *et al.*, 1976; Hochstein, 1978) isolated a number of halobacteria that produced sugars from carbohydrates. *Halobacterium saccharovorum,* the species most studied, metabolizes glucose by a modified Entner–Doudoroff pathway and can produce galactonic acid from galactose and lactobionic acid from lactose. The current *H. maris-mortui* (Halobacterium of the Dead Sea) can also produce acid from sugars (Pundak and Eisenberg, 1981). Several halobacteria isolated by Spanish workers also metabolize sugars (Rodriguez-Valera *et al.*, 1980). Some details of the pathways of lipid synthesis are also known (Kates and Kushwaha, 1978). However, we do not have a general overview of metabolic pathways in the halobacteria.

D. REGULATION OF ENZYME ACTIVITY AND SYNTHESIS

Regulation of the activity of many enzymes depends on interactions between enzyme subunits and the effects of allosteric effectors and inhibitors on these interactions. The studies that have been carried out to date, on impure enzyme preparations, suggest that patterns of regulation of enzyme activity in the halobacteria are similar to those found in other bacteria. In the latter, it should be recalled, patterns of regulation of individual enzymes, e.g., aspartate kinases, can differ greatly between different species (Stanier *et al.*, 1976). Only, the effects of salts on enzyme regulation in the halobacteria are as striking and characteristic as their effects on enzyme activity.

Past work is described in some detail in Kushner (1978). Briefly, the aspartate carbamoyltransferase (ATCase) of *Halobacterium cutirubrum* is subject to feedback inhibition by CTP. Both activity and feedback inhibition depend on high salt concentrations (2.0 M). In fact, regulation (feedback inhibition) is somewhat more salt dependent than activity. Both NaCl and KCl are about equally effective in supporting activity, but regulation is slightly stronger in KCl than in NaCl. The behavior in relation to salt of the ATCase of *H. cutirubrum* is almost exactly opposite that of the same enzyme from the yeast, *Saccharomyces cerevisiae*. This enzyme, which also depends on a multisubunit structure for regulation, is inhibited by moderate NaCl concentrations (e.g., 0.5 M), and such salt concentrations inhibit feedback inhibition by UTP more strongly than enzyme activity.

Other enzymes of halobacteria subject to allosteric regulation include catabolic threonine deaminase (Lieberman and Lanyi, 1972), the malate enzyme (Vidal and Cazzulo, 1976), and citrate synthase (Cazzulo, 1973).

Few studies have been carried out on the factors regulating enzyme formation in halobacteria. The complex growth requirements of the most-studied halobacteria interfere with studies of induction and repression, especially of biosynthetic enzymes. Now that species are available that grow in minimal media (Rodriguez-Valera *et al.*, 1980), nutritional control of enzyme formation can be studied in much more detail and, hopefully, will be.

The studies of Hochstein (1978) with *Halobacterium saccharovorum* showed that the sugar on which these cells are grown can greatly influence the enzymes that are formed. During growth on glucose, a modified Entner–Doudoroff pathway was induced, but not during growth on galactose or on media without sugars. In these bacteria and in *H. salinarium* (Aitken and Brown, 1969), glucose dehydrogenase was formed only after growth in the presence of glucose. In *H. salinarium*, acetate induced isocitrate lyase formation (Aitken and Brown, 1969). Hochstein (1978) described other changes in enzyme content following growth on different sugars; among these, citrate and malate synthase activities were reduced in glucose-grown cells, whereas acetyl-CoA deacetylase activity was reduced in galactose-grown cells. Repression of citrate synthase by glucose has also been observed in nonhalophilic bacteria. Two forms of citrate synthase were found in *H. saccharovorum*, which differed in their stability in low salt concentrations.

Hartmann *et al.* (1980), in their study of anaerobic growth of halobacteria, using light energy or substrate level phosphorylation arising from arginine breakdown (see below), obtained results that suggested that the enzymes responsible for arginine degradation are induced under anaerobic growth conditions.

So far, then, enzyme formation in halobacteria seems subject to metabolic regulation, with both induction and repression taking place. Though the mechanisms involved are not known, there is no reason to suspect that they are fundamentally different from those found in nonhalophilic bacteria.

E. *In Vitro* Protein Synthesis

Much of the work on this subject is summarized by Kushner (1978) and Bayley and Morton (1978, 1979). The ribosomes of *Halobacterium cutirubrum,* and presumably other halobacteria and halococci, require high salt concentrations for stability—salt concentrations that analyses have shown to be present inside the cell. To maintain stable 70 S, 50 S, and 30 S forms, about 3 M KCl and 0.1 Mg^{2+} are needed. Potassium chloride cannot be replaced by NaCl—ribosomes aggregate in high NaCl concentrations. At low concentrations of salt, e.g., 0.001 M MgCl$_2$, which maintains *E. coli* ribosome structure, most of the proteins are lost from the subunits of the halobacterial ribosomes. Most of the proteins of these ribosomes are acidic. The chemistry of some of them and their relations with other archaebacterial ribosomes are described in Yaguchi *et al.* (1982) and in much more detail elsewhere in this book.

Optimal *in vitro* protein synthesis takes place in the presence of 3.8 M KCl, 1.0 M NaCl, 0.4 M NH$_4$Cl, and 0.04 M Mg^{2+}. A number of halobacterial aminoacyl-tRNA synthetases function best in 3.8 M KCl. However, the tRNAs themselves do not require a high salt concentration. Those from *E. coli* can be used in the *Halobacterium cutirubrum* system, and vice versa.

Studies with artificial messenger RNAs showed that codon assignments for the *Halobacterium cutirubrum* system were the same as in nonhalophiles. Synthesis of proteins is initiated by methionine, not *N*-formylmethionine. The initiating tRNA^{Met} and other tRNAs have a pseudouridine or modified pseudouridine instead of ribothymidine in the "common arm" (see Chapter 6 by Gupta). Some of these characteristics resemble the protein synthesizing systems in eukaryotic rather than prokaryotic cells, as does the fact that the elongation factor EF-2 may be ADP-ribosylated by diphtheria toxin (Kessel and Klink, 1980, 1981; see Chapter 8 by Klink). Together with rRNA sequences (Fox *et al.,* 1980; Gupta *et al.,* 1983), they place the halobacteria among the archaebacteria, a subject discussed much more fully elsewhere in this book (see Chapter 5 by Fox).

F. Action of Antibacterial Substances

Early studies showed that halobacteria and halococci were insensitive to standard antibiotics that affected protein synthesis: chloramphenicol, streptomycin, erythromycin, and tetracycline. They were also insensitive to penicillin and related compounds, as would be expected from their lack of peptidoglycan (Kushner, 1978). Since then, tests of other halobacteria with antibiotic disks (Colwell *et al.,* 1979; Soliman and Trüper, 1982; Hilpert *et al.,* 1981) and tube dilutions (Pecher and Böck, 1981) showed that, depending on the strain used and the test employed, chloramphenicol and tetracycline were sometimes inhibitory. Aminoglycosides (e.g., streptomycin) and many inhibitors of protein synthesis that act on 70 S and 80 S ribosomes in other cells were inactive, as was penicillin. Anisomycin, which acts on 80 S ribosomes, was especially active, as was rifampicin, which acts on RNA polymerase (Pecher and Böck, 1981). Novobiocin, a strong magnesium antagonist, was much more active against alkaliphilic than nonalkaliphilic halophiles (Soliman and Trüper, 1982.) A new substance, haloquinone, produced by a *Streptomyces* species was strongly inhibitory against halobacteria (Ewersmeyer-Wenk *et al.*, 1981).

Bacitracin, which inhibits lipid biosynthesis, is active against halobacteria and halococci (Basinger and Oliver, 1979; Hunter and Millar, 1980). Cerulenin, which inhibits fatty acid synthesis, is also effective against *Halobacterium cutirubrum* (Dees and Oliver, 1977), though only traces of fatty acids are found in this organism (see Chapter 12 by Böck and Kandler).

Rodriguez-Valera *et al.* (1982b) found that 39 strains of halobacteria from a Spanish saltern produced substances that inhibited the growth of other halobacteria, especially those from culture collections, but not halococci or moderate halophiles. These appeared to be bacteriocins, the first demonstrated in the archaebacteria. They were called "halocins." That produced by strain R4 (now *Halobacterium mediterranei*; Rodriguez-Valera *et al.*, 1983b) was studied, using *H. halobium* as an indicator bacterium. It is a high molecular weight substance, not a virus, that is destroyed by heat or exposure to low salt concentrations.

G. CELL ENVELOPES AND CELL MEMBRANES OF HALOBACTERIA

The cytoplasm of *Halobacterium halobium* and closely related halobacteria is surrounded by two layers, a cytoplasmic membrane and an external layer that is about 17.5 mm thick, composed of a hexagonal array of glycoprotein particles, which is sometimes called the "cell wall" and which contains no muramic acid, diaminopimelic acid, D-amino acids, or teichoic acids (Kushner, 1978; Mescher and Strominger, 1978; Robertson *et al.*, 1982; Stoeckenius *et al.*, 1979). The presence of glycoproteins on the cell surface is of special interest, because such glycoproteins are common in eukaryotic cells but little (or not at all) known in prokaryotic cells. (See Chapter 2 by Zillig.)

H. CYTOPLASMIC MEMBRANE AND PURPLE MEMBRANE OF HALOBACTERIA

The physiology of halobacterial membranes would naturally be of interest because of the very high external NaCl and internal KCl concentrations at which they function. They require NaCl for structural stability and for active transport of a number of substances. They contain characteristic lipids, diphytanyl ether analogs of phosphatidylglycerol and of a glycolipid sulfate, as well as the red C_{50} carotenoid pigments (bacterioruberins), squalenes, phytoene, lycopene, β-carotene, vitamin MK8, and the visual pigment retinal (reviewed in Kates, 1978; Kates and Kushwaha, 1978; Kushner, 1978). When first described, the ether-linked lipids seemed unique to the biological world; now, of course, they are characteristic of other archaebacteria (see Chapter 10 by Langworthy). Another characteristic of the membranes of halobacteria is their rigidity. Only in a narrow central portion of the bilayer is the environment flexible. This seemed due to the high protein content of the membranes and to strong protein–lipid interactions within the bilayer (reviewed in Kushner, 1978).

By far, the greatest interest in the physiology of membranes of halobacteria has been directed towards a component that, paradoxically, does *not* require high salts for activity and stability: the purple membrane system and its protein, bacteriorhodopsin. The statement by Stoeckenius and Bogomolni (1982) that "bacteriorhodopsin obviously is the simplest and best understood of all known active transport systems" does not seem exaggerated. This system gave us the clearest and most unequivocal evidence supporting Mitchell's (1972) chemiosmotic hypothesis. In recent years, it has been the major focus of interest for research on the halobacteria. More than one-half the published papers deal with the bacteriorhodopsin system, or related systems, from several very active laboratories. Excellent review articles (Stoeckenius, 1976; Lanyi, 1978; Stoeckenius *et al.*, 1979; Stoeckenius and Bogomolni, 1982) and a book (Packer, 1982) give a detailed account of this system's history, architecture, and molecular details. Only a brief account can be given here.

The purple membrane was first identified as a fraction that remained after cells or isolated envelopes of *Halobacterium halobium* were disaggregated by exposure to distilled water; the other fraction, the "red membrane," represents the disaggregated membrane of the rest of the cell. Gas vacuoles of halobacteria also survive exposure to distilled water (Kushner, 1978). In purple membrane studies, to avoid contamination with gas vacuoles, workers commonly use nonvacuolated strains of *H. halobium*.

Isolated purple membranes contain about 25% lipid and 75% protein. All of the cell's sulfate-containing lipids are found in the purple membrane fraction; they account for about 15% of the total lipid. The other lipids, found in the same proportion, are the same as in the rest of the cell's membrane, i.e., the red membrane (Kushwaha *et al.*, 1975; Stoeckenius *et al.*, 1979). The purple membrane contains only one protein of molecular weight 26,000, which has retinal as a prosthetic group and is called bacteriorhodopsin. In cells and in isolated membranes, this protein aggregates to form crystalline patches (Stoeckenius and Bogomolni, 1982), which have a distinctive appearance in freeze-etch electron micrographs of the cell. They are organized in a two-dimensional hexagonal lattice of trimers that are surrounded by lipid molecules. Their orderly crystalline organization is unique among membrane proteins (Stoeckenius *et al.*, 1979).

The importance of the purple membrane was first realized through the observation of Danon and Stoeckenius (1974) that when *Halobacterium halobium* was suspended under anaerobic conditions in the dark the intracellular ATP contents dropped rapidly. If the cells were illuminated by light of wavelength absorbed by the purple membrane or if oxygen were admitted to the suspension, the ATP level rose again to the initial value or higher. Uncouplers or agents that block membrane ATPase prevented both the light and the oxygen effect; inhibitors of either the respiratory transport chain or that found in other photosynthetic organisms (KCN and DCCD) blocked only the oxygen effect.

It was later determined that both light and oxygen induced ATP synthesis by causing an ejection of protons from the cell. The return of protons catalyzes ATP formation, following the mechanism outlined in Mitchell's (1972) chemiosmotic theory. Indeed, the purple membrane has provided the clearest evidence for this hypothesis.

It was also observed that illumination of these cells inhibited respiration. Because both processes act by generating proton gradients, it is not surprising that they counteract each other (Oesterhelt and Krippahl, 1973). The mechanisms for this antagonism are not entirely clear. One possibility is that the proton gradient caused by light reverses electron transport through the electron transport chain. This is supported by the finding that CO_2 incorporation by *H. halobium* is greatly stimulated by light (Danon and Caplan, 1979).

Proton ejection caused either by electron flow through the electron transfer chain or by illuminating the purple membrane may have other consequences than ATP synthesis. The proton motive force set up is involved in maintaining the ion balance of the cells. By a sodium–proton antiport mechanism, Na^+ ions are ejected from the cell. To maintain electroneutrality, K^+ ions enter, and these changes seem responsible for the very high K^+ contents within the halobacterial cell (Lanyi, 1978). The Na^+ gradient that is set up ($Na_{in}^+ < Na_{out}^+$) serves to drive the transport of amino acids, almost all of which depend on a Na^+ symport mechanism.

Transport has been studied in whole cells and also in vesicles prepared by mechanically disrupting cells. These vesicles are usually inside-in, i.e., the orientation of the membrane is the same as in the intact cells (Lanyi, 1978), but inside-out vesicles can be prepared by sonication and sucrose density fractionation. The sidedness of the vesicle preparation can be determined by measuring the activity of NADH-menadione reductase before and after treatment with the detergent Triton X-100. The enzyme, which is located inside the cytoplasmic membrane, is accessible to NADH only after detergent treatment in intact cells or vesicles in the inside-in configuration, but does not require this treatment for activity in inside-out vesicles (Lanyi, 1978; Garty *et al.*, 1980). Inverted vesicles transport protons from the outside in following illumination.

Purple membrane sheets can be incorporated into planar lipid films (''black lipid membranes'') or into liposomes, which are enclosed vesicles made from phospholipids extracted from other biological material. In the latter, it often is situated in the inside-out configuration in respect to the external medium (Lanyi, 1978). The fact that proton pumping is observed with other than halobacterial lipids shows that the protein, not the lipid, fraction of the purple membrane has the special properties needed for the pumping. Though the halobacterial lipids are quite acidic, neutral or basic lipids and even *n*-octane can serve as support for bacteriorhodopsin in active model systems (Stoeckenius *et al.*, 1979). Detailed methods for preparing different artificial membranes are given in the different sections of Packer (1982). The demonstration by Racker and Stoeckenius (1974) that purple membranes (incorporated together with mitochondrial ATPase in liposomes) could, on illumination, catalyze ATP synthesis showed clearly that an electrochemical proton gradient could yield stored energy. The purple membrane in vesicles, liposomes, or planar membranes has been used to study other fundamental aspects of membrane physiology, as outlined in much more detail in the reviews already cited.

1. MOLECULAR ARCHITECTURE OF BACTERIORHODOPSIN AND THE PURPLE MEMBRANE

The polypeptide chain of bacteriorhodopsin is composed of 248 amino acids, 70% of which are hydrophobic. They are arranged in seven α-helical segments, clustered together in a rough crescent. The segments cross the purple membrane, roughly perpendicular to its surface, and are linked by much shorter segments. Three to six N-terminal amino acids are on the exterior surface of the membrane and 17–24 C-terminal amino acids are on the cytoplasmic surface (Stoeckenius and Bogomolni, 1982). The sequence of bacteriorhodopsin is known, as is the base sequence of the bacteriorhodopsin gene (Dunn *et al.*, 1981). This gene codes for a precursor sequence, which has 13 additional amino acids at the N-terminal and one, an aspartate residue, at the C-terminal.

The pigment, retinal, is attached to the ϵ-amino group of lysine at position 216 of bacteriorhodopsin, by means of a Schiff base between its C-terminal and the ϵ-amino group of lysine (retinal-C $=$ N-lysine). The base can exist in an unprotonated form, as shown, or in a protonated form:

$$\text{Retinal-C} = \text{N}_\text{H}^+\text{-lysine}$$

Both forms seem involved in the cyclic events that lead to proton translocation by bacteriorhodopsin. The chains of bacteriorhodopsin are so folded that the retinal appears to be about one-third of the distance from the cytoplasm to the external surface in the last α-helical segment that crosses the purple membrane (Stoeckenius and Bogomolni, 1982; Ovchinnikov, 1982).

Interestingly, the arrangement of the rhodopsin molecule with its attached retinal in photoreceptor membranes of animal cells is somewhat similar to that of the bacteriorhodopsin. Again, there is a pattern of seven α-helices crossing the membrane, with retinal attached to a lysine residue near the center of the last helix. Comparison of the amino acid sequences, however, shows no evidence of strong amino acid homology (Ovchinnikov, 1982). It would be intriguing to know which organism first devised this arrangement, the archaebacterium or the eukaryote.

2. LIGHT-INDUCED CHANGES IN BACTERIORHODOPSIN ABSORPTION SPECTRUM AND THE PHOTOCHEMICAL CYCLE

In the light, the visible absorption peak of bacteriorhodopsin is 568 nm. In the dark, this shifts, slowly and reversibly, to 560 nm. These changes are due to the isomerization of retinal: the dark-adapted form contains an equal mixture of 13-*cis*- and all-*trans*-retinals. In the light, all retinal is in the *trans* form. Under the influence of light, an intermediate, known as the "M" intermediate, is formed, which absorbs maximally at 412 nm. This contains 13-*cis*-retinal. There are a number of other intermediate steps in the photocycle, which lasts about 10 msec. Protons are taken up on the cytoplasmic side of the purple membrane and released on the exterior. Two protons are released per "turn" of the cycle (Stoeckenius and Bogomolni, 1982).

Much work has been done on this cycle, (see reviews cited above for details). Workers are not in agreement on all details of the cycle and the intermediates involved. It seems established that a brief deprotonation of the Schiff base takes place during the cycle, as well as changes in protein conformation. Studies with purple membranes from which retinal has been removed and replaced by retinal analogs (removal is by hydroxylamine treatment or by growth in the presence of nicotine, which prevents retinal and carotenoid synthesis (Kushwaha and Kates, 1979a) confirm that the Schiff base is essential in proton pumping activity (Stoeckenius *et al.,* 1979).

The precise way in which protons are pumped through the purple membrane is still a matter for active investigation and discussion. Proteins may provide excellent means for proton conduction by setting up continuous chains of hydrogen bonds on the amino acid side groups through which protons can pass, the so-called "proton wires" (Nagle and Morowitz, 1978; Nagle and Mille, 1981). The extent to which longer proton pathways, or a number of shorter ones, are involved in the bacteriorhodopsin proton translocation, as well as the means by which the direction of this translocation is determined, is still unclear (Stoeckenius and Bogomolni, 1982).

3. The Halorhodopsin System

Although the action of bacteriorhodopsin was well established as leading to H^+ extrusion from the cell, many investigators noticed that the acidification of the external medium caused by this extrusion was often preceded by alkalization. This was explained, around 1979, as being due to a second light-activated transport system, halorhodopsin, which was considered to be a pump for the direct extrusion of Na^+ ions. This could be distinguished from the Na^+ extrusion that follows bacteriorhodopsin action and is due to a Na^+/H^+ antiport following H^+ extrusion. The indirect Na^+ pumping action by bacteriorhodopsin action was sensitive to uncouplers such as DCCD; the direct Na^+ pumping action of halorhodopsin was not (reviewed in Stoeckenius and Bogomolni, 1982; Lanyi, 1981).

Studies of halorhodopsin were made difficult by the simultaneous presence of bacteriorhodopsin. Some of the properties of the former pigment were studied in mutants of *Halobacterium halobium* that lacked the ability to make bacteriorhodopsin. Halorhodopsin has a maximal absorption at 588 nm; it also passes through intermediate forms on illumination, leading to a cycle that results in Na^+ extrusion. This causes ATP synthesis, probably as a result of increased membrane potential (Wagner *et al.*, 1981), but halorhodopsin alone is not able to support anaerobic growth.

The apoprotein of halorhodopsin has now been isolated (Lanyi and Oesterhelt, 1982). It is distinct from the apoprotein of bacteriorhodopsin and has a slightly lower molecular weight (25,000).

Probably the most striking finding about halorhodopsin is the discovery by Schobert and Lanyi (1982) that it is not an outward-directed Na^+ pump at all, but rather an inward-directed Cl^- pump. Chloride uptake is accompanied by passive H^+ entry and Na^+ extrusion by Na^+/H^+ antiport action. Further recent studies of effects of Cl^- and H^+ on the photochemical cycling of halorhodopsin have been reported (Lanyi and Schobert, 1983).

The biological role of halorhodopsin is still unknown. It does not make a very large energetic contribution to the cell, and as stated, it cannot sustain anaerobic growth (Wagner *et al.*, 1981). However, it may be involved in phototaxis of *Halobacterium halobium*.

I. Phototaxis and Chemotaxis

The halobacteria are proving to be interesting microorganisms for studying these aspects of behavioral physiology. In fact, phototaxis, in which the bacteriorhodopsin system or closely related systems are involved, was discovered before the chemical nature of the purple membrane was worked out (Stoeckenius *et al.*, 1979). Some aspects of work on phototaxis are reviewed by Stoeckenius and Bogomolni (1982).

Halobacterium halobium responds in different ways to changes in light intensity. Cells that are under no chemical or light gradient swim at about 2.5 μm/sec at room temperature and reverse their direction of swimming about every 10 sec. If UV light (maximal effect at 370 nm) is increased or visible light (maximal effect at 565 nm) is decreased, reversal responses occur more frequently, every 2–5 sec. The converse changes, decrease in UV or increase in visible light intensities, lead to a lowering of the reversal frequencies, so that the cells can swim in straight "runs" for up to 50 sec. As a result of these responses, cells swim toward green-yellow light and away from blue light. The sensory photosystems have been called PS 370 and PS 565. The responses to these systems have sometimes been called the "blue" and "red" responses, respectively (Hildebrand and Schimz, 1983; Stoeckenius and Bogomolni, 1982; Traulich *et al.*, 1983).

The absorption spectrum of PS 565 is close to that of bacteriorhodopsin (568 nm). Bacteriorhodopsin, or bacteriorhodopsin-like pigments, certainly seem involved in the phototactic response, because both PS 370 and PS 565 responses can be abolished by growing cells in nicotine, thus preventing retinal synthesis. Adding retinal restores these responses (Stoeckenius *et al.*, 1979). Recent studies with bacteriorhodopsin-deficient mutants of *H. halobium* showed that bacteriorhodopsin itself is not indispensible for photosensory behavior (Hildebrand and Schimz, 1983). It is possible that halorhodopsin is implicated in photosensory behavior, but it does not seem to be a main photosensory pigment (Traulich *et al.*, 1983).

Studies by Baryshev *et al.* (1983) suggested that membrane potential, as affected by the Na^+/K^+ gradient and other energetic determinants, may affect green-yellow light taxis (PS 565), whereas blue light taxis (PS 370) is governed by a specific photoreceptor.

Halobacterium halobium also shows chemotactic responses, as do many other bacteria. These responses mediate behavior toward both attractive and repellant substances (Schimz and Hildebrand, 1979). These systems seem to share intermediate mechanisms with the phototactic systems. Schimz and Hildebrand (1979) found that formation of the PS systems interfered with the formation of systems mediating positive chemotaxis to both glucose and histidine. Methylation, presumably of methyl-accepting chemotaxis proteins (MCPs), accompanied both chemotaxis and green-light (PS 565) taxis in *H. halobium* (Schimz, 1981, 1982; Baryshev *et al.*, 1982.)

J. PHOSPHORYLATION OF PROTEINS

When cells of *Halobacterium halobium* that contain bacteriorhodopsin are exposed to light, there is a rapid dephosphorylation of a membrane phosphoprotein and of two other phosphoproteins. These proteins are rephosphorylated in darkness (Spudich and Stoeckenius, 1980; Spudich and Spudich, 1980, 1981). These dephosphorylations and phosphorylations appear to be mediated by the pmf changes brought about by illumination (Spudich and Spudich, 1981). Spudich and Stoeckenius (1980) pointed out that regulation in response to environmental stimuli (e.g., to hormones or light) is often regulated by protein phosphorylation in eukaryotes, but that it is also regulated in certain prokaryotes, such as *Escherichia coli* and *Salmonella typhi*.

K. Biosynthesis of Bacteriorhodopsin and the Purple Membrane

Some aspects of this subject were reviewed by Stoeckenius *et al.* (1979), Sumper and Herrmann (1978), and Sumper (1982). Retinal is formed by the oxidative cleavage of β-carotene. *Halobacterium halobium* grown under high aeration or in the presence of nicotine forms a "brown membrane," which contains bacterioopsin and which may also contain small amounts of bacteriorhodopsin, possibly in the monomer form. The brown membrane has been shown to be the precursor of the purple membrane. The former quickly binds retinal to form bacteriorhodopsin, and under suitable conditions, purple membrane is formed. Low O_2 concentrations stimulate both bacterioopsin and bacteriorhodopsin synthesis. The level of bacterioopsin synthesis, in turn, has some control on retinal synthesis, so that the two are coordinated.

The brown membrane also contains cytochrome *b*-561, in a 1:3 ratio with bacterioopsin. The cytochrome has not been found in the mature purple membrane, and it has been suggested that it may be located around the periphery of the crystalline domain of the bacteriorhodopsin molecules (Papadopoulos and Cassim, 1981).

Mukohata *et al.* (1981) have described a "white membrane" isolated from a strain of *H. halobium* that lacks the ability to form any isoprenoid pigments. The white membrane seems mainly composed of bacterioopsin, which is readily converted to bacteriorhodopsin by retinal addition. In this strain, bacterioopsin synthesis was not depressed by the absence of retinal. Mukohata *et al.* (1981) suggested that the genetic lesion that prevents retinal synthesis may also remove feedback control of bacterioopsin synthesis. It would seem that all determinants of bacteriorhodopsin synthesis and assembly are still not understood.

L. Occurrence and Ecological Importance of Bacteriorhodopsin and Other Pigments

The purple membrane is not formed under all conditions of growth but appears in cells grown under conditions of limited oxygen. It cannot, however, be formed under completely anaerobic conditions, because O_2 is needed for the formation of retinal from β-carotene (Hartmann et al., 1980). Rogers and Morris (1978) found that, in continuous cultures of Halobacterium halobium, bacteriorhodopsin was formed only when a low growth rate (low dilution rate) accompanied low O_2 tension.

Bacteriorhodopsin has been found in many species of halobacteria, though not in all species. As Pfeifer et al. (1981a) pointed out, the ability to make bacteriorhodopsin is so genetically unstable that it is a doubtful taxonomic trait. Kushwaha et al. (1974) found retinal, and bacteriorhodopsin seem confined to the Halobacteriaceae.

Although no one doubts that bacteriorhodopsin is a fascinating biochemical system, some microbiologists have questioned its importance under natural growth conditions. Is it merely a "laboratory curiosity"? Some of these doubts are now being resolved in favor of the view that bacteriorhodopsin may be as important to the bacteria as to the biochemist.

First, substantial amounts of bacteriorhodopsin may be formed under natural conditions. Oren and Shilo (1981) showed that a dense bloom of red halobacteria in the Dead Sea, at a depth of 10 m, contained large amounts of bacteriorhodopsin, up to 0.4 nmol/mg protein, an amount comparable to that found in laboratory cultures of H. halobium.

Furthermore, bacteriorhodopsin can mediate other effects of visible light on halobacteria, in addition to its possible role in phototaxis. It has long been known that visible light may have a harmful effect on these organisms. This is especially striking in nonpigmented halobacteria, which lack the C_{50} pigments, bacterioruberins, which are responsible for the red color (Kushner, 1978; Kushwaha and Kates, 1979b). The effect of visible light, presumably acting through photosensitizers, may be either inhibitory to growth (Dundas and Larsen, 1962) or lethal (Rodriguez-Valera et al. 1982a).

Kushwaha et al. (1982) have pointed out that different halobacterial species respond differently to salt concentrations, as regards pigmentation. Thus, Halobacterium R4 (H. mediterranei) produces bacterioruberins at its lower salt concentration range, and H. cutirubrum at its higher concentration range. For example, in increasing salt concentration in an evaporating solar saltern, the first microorganisms would be pigmented first and protected from sunlight. They would be expected to predominate, and later to be replaced by bacteria that behave as H. cutirubrum does. By selecting "red halophiles" as those that make red colonies in 25% salt on a complex medium, many microbiologists have, perhaps unwittingly, selected for the H. salinarium group (e.g., Colwell et al., 1979).

Brock and Petersen (1976) found that starving halobacteria that contained bacteriorhodopsin died more rapidly in the light than in the dark under aerobic conditions but that under anaerobic conditions, the reverse was true.

Such experiments suggest that the toxic photodynamic effects of light and oxygen can be counteracted to some extent by bacteriorhodopsin. They are, of course, also counteracted by the bacterioruberin pigments of normal cells, just as bacterial pigments counteract many other photodynamic effects (Stanier et al., 1976).

Because halobacteria are subject to a good deal of UV irradiation from sunlight, it is probably important, in their natural conditions, that they have very active photoreaction systems for DNA repair. Curiously, they may completely lack dark repair (Fitt et al., 1983.)

Light seems to serve as an energy source for growth, through the bacteriorhodopsin system. Hartmann et al. (1980) showed that Halobacterium halobium could grow under completely anaerobic conditions (with retinal added to the medium), using the energy from light and bacteriorhodopsin. In such experiments, it is necessary to add retinal, which is not produced by these bacteria anaerobically. Halobacterium halobium can also grow anaerobically using substrate-level phosphorylation from arginine (Dundas, 1977), but substantial amounts of arginine are probably rarer in nature than is light. Both Rogers and Morris (1978) and Rodriguez-Valera et al. (1983a) found that in continuous culture, light permitted halobacteria to reach higher growth rates and population densities. The latter authors also found some putative bacteriorhodopsin-constitutive mutants, which are now being further studied (F. Rodriguez-Valera, personal communication).

VI. Genetics of Halobacteria

The Halobacteriaceae have begun to present us with as many surprises in their genetic properties as in their physiological and biochemical ones. It has long been known that both halobacteria and halococci contain large amounts of extrachromosomal DNA, called "satellite DNA." For many species the chromosomal DNA has a G + C content of 66–68%, whereas that of the satellite DNA is 57–60%. The satellite DNA can account for 11–36% of the total DNA in some species (Moore and McCarthy, 1969a,b). More recent studies have suggested that the extrachromosomal DNA may consist of plasmids, though all of the functions of such plasmids cannot yet be defined. Simon (1978) found that the presence of a plasmid was correlated with that of gas vacuoles in a Halobacterium species. In the absence of the plasmid, defective gas vesicles with different protein structure and organization were formed (Simon, 1981). Weidinger and colleagues (Weidinger et al., 1979; Pfeifer et al., 1981a,b) found a large (95 megadalton) plasmid, pHHl, in H. halobium that contains genes coding for gas vacuoles and probably bacteriorhodopsin and bacterioruberin. Four to five copies of this plasmid were present in H. halobium, in which the satellite DNA had the same restriction pattern when studied with various nucleases, as were present in plasmid pHHl (Pfeifer *et al.*, 1981b). Similar plasmids to pHHl (e.g., pHCl in *H. cutirubrum*) have been found, and restriction maps, which were made of several of these, show that, though not identical, their maps are similar and contain a number of homologous regions (Weidinger *et al.*, 1982).

Other species of halobacteria lack plasmid pHHl or other plasmids, but they have the same gene functions, presumably in the chromosome that contains sequences homologous with the pHHl plasmid. When the plasmid-containing cells lose the genes in question, they do not lose the plasmids; rather, the changes are thought to arise from insertions within the plasmid, leading to genetic change (Pfeifer *et al.*, 1981a,b).

The spontaneous rate of genetic variation in *Halobacterium halobium* and probably in other halobacteria can be extraordinarily high. Kushwaha *et al.* (1980) found that colorless mutants of *H. halobium* arose with a frequency of about 10^{-3}. One of these mutants formed all *trans* phytoene, which appears to be a common intermediate in bacterioruberin and retinal synthesis. Mutants deficient in gas–vacuole formation arise at a frequency of 10^{-2}; those deficient in bacterioruberin, bacteriorhodopsin, and retinal, at a frequency of about 10^{-4} (Pfeifer *et al.*, 1981b). Reversions were also observed at somewhat lower frequencies. Restriction endonuclease digestion showed that most genetic changes are accompanied by alterations in plasmid pHH1 (Pfeifer *et al.*, 1981b). Another mutant type was observed that showed extremely high variation in the above types of mutation; as a result, sectored colonies often appeared.

Pfeifer *et al.* (1981b) suggested that spontaneous mutations arose by insertions in chromosomal and plasmid DNA; they considered the possibility that there might be "hot spots" in the plasmids, so that one insertion could trigger other insertions.

The frequency of genetic alterations suggested to Sapienza and Doolittle (1982) that "the *Halobacterium* genome may contain many transposable elements or regions of sequence homology promoting recombination in and between chromosomal and plasmid DNAs." In support of this, these workers showed that the genomes of *Halobacterium halobium* and *H. volcanii* contain many repeated sequences. About 50 families of such sequences were found, each of which could have up to 20 or more members. The authors pointed out that this is not necessarily typical of all archaebacteria, since they did not find such repeated sequences in a thermoacidophile. However, repeated sequences are, of course, common in plant and animal cells. The halobacteria, thus far, seem to have the most complex repeated sequences in any prokaryote.

A. Isolation of Mutants

Mutations that lead to gas vacuole formation, pigmentation, or purple membrane formation are usually recognized by changed colonial appearance. More recently, techniques for carrying out spectroscopic measurements on single colonies have been devised (Weber and Bogomolni, 1982). Usually the techniques of mutant isolation are nonselective, but, as previously mentioned, the normal frequency of these mutations makes it relatively easy to find them. Spudich and Spudich (1982) described a method of selecting for mutants that are deficient in generating pmf, especially by the bacteriorhodopsin and halorhodopsin systems. It uses "light shock" in low external pH (5.5) and in the presence of ionophores, so that cells with normal ion fluxes suffer acidification of the cytoplasm. Using these techniques, they were able to isolate, among others, halorhodopsin mutants that were not retinal mutants.

To my knowledge, few other mutants have been isolated. Because the most-studied halobacteria have complex growth requirements—and should not be susceptible to penicillin selection—it is not easy to isolate auxotrophic mutants. Some success in obtaining such mutants may, however, be expected with the newly isolated strains that can grow with glucose or other single-carbon sources.

Because the halobacteria are usually not affected by the antibiotics that act on peptidoglycan synthesis or on eubacterial ribosomes, these substances probably cannot be used to isolate mutants with altered cell envelopes or ribosomal proteins. So far, mutants resistant to those antibiotics to which halobacteria are susceptible do not seem to have been isolated, perhaps for lack of trying. As will be seen below, a phage-resistant strain of *H. halobium* has been isolated (Schnabel *et al.*, 1982a), and we might reasonably expect the discovery of other phages and phage-resistant strains in the future. This might provide means of isolating mutants with altered surfaces.

The great potential for genetic variability naturally poses some question of halobacterial classification (Pfeifer *et al.*, 1981b; Weber and Bogomolni, 1982). As these workers have pointed out, it does not seem advisable to use such highly variable traits as gas vacuole or purple membrane formation to distinguish between species.

However, all traits are not necessarily so variable. One property that seems fixed is salt dependence. To my knowledge, this has never been seen to change, in either direction, in any halobacterial species.

It is also striking that only moderate variation seems to occur in natural populations of halobacteria. When Colwell *et al.* (1979) compared many species of red halophilic rods from culture collections and freshly isolated from salterns, they found that all halobacteria belonged to two closely related species (possibly considered as only one species now, see Section IV,A on classification). It is true that these isolates were chosen by their ability to form red colonies in high salt concentrations, while growing on complex media. Their isolation technique might exclude such bacteria as *Halobacterium mediterranei,* which grows on a simpler medium and which is red-pigmented at lower but not higher NaCl concentration. More recent studies such as those on halophiles isolated by Spanish workers and on the salterns of California and Baja California, using simpler isolation media have revealed newer halobacterial species, but there is no hint of an infinite variability.

B. Phages

A number of double-stranded DNA phages have been described in halobacteria (Torsvik and Dundas, 1974, 1980; Wais *et al.*, 1975; Schnabel *et al.*, 1982a,b). All these phages have isometric heads and tails; all require high salt concentrations for stability. Phage HsI of *Halobacterium salinarium* has a latent period, which may be much longer than the generation time of the host, especially at higher NaCl concentrations (25%). At this concentration, a carrier state is established, in which lysis of the infected bacterium is delayed for several generations. Lowering the NaCl concentration to 17% slows growth and leads to lytic development with phage production (Torsvik and Dundas, 1980).

Schnabel *et al.* (1982a) studied the organization of the linear ds DNA of their phage, φH of *H. halobium* by restriction cleavage and hybridization techniques. A packaging model of the DNA was proposed. A good deal of homology was found between regions of phage and chromosomal DNA, and some evidence was found for a lysogenic relationship. As might be expected from genetic studies already cited, the genome of this phage seems extremely variable. A number of phage variants were found, which differed by insertions, a deletion, and an inversion (Schnabel *et al.*, 1982b).

VII. Some Notes on the Halococci

The halococci have been much less studied than the halobacteria, probably for a very practical reason: they are much more refractile as experimental material. Although the halobacteria disintegrate extensively in distilled water, often leaving behind almost pure preparations of purple membranes and/or gas vacuoles, the halococci possess some of the toughest envelopes known in bacteria. They are very difficult to break, though they can be broken by shaking with glass beads (Forsyth, 1971; Schleifer *et al.*, 1982.)

Analyses of lipids of halococci show that these are similar to those of halobacteria; furthermore, retinal has been detected in one *Halococcus* strain, (Kates *et al.*, 1966; Kushwaha *et al.*, 1974). Purple membranes, however, have not been isolated from these bacteria, and it might be difficult to do so.

Moore and McCarthy (1969b) showed by DNA/DNA and DNA/rRNA hybridization that *Halococcus morrhuae* was related to strains of halobacteria, though not as closely as the halobacteria were related to each other (see also this volume, Chapter 5). A similar conclusion was reached by ribonuclease T_1 fingerprinting of 5 S rRNAs (Nicholson and Fox, 1983).

The rigid cell envelope of halococci is not based on a peptidoglycan structure. Brown and Cho (1970) first showed that the walls lack muramic acid. The shape-determining part of the wall is mainly polysaccharide. In the electron microscope, it consists of a single homogenous layer, 50–60 nm thick (Kocur *et al.*, 1972). Several studies of its chemical composition and attempts to determine its chemical architecture have been reported (Reistad, 1972, 1974; Steber and Schleifer, 1975, 1979; Schleifer *et al.*, 1982.) The major polymer of *H. morrhuae* strain CCM 859 contains neutral sugars, amino sugars, uronic acids, including gulosoaminuronic acid, acetate, glycine, and sulfate. A branched structure has been proposed, with the suggestion that glycine residues may play a role in connecting glycan strands through peptidic linkages between amino groups of glucosamine and the carboxyl groups of uronic acids or gulosaminuronic acid (Schleifer *et al.*, 1982.)

It has long been known that halococci do not lyse in the absence of high salt concentrations and can survive exposure to much lower salt concentrations than haloacteria (Brown and Cho, 1970). Rodriguez-Valera *et al.* (1979) showed, in fact, that living halococci could be isolated from seawater, in which halobacteria would certainly die. Since then, Rodriguez-Valera *et al.* (1982c) have shown that in 3% salts (of the composition of seawater) respiration of a *Halococcus* species was unchanged and amino acid uptake continued at about one-half the normal rate, though the K^+ content fell tenfold. The ability of halococci to survive much lower salt concentrations should give them a competitive advantage over the halobacteria in natural conditions of changing salinity.

VIII. Concluding Remarks: What's Next with the Halobacteriaceae?

The last time I reviewed this subject in depth (Kushner, 1978), I posed much the same question and stated that "the unusual biochemical properties should be prized as possible clues to the taxonomic and evolutionary status of these organisms." Apparently, I wasn't the only one who thought so, and some had been thinking along these lines more deeply than I have as witnessed by the existence of this book.

This is the place to speculate, however, on what future work with the halobacteria and even the halococci or the square or alkaliphilic bacteria, may bring.

1. New understanding of membrane function, and perhaps new ways of using light to produce food and new sources of energy.

2. Employment of an apparently infinitely variable genome for new tech-

niques in genetic engineering and, conversely, for understanding why the possessors of such genomes do not vary phenotypically as much as they might.

3. Insights into mechanisms of salt *tolerance* (as distinct from salt *requirement*), a problem that, given the limited amount of water in the world, and elsewhere, may be as important as any we have.

4. Something quite new and unexpected.

Since I have long regarded microorganisms not only as biochemical tools but as individuals in their own right, I have been amazed for more than 20 years with the mysteries of the red halophiles. A microorganism, even a common one, always "has something new up its sleeve," and now it is only a question with these out-of-the-way creatures, what the next surprise will be.

ACKNOWLEDGMENTS

This article was prepared while the author was the recipient of a grant from the Natural Sciences and Engineering Research Council of Canada. I am grateful for colleagues who sent me copies of their work before publication and who discussed some aspects of this chapter with me. I am especially grateful to Dr. M. Kates and to an unknown reviewer for their critical and constructive comments. In such a large and active field, it is not possible to give credit to all primary authors. In citing reviews, I am well aware of those who did the work reported in the review, and I thank them.

REFERENCES

Aitken, D. M., and Brown, A. D. (1969). Citrate and glyoxylate cycles in the halophil, *Halobacterium salinarium*. *Biochim. Biophys. Acta* **177**, 351–354.

Arshavsky, V., Baryshev, V. A., Brown, I. I., Glagolev, A. N., and Skulachev, V. P. (1981). Transmembrane gradient of K^+ and Na^+ ions as an energy buffer in *Halobacterium halobium* cells. *FEBS Lett.* **133**, 22–26.

Baas-Becking, L. G. M. (1931). Historical notes on salt and salt manufacture. *Sci. Mon.* **32**, 434–446.

Baryshev, V. A., Glagolev, A. N., and Skulachev, V. P. (1982). Interrelationship between calcium and a methionine-requiring step in *Halobacterium halobium* taxis. *FEMS Microbiol. Lett.* **13**, 47–50.

Baryshev, V. A., Glagolev, A. N., and Skulachev, V. P. (1983). The interrelation of phototaxis, membrane potential and K^+/Na^+ gradient in *Halobacterium halobium*. *J. Gen. Microbiol.* **129**, 367–373.

Basinger, G. W., and Oliver, J. D. (1979). Inhibition of *Halobacterium cutirubrum* lipid biosynthesis by bacitracin. *J. Gen. Microbiol.* **111**, 423–427.

Baxter, R. M. (1959). An interpretation of the effects of salts on the lactic dehydrogenase of *Halobacterium salinarium*. *Can. J. Microbiol.* **2**, 599–606.

Bayley, S. T., and Morton, R. A. (1978). Recent developments in the molecular biology of extremely halophilic bacteria. *CRC Crit. Rev. Microbiol.* **6**, 151–205.

Bayley, S. T., and Morton, R. A. (1979). Biochemical evolution of halobacteria. *In* "Strategies of Microbial Life in Extreme Environments" (M. Shilo, ed.), Dahlem Konf. Life Sci. Res. Rep., Vol. 13. pp. 109–124. Verlag Chemie, Weinheim.

Beyth, M. (1980). Recent evolution and present stage of Dead Sea brines. *Dev. Sedimentol.* **28,** 155–166.

Brock, T. D., and Petersen, S. (1976). Some effects of light on the viability of rhodopsin-containing halobacteria. *Arch. Microbiol.* **109,** 199–200.

Brown, A. D. (1976). Microbial water stress. *Bacteriol. Rev.* **40,** 803–846.

Brown, A. D., and Cho, K. Y. (1970). The walls of the extremely halophilic cocci: gram-positive bacteria lacking muramic acid. *J. Gen. Microbiol.* **62,** 267–270.

Brown, A. D., and Duong, A. (1982). State of water in extremely halophilic bacteria: Heat of dilution of *Halobacterium halobium*. *J. Membr. Biol.* **64,** 187–193.

Brown, A. D., and Sturtevant, J. M. (1980). State of water in extremely halophilic bacteria: Freezing transitions of *Halobacterium halobium* observed by differential scanning calorimetry. *J. Membr. Biol.* **54,** 21–30.

Brown, H. J., and Gibbons, N. E. (1955). The effect of magnesium, potassium, and iron on the growth and morphology of red halophilic bacteria. *Can. J. Microbiol.* **1,** 486–494.

Cazzulo, J. J. (1973). On the regulatory properties of a halophilic citrate synthase. *FEBS Lett.* **30,** 339–342.

Cazzulo, J. J., and Vidal, M. C. (1972). Effect of monovalent cations on the malic enzyme from the extreme halophile, *Halobacterium cutirubrum*. *J. Bacteriol.* **109,** 437–439.

Chazan, L. L., and Bayley, S. T. (1973). Some properties of a DNA-dependent RNA polymerase from *Halobacterium cutirubrum*. *Can. J. Biochem.* **51,** 1297–1304.

Colwell, R. R., Litchfield, C. D., Vreeland, R. H., Kiefer, L. A., and Gibbons, N. E. (1979). Taxonomic studies of red halophilic bacteria. *Int. J. Syst. Bacteriol.* **29,** 379–399.

Cope, F. W., and Damadian, R. (1970). Cell potassium by ^{39}K spin echo nuclear magnetic resonance. *Nature (London)* **228,** 76–77.

Danon, A., and Caplan, S. R. (1979). CO_2 fixation by *Halobacterium halobium*. *FEBS Lett.* **74,** 255–258.

Danon, A., and Stoeckenius, W. (1974). Photophosphorylation in *Halobacterium halobium*. *Proc. Natl. Acad. Sci. U.S.A.* **71,** 1234–1238.

Dees, C., and Oliver, J. D. (1977). Growth inhibition of *Halobacterium cutirubrum* by cerulenin, a potent inhibitor of fatty acid synthesis. *Biochem. Biophys. Res. Commun.* **78,** 36–44.

de Medicis, E., Laliberte, J. F., and Vass-Marengo, J. (1982). Purification and properties of pyruvate kinase from *Halobacterium cutirubrum*. *Biochim. Biophys. Acta* **708,** 57–67.

de Rosa, M., Gambacorta, A., Nicolaus, B., Ross, H. N. M., Grant, W. D., and Bu'Lock, J. D. (1982). An asymmetric archaebacterial diether lipid from alkaliphilic halophiles. *J. Gen. Microbiol.* **128,** 343–348.

de Rosa, M., Gambacorta, A., Nicolaus, B., and Grant, W. D. (1983). A $C_{25}C_{25}$ diether core lipid from archaebacterial haloalkaliphiles. *J. Gen. Microbiol.* **129,** 2333–2337.

Dundas, I. E. D. (1977). Physiology of *Halobacteriaceae*. *Adv. Microb. Physiol.* **15,** 85–120.

Dundas, I. E. D., and Larsen, H. (1962). The physiological role of the carotenoid pigments of *Halobacterium salinarium*. *Arch. Mikrobiol.* **44,** 233–239.

Dunn, R., McRoy, J., Simsek, M., Majumdar, A., Chang, S. H., RajBhandary, U. L., and Khorana, H. G. (1981). The bacteriorhodopsin gene. *Proc. Natl. Acad. Sci. U.S.A.* **78,** 6744–6748.

Dussault, H. P. (1955). An improved technique for staining red halophilic bacteria. *J. Bacteriol.* **70,** 484–485.

Edgerton, M. E., and Brimblecombe, P. (1981). Thermodynamics of halobacterial environments. *Can. J. Microbiol.* **27,** 899–909.

Elazari-Volcani, B. (1940). Studies on the microflora of the Dead Sea. Doctoral Thesis, Hebrew University, Jerusalem.

Ewersmeyer-Wenk, B., Zahner, H., Krone, B., and Zeek, A. (1981). Metabolic products of microorganisms. 207. Haloquinone, a new antibiotic active against halobacteria. I. Isolation, characterization and biological properties. *J. Antibiot.* **34**, 1531–1537.

Fitt, P. S. (1978). Nucleic acid enzymology of extreme halophiles. *In* "Energetics and Structure of Halophilic Microorganisms" (S. R. Caplan and M. Ginzburg, eds.), pp. 379–393. Elsevier/North-Holland Biomedical Press, Amsterdam.

Fitt, P. S., Sharma, N., and Castellanos, G. (1983). A comparison of liquid-holding recovery and photoreactivation in halophilic and non-halophilic bacteria. *Biochim. Biophys. Acta* **739**, 73–78.

Forsyth, M. P. (1971). Physiological studies on the halophilic bacteria. Ph.D. Thesis, University of Ottawa, Ottawa, Ontario, Canada.

Fox, G. E., Stackenbrandt, E., Hespell, R. B., Gibson, J., Maniloff, J., Dyer, T. A., Wolfe, R. S., Balch, W. E., Tanner, R. S., Magrum, L. J., Zablen, L. B., BLakemore, R., Gupta, R. Bonen, L., Lewis, B. J., Stahl, D. A., Luehrsen, K. B., Chen, K. N., and Woese, C. R. (1980). The phylogeny of prokaryotes. *Science* **209**, 457–463.

Fujii, T. (1980). Some characteristics of spore-forming halophilic bacteria isolated from Bagoony. *Bull. Jpn. Soc. Sci. Fish.* **46**, 1545.

Galinsky, E. A., and Trüper, H. G. (1982). Betaine, a compatible solute in the extremely halophilic phototrophic bacterium, *Ectothiorhodospira halochloris*. *FEMS Microbiol. Lett.* **13**, 357–360.

Garty, H., Danon, A., and Caplan, S. R. (1980). Preparation and characterization of inverted cell envelopes of *Halobacterium halobium*. *Eur. J. Biochem.* **111**, 411–418.

Gibbons, N. E. (1974). Halobacteriaceae. *In* "Bergey's Manual of Determinative Bacteriology" (R. E. Buchanan and N. E. Gibbons, eds.), 8th ed., pp. 269–273. Williams & Wilkins, Baltimore, Maryland.

Gibbons, N. E., and Payne, J. I. (1961). Relation of temperature and sodium chloride concentration to growth and morphology of some halophilic bacteria. *Can. J. Microbiol.* **7**, 483–489.

Ginzburg, M., Sachs, L., and Ginzburg, B. Z. (1970). Ion metabolism in a *Halobacterium*. I. Influence of age of culture on intracellular concentrations. *J. Gen. Physiol.* **55**, 187–207.

Gochnauer, M., and Kushner, D. J. (1971). Potassium binding, growth and survival of an extremely halophilic bacterium. *Can. J. Microbiol.* **17**, 17–23.

Gupta, R., Lanter, J. M., and Woese, C. R. (1983). Sequence of the 16 S ribosomal RNA from *Halobacterium volcanii*, an archaebacterium. *Science* **221**, 656–659.

Hartmann, R., Sickinger, H.-D., and Oesterhelt, D. (1980). Anaerobic growth of halobacteria. *Proc. Natl. Acad. Sci. U.S.A.* **77**, 3821–3825.

Hildebrand, E., and Schimz, A. (1983). Photosensory behavior of *Halobacterium halobium*. *Photochem. Photobiol.* **37**, 581–584.

Hilpert, R., Winter, J., Hammers, W., and Kandler, O. (1981). The sensitivity of archaebacteria to antibiotics. *Zentralbl. Bakteriol., Mikrobiol. Hyg. Abt. 1, Orig. C* **2**, 11–21.

Hochstein, L. I. (1978). Carbohydrate metabolism in the extremely halophilic bacteria: The role of glucose in the regulation of citrate synthase activity. *In* "Energetics and Structure of Halophilic Microorganisms" (S. R. Caplan and M. Ginzburg, eds.), pp. 397–412. Elsevier/North-Holland Biomedical Press, Amsterdam.

Hochstein, L. I., Dalton, B. P., and Pollock, G. (1976). The metabolism of carbohydrates by extremely halophilic bacteria: Identification of galactonic acid as a product of galactose metabolism. *Can. J. Microbiol.* **22**, 1191–1196.

Holmes, P. K., and Halvorson, H. O. (1965). Purification of a salt-requiring enzyme from an obligately halophilic bacterium. *J. Bacteriol.* **90**, 312–315.

Hubbard, J. S., and Miller, A. B. (1969). Purification and reversible inactivation of the isocitrate dehydrogenase from an obligate halophile. *J. Bacteriol.* **99**, 161–168.

Hunter, M. I. S., and Millar, S. J. W. (1980). Effect of wall antibiotics on the growth of the extremely halophilic coccus, *Sarcina marina* NCMB 778. *J. Gen. Microbiol.* **120**, 255–258.

Imhoff, J. F., Sahl, H. G., Soliman, G. S. H., and Trüper, H. G. (1979). The Wadi Natrun: Chemical composition and microbial mass developments in alkaline brines of eutrophic desert lakes. *Geomicrobiol. J.* **1**, 219–234.

Imhoff, J. F., Kushner, D. J., and Anderson, P. J. (1983). Amino acid composition of proteins in halophilic phototrophic bacteria of the genus *Ectothiorhodospira*. *Can. J. Microbiol.* **29**, 1675–1679.

Ingram, M. (1957). Microorganisms resisting high concentrations of sugars or salt. *Symp. Soc. Gen. Microbiol.* **7**, 90–133.

Javor, B. J. (1983a). Nutrients and ecology of the Western Salt and Exportadora de Sal saltern brines. *Symp. Salt. 6th,* 1982.

Javor, B. J. (1983b). Planktonic standing crop and nutrients in a saltern ecosystem. *Limnol. Oceanogr.* **28**, 153–159.

Javor, B. J., Requadt, C., and Stoeckenius, W. (1982). Box-shaped halophilic bacteria. *J. Bacteriol.* **151**, 1532–1542.

Juez, A., Rodriguez-Valera, F., Ventosa, A., and Kushner, D. J. (1985). In preparation.

Kates, M. (1978). The phytanyl ether-linked polar lipids and isoprenoid neutral lipids of extremely halophilic bacteria. *Prog. Chem. Fats Other Lipids* **15**, 301–342.

Kates, M., and Kushwaha, S. C. (1978). Biochemistry of the lipids of extremely halophilic bacteria. *In* "Energetics and Structure of Halophilic Microorganisms" (S. R. Caplan and M. Ginzburg, eds.), pp. 461–479. Elsevier/North-Holland Biomedical Press, Amsterdam.

Kates, M., Palameta, B., Joo, C. N., Kushner, D. J., and Gibbons, N. E. (1966). Aliphatic diether analogs of glyceride-derived lipids. IV. The occurrence of *di*-O-dihydrophytyl-glycerol ether containing lipids in extremely halophilic bacteria. *Biochemistry* **5**, 4092–4099.

Keradjopoulos, D., and Holldorf, A. W. (1979). Purification and properties of alanine dehydrogenase from *Halobacterium salinarium*. *Biochim. Biophys. Acta* **570**, 1–10.

Keradjopoulos, D., and Holldorf, A. W. (1980). Salt dependent conformational changes of alanine dehydrogenase from *Halobacterium salinarium*. *FEBS Lett.* **112**, 183–185.

Keradjopoulos, D., and Wulff, K. (1974). Thermophilic alanine dehydrogenase from *Halobacterium salinarium*. *Can. J. Biochem.* **52**, 1033–1037.

Kessel, M., and Cohen, Y. (1982). Ultrastructure of square bacteria from a brine pool in southern Sinai. *J. Bacteriol.* **150**, 851–860.

Kessel, M., and Klink, F. (1980). Archaebacterial elongation factor is ADP-ribosylated by diphtheria toxin. *Nature (London)* **287**, 250–251.

Kessel, M., and Klink, F. (1981). Two elongation factors from the extremely halophilic archaebacterium *Halobacterium cutirubrum*. *Eur. J. Biochem.* **114**, 481–486.

Koch-Schmidt, A. C., Mosbach, K., and Werber, M. M. (1979). A comparative study on the stability of immobilized halophilic and nonhalophilic malate dehydrogenases at various ionic strengths. *Eur. J. Biochem.* **100**, 213–218.

Kocur, M., Smid, B., and Martinec, T. (1972). The structure of extreme halophilic cocci. *Microbios* **5**, 101–107.

Kushner, D. J. (1968). Halophilic bacteria. *Adv. Appl. Microbiol.* **10**, 73–99.

Kushner, D. J. (1978). Life in high salt and solute concentrations: Halophilic bacteria. *In* "Microbial Life in Extreme Environments" (D. J. Kushner, ed.), pp. 317–368. Academic Press, London.

Kushner, D. J. (1980). Extreme environments. *In* "Contemporary Microbial Ecology" (D. C. Ellwood, J. N. Hedger, M. J. Latham, J. H. Slater, and J. M. Lynch, eds.), pp. 29–54. Academic Press, London.

Kushner, D. J., Masson, G., and Gibbons, N. E. (1965). Simple method of killing halophilic bacteria in contaminated solar salt. *Appl. Microbiol.* **13**, 288.

Kushwaha, S. C., and Kates, M. (1979a). Effect of nicotine on carotenogenesis in extremely halophilic bacteria. *Phytochemistry* **18**, 2061–2062.

Kushwaha, S. C., and Kates, M. (1979b). Studies of the biosynthesis of 50 carbon carotenoids in *Halobacterium cutirubrum*. *Can. J. Microbiol.* **25**, 1292–1297.

Kushwaha, S. C., Gochnauer, M. B., Kushner, D. J., and Kates, M. (1974). Pigments and isoprenoid compounds in extremely and moderately halophilic bacteria. *Can. J. Microbiol.* **20**, 241–245.

Kushwaha, S. C., Kates, M., and Martin, W. G. (1975). Characterization and composition of the red and purple membrane from *Halobacterium cutirubrum*. *Can. J. Biochem.* **53**, 284–292.

Kushwaha, S. C., Kates, M., and Weber, H. J. (1980). Exclusive formation of all-*trans*-phytoene by a colorless mutant of *Halobacterium halobium*. *Can. J. Microbiol.* **26**, 1011–1014.

Kushwaha, S. C., Juez-Perez, G., Rodriguez-Valera, F., Kates, M., and Kushner, D. J. (1982). Survey of lipids of a new group of extremely halophilic bacteria from salt ponds in Spain. *Can. J. Microbiol.* **28**, 1365–1372.

Lanyi, J. K. (1974). Salt-dependent properties of proteins from extremely halophilic bacteria. *Bacteriol. Rev.* **38**, 272–290.

Lanyi, J. K. (1978). Light energy conversion in *Halobacterium halobium*. *Microbiol. Rev.* **42**, 682–706.

Lanyi, J. K. (1981). Halorhodopsin—a second retinal pigment in *Halobacterium halobium*. *Trends Biochem. Sci.* February, pp. 1–3.

Lanyi, J. K., and Oesterhelt, D. (1982). Identification of the retinal-binding protein in halorhodopsin. *J. Biol. Chem.* **257**, 2674–2677.

Lanyi, J. K., and Schobert, B. (1983). Effects of chloride and pH on the chromophore and photochemical cycling of halorhodopsin. *Biochemistry* **22**, 2763–2769.

Lanyi, J. K., and Silverman, M. P. (1972). The state of binding of intracellular K^+ in *Halobacterium cutirubrum*. *Can. J. Microbiol.* **18**, 993–995.

Larsen, H. (1962). Halophilism. *In* ''The Bacteria'' (I. C. Gunsalus and R. Y. Stanier, eds.), Vol. 4, pp. 297–342. Academic Press, New York.

Larsen, H. (1967). Biochemical aspects of extreme halophilism. *Adv. Microb. Physiol.* **1**, 97–132.

Larsen, H. (1973). The halobacteria's confusion to biology. *Antonie van Leeuwenhoek* **39**, 383–396.

Larsen, H. (1980). Ecology of hypersaline environments. *Dev. Sedimentol.* **28**, 23–39.

Larsen, H. (1981). The family Halobacteriaceae. *In* ''The Prokaryotes: A Handbook on Habitats, Isolation, and Identification of Bacteria'' (M. P. Starr, H. Stolp, H. G. Trüper, A. Balows, and H. G. Schlegel, eds.), pp. 985–994. Springer-Verlag, Berlin and New York.

Larsen, H. (1984). Family Halobacteriaceae Gibbons 269. *In* ''Bergey's Manual of Determinative Bacteriology,'' 9th ed., 261–267.

Leicht, W., and Pundak, S. (1981). Large-scale purification of halophilic enzymes by salting-out mediated chromotography. *Anal. Biochem.* **114**, 186–192.

Lieberman, M. M., and Lanyi, J. K. (1972). Threonine deaminase from extremely halophilic bacteria. Cooperative substrate kinetics and salt dependence. *Biochemistry* **11**, 211–216.

Louis, B. G., and Fitt, P. (1972a). Isolation and properties of highly purified *Halobacterium cutirubrum* deoxyribonucleic acid-dependent-ribonucleic acid polymerase. *Biochem. J.* **127**, 69–80.

Louis, B. G., and Fitt, P. (1972b). The role of *Halobacterium cutirubrum* deoxyribonucleic acid-dependent ribonucleic acid polymerase subunits in initiation and polymerization. *Biochem. J.* **127**, 81–86.

Madon, J., and Zillig, W. (1983). A form of the DNA-dependent RNA polymerase of *Halobacterium halobium*, containing an additional component, is able to transcribe native DNA. *Eur. J. Biochem.* **133**, 471–474.

Magrum, L. J., Luehrsen, K. R., and Woese, C. R. (1978). Are extreme halophiles actually ''bacteria''? *J. Mol. Evol.* **11**, 1–8.

Marquez, E. D., and Brodie, A. F. (1973). The effect of cations on the heat stability of a halophilic nitrate reductase. *Biochim. Biophys. Acta* **321**, 84–89.

Mescher, M. F., and Strominger, J. L. (1978). The cell surface glycoprotein of *Halobacterium salinarium*. *In* "Energetics and Structure of Halophilic Microorganisms" (S. R. Caplan and M. Ginzburg, eds.), pp. 503–511. Elsevier/North-Holland Biomedical Press, Amsterdam.

Mitchell, P. (1972). Chemiosmotic coupling in energy transduction: A logical development of biochemical knowledge. *J. Bioenerg.* **3**, 5–24.

Moore, R. L., and McCarthy, B. J. (1969a). Characterization of the deoxyribonucleic acid of various strains of halophilic bacteria. *J. Bacteriol.* **99**, 248–254.

Moore, R. L., and McCarthy, B. J. (1969b). Base sequence homology and renaturation studies of the deoxyribonucleic acid of extremely halophilic bacteria. *J. Bacteriol.* **99**, 255–262.

Mukohata, Y., Sugiyama, Y., Kaji, Y., Usukura, J., and Yamada, E. (1981). The white membrane of crystalline bacterioopsin in *Halobacterium halobium* strain R_1mW and its conversion into purple membrane by exogenous retinal. *Photochem. Photobiol.* **33**, 593–600.

Mullakhanbhai, M. F., and Larsen, H. (1975). *Halobacterium volcanii* spec. nov., a Dead Sea halobacterium with a moderate salt requirement. *Arch. Microbiol.* **104**, 207–214.

Nagle, J. F., and Mille, M. (1981). Molecular models of proton pumps. *J. Chem. Phys.* **74**, 1367–1372.

Nagle, J. F., and Morowitz, H. J. (1978). Molecular mechanisms for proton transport in membranes. *Proc. Natl. Acad. Sci. U.S.A.* **75**, 298–302.

Nicholson, D. E., and Fox, G. E. (1983). Molecular evidence for a close phylogenetic relationship among box-shaped halophilic bacteria, *Halobacterium vallismortis* and *Halobacterium marismortui*. *Can. J. Microbiol.* **29**, 52–59.

Nissenbaum, A. (1975). The microbiology and biogeochemistry of the Dead Sea. *Microb. Ecol.* **2**, 139–161.

Nissenbaum, A., ed. (1980). Hypersaline brines and evaporitic environments. *Dev. Sedimentol.* **28**, 1–270.

Norberg, P., Kaplan, J. G., and Kushner, D. J. (1973). Kinetics and regulation of the salt-dependent aspartate transcarbamylase of *Halobacterium cutirubrum*. *J. Bacteriol.* **113**, 680–686.

Oesterhelt, D., and Krippahl, G. (1973). Light inhibition of respiration in *Halobacterium halobium*. *FEBS Lett.* **36**, 72–76.

Oren, A. (1983). *Halobacterium sodomense* sp. nov., a Dead Sea halobacterium with an extremely high magnesium requirement. *Int. J. Syst. Bacteriol.* **33**, 381–386.

Oren, A., and Shilo, M. (1981). Bacteriorhodopsin in a bloom of halobacteria in the Dead Sea. *Arch. Microbiol.* **130**, 185–187.

Oren, A., and Shilo, M. (1982). Population dynamics of *Dunaliella parva* in the Dead Sea. *Limnol. Oceanogr.* **27**, 201–211.

Ovchinnikov, Y. A. (1982). Rhodopsin and bacteriorhodpsin: Structure-function relationships. *FEBS Lett.* **148**, 179–191.

Packer, L., ed. (1982). "Methods in Enzymology," Vol. 88 (91 different articles, mainly dealing with the bacteriorhodopsin system; some articles also deal with the halorhodopsin system and with classification, chemistry, genetics and physiology of halobacteria). Academic Press, New York.

Papadopoulos, G. K., and Cassim, J. Y. (1981). Orientation of the retinyl and the heme chromophores in the brown membrane of *Halobacterium halobium*. *J. Mol. Biol.* **152**, 35–47.

Parkes, K., and Walsby, A. E. (1981). Ultrastructure of a gas-vacuolate square bacterium. *J. Gen. Microbiol.* **126**, 503–506.

Pecher, T., and Böck, A. (1981). *In vivo* susceptibility of halophilic and methanogenic organisms to protein synthesis inhibitors. *FEMS Microbiol. Lett.* **10**, 295–297.

Pfeifer, F., Weidinger, G., and Goebel, W. (1981a). Characterization of plasmids in halobacteria. *J. Bacteriol.* **145**, 369–374.

Pfeifer, F., Weidinger, G., and Goebel, W. (1981b). Genetic variability of *Halobacterium halobium*. *J. Bacteriol.* **145**, 375–381.

Post, F. J. (1975). Life in the Great Salt Lake. *Utah Sci.* **36**, 43–47.

Post, F. J. (1977). The microbial ecology of the Great Salt Lake. *Microbial Ecol.* **3**, 143–165.

Pugh, E. L., Wassef, M. K., and Kates, M. (1971). Inhibition of fatty acid synthetase in *Halobacterium cutirubrum* and *Escherichia coli* by high salt concentrations. *Can. J. Biochem.* **49**, 953–958.

Pundak, S., and Eisenberg, H. (1981). Structure and activity of malate dehydrogenase from the extreme halophilic bacteria of the Dead Sea, Israel. 1. Conformation and interaction with water and salt between 5 molar and 1 molar sodium chloride concentration. *Eur. J. Biochem.* **118**, 463–470.

Pundak, S., Aloni, H., and Eisenberg, H. (1981). Structure and activity of malate dehydrogenase from the extreme halophilic bacteria of the Dead Sea, Israel. 2. Inactivation, dissociation and unfolding at sodium chloride concentration below 2 molar salt concentration and temperature dependence of enzyme stability. *Eur. J. Biochem.* **118**, 471–478.

Racker, E., and Stoeckenius, W. (1974). Reconstitution of purple membrane vesicles catalyzing light-driven proton uptake and adenosine triphosphate formation. *J. Biol. Chem.* **249**, 662–663.

Rao, J. K. M., and Argos, P. (1981). Structural stability of halophilic proteins. *Biochemistry* **20**, 6536–6543.

Reich, M. H., Kam, Z., and Eisenberg, H. (1982). Small-angle X-ray scattering study of halophilic malic dehydrogenase. *Biochemistry* **21**, 5189–5195.

Reistad, R. (1972). Cell wall of an extremely halophilic coccus: investigation of ninhydrin-positive compounds. *Arch. Microbiol.* **82**, 24–30.

Reistad, R. (1974). 2-Amino-2-deoxyguluronic acid: a constituent of the cell wall of *Halococcus sp.* strain 24. *Carbohydrate Res.* **36**, 420–423.

Robertson, J. D., Schreil, W., and Reedy, M. (1982). *Halobacterium halobium* I. A thin-sectioning electron microscope study. *J. Ultrastruct. Res.* **80**, 148–162.

Rodriguez-Valera, F., Ruiz-Berraquero, F., and Ramos-Cormenzana, A. (1979). Isolation of extreme halophiles from seawater. *Appl. Environ. Microbiol.* **38**, 164–165.

Rodriguez-Valera, F., Ruiz-Berraquero, F., and Ramos-Cormenzana, A. (1980). Isolation of extremely halophilic bacteria able to grow on defined inorganic media with single carbon sources. *J. Gen. Microbiol.* **119**, 535–538.

Rodriguez-Valera, F., Albert, F. J., and Gibson, J. (1982a). Effect of light on growing and starved populations of extremely halophilic bacteria. *FEMS Microbiol. Lett.* **14**, 155–158.

Rodriguez-Valera, F., Juez, G., and Kushner, D. J. (1982b). Halocins: Salt-dependent bacteriocins produced by extremely halophilic rods *Can. J. Microbiol.* **28**, 151–154.

Rodriguez-Valera, F., Ventosa, A., Quesada, E., and Ruiz-Berraquero, F. (1982c). Some physiological features of *Halococcus* sp. at low salt concentrations. *FEMS Microbiol. Lett.* **15**, 249–252.

Rodriguez-Valera, F., Nieto, J. J., and Ruiz-Berraquero, F. (1983a). Light as an energy source in continuous cultures of bacteriorhodopsin-containing halobacteria. *Appl. Environ. Microbiol.* **45**, 868–871.

Rodriguez-Valera, F., Juez, G., and Kushner, D. J. (1983b). *Halobacterium mediterranei* spec. nov., a new carbohydrate-utilizing extreme halophile. *Syst. Appl. Microbiol.* **4**, 369–381.

Rogers, P. J., and Morris, C. A. (1978). Regulation of bacteriorhodopsin synthesis by growth rate in continuous cultures of *Halobacterium halobium*. *Arch. Microbiol.* **119**, 323–325.

Ross, H. N. M., Collins, M. D., Tindall, B. J., and Grant, W. D. (1981). A rapid procedure for the detection of archaebacterial lipids in halophilic bacteria. *J. Gen. Microbiol.* **123**, 75–80.

Sapienza, C., and Doolittle, W. F. (1982). Unusual physical organization of the *Halobacterium* genome. *Nature (London)* **295**, 384–389.

Schimz, A. (1981). Methylation of membrane proteins is involved in chemosensory and photosensory behavior of *Halobacterium halobium*. *FEBS Lett.* **125**, 205–207.

Schimz, A. (1982). Localisation of the methylation system involved in sensory behaviour of *Halobacterium halobium* and its dependence on calcium. *FEBS Lett.* **139**, 283–286.

Schimz, A., and Hildebrand, E. (1979). Chemosensory responses or *Halobacterium halobium*. *J. Bacteriol.* **140**, 749–753.

Schleifer, K. H., Steber, J., and Mayer, H. (1982). Chemical composition and structure of the cell wall of *Halococcus morrhuae*. *Zentralbl. Bakteriol., Mikrobiol. Hyg., Abt. 1, Orig. C* **3**, 171–178.

Schnabel, H., Zillig, W., Pfaffle, M., Schnabel, R., Michel, H., and Delius, H. (1982a). *Halobacterium halobium* phage φH. *EMBO J.* **1**, 87–92.

Schnabel, H., Schramm, E., Schnabel, R., and Zillig, W. (1982b). Structural variability in the genome of phage φH of *Halobacterium halobium*. *Mol. Gen. Genet.* **188**, 370–377.

Schobert, B., and Lanyi, J. K. (1982). Halorhodopsin is a light-driven chloride pump. *J. Biol. Chem.* **257**, 10306–10313.

Scott, W. J. (1957). Water relations of food spoilage microorganisms. *Adv. Food Res.* **7**, 83–127.

Shporer, M., and Civan, M. M. (1977). Pulsed nuclear magnetic resonance study of ^{39}K within halobacteria. *J. Membr. Biol.* **33**, 385–400.

Simon, R. D. (1978). *Halobacterium* strain 5 contains a plasmid which is correlated with the presence of gas vacuoles. *Nature (London)* **273**, 314–317.

Simon, R. D. (1981). Morphology and protein composition of gas vesicles from wild type and gas vacuole defective strains of *Halobacterium salinarium* Strain 5. *J. Gen. Microbiol.* **125**, 103–112.

Soliman, G. S. H., and Trüper, H. G. (1982). *Halobacterium pharaonis* sp. nov., a new extremely haloalkaliphilic archaebacterium with low magnesium requirements. *Zentralbl. Baktriol., Mikrobiol. Hyg., Abt. 1, Orig. C* **3**, 318–329.

Spudich, E. N., and Spudich, J. L. (1980). Light-regulated retinal-dependent reversible phosphorylation of *Halobacterium* proteins. *J. Biol. Chem.* **255**, 5501–5503.

Spudich, E. N., and Spudich, J. L. (1981). Photosensitive photosphoproteins in halobacteria: Regulatory coupling of transmembrane proton flux and protein dephosphorylation. *J. Cell Biol.* **91**, 895–900.

Spudich, E. N., and Spudich, J. L. (1982). Control of transmembrane ions fluxes to select halorhodopsin-deficient and other energy-transduction mutants of *Halobacterium halobium*. *Proc. Natl. Acad. Sci. U.S.A.* **79**, 4308–4312.

Spudich, J. L., and Stoeckenius, W. (1980). Light-regulated retinal-dependent reversible phosphorylation of *Halobacterium* proteins. *J. Biol. Chem.* **255**, 5501–5503.

Stanier, R. Y., Adelberg, E., and Ingraham, J. (1976). "The Microbial World," 4th ed. Prentice-Hall, Englewood Cliffs, New Jersey.

Steber, J., and Schleifer, K. H. (1975). *Halococcus morrhuae:* a sulfated heteropolysaccharide as the structural component of the bacterial cell wall. *Arch. Microbiol.* **105**, 173–177.

Steber, J., and Schleifer, K. H. (1979). N-Glycyl-glucosamine, a novel constituent in the cell wall of *Halococcus morrhuae*. *Arch. Microbiol.* **123**, 209–212.

Stoeckenius, W. (1976). The purple membrane of salt-loving bacteria. *Sci. Am.* **234**, 38–46.

Stoeckenius, W. (1981). Walsby's square bacterium: Fine structure of an orthogonal procaryote. *J. Bacteriol.* **148**, 352–360.

Stoeckenius, W., and Bogomolni, R. A. (1982). Bacteriorhodopsin and related pigments of halobacteria. *Annu. Rev. Biochem.* **52**, 587–616.

Stoeckenius, W., Lozier, R. H., and Bogomolni, R. A. (1979). Bacteriorhodopsin and the purple membrane of halobacteria. *Biochim. Biophys. Acta* **505**, 215–278.

Sumper, M. (1982). The brown membrane of *Halobacterium halobium;* the biosynthetic precursor of the purple membrane. *Methods Enzymol.* **88**, 391–395.

Sumper, M., and Herrmann, G. (1978). Studies on the biosynthesis of bacterio-opsin. Demonstration of the existence of protein species structurally related to bacterio-opsin. *Eur. J. Biochem.* **89**, 229–235.

Tindall, B. J., Mills, A. A., and Grant, W. D. (1980). An alkalophilic red halophilic bacterium with a low magnesium requirement from a Kenyan soda lake. *J. Gen. Microbiol.* **116,** 257–260.

Tindall, B. J., Ross, H. N. M., and Grant, W. D. (1984). *Natronobacterium* gen. nov. and *Natronococcus,* gen. nov., genera of haloalkalophilic archaebacteria. *Syst. Appl. Microbiol.* **5,** 41–57.

Tomlinson, G. A., Koch, T. K., and Hochstein, L. I. (1974). The metabolism of carbohydrates by extremely halophilic bacteria: Glucose metabolism via a modified Entner-Doudoroff pathway. *Can. J. Microbiol.* **20,** 1085–1091.

Torsvik, T., and Dundas, I. (1974). Bacteriophage of *Halobacterium salinarium. Nature (London)* **248,** 680–681.

Torsvik, T., and Dundas, I. (1980). Persisting phage infection in *Halobacterium salinarium* str. 1. *J. Gen. Virol.* **47,** 29–36.

Torsvik, T., and Dundas, I. (1982). The classification of halobacteria. *In* ''Methods in Enzymology'' (L. Packer, ed.), Vol. 88, pp. 360–368. Academic Press, New York.

Traulich, B., Hildebrand, E., Schimz, A., Wagner, G., and Lanyi, J. K. (1983). Halorhodopsin and photosensory behavior in *Halobacterium halobium* mutant strain L-33. *Photochem. Photobiol.* **37,** 577–579.

Vidal, M. C., and Cazzulo, J. J. (1976). On the regulatory properties of a halophilic malic enzyme from *Halobacterium cutirubrum. Experientia* **32,** 441–442.

Vreeland, R. H., and Martin, E. L. (1980). Growth characteristics, effects of temperature and ion specificity of the halotolerant bacterium, *Halomonas elongata. Can. J. Microbiol.* **26,** 746–752.

Wagner, G., Hartmann, R., and Oesterhelt, D. (1978). Potassium uniport and ATP synthesis in *Halobacterium halobium. Eur. J. Biochem.* **89,** 169–179.

Wagner, G., Oesterhelt, D., Krippahl, G., and Lanyi, J. K. (1981). Bioenergetic role of halorhodopsin in *Halobacterium halobium* cells. *FEBS Lett.* **131,** 341–345.

Wais, A. C., Kon, M., MacDonald, R. E., and Stollar, R. D. (1975). Salt-dependent bacteriophage infecting *Halobacterium cutirubrum* and *Halobacterium halobium. Nature (London)* **256,** 314–315.

Walsby, A. E. (1980). A square bacterium. *Nature (London)* **283,** 69–71.

Weber, H. J., and Bogomolni, R. A. (1982). The isolation of *Halobacterium* mutant strains with defects in pigment synthesis. *In* ''Methods in Enzymology'' (L. Packer, ed.), Vol. 88, pp. 379–395. Academic Press, New York.

Weidinger, G., Klotz, G., and Goebel, W. (1979). A large plasmid from *Halobacterium halobium* carrying genetic information for gas vacuole formation. *Plasmid* **2,** 377–386.

Weidinger, G., Pfeifer, F., and Goebel, W. (1982). Plasmids in halobacteria: Restriction maps. *In* ''Methods in Enzymology'' (L. Packer, ed.), Vol. 88, pp. 374–379. Academic Press, New York.

Werber, M. M., and Mevarech, M. (1978). Induction of a dissimilatory reduction pathway of nitrate in *Halobacterium* of the Dead Sea. A possible role for the 2 Fe-ferredoxin isolated from this organism. *Arch. Biochem. Biophys.* **186,** 60–65.

Werber, M. M., Mevarech, M., Leicht, W., and Eisenberg, H. (1978). Structure-function relationships in proteins and enzymes of *Halobacterium* of the Dead Sea. *In* ''Energetics and Structure of Halophilic Microorganisms'' (S. R. Caplan and M. Ginzburg, eds.), pp. 427–443. Elsevier/North-Holland Biomedical Press, Amsterdam.

Yaguchi, M., Visentin, L. P., Zuker, M., Matheson, A. T., Roy, C., and Strom, A. R. (1982). Amino-terminal sequences of ribosomal proteins from the 30S subunit of archaebacterium *Halobacterium cutirubrum. Zentralbl. Bakteriol., Mikrobiol. Hyg., Abt. 1, Orig. C* **3,** 200–208.

Zillig, W., Stetter, K. O., and Tobien, M. (1978). DNA-dependent RNA polymerase from *Halobacterium halobium. Eur. J. Biochem.* **91,** 193–200.

Chapter 4

Sedimentary Record and Archaebacteria

JÜRGEN HAHN AND PAT HAUG

I. Introduction

It is commonly believed that the earth's crust and the atmosphere began to form about 4.5 billion years ago. Since hydrogen and the noble gases are significantly depleted on earth as compared to solar abundances, it has been proposed that an earlier reducing atmosphere with reduced species such as CH_4, NH_3, H_2O, and H_2 as major constituents was lost during the *t tauri* phase of our sun. A more recent alternative explanation for the depletion in hydrogen and noble gases is that the protoplanet earth did not accumulate an atmosphere. At the time when the earth accreted the upper mantle from residual matter of the solar nebula in or near the Earth's orbit, the differentiation of residual presolar matter had reached such an extent that the bulk of hydrogen and noble gases had already migrated to the outer parts of the solar system (Alfvén and Arrhenius, 1976). The atmosphere, which formed during the outgassing of the upper mantle, was probably weakly reducing or even neutral, consisting of CO_2, N_2, water vapor, and a suite of minor constituents such as CH_4, CO, SO_2, H_2S, H_2, HF, HCl, and others (Schidlowski, 1980). Free oxygen was only a trace constituent with probably an extremely low partial pressure (Walker *et al.*, 1983).

A. Sediments and the Geological Time Scale

The first rocks of the crust were igneous, and at a very early period of earth's history, the crust may have consisted solely of igneous rocks. But more than 4 billion years of interaction of sun-driven erosional and depositional processes

Copyright © 1985 by Academic Press, Inc.
All rights of reproduction in any form reserved.
ISBN 0-12-307208-5

have destroyed most of the primeval rocks together with many of the younger igneous rocks. Through reaction with atmospheric constituents, igneous rocks were granulated, chemically altered, or even totally dissolved. The solid debris was then transported by wind and water to be subsequently deposited as clastic sediments, while the dissolved material was eventually precipitated either inorganically or as a result of biological activities. After burial to a sufficient depth, the sediments were gradually turned into sedimentary rocks. Upon exposure to the atmosphere, these rocks were eroded again and redeposited at another location. Thus, the sedimentary rocks that one finds today have been derived from igneous precursors or from pre-existing sedimentary rocks. They may have completed one or more cycles of weathering, erosion, and redeposition. The cycling time is on the order of several hundreds of millions of years and varies with the type of sedimentary rock.

The geological record begins at about 3.8 billion years B.P. (B.P. $\equiv$ df before present) with the metasediments of the Isua supracrustal belt in Greenland. Note that the oldest known rocks are *sedimentary* rocks. The era earlier than 3.8 billion years B.P. is commonly referred to as the Hadean. From this era, we have no records. The time span from 3.8 to 2.5 billion years B.P. is called Archean, as shown in Table I, followed by the Proterozoic from 2.5 to 0.6 billion years B.P., and the Phanerozoic, which comprises the last 600 million years. Until recently, it was believed that life sprang up in the Archean around 3.3 billion years ago and that the Proterozoic was the era of algae. Recent finds, however, suggest that prokaryotic microorganisms may be as old as 3.8 billion years (Schidlowski *et al.*, 1979; Pflug, 1982). The border line between Proterozoic and Phanerozoic has been defined as the time of the appearance of the first multicellular organisms. Recently, Cloud and Glaessner (1982) have proposed to move this line to 670 million years B.P. in order to include the *Ediacara* fauna, which consisted largely of multicellular organisms, into the Phanerozoic. For more details on Phanerozoic subdivisions, see Odin (1982). While there is some indirect evidence that atmospheric CO_2 has continuously, although irregularly, decreased in the course of earth's history (Walker *et al.*, 1981; Lovelock and Whitfield, 1982), it is now widely accepted that free oxygen did not appear in the atmosphere in appreciable amounts earlier than 2.0 billion years B.P.

Sedimentary rocks may be divided into four major groups namely shales, sandstones, evaporites, and carbonates. Shales are rocks composed of particles which are smaller than those of sandstones, typically in the range of a few microns. The main constituents of shales are the various clay minerals, but fine-grained quartz and feldspar frequently occur. The chief mineral of sandstones is quartz. While sandstones and shales are deposited as sand and mud, respectively, evaporites form as a precipitate when sea or freshwater bodies concentrate due to evaporation of water. In the case of seawater, the chief minerals formed are gypsum, anhydrite, and halite. Evaporite formation from seawater may be ob-

TABLE I

CHIEF DIVISIONS OF GEOLOGICAL TIME

Era	Period	Age	Before the present (millions of years)
Cenozoic	Quaternary	Holocene	Present–0.010
		Pleistocene	0.010–2.0
	Tertiary	Pliocene	2–7
		Miocene	7–26
		Oligocene	26–33
		Eocene	33–54
		Palaeocene	54–65.5
Mesozoic	Cretaceous		65.5–137
	Jurassic		137–195
	Triassic		195–225
Palaeozoic	Permian		225–280
	Carboniferous		280–345
	Devonian		345–395
	Silurian		395–435
	Ordovician		435–500
	Cambrian		500–600
Proterozoic			600–2500
Archean			2500–3800
Hadean			Earlier than 3800 BP

served today, e.g., in the so-called "sabkhas" along the coasts of the Persian Gulf. Carbonate rocks can be grouped into two categories: limestones and dolomites. Limestones are chiefly composed of the minerals calcite and aragonite, whereas dolomites are rich in dolomite, as the name indicates. Carbonates precipitate whenever the solubility product of calcium or calcium/magnesium carbonate in water is exceeded. This may be due to the evaporation of water or to the depletion of dissolved CO_2 as a result of the activity of photosynthesizing organisms. Also, large carbonate deposits have formed directly from the calcareous shells and skeletons of microorganisms such as, for example, foraminifera. Carbonate precipitation from seawater occurs today, e.g., in the area of the Grand Bahama Bank, in Baja, California, and in some places on the Australian coasts, such as Shark Bay.

The compaction of sediments due to burial and the subsequent formation of sedimentary rock is accompanied by a series of chemical and mineralogical changes in their composition. The early stages of this process are called *diagenesis* and are characterized by the squeezing out of pore waters, the cementing or decementing of the original grains, and the formation of small quantities of new minerals from old ones. As a result of deeper burial (to depths of more than

10 km), heating (sometimes short of partial melting), or folding, more drastic changes occur that alter both texture and structure of the sedimentary rocks. These advanced stages of rock transformation are called *metamorphism*. Even if no partial melting occurs, lateral pressure may severely fold the rocks, most of the minerals may undergo recrystallization, and pore space may be greatly reduced, sometimes almost to null. During high-grade metamorphism, any fossils contained in the rock are usually destroyed. Severe heating may alter sedimentary rocks to such an extent that they look like igneous rocks after such an event. The final product of the metamorphic alteration of a sandstone is quartzite, that of a shale is gneiss, and that of a limestone is marble.

B. Organic Matter in Sedimentary Rocks

Organic matter in various forms from light hydrocarbons to complex polymers and elemental carbon are common constituents of sedimentary rocks as far as the geological record goes back. The content of a rock of both organic matter and elemental carbon is usually expressed in terms of percent organic carbon ($\%C_{org}$). Percentages vary from a few hundred ppm to more than 50% (in the case of coal- or oil-bearing rocks). Numerous measurements of organic carbon contents by Trask and Patnode (1942) and by Ronov (1958) and Ronov and Yaroshevski (1969) have provided a fairly large data base, essentially for Phanerozoic sedimentary rocks. Using these data and Hunt's (1972) revised figures for the organic carbon contents of the chief sediment types, Schidlowski (1982) arrived at calculated mean values for total shales, carbonates, and sandstones of 0.77, 0.30, and $0.27\%C_{org}$, respectively. Weighted by the relative abundance of the host rocks, he calculated 0.50–0.60% for the organic carbon content of the "average" sedimentary rock. With a total mass of sedimentary rocks of 2.4×10^{24} g (Garrels and Lerman, 1977), the organic carbon inventory of the sedimentary shell of the earth totals $1.2–1.4 \times 10^{22}$ g C.

The data obtained for the Phanerozoic show a distinct variation of the C_{org} contents of shales and "average" sedimentary rock with time. Carbonates and sandstones exhibit the same temporal trend with a proportionately smaller amplitude (Ronov, 1958). The time curves show four maxima, Ordovician, Lower Carboniferous, Jurassic/Lower Cretaceous, Tertiary, and one pronounced minimum during the Triassic (e.g., Schidlowski, 1980). The principal correspondence of the time curves obtained for the sediments of the American and the Russian platform led Ronov (1958) to conclude that the variations occurred on a global scale.

Mean values obtained for the comparatively small number of C_{org} data from Precambrian sedimentary rocks (Cameron and Garrels, 1980; Ronov, 1980) are within the scatter of data for the Phanerozoic, suggesting that the average organic carbon contents of Precambrian rocks are not too different from those of geologically younger rocks of the same or similar type. The data base is still too limited to reveal any secular C_{org} variations during the Precambrian.

The organic carbon in sedimentary rocks may originate from an extraterrestrial, a terrestrial abiological, and/or a terrestrial biological source. Input of extraterrestrial organic or elemental carbon may have been important during the time period of major bombardment of the earth's surface by comets, meteorites, and interplanetary dust particles, the remains of the presolar nebula and the first stages of planet formation. As can be deduced from the analyses of the material collected on the moon's surface during the various Apollo missions, this bombardment must have been severe during the Hadean, 3.9–4.4 billion years ago. Larger bodies were probably seriously altered during passage through the earth's atmosphere due to frictional heating and destruction and perhaps even evaporation upon impact. Dust particles, however, landed virtually unaltered. From the different types of interplanetary debris that have reached the earth, meteorites are the most thoroughly studied species. Richest in carbon are the coaly chondrites, which may contain up to 4% C in the form of graphite-like coaly material, carbonates, and minor amounts of volatile organic compounds (e.g., Hayatsu and Anders, 1981).

As long as the environment was slightly reducing or neutral, a variety of organic compounds may have formed on the surface of the primitive earth from atmospheric gases such as CO_2, CH_4, N_2, NH_3, H_2, and H_2O with solar UV radiation as the energy source (Urey, 1952; Rubey, 1955; Miller and Urey, 1959; Miller, 1982) or during Fischer–Tropsch processes involving CO and H_2 [see reaction (4)]. Subsequent reactions under the influence of inorganic and organic catalysts (condensing agents) may have led to a variety of amino acids, peptides, and polypeptides in the surface water of the primordial oceans (Ponnamperuma, 1978). It is conceivable that such reactions occurred in the droplets of clouds, even before the oceans had formed. Together with any extraterrestrial organic or reduced carbon, the organic matter, which had formed on the earth's surface in one way or another, may have been slowly cycled through the surface layers of the primitive earth during the early stages of crustal formation as a result of chemical equilibrium processes such as

$$CH_4 + 2H_2O \rightleftharpoons CO_2 + 4H_2 \tag{1}$$

$$6FeCO_3 \rightleftharpoons 2Fe_3O_4 + 5CO_2 + C \tag{2}$$

$$C + CO_2 \rightleftharpoons 2CO \tag{3}$$

$$nCO + (2n+1) H_2 \rightleftharpoons C_nH_{2n+2} + nH_2O \tag{4}$$

$$C_nH_{2n+2} \rightleftharpoons nC + (n+1) H_2 \tag{5}$$

and others.

After life began, available abiogenic organic matter was utilized by the early organisms and thus gradually turned into biological matter. (The chance that there are sedimentary rocks in which significant amounts of abiogenic organic matter have been preserved without previous reworking by organisms is very small.) But this is only a part of the story. It appears that life, very early in its evolutionary history, overcame the restrictions of a limited supply of available organic matter by evolving biosynthetic pathways for the production of organic compounds from CO_2 and H_2. There is indirect evidence that the earliest form of photoautotrophy, i.e., the synthesis of organic matter from CO_2 and H_2 with visible light as the energy source, is at least 3.5 billion years old (e.g., Schidlowski, 1982). A simple calculation shows that by far the largest portion of the existing sedimentary organic matter must have been synthesized by autotrophic organisms from a carbon pool other than that in igneous rocks: Igneous rocks contain an average of 200 ppm by weight reduced or elemental carbon (Hoefs, 1973). This may be compared with the average value of about 150 ppm carbon for moon regolith (e.g., Cadogan et al., 1972). If the sediments had inherited their reduced carbon directly or after modification by chemical or biochemical processes from the weathered igneous rocks, they should show similar or lower organic carbon contents. The average organic carbon content of sedimentary rock is, however, 0.5–0.6% by weight, as was previously mentioned. That means that only a minor fraction of 3–4% of the total sedimentary organic carbon may ultimately originate from the reduced or elemental carbon of weathered igneous rocks. The fact that life very early invented biosynthesis of organic matter from CO_2 shows that, considering the total mass of preserved sedimentary organic carbon, Miller–Urey and Fischer–Tropsch-type processes could not have contributed significantly. Evidence derived from the relative abundances of ^{12}C and ^{13}C isotopes in sedimentary organic matter (see Section III,A) supports this view.

II. Fate of Organic Matter in a Sediment

Figure 1 illustrates in a schematic way the cycle(s) and the fate of biological material in aquatic environments and in younger sediments. A multitude of organic compounds is continuously cycled through the aquatic biota (living systems), which altogether currently represent a pool of organic matter of a size equivalent to about 2×10^{15} g C_{org}. Although the mean cycling time is only 1 month, calculated on the base of the rate of total net primary productivity of the aquatic biota, the residence time of organic matter (living and dead) in the upper layers of the oceans is much longer, on the order of 18 years. From the about 1.7 $\times 10^{15}$ g C_{org} that annually enter the deep oceans, only about 10% reaches the bottom sediments. Estimates of present burial rates of organic carbon range from 0.3 to 2.6×10^{14} g C per year (Junge et al., 1975). Hence, from the about 24 $\times 10^{15}$ g C per year that are fed by photosynthesis from the atmosphere into the

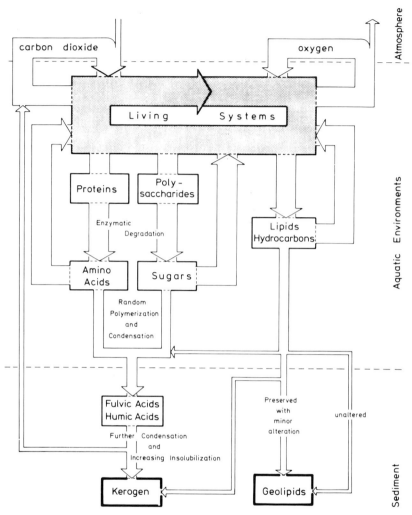

FIG. 1. Fate of biological material during sedimentation and diagenetic alteration in a sediment, resulting in kerogen and geolipids as the two main fractions of sedimentary organic matter. (From Hahn, 1982.)

aquatic biota, less than one-hundredth will end up in the bottom sediments of the oceans. More than 99% is lost by mineralization processes, i.e., microbial oxidation, which turns organic matter back into CO_2 and water. During the Archean, this loss was probably much smaller because there was very little free molecular oxygen available, and the supply of available sulfate limited mineralization. This may be an explanation for the otherwise somewhat surprising observation that the C_{org} values for Archean sedimentary rocks are in the same range as those for geologically younger sediments.

Organic matter may be preserved in a sediment with minor alteration or in a markedly altered form. Chemical and microbial degradation and transformation processes such as hydrolysis, reduction, or decarboxylation leave the carbon skeletons of the biomolecules essentially unchanged. But condensation, cyclization, and polymerization reactions lead to macromolecules with a higher degree of complexity. As shown in Fig. 1, lipids and hydrocarbons, after minor alteration, form the extractable and largely volatile constituents of sedimentary organic matter, called *geolipids*. Proteins and polysaccharides are the chief precursors of the particulate, rather inert, insoluble *kerogen*. Apart from local accumulations, resulting from the migration of geolipids in sedimentary strata, kerogen which may be considered as a cross-linked heteropolycondensate constitutes by far the largest portion of sedimentary organic matter. On the average, about 95% of the total of organic carbon is present in the form of kerogen.

Figure 2 is a schematic that shows what happens when sedimentary organic matter is buried increasingly deeper in the course of time. Increasing depth means growing pressure, but, more important, increasing temperature. After a predominantly biochemical transformation phase during the early stages of diagenesis, in which microbial activity gives rise to the formation of CO_2 and CH_4, the fractions of geolipids and kerogen begin to form. When the kerogen, as a result of increasing temperature, reaches a certain degree of maturation, the so-called "oil-window," a major portion of hydrocarbons is formed (e.g., Tissot and Welte, 1978). As the host rock has been subjected to the temperature and pressure conditions typical of incipient metamorphism, the entrapped organic matter begins to further disproportionate into a hydrogen-rich and a hydrogen-

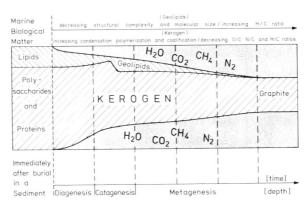

FIG. 2. Alteration from microbial and chemical processes of marine biological matter that had been buried in sediment. Ultimately, the alteration resulted in disproportionate amounts of a hydrogen-rich (methane) and a hydrogen-depleted (graphite) product.

poor fraction. In the process, kerogen looses, in addition to CH_4 and hydrogen, essentially all hetero-atoms such as oxygen, nitrogen, and sulfur. The average molecular weight of the geolipid fraction and the amount of geolipids present decrease. Finally, when the late stages of rock metamorphism are reached, the organic matter in the rock has been more or less completely transformed to graphite and methane, plus some water, CO_2, and nitrogen. The sulfur is lost chiefly in the form of H_2S, which in turn reacts with the mineral matrix to form pyrite. Temperature is the most critical parameter for the maturation and further transformation of sedimentary organic matter. The role of the mineral matrix in these processes is still poorly understood. It looks, however, as if the present state of sedimentary organic matter is not only a function of the chemical nature of its principal biological precursor molecules and its integrated time–temperature history, but also of the oxidation state of its depositional environment and of the mineralogical composition and type of host rock. There is a well-established relationship between burial temperature (depth), duration of heating, and rank (maturity, degree of coalification), which means that, in general, the older a sedimentary rock, the more altered is the syngenetic organic matter (Tissot and Welte, 1978). Thus, a syngenetic Precambrian kerogen, by virtue of its age, is likely to be extensively altered, unless burial has never been deep and the temperatures to which it has been exposed were not much higher than 60°–70°C. In this context, the findings by Price (1982), which are in contrast to this view, may be of interest. According to Price, the organogeochemical investigation of a core from a 7000 m-deep well into cretaceous sediments in South Texas showed that sedimentary rock from a metamorphic terrain of green schist facies (temperatures of more than 300°C) contained a significant amount of geolipids and that the kerogen, although according to reflectance data of high rank, still had hydrocarbon generating potential.

The time-span between burial of organic matter (deposition of the host sediment) and the beginning of major oil formation (principal zone of oil formation) may vary from a few million years to more than 300 million years. Therefore, it is not surprising that the ages of petroleum occurrences were found to cover a wide range from Oligocene to Precambrian, indicating that the rank/time factor for the alteration of sedimentary organic matter varies considerably.

III. Geochemical Fossils: Biological Markers

Usually, the solvent-extractable geolipids are found to form very complex mixtures of various classes of organic compounds such as hydrocarbons, esters, and carboxylic acids. There are also minor quantities of nitrogen and sulfur compounds. Mature geolipids consist to more than 90% of hydrocarbons. Many of the constituents of these mixtures may be considered *geochemical fossils*. Geochemical fossils are individual organic compounds of biological origin that have survived entrapment in a sediment and the processes which accompany lithification and rock transformation with their molecular structure remaining unchanged or little altered from the time when they were part of a living organism. Proteins, amino acids, sugars and amino sugars, porphyrins, and other pigments such as bile pigments (phycobilins) may also form geochemical fossils if they manage to escape degradation or incorporation into kerogen macromolecules.

Modern instrumentation for gas chromatography, liquid chromatography, and particularly for gas chromatography–mass spectrometry permits analysis of sedimentary rocks for geochemical fossils that are present in the lower parts per billion (ppb) range. It is the level of postdepositional organic compounds contaminating the syngenetic organic matter in sedimentary rocks that determines the lower detection limit for geochemical fossils and not the sensitivity of the analytical apparatus. Contamination often makes it difficult to substantiate the assumption that the organic compounds originate from organisms that existed at the time when the sediment was deposited, especially when the rocks contain very little organic matter (McKirdy and Hahn, 1982).

In cases where morphological fossils are absent, geochemical fossils can provide valuable information on ancient life processes and on major evolutionary events. This holds for ancient sediments in which organisms living at the time of their deposition did not leave any morphological traces, where such traces are ambiguous, or where rock transformation processes have destroyed the morphological record. Although the morphological record covers more than 3 billion years, geochemical fossils are in many cases the only remnants of organisms in ancient sedimentary rocks. Their diagnostic value as *biological markers* varies inversely with their structural and stereochemical complexity.

A. Abundance of Stable Carbon Isotopes as Biological Markers

A biological marker may be a discrete compound or a homologous series of compounds with a characteristic distribution pattern or any other feature, organic or inorganic, which can be firmly attributed to biological activity. The isotopic composition of the carbon species constituting the sedimentary organic matter is such a feature. There is a wealth of data on the abundance of stable carbon isotopes, especially for the Phanerozoic, in petroleum, coal, and sedimentary organic matter in general. Using a mass spectrometer, the ratio of the stable isotopes ^{13}C and the much more abundant ^{12}C is measured against a standard to determine the small variations, typically in the per mille range, with the necessary accuracy. The relative deviation of the sample from the standard (a Cre-

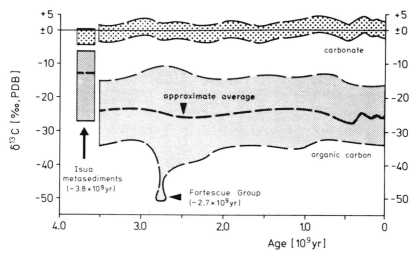

FIG. 3.   Isotopic composition of sedimentary organic carbon and carbonate carbon as a function of geological time. The relatively small variations for carbonate are markedly different from the wide scatter of $\delta^{13}C$ values for organic carbon. (From Schidlowski, 1982.)

taceous belemnite from the Peedee formation in South Carolina) is given as the so-called $\delta^{13}C$ value in ‰ PDB.

$$\delta^{13}C = \left[\frac{(^{13}C/^{12}C)_{sa}}{(^{13}C/^{12}C)_{st}} - 1 \right] \times 10^3 \ [‰ \ PDB]$$

Figure 3 is a compilation of all available data by Schidlowski (1982) for both sedimentary carbonate and organic carbon. Accompanied by a corresponding constancy of the $\delta^{13}C$ values of carbonates, the isotopic composition of sedimentary organic carbon has been remarkably constant. Average values vary within relatively narrow confines around $-25‰$ PDB from 3.5 billion years ago. As one goes from the Phanerozoic to the Precambrian, the spread of individual δ values, as indicated by the dark band, becomes larger. The negative extremes sometimes exceed $-40‰$, in some cases even $-50‰$ PDB.

Uptake and intracellular transport of CO_2, but chiefly the enzymes involved in biological carbon fixation from CO_2 or bicarbonate, discriminate in their preference for the lighter ^{12}C isotope. In particular, ribulose-1,5-bisphosphate carboxylase, which irreversibly incorporates CO_2 into the Calvin cycle to form a C_3 compound produces fractionations in the range of -25 to $-35‰$. The fractionations depend on pH, temperature, and metal ion availability. Phosphoenolpyruvate carboxylase, in contrast, incorporates bicarbonate into C_4 plants with a ^{13}C discrimination of only -2 to $-3‰$. For more details, see Schidlowski (1982).

Total fractionation may vary with plant type and environment. This leads to

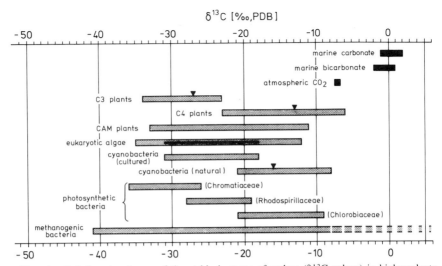

FIG. 4. Relative abundances of the stable isotopes of carbon ($\delta^{13}C$ values) in higher plants, algae, and various autotrophic bacteria, as compared with those in atmospheric carbon dioxide and in marine bicarbonate and carbonate. (From Schidlowski, 1982.)

approximate ranges of $\delta^{13}C$ values for various groups of plants and bacteria as shown in Fig. 4. There are only few isotope data for chemoautotrophic bacteria. Schidlowski (1982) recalculated the fractionations reported by Fuchs *et al.* (1979) for atmospheric CO_2 as a substrate and obtained a range of $+6$ to $-41‰$ for *Methanobacterium thermoautotrophicum* grown on CO_2 and H_2 at different gassing rates. Similar recalculation of data obtained by DeNiro *et al.* (1983) for *M. thermoautotrophicum*, *M. barkeri*, and *M. ivanovii* yielded values from -26.8 to $-37.8‰$. The methane generated by these bacteria is highly enriched in ^{12}C with $\delta^{13}C$ values reaching $-60‰$ or even lower. As a whole, the terrestrial biomass appears to be enriched in ^{12}C by about 25‰ as compared to the inorganic carbon of the crust–ocean system, which is mainly in the form of carbonate and bicarbonate with $\delta^{13}C$ values close to 0‰ PDB.

Both carbonate and organic carbon are initially incorporated into sediments without alteration of their isotopic ratios. However, during burial and diagenesis of the sediments and the concomitant alteration of the sedimentary organic matter, isotope ratios may be shifted by several per mille as a result of preferential loss of isotopically light lipids and biologically and thermally labile functional groups so that the fraction of the geolipids becomes increasingly lighter and the kerogen becomes heavier. During metamorphism, at temperatures of more than 400°C, exchange occurs between carbonate and organic (reduced) carbon leading to isotopically lighter carbonates and heavier kerogen.

With this in mind, the shift in carbon isotope ratios, observed in the metasediments from Isua, Greenland (Fig. 3), can be explained by the high-grade metamorphism these rocks have experienced (Schidlowski *et al.*, 1979). Hence, one may assume that the average $\delta^{13}C$ value of sedimentary organic carbon at 3.8 billion years B.P. was not too different from the value of $-25‰$. This, in turn, leads to the conclusion that life and biosynthesis of organic matter are as old as or even older than 3.8 billion years, since Fischer–Tropsch processes and Miller–Urey spark discharge syntheses just marginally reach the range of carbon isotope fractionations attributed to biological syntheses and their products preserved in sedimentary rocks. The absence of appreciable amounts of comparatively heavy organic matter ($\delta^{13}C$ values near $-15‰$ PDB) may be taken to exclude any significant contribution from such abiological processes to the kerogen of rocks at least as old as 3.5 billion years. In view of the negative extremes found in Late Archean kerogens, such as in the rocks of the Australian Fortescue and the South African Ventersdorp groups, the possibility that the organic matter in Precambrian sedimentary rocks could incorporate an isotopic signature from archaebacteria in the $^{13}C/^{12}C$ ratio is intriguing. Assimilation of ^{13}C-deficient CO_2 produced by microorganisms during the anaerobic oxidation of isotopically light CH_4, which had been released by methanogens, has been postulated to account for these light kerogens (Schoell and Wellmer, 1981; Schidlowski *et al.*, 1983).

B. ORGANIC COMPOUNDS AS BIOLOGICAL MARKERS

The optical activity of organic fossil material and of petroleum was first advanced as evidence of their biological origin by Walden (1906). The ubiquity of *amino acids* in proteins of living organisms led to the search for their preservation as well as to the study of their conversion to the thermodynamically more stable humus in soils, sediments, rocks, and macrofossils (for review of earlier work on amino acids, see Waksman, 1938). The same holds for *carbohydrates* (earlier work on carbohydrates in sediments was reviewed by Vallentyne, 1963). In view of the glycan strand of alternating *N*-acetyl-D-glucosamine and *N*-acetyl-L-talosaminuronic acid in β-1,3 linkages found in archaebacteria by König and Kandler (1979), it is interesting that β-1,3-glucans such as laminarans have been enzymatically detected in Devonian shales and the presence of *N*-acetylglucosane established (Rogers, 1965, as cited in Swain, 1969). Both amino acids and carbohydrates racemize during diagenesis in sediments and become part of the kerogen macromolecules as was explained in Section II. Since they are extremely soluble in water, the sedimentary record may have been contaminated by percolating waters with postdepositional material. Therefore, although they are characteristic of biological material, both amino acids and carbohydrates are of limited value as geochemical fossils.

In 1934, Treibs succeeded in identifying deoxophylloerythrin (DPEP) in bitumina and petroleum. Furthermore, he postulated that this compound arose by degradation from chlorophyll:

Chlorophyll *a*   Vanadyl-DPEP complex

(6)

practically coincident with the establishment of the structure of the latter. Since then, numerous papers have dealt with fossil *porphyrins* (petroporphyrins). The state of knowledge may be summarized as follows: Alkylmetalloporphyrins are known to be present in a wide range of petroleums, coals, and sedimentary rocks as old as 3.4 billion years (Kvenvolden and Hodgson, 1969; Schopf, 1970). They occur mainly as complex mixtures of nickel and vanadyl (VO) chelates of two major skeletal types, DPEP and etio (etioporphyrins). Minor homologous series are formed by a di-DPEP and two different rhodo-type al-kylbenzoporphyrins (Fig. 5) whose precise structures and origins are still not known. It is commonly believed that the major precursor of the two main series, DPEP and etio, is chlorophyll *a*. The range of carbon numbers can extend from C_{27} to higher than C_{39} (both DPEP and etioporphyrin III have 32 carbon atoms). That means that although the alkyl substituents of the porphyrin nucleus (R_1, R_2, . . ., R_8, in Fig. 5) are chiefly methyl and ethyl groups, alkyl side chains with more than 11 carbon atoms can occur. The high carbon number petroporphyrins seem to be generated in the zone of oil formation during thermal cracking of carbon–carbon bonds of kerogen macromolecules, which, at an earlier stage of diagenesis, had incorporated petroporphyrin precursors. Naturally occurring chlorophylls with extended alkyl side chains, such as the *Chlorobium* chlo-rophylls, may be important precursors in specific environments. With increasing burial depth, i.e., increasing temperature, a set of maturational changes occurs. Major effects associated with increasing degree of maturation (rank) appear to be (1) a decrease in the DPEP/etio ratio (which is typically well above unity in shallow sediment samples), possibly as a result of thermal cracking of the iso-cyclic ring in the DPEP compounds, (2) a decrease in the ratio of nickel por-phyrins to vanadyl porphyrins, (3) a decrease in average molecular weight result-ing from carbon–carbon bond cleavage in the side chains, (4) an increase in average carbon number as a result of the generation of high carbon number

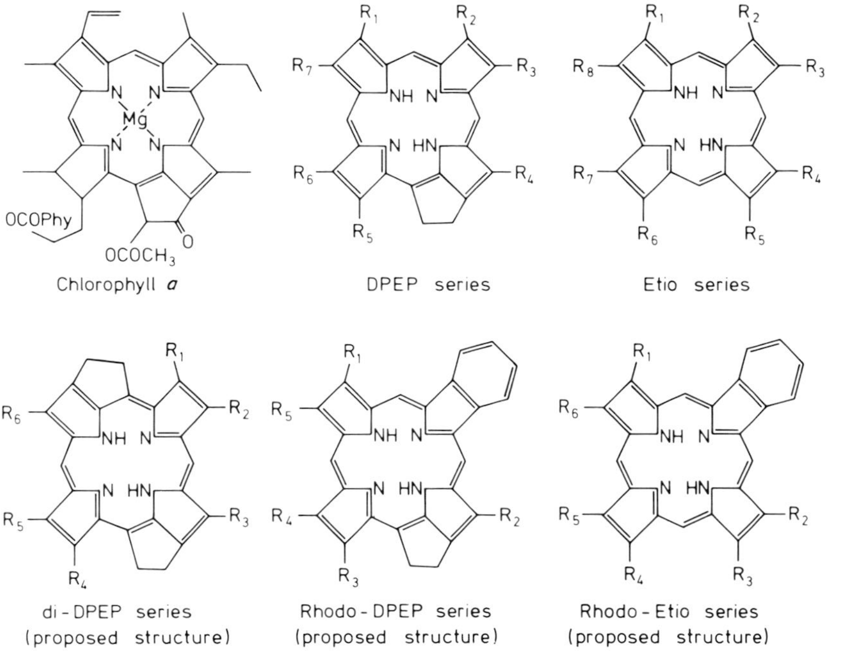

FIG. 5. Skeletons of chlorophyll *a* and various series of petroporphyrins found in petroleum, coal, and sedimentary rocks.

porphyrins during thermal cracking of carbon–carbon bonds in kerogen macromolecules which had previously incorporated petroporphyrin precursors, and, finally, (5) the disappearance of all petroporphyrins during the early stages of metagenesis (see Fig. 2). [For more details, see reviews of the organic geochemistry of porphyrins by Hodgson *et al.* (1968) and by Baker and Palmer (1978) and papers by Eglinton *et al.* (1980) and by Mackenzie *et al.* (1980)].

Recently, factor F_{430}, the prosthetic group of methyl-CoM reductase in methanogenic archaebacteria, has been shown to have a nickel tetrapyrrole structure (Diekert *et al.*, 1980). A careful examination led Pfaltz and colleagues (1982) to propose a structure for the porphinoid component of factor F_{430} as shown in Fig. 6. In analogy to the alteration of chlorophyll, diagenetic processes in a sediment may transform this molecule into a nickel–tetrahydrocorphine (THC) complex, which then could give rise to a suite of nickel porphinoid complexes with various degrees of hydrogenation (Fig. 6: I, II, III). Compounds with type-III structure have great similarity to DPEP compounds and may, therefore, mimic a DPEP series in mass spectral analysis, the only difference being a shift of the mass distribution pattern to higher masses by 14 mass units, which may easily be

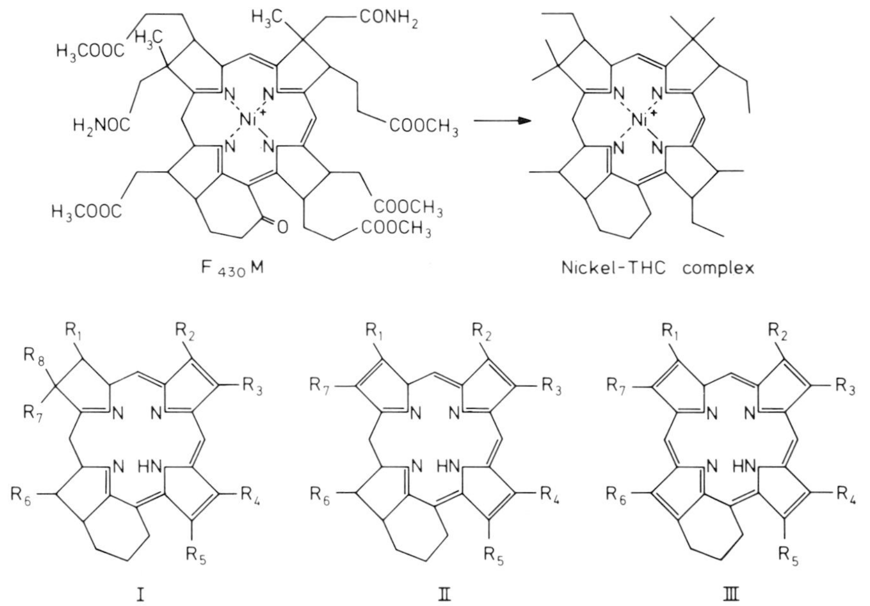

FIG. 6. Skeletons of the porphinoid component of factor F_{430}, the prosthetic group of methyl-CoM reductase in methanogenic archaebacteria, and of possible products of its diagenetic alteration in a sediment.

overlooked. Similarly, compounds with a structure intermediate between structures II and III may mimic an etio series in mass spectral analysis. Although the potential pseudo-DPEP series (Fig. 6, structure III) may be expected to possess a stability against thermal cracking which is quite similar to real DPEPs, the pseudo-etio series should be less stable than the real etioporphyrins.

There are no reports on the occurrence of these pseudo-DPEP and etio series in the literature, but published results of analyses of petroporphyrin mixtures provide indirect evidence. A significant contribution of factor F_{430}-derived porphin complexes to the mixture of nickel and vanadyl petroporphyrins found in sediments or petroleum should be apparent when samples of increasing rank are compared. From stability and diagenetic considerations, one would expect a clear predominance of DPEPs over etio series petroporphyrins in samples from shallow depths. The distribution pattern of demetalled DPEPs with respect to carbon number should be narrow with a steep maximum at mass 476 (DPEP). The distribution pattern of the demetalled etios should exhibit a flat maximum in the mass range 478–492 ($C_{32}H_{38}N_4$–$C_{33}H_{40}N_4$). With increasing sampling depth or rank, respectively, the amount of etio's would be expected to increase at the expense of DPEP compounds. The distribution pattern of the various DPEPs should grow wider due to the contribution of pseudo-DPEPs, and the distribution maximum (the high mass flank in particular) should become flatter. The opposite would be expected for the distribution pattern of the etio series. Here, the high mass flank of the distribution maximum should grow steeper because the less stable pseudo-etios disappear. Both the distribution maximum of the DPEP and that of the etio series shift toward lower masses (carbon numbers). Then with a further increase in rank, there will presumably be an increase in DPEPs due to liberation of high carbon number DPEPs from kerogen. The DPEP distribution pattern should grow increasingly flatter. The distribution maximum might even shift back to higher masses. A slight further increase in rank would then lead to the complete disappearance of DPEPs, soon followed by the etio series petroporphyrins. These postulations correlate very well with the observations of Baker and associates (1967) and of Mackenzie and associates (1980).

Additional indirect evidence for the occurrence of factor F_{430}-derived petroporphyrins comes from Eglinton and associates (1980) who found that the nickel-petroporphyrin series (which was isolated from gilsonite, a natural bitumen of Eocene age found in the Uinta basin in Utah) could not be fully separated by chromatographic techniques into DPEP and etioporphyrins. The authors obtained three fractions. All three of them were found to contain both DPEP and etio series porphyrins, suggesting that there were more than one series of both DPEP and etioporphyrins with overlapping chromatographic separation characteristics.

Radchenko and Sheshina (1955) reported that, in various crude oils, vanadium–porphyrin complexes were chiefly associated with asphaltic substances, whereas nickel–porphyrin complexes were associated with the oleaginous components (geolipids). Although this observation cannot be generalized and was even dismissed by later reviewers (e.g., Dunning, 1963), it appears interesting in the light of this discussion. Admittedly, none of the observations can be interpreted as convincing evidence for the occurrence of factor F_{430}-derived petroporphyrins, but it seems as if there is a good chance that one will find them once the analysis is directed toward these compounds. The effort appears worthwhile, since factor F_{430}-derived porphyrins, in addition to their general potential as biological markers, may provide information on the occurrence of methanogenic archaebacteria during the geological past.

A number of other natural pigments (such as various bile pigments, flavones, anthocyanins, carotenoids, and others) occur in both the plant and the animal kingdoms (Dunning, 1963). However, they are not stable enough to withstand diagenesis without major alteration because they do not form metal chelates. Furthermore, many are water-soluble and can migrate with percolating waters. *Carotenoids,* which consist mainly of long hydrocarbon chains forming an uninterrupted system of conjugated double bonds, are generally insoluble in water but dissolve easily in hydrocarbon mixtures and fats. Carotenoids occur in practically all types of plants. Several of these compounds have also been found in

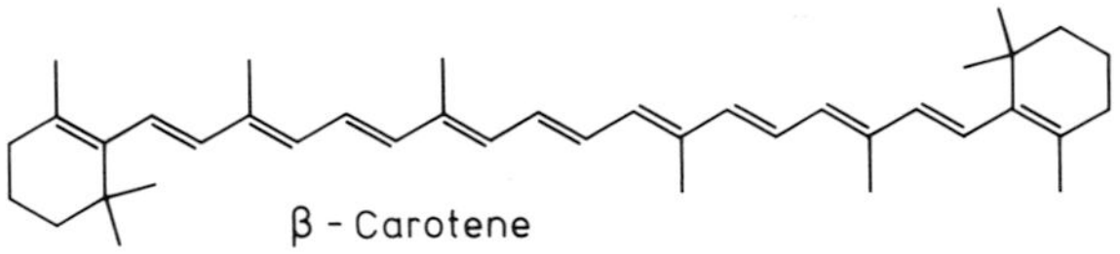

β - Carotene

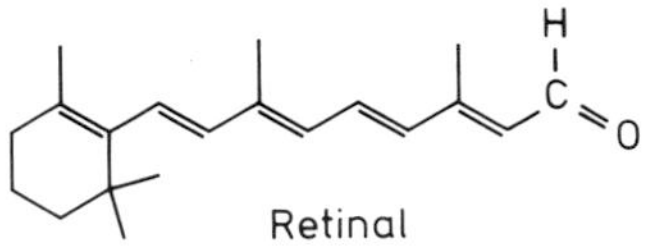

Retinal

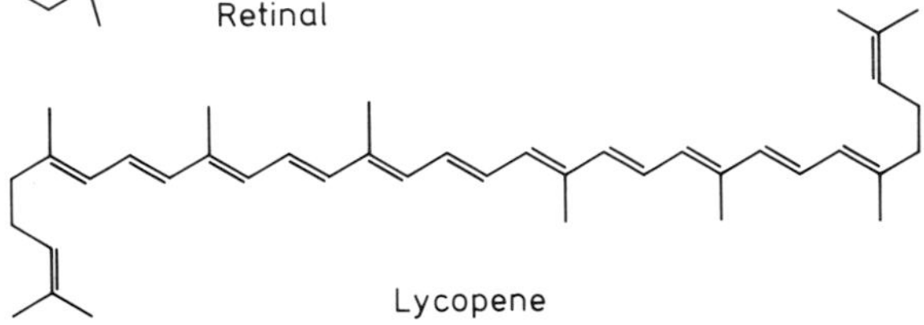

Lycopene

purple bacteria (Bogert, 1938). All known algae contain carotenes and most of them also carotenols (Rabinowitch, 1945). Hydrolysis of the central double bond in β-carotene leads to vitamin A (retinol). The analogous aldehyde, retinal, linked through a Schiff base to the ε-amino group of a lysine residue is the prosthetic group of retinal pigments. In animals, their only known function is that of light sensors. In extreme halophilic bacteria, however, they serve as both signal and energy transducers. Bacteriorhodopsin is the best characterized of these pigments (Stoeckenius *et al.,* 1979). Besides, there are at least two other retinal pigments in the same halobacteria cells, with maximum absorbance in the blue and near UV and in the green region of the spectrum (Lanyi, 1981). The presence of these pigments enables halobacteria to use sunlight as an energy source.

Carotenoids have been found in a number of recent sediments (e.g., Vallentyne, 1960; Schwendinger, 1969). The oldest sediments that have been shown to contain fully preserved carotenoids are an interglacial gyttja about 100,000 years old (Andersen and Gundersen, 1955) and the approximately 340,000 year-old marine sediments in the Cariaco trench off Venezuela (Watts and Maxwell, 1977). No carotenoids have been identified so far in lithified sediments, i.e., sedimentary rocks. Despite their resonance-stabilized structure, they are less stable than porphyrins. Many carotenoids are sensitive to heat and light or other forms of radiation. They are also extremely sensitive to oxidation. In addition, they are subject to reduction processes. The occurrence of partly hydrogenated carotenoids in Cariaco trench sediments (Watts and Maxwell, 1977), in deep-sea sediments (Simoneit and Burlingame, 1972), and in the 50 million-year-old Green River (Gallegos, 1971) and Messel shales (Kimble *et al.,* 1974), together with C_{40} hydrocarbons having the carbon skeleton of carotenoids, may be taken as evidence for the reduction of carotenoids during the early stages of diagenesis.

Perhydrated carotene, carotane, has been found in the Green River shale (Murphy *et al.,* 1967) and perhydrated lycopene, lycopane, in the Messel shale (Kimble *et al.,* 1974).

The time scale of reduction of carotenoids in a sediment seems to depend on the particular carotenoid. According to Watts and Maxwell (1977), canthaxanthin (a diketone derived from β-carotene) evidently lost up to five double bonds in the sediments of the Cariaco trench over a time-span of about 340,000 years, whereas the analogous diol, zeaxanthin, lost only up to two double bonds, and β-carotene was fully preserved. Heating experiments with pure carotenoids showed that toluene, *m*-xylene, 2,6-dimethylnaphthalene, and ionene are thermal degradation products of β-carotene (Zechmeister and von Cholnoky, 1930; Kuhn and Winterstein, 1933; Day and Erdman, 1963). The same products were observed in thermal degradation experiments at temperatures between 65° and 150°C with recent marine sediments containing β-carotene (Day and Erdman, 1963; Ikan *et al.,* 1975). The structures of the mono- and dimethylalkylbenzenes with alkyl side chains consisting of seven and more carbon atoms (see Fig. 7) found in petroleum (e.g., Mair and Barnewall, 1964) and in sediments such as the above-mentioned Green River shale (Gallegos, 1973) strongly suggest that their precursors were carotenoids. However, distinction of carotenoid fossils of (halophilic) archaebacteria from other carotenoid fossils must await more detailed studies.

A slightly different type of alkylbenzenes has recently been identified in the nonpolar lipids of the thermoacidophilic archaebacteria *Thermoplasma* and *Sulfolobus* (Langworthy *et al.,* 1982). All compounds were found to be monosubstituted alkylbenzenes with a mono- or dimethyl-branched isoalkane chain consisting of between 6 and 14 carbon atoms.

Because *fatty acids* are widely distributed in nature, particularly in the form of ester constituents of biological membranes, their presence in sediments has been widely studied. A marked predominance has been observed of normal (i.e., unbranched straight-chain) acids containing an even number of carbon atoms as compared with those with an odd number. This observation is in accord with the biochemical synthesis of these compounds involving malonyl coenzyme A to produce a predominance of the normal saturated C_{16} acid (palmitic acid) and the monounsaturated C_{18} acid (oleic acid). With increasing depth and sediment age, the even/odd carbon preference becomes less pronounced. Bacteria are believed to account for the 2-methyl (iso) and 3-methyl (anteiso) acids, particularly iso and ante/isopentadecanoic acid, found in sediments (Leo and Parker, 1966), and possibly for the *dicarboxylic acids* as well (Haug *et al.,* 1967). As far as is known, higher organisms and marine plankton do not contain enough branched-chain fatty acids to account for the relative proportions of these compounds observed in sediments without postulating selective preservation, which does not seem very likely. Therefore, one may take the ratio of methyl-branched to normal saturated C_{15} fatty acids as a fairly sensitive paleobiological indicator of the depositional environment. The isolation of dicarboxylic acids from the freshwater alga *Botryococcus braunii* (Douglas *et al.*, 1969a) suggests that, at least in the case of Carboniferous Scottish torbanite, this organism may have been the source of the dicarboxylic acids found (Douglas *et al.*, 1969b).

Cason and Graham (1965) were the first to isolate C_{14}, C_{15}, C_{19}, and C_{20} *isoprenoid acids* from a California petroleum. Just 1 year later, the C_{19} and C_{20}

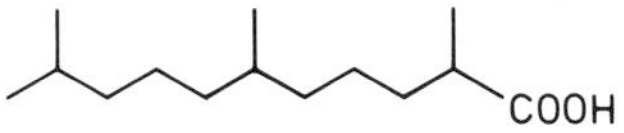

C_{14} isoprenoid acid

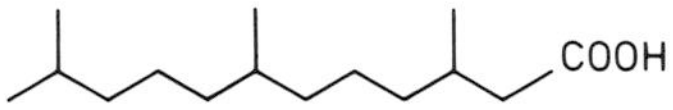

C_{15} isoprenoid acid

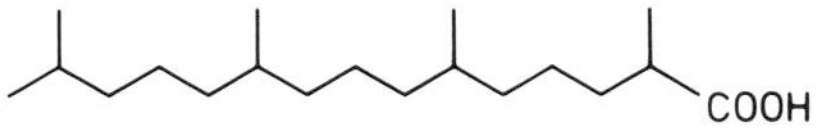

Pristanic acid

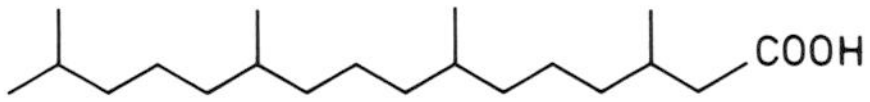

Phytanic acid

isoprenoid acids were isolated from Eocene Green River shale (Eglinton *et al.*, 1966). It is interesting that as early as 1966, Kates (as documented in Eglinton *et al.*, 1966) suggested that bacteria such as the halophiles could be responsible for some of the isoprenoid acids in sediments on the basis of the C_{20} isoprenoid diether found in *Halobacterium cutirubrum* (Kates *et al.*, 1965). In 1977, Anderson and colleagues isolated free and esterified C_{20} isoprenoid acid (phytanic acid) from Dead Sea sediments and determined its stereoisomeric composition to be 3R, 7R, 11R. In both the isolated ester-bound phytanic acid and the isolated free and phospholipid-bound di-O-phytanylglycerol, only the R,R,R-isomer was detected. The authors thus concluded that the various phytanyl compounds isolated from Dead Sea sediments are most likely derived from extremely halophilic bacteria rather than from phytol of chlorophyll origin. They observed that their results provide further evidence that the mixture of R,R,R- and S,R,R-phytanic acid isomers found in the lacustrine Green River shale (Maxwell *et al.*, 1973) were most likely derived from chlorophyll.

Organic acids have been found in sediments and sedimentary rocks as old as Cambrian, and the previous discussion shows that they would make very useful geochemical fossils. Since some fatty acids may be transported by seep waters, the geological record of these compounds must be evaluated with some reservations.

Decarboxylation of fatty acids leads to the more stable hydrocarbons so that for the *normal alkanes* (i.e., saturated straight-chain hydrocarbons), odd carbon numbers consequently predominate in recent and, to a lesser degree, in ancient sediments in accordance with the free radical decarboxylation mechanism proposed by Cooper and Bray (1963). Other processes which are important for hydrocarbon formation and transformation in sediments are reduction, cracking, cyclization, and aromatization of ring structures (see also Section II). The types of hydrocarbon geochemical fossils which have been found in ancient sediments are shown in Fig. 7. Due to the relative stability of the carbon–carbon bond against thermal alteration, the alkanes have the greatest chance to survive in a sediment or sedimentary rock from early periods of earth's history. Because normal alkanes may principally be synthezised by abiogenic Fischer–Tropsch processes, their value as biological markers is limited to characteristic ranges, maxima and odd/even carbon preferences. Besides *n*-alkanes, series of *2- and 3-methyl-branched alkanes* were also isolated from a number of sediments. The broad distribution of these compounds in the biosphere appeared adequate to account for their equally broad distribution in sediments. The *n-alkylcyclohexanes,* which have been found in sediments, were postulated to arise by cyclization and decarboxylation of 6,7-unsaturated fatty acids (Johns *et al.,* 1966).

The 50:50 mixture of 7- and 8-methylheptadecanes found in Bell Creek crude

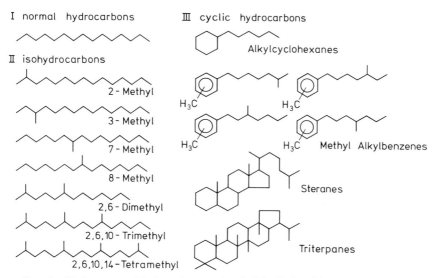

FIG. 7.   Various types of hydrocarbon geochemical fossils found in ancient sediments.

oil appears to be a biological marker for and geochemical fossils of blue-green algae from which they have also been isolated (Han *et al.*, 1968a,b; Han and Calvin, 1969a). In 1968, an unusual C_{34}-branched alkane, botryococcane, and

Botryococcane

analogs were isolated from the alga *Botryococcus braunii*. Later, they were also found in sediments and petroleum (Cox *et al.*, 1973) and have been used to trace the source of crude oils (Moldowan and Seifert, 1979; Seifert and Moldowan, 1981).

In 1965, Burlingame and colleagues reported the isolation and mass spectral characterization of $C_{27}-C_{29}$ steranes as well as a C_{30} triterpane from Green River shale, which have subsequently been extensively studied and further characterized (Murphy *et al.*, 1967; Hills *et al.*, 1966; Anderson *et al.*, 1969; Henderson *et al.*, 1969; Gallegos, 1971). It is believed that sedimentary *steranes* derive from sterols and *triterpanes* from triterpenes. Most eukaryotic organisms contain sterols, whereas prokaryotes have triterpanes (hopanoids) instead, and archaebacteria have neither sterols nor hopanoids. Ourisson and colleagues (1979) have given specific examples from *Methylococcus capsulatus* of how the early prokaryotic hopanoid enzyme system could have evolved to the more complex eukaryotic steroid system. They proposed that the hopanoids in ferns and few plants represent the obsolete products of captured cyanobacteria, which were captured by pre-eukaryotic organisms and subsequently converted into chloroplasts or other organelles. Before squalene cyclization enzymes had evolved, it appears likely that the function of membrane stabilization was filled by available acyclic isoprenoids.

Considerable importance was attached to the isolation of the regular *isoprenoids* (i.e., head-to-tail linked polymers of isoprene) and to the isolation of pristane and phytane from a wide variety of sediments (Dean and Whitehead, 1961; Bendoraitis *et al.*, 1962, 1963; Robinson *et al.*, 1965; Eglinton *et al.*, 1965; Johns *et al.*, 1966). Blumer and Thomas (1965) proposed that zooplankton converted the phytol group of chlorophyll through dihydrophytenol and phytanic acid to pristane and succeeded in isolating a number of pristenes as well. In a very elaborate synthetic work, Cox and colleagues (1970) and Maxwell and colleagues (1972, 1973) demonstrated that the tertiary carbons had stereospecificity corresponding to that of the biologically occurring precursors. More recently, it has been shown that, with increasing depth and temperature, pristane isomerizes at its optically active chiral center to a 1:1 mixture of its mesoform versus two enantiomers (Patience *et al.*, 1978). Using radioactive markers, Didyk and colleagues (1978) were able to demonstrate the biological conversion of phytol through phytanic acid and various phytenes and pristenes to pristane and phytane. An abbreviated scheme for this conversion is shown in Fig. 8. The accompanying C_{15}, C_{16}, and C_{18} isoprenoids were postulated to be derived from the phytyl side chain of chlorophyll by a complex sequence of reactions involving oxidation, decarboxylation, reduction, and cracking. Bayliss demonstrated in 1968 their production from chlorophyll in the presence of hydrogen at elevated temperature. With the possible exception of the 2 billion-year-old Antrim shale, the unique absence of the C_{17} isoprenoid, which required cleavage of two bonds to the same carbon atom, was taken as additional support for this mechanism (McCarthy and Calvin, 1967).

The occurrence in sediments of larger, regular (head-to-tail) isoprenoids up to C_{30} required a not-yet discovered precursor, because the only known potential precursors are produced by organisms which are too highly evolved and are too limited in occurrence to be considered. Waples *et al.* (1974) found the regular C_{25} isoprenoid alkane to be the major organic component of a saline, lagoonal Tertiary sediment, and they correctly predicted that eventually an adequate biological precursor would be found.

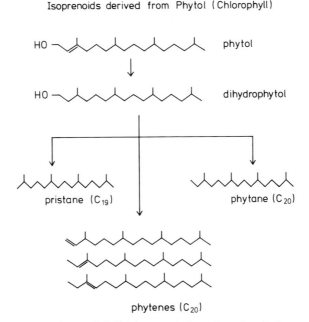

FIG. 8. Formation of pristane (2,6,10,14-tetramethylpentadecane) and phytane (2,6,10,14-tetramethylhexadecane) from phytol (chlorophyll). (From Hahn, 1982.)

In 1972, Hoering examined the organic matter from a highly reduced core from the Cariaco trench using lithium aluminum deuteride. He presented strong mass spectral evidence for the presence of a head-to-head diphytyl compound, which he suggested was originally an alcohol (although his procedure would not exclude a monoether). He considered this unique compound to be closely related to the red plant pigment lycopene ($C_{40}H_{56}$), but noted that, in contrast to the compound that he isolated, lycopene consists of two C_{20} units linked head-to-tail about the point of symmetry of the molecule. The relationship to the diphytanylglycerol ether compounds isolated by Kates et al. (1965, 1966) from the lipids of extreme halophiles of the genus *Halobacterium* and by Kaplan and Baedecker (1970) from Dead Sea sediments was not apparent at that time. It was not until several years later that it was recognized that archaebacteria are the most likely source of head-to-head linked isoprenoids.

After Kates and colleagues (see review by Kates, 1978) had established that halophilic archaebacteria contain only diphytanylglycerol diethers in their glycerolipid fraction, diglycerol tetraethers with two biphytanyl chains (consisting of two head-to-head linked C_{20} units each; see Fig. 9) were recognized more recently as major constituents of the glycerolipids of the thermoacidophilic archaebacteria *Thermoplasma* and *Sulfolobus* and of several methanogens. The distribution of diether and tetraether glycerolipids has been found to vary within the different archaebacterial subgroups from 100% diethers in halophiles, for example, to only 5% diethers and 95% tetraethers in some thermoacidophiles (see, e.g., review by Langworthy et al., 1982). The tetraethers of thermoacidophiles can contain between one and four cyclopentyl rings (De Rosa et al., 1977, 1980a,b,c). The degree of cyclization is different for thermoacidophiles and methanogens. While *Sulfolobus* possesses diglycerol tetraethers with biphytanyl chains, which contain up to four cyclopentyl rings and *Thermoplasma* up to two of such rings, the tetraethers of methanogens contain only acyclic biphytanyl chains as indicated in Fig. 9.

In the nonpolar lipids of methanogens and thermoacidophiles, regular C_{15}–C_{30} isoprenoid alkanes, squalene, a C_{30} isoprenoid consisting of two regular C_{15} isoprenoid units linked tail-to-tail, and a C_{25} squalene homologue have been identified (Tornabene et al., 1979; Holzer et al., 1979). The structures of the isoprenoids found in the nonpolar lipid fraction of archaebacteria are shown in Fig. 10.

In 1979, Moldowan and Seifert identified a series of head-to-head linked C_{32}–C_{40} isoprenoid alkanes, the largest homolog being biphytanyl, together with a series of regular C_{18}–C_{36} isoprenoid alkanes in a Miocene crude oil from California. They concluded that the biphytanyl and its lower homologs represent "diagenetic debris" from one or more precursors containing the head-to-head linked C_{20}–C_{20} isoprenoid chain. They cited the head-to-head linked "biphytanediol constituent" found by De Rosa et al. (1977) in the cell-wall mem-

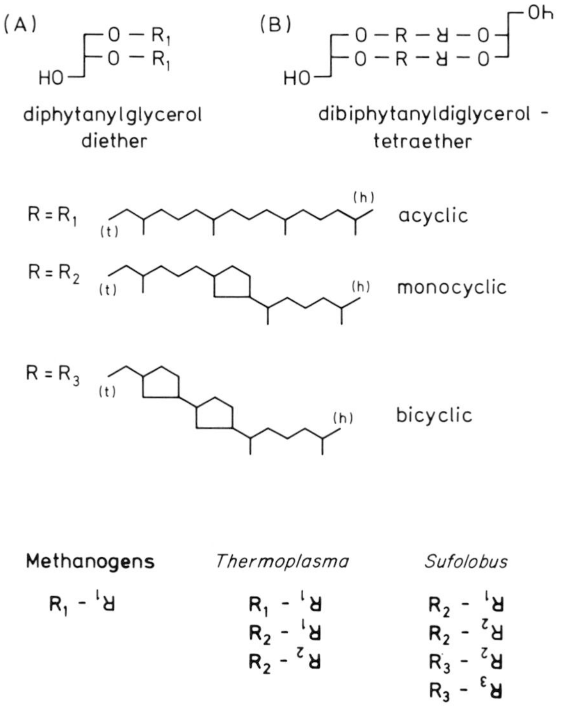

(A)

$$-O - R_1$$
$$-O - R_1$$
HO—

diphytanylglycerol
diether

(B)

$$-O - R - Я - O$$
$$-O - R - Я - O$$
HO—
$$—Oh$$

dibiphytanyldiglycerol –
tetraether

R = R₁   (h) (t)   acyclic

R = R₂   (t) (h)   monocyclic

R = R₃   (t) (h)   bicyclic

Methanogens	Thermoplasma	Sufolobus
$R_1 - Я^1$	$R_1 - Я^1$	$R_2 - Я^1$
	$R_2 - Я^1$	$R_2 - Я^2$
	$R_2 - Я^2$	$R_3 - Я^2$
		$R_3 - Я^3$

FIG. 9. Various isoprenoid glycerol ethers from the polar cell-wall lipids of archaebacteria.

branes of archaebacteria of the *Caldariella* group (i.e., *Sulfolobus*) as a potential precursor prototype. In the same year, Albrecht and colleagues (Michaelis and Albrecht, 1979; Chappe *et al.*, 1979) brominated the ether linkages (chemical knife) in the Eocene Messel oil shale kerogen and converted the resulting bromides to hydrocarbons with lithium aluminum deuteride. In this way, they were able to establish the presence of the ω,ω-diether of a C_{40} head-to-head linked isoprenoid. Also identified was a diether with a 1,4-cyclopentyl ring, as well as the monoethers of phytane and 2-methyltetradecane and the diether of a head-to-head linked dimer of 2-methyltetradecane. The first two compounds were shown to be identical with those observed by De Rosa *et al.* (1977) for the thermoacidophiles of the *Caldariella* (*Sulfolobus*) group. Since the depositional environment of the Messel oil shale could not, at any time in its geological history, have supported thermophiles, thermoacidophiles, or halophiles, the authors concluded that methanogens had produced these analogs.

Squalane was identified in a Libyan crude oil (Hills *et al.*, 1970) and in a Nigerian petroleum (Gardner and Whitehead, 1972). More recently, squalane and a C_{25} homolog (2,6,10,15,19-pentamethyleicosane) were found in Cre-

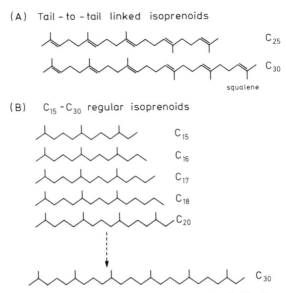

(A) Tail - to - tail linked isoprenoids

C_{25}

C_{30}

squalene

(B) C_{15} - C_{30} regular isoprenoids

C_{15}

C_{16}

C_{17}

C_{18}

C_{20}

C_{30}

FIG. 10. Isoprenoid alkenes and alkanes from the nonpolar lipids of archaebacteria.

taceous marine sediments. Because the presence of methanogens and/or high methane concentrations were found in those sediments and 2,6,10,15,19-pentamethyleicosane has only been found in methanogens, this tail-to-tail linked C_{25} isoprenoid may be a biological marker for methanogenic archaebacteria (Brassell *et al.*, 1981).

Geochemical studies of evaporitic source rocks and crude oils [Mesozoic and Cenozoic (Miocene) age] from offshore drillings in the Mediterranean Sea near Spain have shown a predominance of even carbon numbers for the *n*-alkanes in the C_{20}–C_{30} range (Albaiges and Torradas, 1977). There was also a predominance of the C_{35} 17α-H-hopane (triterpane, see Fig. 7) over the C_{31}–C_{34} homologs. Such features are generally associated with bacterial activity in highly saline carbonate environments (Dembicki *et al.*, 1976; Van Dorsselaer *et al.*, 1977). More recently, three different types of acyclic isoprenoid alkanes have been identified in these oils and source rocks, namely a regular series up to C_{45}, a head-to-head linked series up to C_{40}, and a new C_{21}–C_{39} series with methyl branches in the 3,7,11,15, . . ., positions (Albaiges *et al.*, 1978; Albaiges, 1980). The new series requires an isoprenoid precursor with more than 40 carbon atoms.

McKirdy and Kantsler (1980) investigated crude oils and their source rocks of Cambrian age from several drillings in the Officer Basin in western South Australia. The oils occupy vugs and partly healed fractures in an alkaline playa–lake sequence of bituminous carbonate and argillite (clay), which contain calcite pseudomorphs of sodium carbonate–bicarbonate evaporitic minerals (lacustrine evaporitic facies). The oils were found to be immature and rich in the regular C_{13}–C_{25} acyclic isoprenoid alkanes and squalane. The authors cited lipids of benthic algae, probably blue-greens, heterotrophic sulfate-reducing, and halophilic bacteria as likely biological precursor materials. They noted that the regular C_{13}–C_{25} isoprenoids and squalane present in relatively high concentrations both in the crude oils and in their source rocks may be biological markers of "autotrophic halophilic and/or methanogenic bacteria."

The discussion in this section on organic compounds as biological markers is summarized in Tables II-IV. Table II deals with sedimentary hydrocarbons as potential biological markers. Items c and d in section B of Table II are shown in more detail in Table III. Information on sedimentary organic compounds other than hydrocarbons is compiled in Table IV. The emphasis is on facts and findings that are relevant for tracing archaebacteria in ancient sediments.

TABLE II

SEDIMENTARY HYDROCARBONS WITH POTENTIAL AS BIOLOGICAL MARKERS

Characteristic feature	Likely precursor	Biological source(s)
A. Normal alkanes		
(a) C_{12}–C_{21} with odd carbon-number preference; C_{15} or C_{17} dominant	n-Fatty acids, n-fatty acid esters, free normal alkenes, and alkanes	Algae, bacteria, cyanobacteria
(b) C_{22}–C_{35} with odd carbon-number preference	n-Fatty acids, n-fatty acid esters, free normal alkenes, and alkanes	*Botryococcus* type chlorophytes, certain cyanobacteria
(c) C_{20}–C_{32} with even carbon-number preference	n-Fatty acids, n-fatty acid esters, free normal alkenes, and alkanes	Anaerobic bacteria in highly saline carbonate environments (e.g., sabkhas)
(d) C_{22} prominent	Unsaturated $C_{22:6}$ fatty acid, free normal C_{22} alkane	Dinoflagellates, zooplankton
B. Branched alkanes		
(a) C_{16}–C_{30} iso- and ante-isoalkanes	Fatty acids, fatty acid esters, free alkenes, and alkanes	Bacteria
(b) 7- and 8-methylheptadecane	Free alkanes	Cyanobacteria
(c) C_{13}–C_{20} regular isoprenoids	Phytol from chlorophylls, diphytanylglycerol diether	Green plants and algae, phytoplankton, halophilic archaebacteria
(d) C_{21}–C_{40} regular and	Free isoprenoids, di-	Archaebacteria

(continued)

TABLE II (*Continued*)

Characteristic feature	Likely precursor	Biological source(s)
irregular isoprenoids	biphytanyldiglycerol tetraether	
(e) Botryococcane	Free alkene/alkane	*Botryococcus braunii*
(f) Carotane, lycopane	Carotenoids	Plants, algae, purple bacteria
C. Cyclic alkanes		
(a) Steranes	Steroids	Animals, plants, algae, fungi, cyanobacteria
(b) 4-Methylsteranes	4-Methylsteroids	Dinoflagellates, methanotrophic bacteria
(c) C_{27}–C_{35} pentacyclic triterpanes (hopanes)	Triterpenoids	Prokaryotes
(d) *n*-Alkylcyclohexanes	Unsaturated fatty acids, free alkanes	Algae, thermoacidophilic bacteria
D. Aromatic hydrocarbons		
(a) Mono- and dimethyl-alkylbenzenes	Carotenoids	Plants, algae, purple bacteria
(b) Mono- and dimethyl-branched alkylbenzenes	Free hydrocarbons	Thermoacidophilic archaebacteria
(c) Mono- and poly-aromatic steranes and triterpanes	Steroids and triterpenoids	Animals, plants, fungi, algae, bacteria
(d) Other polyaromatic hydrocarbons (e.g., perylene)	Lignin (?)	Higher plants

From all sedimentary organic compounds which may represent geochemical fossils of archaebacteria, the isoprenoid alkanes with carbon skeletons $>C_{20}$, especially the head-to-head and the tail-to-tail linked isoprenoid alkanes, are the least ambiguous biological markers. How much of the isoprenoid alkanes $\leqslant C_{20}$ found in ancient sediments and petroleums stems from archaebacteria and how much is derived from phytol or chlorophyll, respectively, is an open question, which may be answered in cases where sufficient geochemical and sedimentological background information is available. The stereospecificity of phytanic acid, phytane, or pristane is a very useful parameter in this respect. Unfortunately, it gradually disappears with increasing diagenesis. Table V lists ancient sediments and petroleums, which were found to contain pristane and phytane as well as isoprenoid alkanes other than pristane and phytane. Although certainly not complete, this listing indicates how far the sedimentary record of occurrences of archaebacteria reaches back in time. There is evidence that the geolipids extracted from the 2.7 billion-year-old Soudan Shale are geologically younger

TABLE III

Sᴇᴅɪᴍᴇɴᴛᴀʀʏ Iꜱᴏᴘʀᴇɴᴏɪᴅ Aʟᴋᴀɴᴇꜱ ᴡɪᴛʜ Pᴏᴛᴇɴᴛɪᴀʟ Aꜱ Bɪᴏʟᴏɢɪᴄᴀʟ Mᴀʀᴋᴇʀꜱ

Characteristic feature	Likely precursor	Biological source(s)
(a) C_{13}–C_{20} regular isoprenoids	Phytol from chlorophylls, diphytanylglycerol diether from polar lipids	Green plants and algae, phytoplankton, halophilic archaebacteria
(b) C_{15}–C_{30} regular isoprenoids	Free isoprenoid alkenes/alkanes from nonpolar lipids	Methanogenic and thermoacidophilic archaebacteria
(c) C_{21}–C_{40} head-to-head linked isoprenoids (acyclic)	Dibiphytanyldiglycerol tetraether from polar lipids	Methanogenic and thermoacidophilic archaebacteria
(d) C_{21}–C_{40} head-to-head linked isoprenoids (monocyclic)	Dibiphytanyldiglycerol tetraethers from polar lipids	Thermoacidophilic archaebacteria
(e) C_{21}–C_{40} head-to-head linked isoprenoids (bicyclic)	Dibiphytanyldiglycerol tetraethers from polar lipids	*Sulfolobus*
(f) C_{25} tail-to-tail linked isoprenoids	Free isoprenoid alkane/alkenes from nonpolar lipids	Methanogenic archaebacteria
(g) C_{30} tail-to-tail linked isoprenoids (squalane/squalenes)	Free isoprenoid alkane/alkenes from nonpolar lipids	Most archaebacteria

than the host rock. Certainly, the sedimentary record of occurrences of archaebacteria extends well into the Precambrian.

The similar structural distribution of the isoprenoids (alkanes, alkenes, ethers, and acids) from both the geolipid and the kerogen fractions of ancient sediments and from petroleum with that found in the lipids of archaebacteria suggests that a major amount of sedimentary organic matter has derived from the latter. Approximation of the relative contribution of chemotrophic archaebacteria as opposed to that of phototrophic bacteria and algae is difficult. However, it is apparent that a major portion of the organic matter orginally assimilated by cyanobacteria (and phytoplankton) has been reworked by archaebacteria (methanogens), at least as far back as 1 billion years and possibly as far back as 2.7 billion years ago.

The Archean may have been the golden age of the methanogens. During this early period of earth's history, they were probably not restricted to special environments because free molecular oxygen, which is poisonous to them, existed at most in trace concentrations in the global environment. Since they are not dependent on sunlight, the methanogens probably had an advantage over phototrophic bacteria in the competition for available molecular hydrogen as an

TABLE IV

SEDIMENTARY ORGANIC COMPOUNDS OTHER THAN HYDROCARBONS
WITH POTENTIAL AS BIOLOGICAL MARKERS

Characteristic feature	Likely precursor	Biological source(s)
A. Porphyrins		
(a) $C_{27}–C_{32}$ DPEP and etiotype porphyrins (Ni and vanadyl complexes)	Chlorophyll a, bacteriochlorophyll a	Green plants and algae, phytoplankton, photosynthetic bacteria
(b) $>C_{32}$ DPEP and etiotype porphyrins (Ni and vanadyl complexes)	Chlorophyll a, bacteriochlorophyll a, *Chlorobium* chlorophylls	Green plants and algae, phytoplankton, photosynthetic bacteria (especially green sulfur bacteria)
(c) $C_{27}–C_{33}$ pseudo DPEP and etio-type porphyrins	Factor F_{430}	Methanogenic archaebacteria
B. Fatty acids		
(a) n-$C_{12}–C_{22}$ fatty acids with even carbon-number preference	Free fatty acids, fatty acid esters	Lacustrine and marine plants and algae, plankton
(b) n-$C_{22}–C_{32}$ fatty acids with even carbon-number preference	Plant–wax esters	Land plants
(c) iso-$C_{14}–C_{18}$ fatty acids, iso-C_{15} dominant	Free fatty acids, fatty acid esters	Bacteria
C. Isoprenoid acids		
(a) C_{14}, C_{15}, C_{19}, and C_{20} regular isoprenoid acids	Phytol from chlorophylls, diphytanylglycerol diether	Green plants and algae, phytoplankton, halophilic archaebacteria
(b) Mixture of RRR- and SRR-phytanic acid stereoisomers	Chlorophyll a	Green plants and algae, phytoplankton
(c) Only RRR-phytanic acid	Diphytanylglycerol diether	Halophilic archaebacteria

electron-donor for the synthesis of organic matter from CO_2. Furthermore, they were able to subsist everywhere on the excreta and remains of other organisms, including those of their competitors. Conditions, however, changed when cyanobacteria evolved, because the cyanobacteria were able to use water as an electron-donor. For the methanogens, one may speculate that the spread of water-splitting photosynthesis led to the pollution of the global habitat with toxic oxygen, forcing them to retreat into (anoxic) ecological niches.

TABLE V

SEDIMENTARY RECORD OF C_{14}–C_{45} ISOPRENOID HYDROCARBON GEOCHEMICAL FOSSILS

Sediment/petroleum	Geological period/age	Carbon-number range	Linkage of major isoprenoid elements	Reference
Abott Rock Oil (United States)	3×10^6 years	C_{18}–C_{21}	Head-to-tail	Johns et al., 1966
Californian Petroleum (United States)	Miocene	C_{18}–C_{36}	Head-to-tail	Moldowan and Seifert, 1979
Castellon Crude Oil (Spain)	Miocene	C_{32}–C_{40}	Head-to-head	Albaiges, 1980
		C_{20}–C_{45}	Head-to-tail	
		C_{32}–C_{40}	Head-to-head	
		C_{21}–C_{39}	Head-to-tail (new series)	
Rheingraben Dolomite (Corbicula facies, W. Germany)	Tertiary	C_{25}	Head-to-tail	Waples et al., 1974
San Joaquin Valley Oil (United States)	30×10^6 years	C_{16}, C_{18}–C_{20}	Head-to-tail	Johns et al., 1966
Green River Shale (United States)	50×10^6 years	C_{15}, C_{16}, C_{18}–C_{20}	Head-to-tail	Robinson et al., 1965
Messel Oil Shale (W. Germany)	50×10^6 years	C_{20}–C_{40}	Head-to-tail	Michaelis and Albrecht, 1979
		C_{20}–C_{40}	Head-to-head	
Costa Rican Seep Oil (Costa Rica)	?	C_{24}, C_{25}, C_{28}, C_{30}	Head-to-tail	Haug and Curry, 1974
Nigerian Petroleum (Nigeria)	?	C_{30}	Tail-to-tail	Gardner and Whitehead, 1972
Moroccan Basin Turbidite (Morocco)	130×10^6 years	C_{25}, C_{30}	Tail-to-tail	Brassell et al., 1981

(continued)

TABLE V (*Continued*)

Sediment/petroleum	Geological period/age	Carbon-number range	Linkage of major isoprenoid elements	Reference
Bell Creek Crude Oil (United States)	135×10^6 years	$C_{22}-C_{25}$	Head-to-tail	Han and Calvin, 1969b
Moonie Crude Oil (Australia)	200×10^6 years	C_{15}, C_{16}, $C_{18}-C_{21}$	Head-to-tail	Van Hoeven et al., 1966
Antrim Shale (United States)	265×10^6 years	C_{16}, $C_{18}-C_{21}$	Head-to-tail	McCarthy et al., 1967; McCarthy and Calvin, 1967
Windy Knoll Bitumen (England)	Lower Carboniferous	$C_{14}-C_{23}$	Head-to-tail	Nooner et al., 1973
Namibian Metasediments (Quartz inclusions; southwestern Africa)	700×10^6 years	$C_{14}-C_{20}$	Head-to-tail	Kvenvolden and Roedder, 1971
Observatory Hill Beds (Carbonates, Evaporites; South Australia)	Early Cambrian	$C_{13}-C_{25}$, C_{30}	Head-to-tail Tail-to-tail	McKirdy and Kantsler, 1980
Observatory Hill Bed Oil (South Australia)	Early Cambrian	$C_{13}-C_{25}$, C_{30}	Head-to-tail Tail-to-tail	McKirdy and Kantsler, 1980
Nonesuch Shale (United States)	1000×10^6 years	C_{16}, C_{19}	Head-to-tail	Eglinton et al., 1965
Nonesuch Seep Oil (United States)	1000×10^6 years	C_{15}, C_{16}, C_{19}, C_{20}, C_{21}	Head-to-tail	Johns et al., 1966
Soudan Shale (United States)	2700×10^6 years	$C_{18}-C_{21}$	Head-to-tail	Johns et al., 1966

REFERENCES

Albaiges, J. (1980). Identification and geochemical significance of long chain acyclic isoprenoid hydrocarbons in crude oils. *In* "Advances in Organic Geochemistry, 1979" (A. G. Douglas and J. R. Maxwell, eds.), pp. 19–28. Pergamon, Oxford.

Albaiges, J., and Torradas, J. (1977). Geochemical characterization of Spanish crude oils. *In* "Advances in Organic Geochemistry, 1975" (R. Campos and J. Goni, eds.), pp. 99–115. ENADISMA, Madrid.

Albaiges, J., Borbon, J., and Salagre, P. (1978). Identification of a series of C_{25}-C_{40} acyclic isoprenoid hydrocarbons in crude oils. *Tetrahedron Lett.* **6**, 595–598.

Alfvén, H., and Arrhenius, G. (1976). Earth's ocean and the formation of the solar system. *In* "Evolution of the Solar System," pp. 483–503. NASA Sci. Tech. Inf. Office, Washington, D.C.

Andersen, S. T., and Gundersen, K. (1955). Ether soluble pigments in interglacial gyttja. *Experientia* **11**, 345–348.

Anderson, P. C., Gardner, P. M., Whitehead, E. V., Anders, D. E., and Robinson, W. E. (1969). The isolation of steranes from Green River oil shale. *Geochim. Cosmochim. Acta* **33**, 1304–1307.

Anderson, R., Kates, M., Baedecker, M. J., Kaplan, I. R., and Ackman, R. G. (1977). The stereoisomeric composition of phytanyl chains in lipids of Dead Sea sediments. *Geochim. Cosmochim. Acta* **41**, 1381–1390.

Baker, E. W., and Palmer, S. E. (1978). Geochemistry of porphyrins. *In* "The Porphyrins" (D. Dolphin, ed.), Vol. 1, pp. 485–551. Academic Press, New York.

Baker, E. W., Yen, T. F., Dickie, J. P., Rhodes, R. E., and Clark, L. F. (1967). Mass spectrometry of porphyrins. II. Characterization of petroporphyrins. *J. Am. Chem. Soc.* **89**, 3631–3639.

Bayliss, G. S. (1968). The formation of pristane, phytane and related hydrocarbons by thermal degradation of chlorophyll. *Pap. 155th Natl. Meet. Am. Chem. Soc.*

Bendoraitis, J. G., Brown, B. L., and Hepner, L. S. (1962). Isoprenoid hydrocarbons in petroleum: Isolation of 2,6,10,14-tetramethyl-pentadecane by high temperature gas-liquid chromatography. *Anal. Chem.* **34**, 49–53.

Bendoraitis, J. G., Brown, B. L., and Hepner, L. S. (1963). Isolation and identification of isoprenoids in petroleum. *World Pet.* **34**, No. 6, 13–29.

Blumer, M., and Thomas, D. W. (1965). Phytadienes in zooplankton. *Science* **147**, 1148.

Bogert, M. T. (1938). Carotenoids, the polyene pigments of plants and animals. *In* "Organic Chemistry" (H. Gilman, ed.), Vol. II, Chapter 14. Wiley, New York.

Brassell, S. C., Wardroper, A. M. K., Thomson, I. D., Maxwell, J. R., and Eglinton, G. (1981). Specific acyclic isoprenoids as biological markers of methanogenic bacteria in marine sediments. *Nature (London)* **290**, 693–696.

Burlingame, A. L., Haug, P., Belsky, T., and Calvin, M. (1965). Occurrence of biogenic steranes and pentacyclic triterpanes in an Eocene shale (52 × million years) and in an Early Precambrian shale (2.7 × billion years): A preliminary report. *Proc. Natl. Acad. Sci. U.S.A.* **54**, 1406–1412.

Cadogan, P. H., Eglinton, G., Firth, J. N. M., Maxwell, J. R., Mays, B. J., and Pillinger, C. T. (1972). Survey of lunar carbon compounds. II. The carbon chemistry of Apollo 11, 12, 14, and 15 samples. *Geochim. Cosmochim Acta, Suppl.* **3**, 2069–2090.

Cameron, E. M., and Garrels, R. M. (1980). Geochemical compositions of some Precambrian shales from the Canadian Shield. *Chem. Geol.* **28**, 181–197.

Cason, J., and Graham, D. W. (1965). Isolated isoprenoid acids from a California petroleum. *Tetrahedron* **21**, 471–483.

Chappe, B., Michaelis, W., Albrecht, P., and Ourisson, G. (1979). Fossil evidence for a novel series of archaebacterial lipids. *Naturwissenschaften* **66**, 522–523.

Cloud, P., and Glaessner, M. (1982). The Ediacaran Period and System: Metazoa inherit the earth. *Science* **218**, 783–792.

Cooper, J. E., and Bray, E. E. (1963). A postulated role of fatty acids in petroleum formation. *Geochim. Cosmochim. Acta* **27**, 1113–1127.

Cox, R. E., Maxwell, J. R., Eglinton, G., Pillinger, C. T., Ackman, R. G., and Hooper, S. N. (1970). The geological fate of chlorophyll: The absolute stereochemistries of the series of acyclic isoprenoid acids in a 50 million year old lacustrine sediment. *Chem. Commun.*, pp. 1639–1641.

Cox, R. E., Burlingame, A. L., Wilson, D. M., Eglinton, G., and Maxwell, J. R. (1973). Botryococcene—a tetramethylated acyclic triterpenoid of algal origin. *Chem. Commun.*, pp. 284–285.

Day, W. C., and Erdman, J. G. (1963). Ionene: A thermal degradation product of β-carotene. *Science* **141**, 808.

Dean, R. A., and Whitehead, E. V. (1961). The occurrence of phytane in petroleum. *Tetrahedron Lett.* **21**, 768–770.

Dembicki, H., Meinschein, W. G., and Hattin, D. E. (1976). Possible ecological and environmental significance of the predominance of even-carbon number C_{20}-C_{30} alkanes. *Geochim. Cosmochim. Acta* **40**, 203–208.

DeNiro, M. J., Epstein, S., Kenealy, W. R., Belyaev, S. S., Wolkin, R., and Zeikus, G. J. (1983). Isotopic fractionation during carbon dioxide reduction by three methanogens (unpublished manuscript).

De Rosa, M., De Rosa, S., Gambacorta, A., Minale, L., and Bu'Lock, J. D. (1977). Chemical structure of the ether lipids of thermophilic acidophilic bacteria of the *Caldariella* group. *Phytochemistry* **16**, 1961–1965.

De Rosa, M., Gambacorta, A., Nicolaus, B., and Bu'Lock, J. D. (1980a) Complex lipids of *Caldariella acidophila,* a thermoacidophile archaebacterium. *Phytochemistry* **19**, 821–825.

De Rosa, M., Esposito, E., Gambacorta, A., Nicolaus, B., and Bu'Lock, J. D. (1980b). Effects of temperature on the ether lipid composition of *Caldariella acidophila. Phytochemistry* **19**, 827–831.

De Rosa, M., Gambacorta, A., Nicolaus, B., Sodana, S., and Bu'Lock, J. D. (1980c). Structural regularities in tetraether lipids of *Caldariella* and their biosynthetic and phyletic implications. *Phytochemistry* **19**, 833–836.

Didyk, B. M., Simoneit, B. R., Brassel, S. C., and Eglinton, G. (1978). Organic geochemical indicators of paleoenvironmental conditions of sedimentation. *Nature (London)* **272**, 216–222.

Diekert, G., Jaenchen, R., and Thauer, R. K. (1980). Biosynthetic evidence for a nickel tetrapyrrole structure factor F_{430} from *Methanobacterium thermoautotrophicum. FEBS Lett.* **119**, 118–120.

Douglas, A. G., Eglinton, G., and Maxwell, J. R. (1969a). The hydrocarbons of coorongite. *Geochim. Cosmochim. Acta* **33**, 569–577.

Douglas, A. G., Eglinton, G., and Maxwell, J. R. (1969b). The organic geochemistry of certain samples from the Scottish Carboniferous formation. *Geochim. Cosmochim. Acta* **33**, 579–590.

Dunning, H. N. (1963). Geochemistry of organic pigments. *In* "Organic Geochemistry" (I. A. Breger, ed.), pp. 367–430. Pergamon, Oxford (distributed by The Macmillan Company, New York.

Eglinton, G., Scott, P. M., Belsky, T., Burlingame, A. L., Richter, W., and Calvin, M. (1965). Occurrence of isoprenoid alkanes in a Precambrian sediment. *Space Sci. Lab., Univ. Calif. Tech. Rep. Ser.*, No. 6, Issue No. 9, January.

Eglinton, G., Douglas, A. G., Maxwell, J. R., and Ramsay, J. N. (1966). Occurrence of isoprenoid fatty acids in the Green River shale. *Science* **153**, 1133–1135.

Eglinton, G., Hajibrahim, S. K., Maxwell, J. R., and Quirke, J. M. E. (1980). Petroporphyrins: Structural elucidation and the application of HPLC fingerprinting to geochemical problems. *In*

"Advances in Organic Geochemistry, 1979" (A. G. Douglas and J. R. Maxwell, eds.), pp. 193–203. Pergamon, Oxford.

Fuchs, G., Thauer, R. K., Ziegler, H., and Stichler, W. (1979). Carbon isotope fractionation by *Methanobacterium thermoautotrophicum*. Arch. Microbiol. **120**, 135–139.

Gallegos, E. J. (1971). Identification of new steranes, terpanes, and branched paraffins in Green River shale by combined capillary gas chromatography and mass spectrometry. *Anal. Chem.* **43**, 1151–1160.

Gallegos, E. J. (1973). Identification of phenylcycloparaffin alkanes and other monoaromatics in Green River shale by gas chromatography - mass spectrometry. *Anal. Chem.* **45**, 1399–1403.

Gardner, P. M., and Whitehead, E. V. (1972). The isolation of squalane from a Nigerian petroleum. *Geochim. Cosmochim. Acta* **36**, 259–263.

Garrels, R. M., and Lerman, A. (1977). The exogenic cycle: Reservoirs, fluxes and problems. *In* "Global Chemical Cycles and their Alterations by Man" (W. Stumm, ed.), pp. 23–31. Abakon, Berlin.

Hahn, J. (1982). Geochemical fossils of a possibly archaebacterial orgin in ancient sediments. *Zentralbl. Bakteriol., Mikrobiol. Hyg., Abt. 1, Orig. C* **3**, 40–52.

Han, J., and Calvin, M. (1969a). Hydrocarbon distribution of algae and bacteria, and microbiological activity in sediments. *Proc. Natl. Acad. Sci. U.S.A.* **64**, 436–443.

Han, J., and Calvin, M. (1969b). Occurrence of C_{22}-C_{25} isoprenoids in Bell Creek crude oil. *Geochim. Cosmochim. Acta* **33**, 733–742.

Han, J., McCarthy, E. D., Calvin, M., and Benn, M. H. (1968a). The hydrocarbon constituents of the blue-green algae, *Nostoc muscorum, Anacystis nidulans, Phormidium loridum,* and *Chlorogloea fritschii. J. Chem. Soc. C,* pp. 2785–2791.

Han, J., McCarthy, E. D., Van Hoeven, W., Calvin, M., and Bradley, W. H. (1968b). Organic geochemical studies. II. A preliminary report on the distribution of aliphatic hydrocarbons in algae, in bacteria, and in a Recent lake sediment. *Proc. Natl. Acad. Sci. U.S.A.* **59**, 29–33.

Haug, P., and Curry, D. J. (1974). Isoprenoids in a Costa Rican seep oil. *Geochim. Cosmochim. Acta* **38**, 601–610.

Haug, P., Schnoes, H. K., and Burlingame, A. L. (1967). Isoprenoid and dicarboxylic acids isolated from Colorado Green River shale (Eocene). *Science* **158**, 772.

Hayatsu, R., and Anders, E. (1981). Organic compounds in meteorites and their origins. *Top. Curr. Chem.* **99**, 1–37.

Henderson, W., Wollrab, V., and Eglinton, G. (1969). Identification of steroids and triterpenes from a geological source by capillary gas-liquid chromatography and mass spectrometry. *In* "Advances in Organic Geochemistry, 1968" (P. A. Schenck and I. Havenaar, eds.), pp. 181–207. Pergamon, Oxford.

Hills, I. R., Whitehead, E. V., Anders, D. E., Cummins, J. J., Robinson, W. E. (1966). An optically active triterpane, gammacerane, in Green River, Colorado, oil-shale bitumen. *Chem. Commun.,* pp. 752–754.

Hills, I. R., Smith, G. W., and Whitehead, E. V. (1970). Hydrocarbons from fossil fuels and their relationship with living organisms. *J. Inst. Pet.* **56**, 127–137.

Hodgson, G. W., Hitchon, B., Taguchi, K., Baker, B. L., and Peake, E. (1968). Geochemistry of porphyrins, chlorins and polycyclic aromatics in soils, sediments and sedimentary rocks. *Geochim. Cosmochim. Acta* **32**, 737–772.

Hoefs, J. (1973). Ein Beitrag zur Isotopengeochemie des Kohlenstoffs in magmatischen Gesteinen. *Contrib. Mineral. Petrol.* **41**, 277–300.

Hoering, T. C. (1972). The benzene-soluble organic matter, humic acids, and insoluble organic matter in a core from the Cariaco Trench. *Year Book—Carnegie Inst. Washington* **71**, 585–592.

Holzer, G., Oró, J., and Tornabene, T. G. (1979). Gas chromatographic mass spectrometric analysis of neutral lipids from methanogenic and thermoacidophilic bacteria. *J. Chromatogr.* **186,** 795–809.

Hunt, J. M. (1972). Distribution of carbon in crust of earth. *Am. Assoc. Pet. Geol. Bull.* **56,** 2273–2277.

Ikan, R., Aizenshtat, Z., Baedecker, M. J., and Kaplan, I. R. (1975). Thermal alteration experiments on organic matter in recent marine sediment. I. Pigments. *Geochim. Cosmochim. Acta* **39,** 173–185.

Johns, R. B., Belsky, T., McCarthy, E. D., Burlingame, A. L., Haug, P., Schnoes, H. K., Richter, W., and Calvin, M. (1966). The organic geochemistry of ancient sediments. Part II. *Geochim. Cosmochim. Acta* **30,** 1191–1222.

Junge, C. E., Schidlowski, M., Eichmann, R., and Pietrek, H. (1975). Model calculations for the terrestrial carbon cycle: Carbon isotope geochemistry and evolution of photosynthetic oxygen. *JGR, J. Geophys. Res.* **80,** 4542–4552.

Kaplan, I. R., and Baedecker, M. J. (1970). Biological productivity in the Dead Sea. Part II. Evidence for phosphatidyl glycerophosphate in sediment. *Isr. J. Chem.* **8,** 529–535.

Kates, M. (1978). The phytanyl ether-linked polar lipids and isoprenoid neutral lipids of extremely halophilic bacteria. *Prog. Chem. Fats Other Lipids* **15,** 301–342.

Kates, M., Yengoyan, L. S., and Sastry, P. S. (1965). Diether analog of phosphatidyl glycerophosphate in *Halobacterium cutirubrum*. *Biochim. Biophys. Acta* **98,** 252–268.

Kates, M., Palameta, B., Joo, C. N., Kushner, D. J., and Gibbons, N. E. (1966). Aliphatic diether analogs of glyceride-derived lipids. IV. The occurrence of di-O-phytanylglycerol ether containing lipids in extremely halophilic bacteria. *Biochemistry* **5,** 4092–4099.

Kimble, B. J., Maxwell, J. R., Philps, R. P., and Eglinton, G. (1974). Tri- and tetraterpenoid hydrocarbons in the Messel oil shale. *Geochim. Cosmochim. Acta* **38,** 1165–1181.

König, H., and Kandler, O. (1979). N-Acetyltalosaminuronic acid a constituent of the pseudomurein of the genus *Methanobacterium*. *Arch. Microbiol.* **123,** 295–299.

Kuhn, R., and Winterstein, A. (1933). Kettenverkürzung und Cyclisierung beim thermischen Abbau natürlicher Polyen-Farbstoffe. *Ber. Dsch. Chem. Ges.* **66,** 1733–1741.

Kvenvolden, K. A., and Hodgson, G. W. (1969). Evidence for porphyrins in Early Precambrian Swaziland System sediments. *Geochim. Cosmochim. Acta* **33,** 1195–1202.

Kvenvolden, K. A., and Roedder, E. (1971). Fluid inclusions in quartz crystals from South-West Africa. *Geochim. Cosmochim Acta* **35,** 1209–1229.

Langworthy, T. A., Tornabene, T. G., and Holzer, G. (1982). Lipids of archaebacteria. *Zentralbl. Bakteriol., Microbiol. Hyg., Abt. 1, Orig. C* **3,** 228–244.

Lanyi, J. K. (1981). Halorhodopsin—a second retinal pigment in *Halobacterium halobium*. *Trends Biochem. Sci.* **6,** 60–62.

Leo, R. F., and Parker, P. L. (1966). Branched-chain fatty acids in sediments. *Science* **152,** 649–650.

Lovelock, J. E., and Whitfield, M. (1982). Life span of the biosphere. *Nature (London)* **296,** 561–563.

McCarthy, E. D., and Calvin, M. (1967). The isolation and identification of the C_{17} isoprenoid 2,6,10-Trimethyltetradecane from a Devonian shale: The role of squalene as a possible precursor. *Tetrahedron* **23,** 2609–2619.

McCarthy, E. D., Van Hoeven, W., and Calvin, M. (1967). The synthesis of standards in the characterization of a C_{21} isoprenoid alkane isolated from Precambrian sediments. *Tetrahedron Lett.*, pp. 4437–4442.

Mackenzie, A. S., Quirke, J. M. E., and Maxwell, J. R. (1980). Molecular parameters of maturation in the Toarcian shales, Paris Basin, France. II. Evolution of metalloporphyrins. *In* "Advances in Org. Geochemistry, 1979" (A. G. Douglas and J. R. Maxwell, eds.), pp. 239–248. Pergamon, Oxford.

McKirdy, D. M., and Hahn, J. (1982). The composition of kerogen and hydrocarbons in Precambrian rocks. *In* "Mineral Deposits and the Evolution of the Biosphere" (H. D. Holland and M. Schidlowski, eds.), Dahlem Konf., pp. 123–154. Springer-Verlag, Berlin and New York.

McKirdy, D. M., and Kantsler, A. J. (1980). Oil geochemistry and potential source rocks of the Officer Basin, South Australia. *APEA J.* **20**, 68–86.

Mair, B. J., and Barnewall, J. M. (1964). Composition of the mononuclear aromatic material in the light gas oil range, low refractive index portion, 230° to 305°C. *J. Chem. Eng. Data* **9**, 282–292.

Maxwell, J. R., Cox, R. E., Ackman, R. G., and Hooper, S. N. (1972). The diagenesis and maturation of phytol. The stereochemistry of 2,6,10,14-Tetramethylpentadecane from an ancient sediment. *In* "Advances in Organic Geochemistry, 1971" (H. R. von Gaertner and H. Wehner, eds.), pp. 277–291. Pergamon, Oxford.

Maxwell, J. R., Cox, R. E., Eglinton, G., Pillinger, C. T., Achman, R. G., and Hooper, S. N. (1973). Stereochemical studies of acyclic compounds. II. The role of chlorophyll in the derivation of isoprenoid-type acids in a lacustrine sediment. *Geochim. Cosmochim. Acta* **37**, 297–313.

Michaelis, W., and Albrecht, P. (1979). Molecular fossils of archaebacteria in kerogen. *Naturwissenschaften* **66**, 420–422.

Miller, S. L. (1982). Prebiotic synthesis of organic compounds. *In* "Mineral Deposits and the Evolution of the Biosphere" (H. D. Holland and M. Schidlowski, eds.), Dahlem Konf., pp. 155–176. Springer-Verlag, Berlin and New York.

Miller, S. L., and Urey, H. C. (1959). Organic compound synthesis on the primitive earth. *Science* **130**, 245–251.

Moldowan, J. M., and Seifert, W. K. (1979). Head-to-head linked isoprenoid hydrocarbons in petroleum. *Science* **204**, 169–171.

Murphy, M. T. J., McCormick, A., and Eglinton, G. (1967). Perhydro-β-Carotene in the Green River shale. *Science* **157**, 1040–1042.

Nooner, D. W., Updegrove, W. S., Flory, D. A., Oro, J., and Mueller, G. (1973). Isotopic and chemical data of bitumens associated with hydrothermal veins from Windy Knoll, Derbyshire, England. *Chem. Geol.* **11**, 189–202.

Odin, G. S. (1982). The Phanerozoic time scale revisited. *Episodes,* pp. 3–9.

Ourisson, G., Albrecht, P., and Rohmer, M. (1979). The hopanoids. Paleochemistry and biochemistry of a group of natural products. *Pure Appl. Chem.* **51**, 709–729.

Patience, R. L., Rowland, S. J., and Maxwell, J. R. (1978). The effect of maturation on the configuration of pristane in sediments and petroleum. *Geochim. Cosmochim. Acta* **42**, 1871–1875.

Pfaltz, A., Jaun, B., Fässler, A., Eschenmoser, A., Jaenchen, R., Gilles, H. H., Diekert, G., and Thauer, R. K. (1982). Zur Kenntnis des Faktors F430 aus methanogenen Bakterien: Struktur des porphinoiden Ligandsystems. *Helv. Chim. Acta* **65**, Fasc. 3, 828–865.

Pflug, H. D. (1982). Early diversification of life in the Archean. *Zentralbl. Bakteriol., Mikrobiol. Hyg., Abt. 1, Orig. C* **3**, 53–64.

Ponnamperuma, C. (1978). Prebiotic molecular evolution. *In* "Origin of Life, Proceedings of the Second ISSOL Meeting" (H. Noda, ed.), pp. 67–81. Bus Cent. Acad. Soc. Jpn., Jpn, Sci. Soc. Press, Tokyo.

Price, L. C. (1982). Organic geochemistry of core samples from an ultra-deep hot well (300 °C, 7 km). *Chem. Geol.* **37**, 215–228.

Rabinowitch, E. I. (1945). "Photosynthesis and Related Processes," Vol. I, Chapter D. Wiley (Interscience), New York.

Radchenko, O. A., and Sheshina, L. S. (1955). The genesis of porphyrins in crude oils. *Dokl. Akad. Nauk SSSR* **105**, 1285.

Robinson, W. E., Cummins, J. J., and Dineen, G. U. (1965). Changes in Green River oil shale paraffins with depth. *Geochim. Cosmochim. Acta* **29**, 249–258.

Ronov, A. B. (1958). Organic carbon in sedimentary rocks (in relation to the presence of petroleum). *Geochemistry (USSR)*, pp. 510–536.

Ronov, A. B. (1980). "Osadotchnaya obolotchka zemli" (20th Vernadski lecture). Izd. Nauka, Moscow.

Ronov, A. B., and Yaroshevski, A. A. (1969). Chemical composition of the earth's crust. *Geophys. Monogr., Am. Geophys. Union.* **13**, 37–57.

Rubey, W. W. (1955). Development of the hydrosphere and atmosphere with special reference to the probable composition of the early atmosphere. *Spec. Pap.—Geol. Soc. Am.* **62**, 631–650.

Schidlowski, M. (1980). Composition and origin of the atmosphere. *In* "Handbook of Environmental Chemistry" (O. Hutzinger, ed.), pp. 1–16. Springer-Verlag, Berlin and New York.

Schidlowski, M. (1982). Content and isotopic composition of reduced carbon in sediments. *In* "Mineral Deposits and the Evolution of the Biosphere" (H. D. Holland and M. Schidlowski, eds.), Dahlem Konf., pp. 103–122. Springer-Verlag, Berlin and New York.

Schidlowski, M., Appel, P. W. U., Eichmann, R., and Junge, C. E. (1979). Carbon isotope geochemistry of the 3.7×10^9 yr old Isua sediments, West Greenland: Implications for the Archean carbon and oxygen cycles. *Geochim. Cosmochim. Acta* **43**, 189–199.

Schidlowski, M., Hayes, J. M., and Kaplan, I. R. (1983). Isotopic inferences of ancient biochemistries: Carbon, sulfur, hydrogen and nitrogen. *In* "The Earth's Earliest Bioshpere: Its Origin and Evolution" (J. W. Schopf, ed.), Chapter 7. Princeton Univ. Press, Princeton, New Jersey.

Schoell, M., and Wellmer, F.-W. (1981). Anomalous ^{13}C depletion in early Precambrian graphites from Superior Province, Canada. *Nature (London)* **290**, 696–699.

Schopf, J. W. (1970). Precambrian microorganisms and evolutionary events prior to the origin of vascular plants. *Biol. Rev. Cambridge Philos. Soc.* **45**, 319–352.

Schwendinger, R. B. (1969). Carotenoids. *In* "Organic Geochemistry" (G. Eglinton and M. T. J. Murphy, eds.), pp. 425–437. Springer-Verlag, Berlin and New York.

Seifert, W. K., and Moldowan, J. M. (1981). Paleoreconstruction by biological markers. *Geochim. Cosmochim. Acta* **45**, 783–794.

Simoneit, R. B., and Burlingame, A. L. (1972). Preliminary organic analyses of the DSDP (JOIDES) cores, legs V-IX. *In* "Advances in Organic Geochemistry, 1971" (H. R. von Gaertner and H. Wehner, eds.), pp. 189–229. Pergamon, Oxford.

Stoeckenius, W., Lozier, R. H., and Bogomolni, R. A. (1979). Bacteriorhodopsin and the purple membrane of halobacteria. *Biochim. Biophys. Acta* **505**, 215–278.

Swain, F. M. (1969). Fossil carbohydrates. *In* "Organic Geochemistry" (G. Eglinton and M. T. J. Murphy, eds.), pp. 374–400. Springer-Verlag, Berlin and New York.

Tissot, B. P., and Welte, D. H. (1978). "Petroleum Formation and Occurrence." Springer-Verlag, Berlin and New York.

Tornabene, T. G., Langworthy, T. A., Holzer, G., and Oró, J. (1979). Squalenes, phytanes and other isoprenoids as major neutral lipids of methanogenic and thermoacidophilic "archaebacteria." *J. Mol. Evol.* **13**, 73–83.

Trask, P. D., and Patnode, H. W. (1942). "Source Beds of Petroleum." Am. Assoc. Pet. Geol., Tulsa, Oklahoma.

Treibs, A. (1934). Chlorophyll- und Häminderivate in bituminösen Gesteinen, Erdölen, Erdwachsen, und Asphalten. *Justus Liebig's Ann. Chem.* **510**, 42–62.

Urey, H. C. (1952). "The Planets." Yale Univ. Press, New Haven, Connecticut.

Vallentyne, J. R. (1960). Fossil pigments: The fate of carotenoids. In "Symposia on comparative biology." Vol. I. Comparative biochemistry of photoreactive systems. (M. B. Allen, ed.). Academic Press, New York.

Vallentyne, J. R. (1963). Geochemistry of carbohydrates. *In* "Organic Geochemistry" (I. A.

Breger, ed.), pp. 456–502. Pergamon, Oxford (distributed by The Macmillan Company, New York).

Van Dorsselaer, A., Albrecht, P., and Ourisson, G. (1977). Identification of novel (17αH)-hopanes in shales, coals, lignites, sediments and petroleum. *Bull. Soc. Chim. Fr.,* pp. 165–170.

Van Hoeven, W., Haug, P., Burlingame, A. L., and Calvin, M. (1966). Hydrocarbons from Australian oil, two hundred million years old. *Nature (London)* **211,** 1361–1366.

Waksman, S. A. (1938). "Humus—Origin, Chemical Composition, and Importance in Nature, 2nd ed. Williams & Wilkins, Baltimore, Maryland.

Walden, P. (1906). Optische Aktivität und die Entstehung des Erdöls. *Chem.-Ztg.* **30,** 391–393.

Walker, J. C. G., Hays, P. B., and Kasting, J. F. (1981). A negative feedback mechanism for the long-term stabilization of Earth's surface temperature. *JGR, J. Geophys. Res.* **86,** 9776–9782.

Walker, J. C. G., Klein, C., Schidlowski, M., Schopf, J. W., Stevenson, D. J., and Walter, M. R. (1983). Environmental evolution of the Archean-Early Proterozoic earth. *In* "The Earth's Earliest Biosphere: Its Origin and Evolution" (J. W. Schopf, ed.), Chapter 10. Princeton Univ. Press, Princeton, New Jersey.

Waples, D. W., Haug, P., and Welte, D. H. (1974). Occurrence of a regular C_{25} isoprenoid hydrocarbon in Tertiary sediments representing a lagoonal-type, saline environment. *Geochim. Cosmochim. Acta* **38,** 381–387.

Watts, C. D., and Maxwell, J. R. (1977). Carotenoid diagenesis in a marine sediment. *Geochim. Cosmochim. Acta* **41,** 493–497.

Zechmeister, L., and von Cholnoky, L. (1930). Untersuchungen über den Paprikafarbstoff. IV. Einige Umwandlungen des Capsanthins. *Justus Liebig's Ann. Chem.* **478,** 95–112.

PART II

Translation Apparatus of Archaebacteria

Chapter 5

The Structure and Evolution of Archaebacterial Ribosomal RNA

George E. Fox

I. Introduction

The ribosomal RNAs of the archaebacteria have been studied almost exclusively from the perspective of establishing evolutionary relationships. Consequently, the data available are primarily derived from sequencing experiments. The underlying theme of this chapter will necessarily be an evolutionary one, but much of the emphasis will be on secondary structure. A brief review of the ideas that led up to the articulation of the archaebacterial concept is given, followed by some description of the 16 S rRNA cataloging procedure and the associated data analysis. An outline of our current knowledge of rRNA gene organization and how it relates to the more extensive knowledge available from eukaryotic and eubacterial systems is given next. A foundation for the detailed discussion of the 5 S rRNA and 16 S rRNA data is then provided by discussing in depth the utility of the comparative approach for obtaining insight to RNA secondary structure. Direct experimental investigations of rRNA structure are largely ignored here because the unavailability of data of this type for the archaebacterial rRNAs only allows meaningful discussion in terms of a comparative approach. The presentation of the archaebacterial 5 S rRNA data is preceded by a lengthy analysis of current views on eubacterial and eukaryotic 5 S rRNA secondary structure. This provides a framework for examining the likely archaebacterial 5 S rRNA struc-

Copyright © 1985 by Academic Press, Inc.
All rights of reproduction in any form reserved.
ISBN 0-12-307208-5

tures. The 16 S rRNA discussion follows similar lines but is greatly restricted by the limited amount of primary sequence data. The final section attempts to define the issues surrounding the evolutionary position of the archaebacteria. The most interesting questions cannot yet be answered with certainty, but at least some of them can be articulated.

II. Origins of the Archaebacterial Concept

The archaebacterial concept emerged from studies of 16 S rRNA primary sequences. By 1965, it had become apparent that macromolecules held a trace of evolutionary history in their primary structures (Zuckerkandl and Pauling, 1965). With the demonstration of the potential value of cytochrome c as a molecular chronometer (Fitch and Margoliash, 1967), the stage was set for the widespread application of molecular techniques in phylogenetics. Nevertheless, attempts to resolve the evolutionary relationships among the bacteria were not immediately forthcoming. In part, this delay may have reflected the discouragement of micro-biologists who had for years been frustrated in their attempts to uncover such natural relationships (Woese, 1982). By the early 1970s, Carl Woese had set the elucidation of bacterial evolutionary history as the primary objective of his research effort.

The first crucial decision was to select a macromolecule. The essentials of a good choice were (a) ubiquitous occurrence of equivalent macromolecules throughout the prokaryotic world, (b) ease in isolation, and (c) technical feasi-bility of obtaining substantial primary sequence information (Sogin *et al.*, 1972). A promising molecule seemed to be ribosomal RNA. At the time, it could be readily isolated and was known to be distributed throughout the living world. That the various RNAs were likely to be functionally equivalent was supported by reconstitution studies (Nomura *et al.*, 1968; Bellemare *et al.*, 1973; Wrede and Erdmann, 1973). Such interchangeability implied conservation of function and, by inference, structure. It is in this kind of situation that one could best hope to obtain meaningful phylogenetic information. Early efforts employing RNA–DNA hybridization had been made (Moore and McCarthy, 1967; Pace and Campbell, 1971) and strongly supported the notion that the RNA sequences would be sufficiently conserved to reveal deep evolutionary divisions. The exist-ing knowledge thus suggested the selection of rRNA.

But which species should be employed? At the time, RNA sequencing was rudimentary and even 5 S rRNAs would have been difficult to examine in large numbers. 16 S and 23 S rRNA offered far more information, but could they be rationally approached in 1970? Even with current rapid gel sequencing tech-nology, it would have been very arduous to complete the substantial number of sequences needed for comprehensive phylogenetic studies. Woese and his colleagues decided to consider a sequence characterization approach. The idea was to digest the RNA with a specific endonuclease and to separate and completely sequence all of the products. The subsequent comparison of these "oligonucleotide catalogs" would presumably give quantitative information about the amount of similarity between the two unknown sequences. The first attempt utilized pancreatic ribonuclease for the primary digestion and tedious column methods for the separation and sequencing (McNamara, 1969). The initial results for *Bacillus subtilis* and *Escherichia coli* indicated that the purine stretches in 16 S rRNA were not strongly conserved. Coupled with the two-base specificity of pancreatic ribonuclease, it was obvious that the desired phylogenetic information would not be forthcoming. Subsequent efforts focused on ribonuclease T_1 digestions of both 5 S and 16 S rRNA (Sogin *et al.*, 1972; Pechman and Woese, 1972) using the Sanger fingerprinting methodology (Sanger *et al.*, 1965). Being specific for one nucleotide (guanylic acid) only, this enzyme gave larger oligonucleotides, which carried significant phylogenetic information. All problems were not yet overcome, as it was soon found that not only were the largest oligomers difficult to sequence but certain systematic errors occurred in the more traditional procedures. Subsequently, the two-dimensional paper methodology was fine-tuned for 5 S and 16 S rRNA cataloging. The need for spot elution was eliminated (Uchida *et al.*, 1974), and an array of secondary and tertiary digestion procedures were devised (Uchida *et al.*, 1974; Woese *et al.*, 1976), which allowed complete sequencing of even the largest oligonucleotide products. Other investigators worked with the initial separation and devised effective procedures utilizing two-dimensional gels (De Wachter and Fiers, 1972, 1982) or homochromatography (Brownlee and Sanger, 1969). A gradient thin layer chromatography approach has been combined with *in vitro* end-labeling procedures and mobility shift methods (Jay *et al.*, 1974; Silberklang *et al.*, 1977, 1979) in a format that is specifically designed for rapid 16 S rRNA cataloging (Stackebrandt *et al.*, 1981).

With the experimental capabilities in hand the 16 S rRNA cataloging procedure was applied to large numbers of phylogenetically interesting strains. The next step was to construct trees reflecting the evolutionary origins of the various strains. In order to accomplish this, it would be useful to know the actual percentage of homology among the underlying 16 S rRNA sequences. Then one can use any of several algorithms for tree construction. In actual practice, what is readily determinable is the total number of well-resolved T_1 oligonucleotides held in common between any two 16 S rRNA species. This allows calculation of an association coefficient, S_{AB}, for each binary comparison of catalogs. This coefficient is defined as follows (Fox *et al.*, 1977b).

$$S_{AB} = 2N_{AB}/(N_A + N_B)$$

where N_A is the total number of residues of at least length L in catalog A, N_B is the total number of residues of at least length L in catalog B, and N_{AB} is the total number of residues represented by all the coincident oligomers between the two catalogs, A and B, of at least length L. Unfortunately this association coefficient is only related to the true similarity in an unknown, nonlinear fashion. Thus, if it were used in tree construction, the branch lengths produced would possibly be meaningless. The usual approach has been to be conservative by only utilizing the values of the association coefficient, S_{AB}, to perform cluster analysis. This results in dendrograms reflecting levels of similarity between various extant strains. Since average linkage clustering is the starting point for many cladistic methods, these dendrograms can with care be interpreted as reflecting evolutionary relationships.

If a relationship could be established between the S_{AB} values and true sequence homology, then cladistic techniques could be utilized with only the usual reservations. To this end, relationships have been developed based on probabilistic arguments (Pechman and Woese, 1972; Bonen and Doolittle, 1975), and Markov simulations of the cataloging procedure have been conducted (Young et al., 1981; Aaronson et al., 1982; Hori et al., 1982). Since the actual relationship is only known in a small number of cases, these predictions are not readily evaluated. In two instances where a test could be made, the estimates proved to be too high (Young et al., 1981). The essential assumption in these studies is that each position is equally mutable. This is not a good assumption as is revealed by comparisons of the actual number of oligomers found in each size with the numbers expected (Pechman and Woese, 1972). Simulations that allow for conserved positions and established 16 S rRNA secondary structural features reveal substantial deviations from the results of the simpler simulations (Chen, 1980). The nature of the deviation is such that the straightforward simulations always give homology values that are high, with the inaccuracy increasing as the actual homology decreases. If the S_{AB} values themselves are viewed as estimates of the true homology, it is clear that they must be under estimates, as they totally ignore correspondences between homologous oligomers of differing sequence. The actual homology is between these two extremes and is undoubtedly closer to the estimates produced by the simplistic probabilistic calculations than to that provided by the S_{AB} values. This is a consequence of the fact that the S_{AB} approach discards information that is available from related but nonidentical oligomers. It is in fact possible to identify with considerable reliability many, but not all, such related nonidentical oligomers in any binary comparison of catalogs (Fox et al., 1977b). However, this has not yet led to improved methodology for analyzing the 16 S rRNA catalog data. Whether or not theoretical improvements in the analysis procedures are forthcoming remains to be seen. Nevertheless, it will at the least soon be feasible to develop empirical insight into the relationship between the S_{AB} value and true homology as more complete 16 S rRNA sequences are published.

One aspect of the relationship between S_{AB} values and actual homology that is not well known is the absence of a one-to-one correspondence between them. Consider that we know two sequences, X and Y, and both are equidistant from (i.e., share the same homology with) a third known sequence, A. Then, it is possible to construct the ribonuclease T_1 catalogs from all three sequences and calculate S_{AX} and S_{AY}. Will $S_{AX} = S_{AY}$? Almost certainly not! Although globally the number of base substitutions between A and X and A and Y will be the same, the actual sites of substitution will be different. As a result, the individual guanine residues will be affected differently so that at the catalog level S_{AX} will not in general be the same as S_{AY}. A useful question, then, is to ask how much variation can be expected. The simulations discussed above are able to address this issue and indicate that over most of the homology range the majority of S_{AB} values that correspond to any particular homology level will fall within $\pm$ 0.05 units of a mean value. It is the relationship between this mean value and the true homology that the simulations have actually attempted to establish.

Determinations of S_{AB} values are also subject to errors in sequencing and recording of the data. These will have a detrimental effect on calculated S_{AB} values when actual oligomer identities are overlooked or when identities are indicated which do not in fact exist. Our experience has been that false identities leading to overstatement of S_{AB} values seldom occur. Far more likely is understatement. The vast majority of experimental difficulties arise with very large oligomers. In many cases, even when these can not be sequenced exactly, it can still be determined that they are certainly different from or probably identical to a similar oligomer in the 16 S rRNA of another organism. In addition, the effect of sequencing errors among the largest oligomers is mediated by the fact that these oligomers are typically unique to the individual organism or at worst of very local phylogenetic distribution. Hence, in those instances where preliminary catalogs containing a significant number of unsequenced large oligomers have been repeated, no change occurred in the organism's overall placement. When in fact any meaningful change occurs, it is local and almost always is in terms of a significant increase in the apparent homology to the most closely related strains. The number of catalogs containing enough sequence uncertainty to warrant concern is actually very small, and it is known which ones these are. Overall then, the effect of experimental errors, with the exception of a few local comparisons, is quite negligible in contrast to the implicit uncertainty seen in the relationship between S_{AB} and true homology.

Although the origin of the implicit uncertainty associated with the S_{AB} values is formally different than experimental error, it seems reasonable to consider its likely effect in an analogous fashion. Thus, the following procedure (K.-N. B. Chen and G. E. Fox, unpublished results) has been devised. Following clustering of the original matrix of S_{AB} values, a new matrix of binary association coefficients is generated by individually permuting each entry to a new value, which is within $\pm$ 0.05 units of the original value in accordance with a random number generator. The resulting matrix is then used to calculate a second dendrogram. One next returns to the original matrix again and repeats the permutation process. After 20 such variant dendrograms have been produced, it is possible to evaluate which branchings in the original dendrogram are most subject to change because of the implicit uncertainty in the S_{AB} values. In addition, the results indicate what reasonable alternative placements might be. Under these conditions, the major branchings are seldom affected. This is because the earliest branchings on the dendrograms reflect averages of numerous S_{AB} values so that even relatively large fluctuations associated with individual S_{AB} values have little consequence to the branching pattern.

Application of the S_{AB} approach to large numbers of 16 S rRNA sequences revealed that one group of organisms, the methanogens, was widely separated from the rest of the prokaryotes (Balch *et al.*, 1977; Fox *et al.*, 1977a). Additional 16 S rRNA catalogs established that the methanogens were specifically related to the extreme halophiles, *Halobacterium* and *Halococcus,* (Magrum *et al.*, 1978; Fox *et al.*, 1980). This grouping was termed the archaebacteria (Woese and Fox, 1977a; Woese *et al.*, 1978), and subsequently many additional strains have been examined by the cataloging approach. In the case of the methanogens, this led to a new taxonomy that was based primarily on 16 S rRNA cataloging results (Balch *et al.*, 1977). A more current view of archaebacterial phylogeny based on 16 S rRNA cataloging is shown in Fig. 1. The majority of the strains are methanogens or extreme halophiles, and it is these, along with *Thermoplasma acidophilum* that form the first of two major clusters. This first grouping is divided into five obvious subclusters, which are designated on Fig. 1 as 1A, 1B, 2, 3, and 4. Three of these, designated 1A, 1B, and 3, contain methanogens exclusively, and one, designated 2, encompasses the extreme halophiles. The final cluster, designated 4, is represented by *T. acidophilum* alone. The halophiles share a common ancestry with the methanogen groups, but it cannot be determined whether they are specifically descended from a methanogen phenotype. As a group, the halophiles exhibit far more phylogenetic variability than has previously been attributed to them. The second major cluster includes the thermoacidophiles *Sulfolobus, Thermoproteus,* and *Thermodiscus.*

When the stability of these two superclusters is examined by the permutation procedure described above, all seem quite stable. The only exception involves *Methanothermus*, which, rather than being an outside member of the group 1B methanogens, was frequently seen to be a specific relative of *Methanococcus jannaschii* in the *Methanococcus* subcluster, group 1A. Within the individual

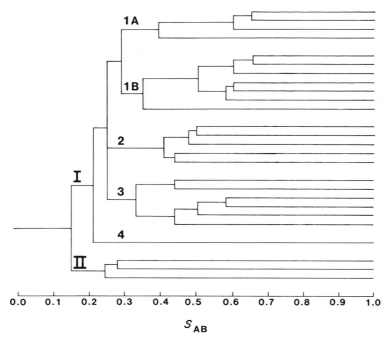

FIG. 1. Archaebacterial phylogeny. Dendrogram calculated by average linkage clustering from a matrix of S_{AB} values from 33 archaebacterial strains. The more closely related strains including five extreme halophiles have been omitted from the figure to improve clarity. A fundamental separation is seen at $S_{AB} = 0.15$, which divides the archaebacteria into two major clusters with the methanogens, extreme halophiles, and *Thermoplasma acidophilum* in Division I and the major group of thermoacidophiles in Division II. Many more members of the first division have been examined so that four clusters are readily detected. These are: 1, Methanococcales (1A)–Methanobacteriales (1B); 2, Halobacteriales; 3, Methanomicrobiales; 4, Thermoplasmales. The individual species shown on the figure are from the top down as follows: 1A, *Methanococcus maripaludis; M. vannielii; M. voltae; M. jannaschii;* 1B, *Methanobrevibacter arboriphilus; M. smithii; M. ruminantium; Methanobacterium bryantii* strain M.o.H.; *M. formicicum; M. thermoautotrophicum; Methanothermus fervidus;* 2, *Halobacterium volcanii;* alkaliphilic halophile strain SP-1; *H. saccharovorum; H. vallismortis; Halococcus morrhuae;* 3, *Methanosarcina barkeri; Methanothrix soehngenii; Methanogenium cariaci; M. marisnigri; Methanomicrobium mobile; Methanospirillum hungatei;* 4, *Thermoplasma acidophilum; Sulfolobus acidocaldarius; Thermoproteus tenax; Thermodiscus maritimus.*

subclusters, the most uncertainty is found among the *Halobacterium* strains. In fact, the 20 variant dendrograms included 10 topologically different trees for the halophiles. The specific relationship between *Sulfolobus* and its two relatives seems to be quite uncertain. In none of the various clusterings, however, did *Thermoplasma* merge with the other thermoacidophiles. Other minor uncertainties are as follows. (1) *Methanococcus vannielii* and *Methanococcus voltae* may be specifically related; (2) the relative branching order of *Methanobacterium formicicum* and *M. thermoautotrophicum* possibly should be reversed; and (3) the relative branching order of *Methanospirillum hungatei* and *Methanobrevi-bacter mobile* is subject to some uncertainty.

The historically minded will find it interesting to learn that it was not this kind of dendrogram approach that led immediately to the idea of the archaebacteria. Rather, it was the stark contrasts in the raw data itself. In fact, the more discontinuous nature of catalog data as compared to sequence data amplified these contrasts. By 1976, 16 S rRNA catalog data had been obtained from diverse bacterial species including *Escherichia coli* (Uchida *et al.*, 1974), *Anacystis nidulans* (Doolittle *et al.*, 1975), *Rhodopseudomonas sphaeroides* (Zablen and Woese, 1975), *Bacillus stearothermophilus* (Woese *et al.*, 1976), *Euglena gracilis* chloroplast (Zablen *et al.*, 1975), and *Porphyridium cruetum* chloroplast (Bonen and Doolittle, 1975). Examination of 27 such catalogs had produced a strong prejudice as to conserved features (Woese *et al.*, 1975) that characterized eubacterial 16 S rRNA. This included not only individual oligonucleotides that were found in every major bacterial group but also patterns of posttranscriptional modification that were repeated over and over again. The sequences that contained the modifications were typically conserved in terms of both sequence and electrophoretic mobility. It thus was a considerable shock when the first 16 S rRNA fingerprint of a methanogen was analyzed and was found to lack so many of these features that one briefly had to wonder if some other RNA had been isolated by mistake!

At present, with over 300 catalogs known, one can still see this aspect in the data. Table I lists the 22 most common oligonucleotides in the eubacterial 16 S rRNAs and indicates the number of eubacterial and archaebacterial catalogs containing them. Only three of these oligomers are present in significant numbers of archaebacteria. Such an intuitive analysis of the data suggests that the methanogens and their relatives are indeed most unusual bacteria but by themselves, the data did not justify the claims ultimately made. It was only after several eukaryotic 18 S rRNAs were examined that one could put the data in a context (see Table II) wherein the archaebacterial/eubacterial divergence could be contrasted with the eukaryotic/eubacterial divergence. It was on these comparisons that the claim of major evolutionary lineage could be based (Woese and Fox, 1977a). That this has subsequently proven to be a useful concept for organizing data is the subject of this volume.

III. Ribosomal RNA Gene Organization

Significant interspecies differences in the organization of ribosomal RNA are known to exist. In *Escherichia coli* (see Morgan, 1982, for a fully documented and recent review), there are known to be seven transcription units containing all

TABLE I

Most Highly Conserved Oligonucleotides in 16 S rRNA

Oligomer	Eubacterial occurrences	Archaebacterial occurrences
UACACACCG	289	0
UAACAAG	289	28
AUACCCUG	265	0
CAACUCG	259	6
UAAUACG	247	0
CUACAAUG	244	28
ACUCCUACG	242	0
UUUAAUUCG	206	0
CUACACACG	201	0
CCACACUG	186	0
CUAACUACG	170	0
CAAACAG	160	0
CUAAUACCG	151	1
UAACACG	148	28
CCCCUUAUG	145	0
AAUAUUG	141	0
CAUUAAG	138	0
AACACCAG	114	0
CCUACCAAG	104	2
CAACCCUUG	103	0
ACAAACCG	101	0
AACACCG	100	0
CATALOGS AVAILABLE	289	28

three ribosomal RNAs in the order 16S-tRNA(s)-23S-5S. In at least three of the units, there is one tRNA gene at the distal end of the 5 S rRNA gene. In other eubacteria, e.g., *Bacillus subtilis* (Zingales and Colli, 1977), *Anacystis nidulans* (Williamson and Doolittle, 1983), *Rhodopseudomonas capsulata* (Yu *et al.*, 1982), and the chloroplasts of *Euglena gracilis* (Orozco *et al.*, 1980), the same 16S-23S-5S gene arrangement is seen. The number of transcription units is variable, however, with three in *Euglena* chloroplasts (Orozco *et al.*, 1980) and two in *A. nidulans* (Williamson and Doolittle, 1983).

The situation in eukaryotes has been recently reviewed by Long and Dawid (1980) in general and for fungi specifically by Bollon (1982). Here it is usual to find one or more blocks of tandomly repeated transcription units arranged in the order 18S-5.8S-28S. The total number of individual genes can be very large. With regard to the 5 S rRNA genes, the eukaryotes exhibit several different arrangements. In some fungi, *Saccharomyces* (Bell *et al.*, 1977; Nath and Bollon, 1977; Philippsen *et al.*, 1978; Kramer *et al.*, 1978), *Torulopsis* (Tabata,

TABLE II

Association Coefficient (S_{AB}) between Representative Members of the Three Primary Kingdoms[a]

Organism	Association coefficient (S_{AB})												
	1	2	3	4	5	6	7	8	9	10	11	12	13
Saccharomyces cerevisiae, 18 S	—	0.29	0.33	0.05	0.06	0.08	0.09	0.11	0.08	0.11	0.11	0.08	0.08
Lemna minor, 18 S	0.29	—	0.36	0.10	0.05	0.06	0.10	0.09	0.11	0.10	0.10	0.13	0.07
L cell, 18 S	0.33	0.36	—	0.06	0.06	0.07	0.07	0.09	0.06	0.10	0.10	0.09	0.07
Escherichia coli	0.05	0.10	0.06	—	0.24	0.25	0.28	0.26	0.21	0.11	0.12	0.07	0.12
Chlorobium vibrioforme	0.06	0.05	0.06	0.24	—	0.22	0.22	0.20	0.19	0.06	0.07	0.06	0.09
Bacillus firmus	0.08	0.06	0.07	0.25	0.22	—	0.34	0.26	0.20	0.11	0.13	0.06	0.12
Corynebacterium diphtheriae	0.09	0.10	0.07	0.28	0.22	0.34	—	0.23	0.21	0.12	0.12	0.09	0.10
Aphanocapsa 6714	0.11	0.09	0.09	0.26	0.20	0.26	0.23	—	0.31	0.11	0.11	0.10	0.10
Chloroplast (*Lemna*)	0.08	0.11	0.06	0.21	0.19	0.20	0.21	0.31	—	0.14	0.12	0.10	0.12
Methanobacterium thermoautotrophicum	0.11	0.10	0.10	0.11	0.06	0.11	0.12	0.11	0.14	—	0.51	0.25	0.30
Methanobacterium ruminantium strain M-1	0.11	0.10	0.10	0.12	0.07	0.13	0.12	0.11	0.12	0.51	—	0.25	0.24
Methanogenium cariaci	0.08	0.13	0.09	0.07	0.06	0.06	0.09	0.10	0.10	0.25	0.25	—	0.32
Methanosarcina barkeri	0.08	0.07	0.07	0.12	0.09	0.12	0.10	0.10	0.12	0.30	0.24	0.32	—

[a] From Woese and Fox (1977a).

1980), *Mucor* (Cihlar and Sypherd, 1980), and *Achlya* (Rozek and Timberlake, 1979), the 5 S genes are at the 5' end of the 18S-5.8S-23S tandem repeat. Consistent with their separate transcription by RNA polymerase III, the 5 S rRNA genes are known to be located on the opposite strand in at least one of these cases, i.e., *Saccharomyces.* In *Neurospora* (Selker *et al.,* 1981), *Aspergillus* (Borsuk *et al.,* 1982), and *Schizosaccharomyces* (Tabata, 1981; Mao *et al.,* 1982), the 5 S rRNA genes are dispersed throughout the genome in several clusters. In still other eukaryotes, the 5 S rRNA genes themselves are arranged in tandemly repeating transcription units. This is known to be the case in *Drosophila* (Procunier and Tartof, 1975; Wimber and Steffensen, 1970), *Vicia* (Kaina *et al.,* 1978), *Tetrahymena* (Kimmel and Gorovsky, 1978), and *Xenopus* (Brown and Gurdon, 1978; Doering and Brown, 1978; Fedoroff and Brown, 1978; Korn and Brown, 1978; Miller *et al.,* 1978). In *Xenopus laevis,* there are known to be three repeat families, two of which are oocyte specific. The somatic 5 S rRNA genes are present in several hundred copies, whereas the oocyte-specific members are estimated at 25,000. One of the oocyte repeats contains both a functional gene and a nonfunctional pseudogene. In *X. borealis,* there are two known repeat units. One of these multiple gene units contains two functional 5 S rRNA genes.

The organization of ribosomal RNA genes has been examined in three archaebacteria, *Halobacterium halobium* (Hofman *et al.,* 1979), *Thermoplasma acidophilum* (Tu and Zillig, 1982), and *Methanococcus vannielii* (Jarsch *et al.,* 1983). The halophile seems to resemble the eubacteria in that the genes are physically linked in the order 16S-23S-5S and are presumably transcribed from a common promoter. As is the case with the mycoplasmas (Ryan and Morowitz, 1969), *Halobacterium halobium* contains just a single gene cluster for the three RNAs.

It would be inappropriate, however, to assume that this is generally true even for the halophiles, as internal microheterogeneity has been detected in the 5 S rRNA populations of *Halobacterium vallismortis* and *Halobacterium marismortui* (Nicholson and Fox, 1983). In *Methanococcus vanniellii,* four clusters of physically linked rRNA genes were isolated with the order again 16S-23S-5S. It is considered likely that the four transcription units are well separated on the chromosome. In addition, a DNA fragment was found that hybridized with 5 S rRNA but not 16 S or 23 S rRNA. This 5 S rRNA hybridizing fragment is well isolated from the four gene clusters (Jarsch *et al.,* 1983). The situation in *Thermoplasma* is rather more novel in that the genes for the three RNAs were not found to be physically linked, and in addition the 5 S rRNA gene is flanked by the 16 S and 23 S RNA genes. The distance between the 16 S and 23 S RNA genes exceeds the size of the spacer region, which separates the 19 S and 25 S rRNA genes of *Neurospora* mitochondria (Heckman and RajBhandary, 1979). Since these latter genes are known to be separately transcribed (Grimm and Lambowitz, 1979; Green *et al.,* 1981), it has been argued (Tu and Zillig, 1982) that one must also seriously question the existence of a single promoter for the three *Thermoplasma* genes. Although considerably more data is needed, it is already clear that further information on rRNA gene organization of the various archaebacteria may be a key element in deciphering the evolutionary history of the group.

IV. RNA Secondary Structure: General Considerations

In the absence of atomic resolution structural information, the molecular biologist has relied on quasitheoretical, direct experimental and comparative techniques for obtaining insight to RNA structure (see discussion in Noller, 1984 for example). The quasitheoretical approach had its origins in the observation (Tinoco *et al.,* 1971) that potential helical regions in any RNA molecule could be systematically enumerated by a diagonalization procedure similar to that which had previously been used in searching for sequence homology in proteins. Once one has enumerated all possible helices under appropriate constraints (i.e., an operational definition of secondary structure that addresses such issues as minimum helix length, minimum loop size, and allowable pairs), it is then necessary to decide which helices are actually biologically significant. In order to make these decisions, thermodynamic data from melting studies on model compounds were incorporated by developing empirical rules that allowed calculation of the contribution to overall free energy of various loop and paired regions (Tinoco *et al.,* 1971, 1973). The calculational task is to construct all feasible combinations of the various possible helical regions and for each case to calculate an estimate of overall free energy, with the ultimate objective being identification of the minimum free energy structure. Even for a small molecule, such as 5 S rRNA, in excess of 100 individual helical regions typically may be consistent with a primary sequence, so that the calculations quickly become enormous. Efficient algorithms have been developed (Comay *et al.,* 1984; Jacobson *et al.,* 1984), and an attempt has been made to improve the empirical parameters by selecting values that optimize the predictability of the well-established cloverleaf structure for tRNA (Ninio, 1979) and, more recently, the consensus 5 S rRNA structure (Papanicolaou *et al.,* 1984). In spite of the considerable progress that has been made, these predictive approaches can not at present be considered definitive. They do, however, provide good insight to reasonable alternatives that can be further explored by other methods and can in fact be readily incorporated into a more comprehensive scheme that also utilizes direct experimental and comparative data (Auron *et al.,* 1982).

Traditionally, the majority of biochemists and molecular biologists, being experimentalists, have favored a direct experimental approach for elucidating secondary structure. One common idea is to use a site-specific agent, either chemical or enzyme, to specifically attack either exposed, e.g., single-stranded residues or, as in the case of the cobra venom nuclease, double-stranded regions. Alternatively, more specialized techniques, such as slow tritium exchange and oligonucleotide binding, or physical techniques, such as Raman spectroscopy, high resolution NMR, and small angle X-ray scattering, have been employed. In the case of tRNA where the structure is known at atomic resolution, it has been possible to examine how well these direct approaches work (see Kim, 1976, for example). In fact, the overall agreement is good, though some individual studies present problems. It nevertheless has been very difficult to make an *a priori* assessment of what a structure is from this kind of data alone. In the case of the rRNAs, this problem is compounded by increased experimental difficulties due to the larger molecule sizes (especially in 16 S and 23 S rRNA) and to the uncertainty about the validity of solution studies in the absence of ribosomal proteins. Recently this outlook has improved dramatically with increasing use of the cobra venom nuclease and advances in NMR instrumentation and theory that have made it possible to utilize the nuclear Overhauser effect in studies of molecules as large as 5 S rRNA.

For purely pragmatic reasons, the primary tool used in deducing secondary structure in rRNAs has been "comparative" or "phylogenetic" analysis. This was first applied to tRNA following the suggestion of the cloverleaf by Holley *et al.* in 1965. The underlying assumption is that functionally equivalent RNAs from phylogenetically diverse sources should have equivalent structures. This does not, of course, mean chemically identical structures, since the sequences differ. Rather, it implies that the structures will be identical at the level of resolution required for biological function. Thus, biologically important aspects of structure are assumed to be conserved as long as function is conserved.

In actual practice, at the level of secondary structure, the main biological constraint is a requirement for helicity, i.e., Watson–Crick pairing, in certain regions of the molecule. Typically, both the location and length of such regions are strongly conserved because variations in distance would be disruptive to the three-dimensional arrangements needed for functionality. Thus, one can identify biologically significant regions of base pairing by comparison of sequences. One looks for helices that are rather precisely conserved in length and location among all members of a group of sequences that exhibit significant primary structure variation. The changes in sequence provide the crucial test without which no final inference can be drawn. Stated another way, conservation of a helix in location, length, and sequence is not sufficient. A seldom-recognized aspect of the comparative approach is that when multiple secondary structural features are found there is no guarantee that they all coexist. If the molecule exists in more than one biologically important conformation, then a given helical region may not be found in all the meaningful structural forms. A possible example of this is the potential pairing between the termini of *E. coli* 23 S rRNA, which may be required for proper processing of the original transcript and thus need not be part of the mature molecule's secondary structure as exhibited in the 50 S particle.

To illustrate how the comparative approach works consider the region near position 1315 in the *E. coli* 16 S rRNA sequence as shown in Table III. The sequence for this region is shown for five ribosomes: (A) *E. coli* (Brosius *et al.*, 1978), (B) corn chloroplast (Schwarz and Kössel, 1980), (C) *Bacillus brevis* (C. R. Woese and H. F. Noller, personal communication), (D) *B. stearothermophilus* (H. F. Noller and C. R. Woese, personal communication), and (E) *Halobacterium volcanii* (Gupta *et al.*, 1983). The alignment of the sequences is established by sequence conservation in areas surrounding the putative paired region, which is boxed. In *E. coli,* the pairing would consist of seven normal Watson–Crick pairs. In the equivalent region in each of the other organisms, seven base pairs can also be made. In each case, the evolutionary process has resulted in primary sequence changes, which provide a total of five different tests of structure, all of which are positive, implying that it is biologically significant.

Studnicka *et al.*(1981) combined these comparative ideas with the helix enumeration technique (Tinoco *et al.*, 1971) to produce an automated methodology, which was termed helical filtering. Their approach is to align primary sequences from diverse sources so that in a particular region of interest conserved residues occur at equivalent positions. A matrix of the helix possibilities is then generated for each individual case, and these are simultaneously compared. Only those helices that consistently show up remain as viable candidates for inclusion in any secondary structure.

A formal criterion of how many favorable tests are required to establish a structure has never been proposed. Without insistence on any actual evolutionary tests, one produces secondary structure models which are "maximal," in the sense that some of their features may later have to be removed due to negative tests when further sequence data becomes available. If, on the other hand, one is

TABLE III

Escherichia coli **16 S rRNA** Sequence

A	GAU	UGGAGUC	UGCAACUC	GACUCCA	UGA
B	GAU	UGCAGGC	UGCAACUC	GCCUGCA	UGA
C	GAU	UGUAGGC	UGCAACUC	GCCUACA	UGA
D	GAU	UGCAGGC	UGCAACUC	GCCUGCA	UGA
E	GAU	UGAGGGC	UGAAACUC	GCCCUCA	UGA

conservative and only includes in a model those features that are supported by large numbers of tests, then a minimal model is produced, which may have to be added to later on. Noise is introduced into the process in that as sequence numbers increase, negative tests will inevitably be encountered either due to experimental errors or to occasional mispairings, which the evolutionary process is able to tolerate. Inevitably then, the judgment-based aspects of the comparative approach, coupled with minor variations in what different investigators consider to be secondary structure, leave room for differences of opinion.

V. 5 S rRNA Primary and Secondary Structure

A. EUKARYOTIC AND EUBACTERIAL 5 S rRNA

The matter of 5 S secondary structure has been actively debated since the first sequence was published (Brownlee *et al.*, 1967). The number of alternatives that have been presented is now legion, perhaps even approaching the number of papers discussing the origins of the genetic code. Although a detailed review of the models will not be attempted, it should be said that in retrospect much of the early confusion can be attributed to sequencing errors. Indeed, by 1975, a collection of 10 phylogenetically diverse sequences was available, five of which have ultimately been found to contain sequencing errors. Currently, this investigator feels the confusion is more apparent than real, an illusion created by numerous investigators drawing structures slightly differently and using different nomenclature. There is controversy left, but far less than workers outside the field perceive.

Nishikawa and Takemura (1974a,b) developed a model by an informal intuitive approach, which contained five helical regions, and suggested that with minor modifications it was universally applicable to all 5 S rRNAs. In actuality, it was not strictly consistent with all the data available at that time, so it tended to be ignored. It did improve, however, as errors were discovered! Although this model contained some pairings that are now considered incorrect, it was largely in accordance with current views. Shortly thereafter, a minimal model based on the comparative approach was introduced (Fox and Woese, 1975a,b), which focused on eubacterial 5 S rRNAs and contained four major helical regions. It was again pointed out that a similar model could be drawn for eukaryotic 5 S rRNAs. In this case, however, the authors stressed the differences, relying heavily on the lack of interchangeability between prokaryotic and eukaryotic 5 S rRNAs in reconstitution assays (Bellemare *et al.*, 1973; Wrede and Erdmann, 1973).

Subsequently, the Fox and Woese model proved to be more popular because it was strongly supported by all the existing comparative data and being a minimal model (a point not explicitly stated at the time), it could be added to readily. Indeed, this process soon began, as 2 four-helix models with differing fourth helices were proposed for eukaryotic 5 S rRNAs (Hori, 1976; Vigne and Jordan, 1977). Later, as sequencing errors were clarified and new sequences determined, it was recognized that both fourth helices were strongly supported by the comparative data, so that a five-helix model for eukaryotic 5 S rRNAs was reasonable (Luehrsen and Fox, 1981a). In a rapid succession of essentially simultaneous papers, the 5 S rRNA community as a whole (Böhm *et al.*, *1981, 1982; Butler et al.,* 1981; Douthwaite and Garrett, 1981; Garrett *et al.,* 1981; Hinnebusch *et al.,* 1981; Peattie *et al.,* 1981; Stahl *et al.,* 1981; Studnicka *et al.,* 1981; Delihas and Andersen, 1982; De Wachter *et al.,* 1982; MacKay *et al.,* 1982; Troutt *et al.,* 1982; Küntzel *et al.,* 1983) realized that the definitions of secondary structure used in developing the Fox and Woese model were too restrictive. In particular, although the possibility of bulge loops was well known, Fox and Woese (1975a,b) reasoned that since bulge loops were not seen in tRNA they should not be proposed unless flanked on each side by a least three nucleotide pairs whose existence could be established by comparative data. When this minimum helix length was relaxed to two nucleotides (Studnicka *et al.*,1981), strong comparative support for extensions of several of these helices by short bulge loops was immediately apparent. Thus, rather belatedly, these features, which had been included in the Nishikawa–Takemura model of 1974, became generally accepted.

There is now considerable consensus among investigators as to 5 S rRNA secondary structures. In Figs. 2 and 3, these consensus structures are displayed for eubacteria and eukaryotes. It is to be emphasized that these are consensus structures in the sense that only those features which appear in essentially every recent model are included. One notable exception is the extension of helix III in eukaryotes, which one group (Lim *et al.,* 1983) has apparently not yet accepted. These two structures should be regarded as the current view of a minimum model for the eukaryotic and eubacterial 5 S rRNA secondary structure. The bacterial model probably does not apply exactly to either the chloroplasts (Delihas and Andersen, 1982) or to the mycoplasmas (Hori *et al.,* 1981; Walker *et al.,* 1982). The issue of whether or not these two models are also complete is still subject to significant differences of opinion, some of which will be discussed in detail.

Before proceeding with this discussion, some nomenclature needs clarification. It is usual to label the helices with either Roman numerals (Nishikawa and Takemura, 1974b) or capital letters (Fox and Woese 1975a,b). We now prefer the former, but the community is leaning to the latter! At present, those investigators that utilize Roman numerals (except Delihas) proceed with the labeling from the 5′ terminus, so that the hairpin between positions 70 and 90 is desig-

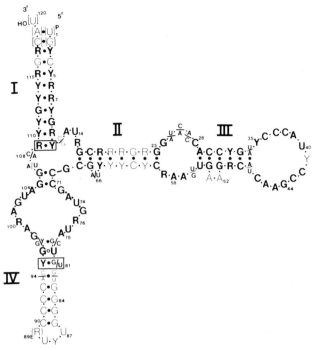

FIG. 2. Consensus secondary structure for eubacterial 5 S rRNA. The figure displays the generally accepted secondary structure features in typical eubacterial 5 S rRNA sequences (excluding organelles and mycoplasmas) and indicates the most common base at each position. Individual base pairs of the Watson–Crick type are indicated by black circles between the bases that are presumed to interact. Pairs that are usually of the wobble type (G-U) are indicated by open circles. The major helical regions are labeled by large Roman numerals. Interactions that are not considered proven by a large number of workers in the field are omitted. Conservation is displayed as follows. Dark letters are used where one particular base occurs in more than 90% of the known sequences. If the two pyrimidines together account for 90% of the known cases, a dark Y is used. Likewise a dark R represents purine positions. In those cases, where some different combination of two bases accounts for 90% of the examples, they are both shown as small dark letters. Light letters are used in an analogous way with a 50% criterion. Individual positions enclosed by brackets are actually not filled in the majority of cases. The two paired positions, 11 with 109 and 81 with 95, which are enclosed in boxes, are actually mispaired in a significant number of sequences. The numbering system is that of the *E. coli* sequence. Many eubacterial sequences are shorter and typically lack positions 1, 2, and 118–120. By beginning the alignment at position 3 rather than position 1, it is straightforward to obtain a universal numbering system for eubacteria that directly relates all of the sequences to the structural studies that have been done primarily on *E. coli*. The only place that a large number of eubacterial sequences present a problem is in the single-stranded region defined by helix IV, which sometimes contains a fourth base in the loop between *E. coli* positions 89 and 90. Since this is the usual situation in eukaryotic cytoplasmic 5 S rRNA, this position is numbered 89E.

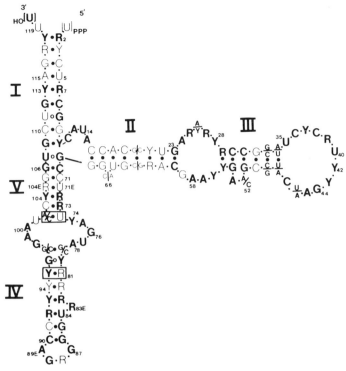

Fig. 3. Consensus secondary structure for eukaryotic cytoplasmic 5 S rRNAs. This figure displays the generally accepted secondary structural features of eukaryotic cytoplasmic 5 S rRNAs. The conventions followed are as in Fig. 2. The numbering system is again essentially that of *E. coli*. This is accomplished by recognizing certain insertions and deletions, which distinguish eubacterial and eukaryotic cytoplasmic 5 S rRNAs. There are three deletions that occur at positions 6, 41, and 114 along with four insertions which are between positions 71 and 72 (71E), 83 and 84 (83E), 89 and 90 (89E), and 104 and 105 (104E). Taken together, Figs. 2 and 3 allow an examination of sequence conservation. When such an evaluation is made it is found that the conservation is greater in the single-stranded regions and that the pattern of conservation is somewhat different in each case. These points are considerably clearer on Fig. 4, which displays only the essentially universal positions.

nated as helix IV, which allows for nomenclatural consistency between the two structures of Figs. 2 and 3. Authors that employ capital letters now usually refer to helix IV of Figs. 2 and 3 as E and helix V of Fig. 3 as D. This creates potential confusion, as in the earlier literature helix V was E and helix IV was D.

Even less consensus exists on a format for displaying the secondary structure, though the largest number of investigators use a format that closely resembles the original Fox and Woese model. In Figs. 2 and 3, we employ a modified arrangement of the Fox and Woese model that places helix I vertically rather than horizontally. The preference for this arrangement stems from the fact that it emphasizes the possibility that helices I and V may be coaxial in the tertiary structure of eukaryotic 5 S rRNAs. In actuality, eukaryotic sequences are also consistent with helices II and V being coaxial. It has been noted that an interchange between these alternative stacking arrangements may be possible (Luehrsen and Fox, 1981a; Studnicka *et al.*, 1981; Stahl *et al.*, 1981), though no effort has been made to pursue this matter experimentally. In order to further conform to the implied suggestion that secondary structure drawings should attempt to conform to actual tertiary structure, one can incorporate some bend into Figs. 2 and 3 where the extensions and main helices join (Troutt *et al.*, 1982).

In order to discuss experimental results pertaining to 5 S rRNA structure and function rationally, it is important to have a universal numbering system. Because most such experiments have been done on *E. coli* 5 S rRNA, the most popular system might be one that preserves the *E. coli* numbers. This approach has been utilized (Delihas and Andersen, 1982) and is formalized here in the following way. First, all eubacterial sequences and eukaryotic cytoplasmic sequences were separately aligned (a straightforward task), and a consensus sequence was determined for each category. These two consensus sequences were next aligned taking the structural features of Figs. 2 and 3 into account. Each position in the eubacterial consensus sequence was assigned its *E. coli* number. In the eukaryotic case, if a position were deleted relative to the eubacteria, the number for that position is skipped. Where insertions relative to the eubacteria occur, the inserted positions are assigned the number of the preceding homologous position followed by a capital letter E indicating that they are uniquely eukaryotic positions, e.g., 71E. Various individual sequences also present special problems, and those can be addressed in an analogous fashion using decimal points. Thus the insertion(s) in the loop defined by helix III, which occurs in *Vibrio harveyi* (Luehrsen and Fox, 1981b), is at position 34.1. The 5' terminus presents a special problem as some sequences have bases that precede position 1 of *E. coli*. These can be numbered toward 0 in decimal point fashion, descending as one goes to the left. In the context of this numbering system, then positions 1, 6, and 42 are deleted in eukaryotes. New positions in the eukaryotes are 71E, 74E, 84E, and 89E. If one is less chauvinistic about the importance of *E. coli*, then a more natural universal numbering system can be created, which eliminates the need for letter designations where insertions relative to the *E. coli* sequence occur (Manske, 1983).

It is an attractive notion to display constant positions, as in Fig. 4. This practice is far more subtle than the reader may realize and considerable caution is essential in interpreting such designations. To begin with, the data set is inherently biased by the presence of very similar sequences, which weight in favor of characteristics of particular phylogenetic groups. Ultimately, this can be corrected for, but no agreement yet exists on the methodology to be used. Next, we

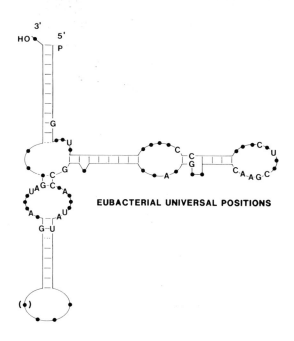

EUBACTERIAL UNIVERSAL POSITIONS

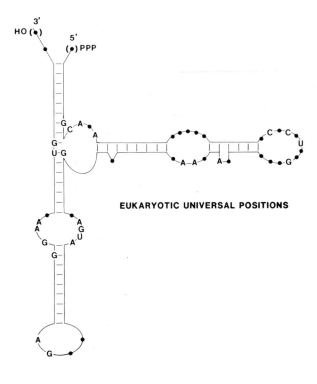

EUKARYOTIC UNIVERSAL POSITIONS

find that the definitions of conserved and universal positions employed by the various authors have been arbitrary. Thus, a position that is an adenine in perhaps 91% of the sequences may be regarded as conserved in terms of a 90% standard, whereas a guanine, which occurs at another position in 89% of the sequences, would be regarded as variable. Even more insidious is the matter of conserved purines and pyrimidines. At first, things seem straightforward, so that if one considers position 3 of the eubacteria one finds cytosine in 32 of 66 known cases and uracil in the other 34. However, what about position 1, which is cytosine in 64 of 69 cases and uracil in the other 5? Is position 1 a conserved cytosine or a universal pyrimidine? Also of concern is that the designation of purine/pyrimidine positions, as usually employed, tends to obscure other either/or positions. For example, eubacterial position 27 is adenine in 36 of 69 cases and cytosine in the other 33.

With these reservations in mind, the author suspects that the best existing approach is the one suggested by Manske (1983) in which information theory is used to define a relative nucleotide variability function, which is then calculated for each position instead of a simple percentage. A simpler alternative (Manske, 1983), which is used in Figs. 2 and 3, is to indicate several ranges of sequence conservation. One excellent feature of this approach is that every position is shown as having some base preference. This is a valid reflection of the fact that in over 95% of the positions one base occurs in at least 50% of the known sequences. There is considerable merit in simply examining the raw data and reaching one's own conclusions. Such a tabulation is available (Singhal and Shaw, 1983). In spite of difficulties described here, there is at least general agreement that the single-stranded regions in all ribosomal RNAs tend to be more conserved than the paired regions.

One advantage of defining universal positions in 5 S rRNA is that it allows better definition of differences between the structural models of Figs. 2 and 3 (Delihas and Andersen, 1982). Thus, one observes that in eukaryotes U_{40} and G_{44} are separated by only two nucleotides, whereas in eubacteria there are three intervening residues. This change in a distance constraint, which necessarily must reflect at least a local perturbation in the three-dimensional structure, is accompanied by the occurrence of the sequence segment $-CG_{44}AAC-$, which is essentially universal in eubacteria but rather variable in eukaryotes. Another characteristic distance is between the conserved guanine at position 69 and the universal UA at positions 77 and 78. In eubacteria, these residues are always

---

Fig. 4. Universal positions in 5 S rRNA. The 5 S rRNAs of eubacteria and the eukaryotic cytoplasm are displayed in the consensus structure of Figs. 2 and 3. Nucleotides, which are essentially universal (no more than two exceptions) in each 5 S rRNA category, are displayed explicitly. In both cases, it is clear that most of these universal positions are in single-stranded segments. It is also of interest that, although the numbers of universal positions are essentially the same, 25 and 24 respectively, only 10 of them are common to both 5 S rRNA types.

separated by seven nucleotides, whereas in eukaryotes, the separation is at least nine residues. Helix IV contains a G-U base pair, positions 80 and 96, which is found universally in all eubacterial and in most eukaryotic cytoplasmic 5 S rRNAs. In all eukaryotes, there are at least 17 bases between these residues, whereas in the eubacteria, the number is typically 15 or 16. This difference reflects meaningful redesign in the helix IV region. The hallmarks of this change are a looped-out residue, position 83E, and the presence of significant sequence conservation in the single-stranded loop. The looped-out residue is likely to alter the geometry in such a way that the eukaryotic version of helix IV is unlikely to be exactly superimposable with the eubacterial version when three-dimensional information becomes available. The eukaryotic version of helix IV frequently contains a U-U mispair at positions 81 and 95. This mispair, which is especially common in higher eukaryotic sequences, can be avoided by using an alternative configuration for helix IV (Luehrsen and Fox, 1981a). When this approach is used, the three-base extension of helix IV can no longer be made. Because the comparative evidence for the extension is overwhelming, the usual format is as in Fig. 3. It is not inconceivable, however, that an alternation between the two forms may be functionally significant (MacKay and Doolittle, 1981). The most obvious difference between eukaryotic and eubacterial 5 S rRNAs is the presence of the six-base helix V in eukaryotes, whose only counterpart in the eubacterial figure is two lone pairs (positions 70 and 71 with positions 105 and 106). Although widely accepted, these two pairs are rather conserved in eubacteria, so the number of true comparative tests is limited. These differences between eubacterial and eukaryotic 5 S rRNA will be central to the discussion of the archaebacterial 5 S rRNAs.

Beyond the consensus minimal structures of Figs. 2 and 3, the various authors argue for a variety of additional structural features. Some of these may be regarded as tertiary interactions that might not be readily amenable to evaluation by the comparative approach. Other examples are clearly secondary structure, but are difficult to evaluate. Only two of these will be discussed in detail here. Especially problematic are individual sequences that allow further pairings that could extend the helices of the consensus structures. The most notable example is the potential addition of a third base pair (between positions 28 and 56) to the helix III extension, which is possible in a large number of 5 S rRNAs. From a thermodynamic view, one would expect such an extra pair to be formed as it would clearly be favored over the apparent alternative of single strandedness. However, recent NMR studies on *E. coli* 5 S rRNA force a tempering of this view as they suggest that the equally obvious and widely accepted extension of helix I is not in fact formed (Kime *et al.*, 1984). From a comparative view, there is no evidence for a pairing between positions 28 and 56. The cytosine at position 28 is conserved, and in those few cases where it does vary there is not a compensating change at position 56. From this view, the large number of cases where the pairing can be made are simply the result of serendipitous occurrences of guanine at position 56. Proponents of the third pair might argue that those 5 S rRNAs that have it exhibit a slightly different structure. Opponents might contend that one of the two residues is actually involved in some other interaction either in 5 S rRNA tertiary structure or with another ribosomal component. Regardless of the correct interpretation, it is likely to be some time before this kind of detail can be finally resolved.

Of the various differences in opinion, by far the most important to the discussion of the archaebacterial data is that raised by investigators (Stahl *et al.*, 1981; De Wachter *et al.*, 1982; Küntzel *et al.*, 1983) who argue that a single consensus model of 5 S rRNA structure applies to all available sequences. This is in contrast to Figs. 2 and 3, which imply that the eubacterial and eukaryotic 5 S rRNAs are fundamentally different. De Wachter and colleagues (1982) do not disagree with the structural information of Figs. 2 and 3. Instead, they contend that Fig. 2, the eubacterial consensus structure, is incomplete in the sense that a fifth helix comparable to helix V of the eukaryotes does exist. The basis for this is the addition of two pairs (positions 73 and 74 with 102 and 103), to the consensus doublet (boxed in Table IV). The result is a helix containing four Watson–Crick pairs and one mispair as shown in Table IV.

The arguments for this uniform view of 5 S rRNA structure (De Wachter *et al.*, 1982) are principally the following. (1) It is universal; (2) it leads to lower estimates of free energy than those obtained with either the Fox and Woese (1975a) model (eubacteria) or the Hori and Osawa (1979) model (eukaryotes); and (3) unusual base pairs are acceptable. The first point is a matter of judgment in applying comparative ideas. In point of fact, since positions 73 and 74 as well as 102 and 103 are highly conserved in eubacteria, the validity of the pairing between these positions is very seldom tested. The central pair is in almost all cases a mispair of the adenine–guanine type because the positions involved are

TABLE IV

TYPICAL HELIX V DESIGNS

Eukaryotes		Eubacteria	
Candida	G[CC]UGA U[GG]ACU	*E. coli*	[CC]GAU [GG]AUG
Euglena	G[CC]CAG U[GG]GUC	*Rh. tenue*	[CC]AAU [GG]AUG
Human	G[CC]UGG U[GG]GCC	*B. subtilis*	[CC]GAU [GG]AUG

conserved. In the eukaryotic cases, a significant number of evolutionary tests occur, and they are essentially all consistent with the pairing. In the eubacterial case, the relevant sequence region is essentially unchanged. There are a small number of examples where some sequence variation does occur, and these are shown in Table V. All of the base changes that pertain to the consensus two base pair segment (boxed) provide positive tests of that pairing. Those changes that test the final pair of the putative helix are all negative. The very limited comparative evidence that is available does not support the presence of a true helix V.

The thermodynamic argument for a universal helix V is likewise not convincing. The energetic comparisons were typically not made to the models of Figs. 2 and 3 but were made, instead, to the earlier models that lacked the various helix extensions (De Wachter *et al.*, 1982). In the presence of these extensions, the calculated free energies would be lowered because the loop sizes are greatly reduced and would certainly approach those obtained by the uniform five helix model. In fact, this was the case for those examples where the Studnicka, Eiserling, and Lake model (1981) (essentially Fig. 2) was compared to the uniform five-helix model. Attempting to decide among the models on the basis of the resulting small differences would be very risky. The numbers used in the calculations are approximate, and there are quite possibly interactions between the 5 S rRNA and other ribosomal components. It would not be hard to envision such additional interactions involving some of the bases of the putative helix V region. Thus the model of Fig. 2 may be consistent with an actual three-dimensional structure in the eubacterial ribosomes that is thermodynamically more stable than any that could be based on the uniform model.

TABLE V

COMPARATIVE TESTS OF HELIX V

L. viridescene	C C G A A G G A U G
Cl. pasterianum	C U G A U G G A U G
My. capricolum	G U G A A C G A U A
Prochloron	G C U A A C G A U A
An. nidulans	G C A A C C G A U A

Finally, it is argued (De Wachter *et al.*, 1982) that occasional mispairs are seen in tRNA and 5 S rRNA, where sequencing errors can be ruled out. This is of course true. However, the instance presented by the putative helix V is another matter altogether. Rather than being an evolutionary transient, this hypothesized guanine–adenine pair is conserved. This would appear to be a major detracting point to the argument for helix V, but, as we shall see, it is not!

The suggestion that a guanine–adenine pair might fit into a helical region if the adenine were in the syn configuration was made over 20 years ago (Donohue and Trueblood, 1960). Model studies later suggested this was unlikely (Traub and Elson, 1966), and it became the dominant view that this pair could only be found at the ends of helices and not in their interior (Traub and Sussman, 1982). Recent high resolution NMR studies on a decadeoxyribonucleotide (Patel *et al.*, 1984) have revised this thinking, with the demonstration of the existence of a guanine–adenine pair, which is of the anti-anti type and resembles (Kan *et al.*, 1983) the $2\text{-m}^2\text{G}(26)\cdot\text{A}(44)$ pairing at the end of the anticodon stem in yeast phenylalanine tRNA (Rich, 1977). The NMR data also establish (Kan *et al.*, 1983; Patel *et al.*, 1984) that the phosphodiester backbone is in fact distorted, which is intuitively reasonable, since two opposing purines have to be accommodated.

These new results with model DNA compounds clearly require that the hypothesis of a eubacterial helix V based on a central guanine–adenine pair be taken seriously. Other evidence is difficult to interpret. Recent experiments with single-strand-specific probes have been negative. These include slow tritium exchange (Farber and Cantor, 1981), chemical modification (Silberklang *et al.*, 1983), and enzymatic digestion (Pieler and Erdmann, 1982). Of particular relevance are studies on an *E. coli* 5 S rRNA half molecule, positions 1–11 and 69–120, which is isolated by limited digestion with pancreatic ribonuclease (Douthwaite *et al.*, 1979). By high resolution NMR this fragment has been shown to exhibit a structure that is very similar to that of the same sequences in the whole molecule (Kime and Moore, 1983a). The downfield proton NMR spectrum of this fragment has been studied by nuclear Overhauser methods (Kime and Moore, 1983b, Kime *et al.*, 1984). Strong evidence was found for helices I and IV, with several resonances left unaccounted for. Indeed, the fact that the fragment was isolated at all suggests the five pyrimidines located between helices I and IV in *E. coli* 5 S rRNA were protected in some way. The two base pair helix of the consensus structure would account for protection of the cytosines. If the three pair extension needed to produce helix V is also included, one can then account for protection of two of the three uracil residues. Although less clearcut, NMR studies on *Bacillus licheniformis* (Salemink *et al.*, 1981) 5 S rRNA likewise suggest more interactions, i.e., 36, than the minimal model of Fig. 2, which only has 32.

At this stage the status of the fifth helix in eubacteria must then be regarded as unresolved. It seems very likely that additional structure exists in the region between helices I and IV, but the particular proposal made by De Wachter *et al.*, (1982) may not be the correct one. One thing that does seem safe to conclude is that resolution of this problem will enhance our insight to the principles governing RNA structure.

An independent and, for the purposes of this discussion, crucial issue is whether eukaryotic and eubacterial 5 S rRNAs differ in a meaningful way. If the universal occurrence of helix V is accepted, then one is still left with the other differences discussed previously. In fact, the presence of a helical region whose precise sequence seldom varies, whose length differs from the eukaryotic equivalent, and which characteristically contains a guanine–adenine pair, would almost certainly represent a commitment to a particular design of the ribosomal machinery, and thus would itself be properly regarded as a characteristic feature of the eubacterial 5 S rRNA! It is suggested that the still skeptical reader make an empirical study of the problem by selecting a few 5 S rRNA sequences at random from each category and attempt to align them. (This will be easier if one takes into account the four universally accepted helices.) It will quickly be discovered that there is no problem in aligning the eukaryotic sequences with one another or the eubacterial sequences with one another. It is only when one mixes them that difficulties arise. These are relatively minor difficulties, admittedly, but quite real nevertheless. So in conclusion, although the similarities between eukaryotic and eubacterial 5 S rRNAs are greater than was once thought, there are still significant differences. If we resist the temptation to produce universal models with appended lists of exceptions that tend to conceal these differences, it will be possible to address the question of their evolutionary origin. With that understanding, we may ultimately be able to develop a theoretical framework in which one will be able to predict possible pathways in the evolution of 5 S rRNA as a macromolecule.

B. ARCHAEBACTERIAL 5 S rRNA

Complete sequences are available for 5 S rRNAs from *Methanobrevibacter ruminantium* and *Methanospirillum hungatei* (Fox *et al.*, 1982), *Halobacterium volcanii* (C. Daniels *et al.*, in preparation), *Halobacterium cutirubrum* (Nazar *et al.*, 1978), *H. halobium* (Mankin *et al.*, 1982), *Halococcus morrhuae* (Luehrsen *et al.*, 1981a), *Thermoplasma acidophilum* (Luehrsen *et al.*, 1981b), *Sulfolobus acidocaldarius* (Stahl *et al.*, 1981), and *S. solfataricus* (Dams *et al.*, 1983b). The published sequence of *H. cutirubrum* (Nazar *et al.*, 1978) contains a minor sequencing error, so that the actual sequence is identical to that reported for *H. halobium* (K. R. Luehrsen *et al.*, unpublished results). Likewise the *S. acidocaldarius* and *S. solfataricus* sequences are identical. The reported sequence for *Methanobrevibacter smithii* (Fox *et al.*, 1982) was mislabeled and actually is from *M. ruminantium* and contained recording errors. The corrected version of this sequence is shown in Fig. 5. The author also has access to several unpublished methanogen sequences (Luehrsen *et al.*, unpublished). Partial sequence data in the form of ribonuclease T_1 catalog data is available for four closely related halophilic stains, which include *Halobacterium marismortui* and *H. vallismortis* (Nicholson and Fox, 1983).

Alignment of these sequences allows calculation of all possible binary differences. The extreme diversity seen among these RNAs is consistent with the early divergence of the archaebacterial ribosomes from those of the eubacteria and the eukaryotic cytoplasm. The difference matrix allows construction of a phylogenetic tree, Fig. 6, which in this case was accomplished by utilizing a modified version of the unweighted pair group method (Li, 1981). This tree reveals three main clusters of archaebacteria. Group 1 contains members of the genera *Methanobacterium, Methanobrevibacter,* and possibly *Methanococcus.* The extreme halophiles, *Halobacterium* and *Halococcus,* as well as the genus *Methanosarcina* comprise group 2. Group 3 includes *Methanomicrobium, Methanospirillum,* and *Methanogenium. Thermoplasma* marginally clusters with group 2 in Fig. 6. *Sulfolobus,* which has been omitted due to alignment problems, quite clearly represents a fourth major grouping. The relationships in Fig. 6 are strongly supported by the secondary structural considerations discussed below and are in reasonable agreement with the phylogeny obtained from 16 S rRNA catalog data. Further discussion of phylogenetic issues will be deferred to the final section.

In terms of secondary structure, the obvious question is whether the archaebacterial 5 S rRNAs conform to the eubacterial model of Fig. 2, the eukaryotic model of Fig. 3, or exhibit some third structure that is uniquely their own. In fact, it is found that none of these are correct. Instead, the various archaebacterial groups exhibit a mixture of unique, eubacterial and eukaryotic features built upon the same structural themes seen in all 5 S rRNAs.

Table VI provides a detailed summary of the structural features seen in specific archaebacterial strains as well as in the usual eubacterial and eukaryotic cytoplasmic 5 S rRNAs of Figs. 2 and 3. The group 1 archaebacterial 5 S rRNAs are readily distinguished from the group 2 and 3 structures. First, they are unique in that in all cases they contain a characteristic looped-out residue in helix I. Next, like eukaryotes, they lack the conserved segment -$CG_{44}AAC$-, though they do have three residues between U_{40} and G_{44}. Finally, helix IV lacks the looped-out nucleotide characteristic of eukaryotic 5 S rRNAs and hence is eubacteria-like. A typical group 1 5S rRNA is that of *Methanobrevibacter ruminantium* shown in Fig. 5. An extended helix V is essentially ruled out in this case due to the high uracil content of the relevant regions, which is not consistent with pairing. Even the two guanine–cytosine interactions that typify the eubacteria are only preserved, dotted lines on Fig. 5, if their location is allowed to vary relative

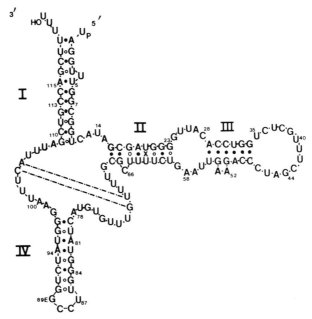

FIG. 5. *Methanobrevibacter ruminantium* 5 S rRNA. The *M. ruminantium* 5 S rRNA is shown to illustrate the features that are common to the 5 S rRNAs of the Division 1, group 1 archaebacteria. An earlier report (Fox *et al.,* 1982) erroneously indicated that this sequence was from *M. smithii* and contained recording errors, which inverted the uracil and adenine at positions 26 and 27 and reported a guanine at position 115 instead of a cytosine. The group 1 5 S rRNAs typically have a looped-out nucleotide preceding residue 5 in helix I and lack the common eubacterial sequence segment - CGAAC- which is replaced by -CGAUC-. This group of archaebacteria lacks the looped-out base in helix IV that is characteristic of all other archaebacteria. Helix IV in this figure is drawn so that it includes two terminal guanine–uracil pairs, which produces a normal-looking helix IV loop. The thermodynamic reasonableness of this is not obvious. In some instances, such as *Methanobrevibacter ruminantium*, it is impossible to make a helix V region of any consequence due to the high uracil content of the two regions, which would be paired. In such cases, it is usual that a two base pair segment such as that indicated on the figure by the dotted lines can be produced. It is not known whether such a two base pair interaction is of any biological relevance as it is not located in a position that is analogous to the two pair segment usually included in eubacterial 5 S rRNA secondary structure. The group 1 archaebacterial 5 S rRNAs resemble the consensus eubacterial 5 S rRNA structure more than any of the other archaebacterial 5 S rRNAs do. The numbering on this figure relates the regions that show obvious structural homology to the eubacterial or eukaryotic structures to *E. coli* 5 S rRNA. In areas such as the putative helix V region where it is not entirely obvious how to best make a correspondence, numbers are not assigned. The reader should thus not regard the numbering as a linear count from one end of the molecule to the other. Its purpose is to provide a nomenclatural framework for those who are attempting to compare various experimental studies on 5 S rRNA structure.

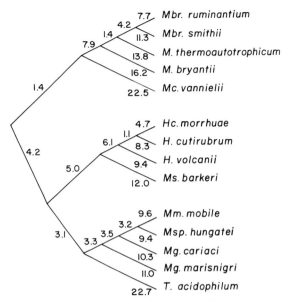

FIG. 6. Archaebacterial phylogeny as deduced from 5 S rRNA sequence data. A phylogenetic tree made by the method of Li (198?) is shown. Branch lengths are given as the number of nucleotide changes per 100 nucleotides. Abbreviations: *H., Halobacterium; Hc., Halococcus; M., Methanobacterium; Mbr., Methanobrevibacter, Mc., Methanococcus; Mg., Methanogenium; Mm., Methanomicrobium; Ms., Methanosarcina; Msp., Methanospirillum; T., Thermoplasma.* From Fox *et al.* (1982).

to the eubacteria. Some of the group 1 strains, e.g., *Methanobacterium thermoautotrophicum, M. bryantii,* and perhaps *Methanococcus vannielli* can form extended Watson–Crick helices in region V, but these differ from the eukaryotes in that the obvious potential for coaxial stacking between helices I and II or II and V is not present.

The remaining archaebacteria of Fig. 6 are divided into two major groups. This split is reflected at the 5 S rRNA structural level by the need to hypothesize a nonstandard base pair in order to preserve the helix II extension in *Methanospirillum hungatei* (Fig. 7) and its relatives, which is not the case for *Halobacterium cutirubrum* (Fig. 8), and its relatives. In other respects, most group 2 strains are in the same structural category. They all contain the sequence segment -CG$_{44}$AAC- with three nucleotides between U$_{40}$ and G$_{44}$, as is typical of eubacteria. In the helix IV region, a characteristic looped-out nucleotide defines a two base pair extension of the helix. This feature is strongly supported by comparative evidence and so can be considered established. Although this archaebacterial version of helix IV appears to resemble the eukaryotic version, it should be considered distinct because the location of the looped-out base, 84A, is different in the two cases. In the helix V region, all the group 2 and 3 archaebacterial 5 S

TABLE VI

Summary of Characteristic Structural Features in Various 5 S rRNAs[a,b]

Structural property	Eb	Group I					Group II				Group III				Ec	14	15
		1	2	3	4	5	6	7	8	9[h]	10	11	12	13	Ec	14	15
A. Helix I looped-out base[c]	–	+	+	+	+	+	–	–	–	–	–	–	–	–	–	–	–
B. Helix II extension[d]	+	+	+	+	+	+	+	+	+	+	–	–	–	–	+	+	+
C. Helix III loop length	13	13	13	13	13	13	13	13	13	13	13	13	13	13	12	13	13
D. -CGAAC- sequence	+	–	–	–	–	–	+	+	+	+	+	+	+	+	–	+	+
E. Helix IV looped-out base[e]	–	–	–	+	–	–	–	+	+	+	+	+	+	+	+	+	+
F. Extended helix V obviously feasible[f,g]	–	–	–	+	+	+	–	+	+	–	–	+	+	+	+	+	+
G. Helices I/V or II/V may be coaxial	–	–	–	–	–	–	–	–	–	–	–	–	–	–	+	+	+
H. Number of bases between main portion of helix II and beginning of helix IV	13	17	17	17	16	16	16	16	16	16	16	16	16	16	15	15	14
I. Total number of bases in helix IV loop and stem	19/20	20	16	19	18	16	21	21	21	21	21	21	21	21	21	21	21

[a] Modified from Fox et al. (1982).

[b] The 5 S rRNAs are as follows: Eb, consensus eubacterial 5 S rRNA (excluding organelles and mycoplasmas); 1, *Methanobrevibacter ruminantium*; 2, *Methanobrevibacter smithii*; 3, *Methanobacterium thermoautotrophicum*; 4, *Methanobacterium bryantii*; 5, *Methanococcus vannielii*; 6, *Halococcus morrhuae*; 7, *Halobacterium cutirubrum*; 8, *Halobacterium volcanii*; 9, *Methanosarcina barkeri*; 10, *Methanogenium marisnigri*; 11, *Methanogenium cariaci*; 12, *Methanospirillum hungatei*; 13, *Methanomicrobium mobile*; Ec, consensus eukaryotic 5 S rRNA; 14, *Thermoplasma acidophilum*; 15, *Sulfolobus acidocaldarius*. Structural property H refers to the distance between positions 66 and 78 in the eubacterial consensus sequence.

[c] The looped-out base in *S. acidocaldarius* is not in the usual location.

[d] A few eubacteria resemble strains 10–13 in this table in that they have a cytosine–adenine mispair at positions 17 and 67.

[e] The looped-out base in eukaryotes occurs one position earlier.

[f] *Halococcus morrhuae* 5 S rRNA contains a large insertion that disrupts the helix V region making the situation unclear.

[g] A three-base helix V can be made in strains 9 and 10.

[h] *Methanosarcina barkeri* is in group II by 5 S rRNA data but group III by the 16 S rRNA data.

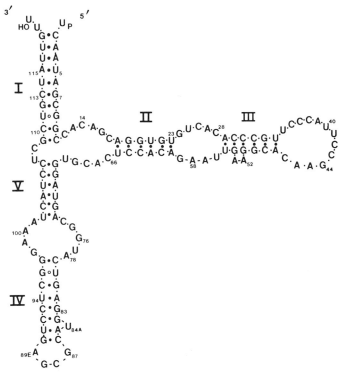

FIG. 7. *Methanospirillum hungatei* 5 S rRNA. In this representative group 3 archaebacterial 5 S rRNA, one sees that the first 66 positions are quite typically eubacterial in their structural features with the exception of the helix II extension, which cannot be made unless one invokes a nonstandard adenine–guanine base pair. Helix IV is the same archaebacterial analog of the eukaryotic helix IV that was described for *Halobacterium cutirubrum*. In this instance, a six base pair helix V is possible, but this is not established by existing comparative data. If it is properly included in the secondary structure, it would not seem to support a coaxial stacking between either helices I and V or II and V. Again, the position numbers on the figure serve to provide a reference to clearly equivalent positions in the *E. coli* sequence and should not be taken as an end-to-end numbering. As in most of the archaebacteria, these numbers are not well defined in the helix V region, which clearly differs structurally from its *E. coli* analog. From Fox *et al.* (1982).

rRNA sequences can make a helix of at least three contiguous base pairs, but without any obvious potential for coaxial stacking with helix I or II. Because the length and location of the putative helix V region is variable and because the number of known sequences is small, the comparative data are at present insufficient to decide the proper status of helix V except in the case of the halophiles, where it is almost certainly present.

The 5 S rRNAs from the (unrelated) thermoacidophiles, *Thermoplasma acidophilum* (Fig. 9) and *Sulfolobus acidocaldarius* (Fig. 10), are anomalous in

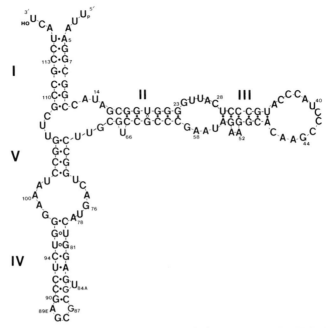

FIG. 8. *Halobacterium cutirubrum* 5S rRNA. A typical group 2 5 S rRNA. Helix I has a larger number of unpaired bases at its terminii than is usual, but the molecule is totally eubacteria-like in its structural features through position 69. Helix IV has a four-base loop as is typical of eukaryotes. The looped-out base in helix IV is followed by two pairs instead of the three that characterize the eukaryotes in this region. This feature is best regarded as uniquely archaebacterial and the relevant position is accordingly numbered 84A, which distinguishes it from the eukaryotic analog, which is 83E. A four-base helix V is possible and is supported by enough comparative data from other extreme halophiles that it should be regarded as likely or perhaps even proven. This helix V does not suggest coaxial stacking of helices I and V or II and V as the eukaryotic version does. *Escherichia coli* equivalent positions are again included and since these are only defined in some regions of the molecule they should not be mistaken for an end-to-end numbering. From Fox *et al.* (1982).

many respects, but do exhibit the structural hallmarks of the group 2 and 3 archaebacteria; the presence of the sequence segment -CG$_{44}$AAC- and the characteristic two base extension of helix IV. Both of these sequences exhibit an extremely stable helix V that, on thermodynamic grounds alone, is likely to be present in the active molecule, and each contains an eight base pair helix II without the characteristic single nucleotide bulge. These modifications of helices II and V make it easy to envision the same type of alternative stacking arrangements that characterizes the eukaryotes. Here the similarities end. The two thermoacidophiles differ extensively from one another in primary sequence, and

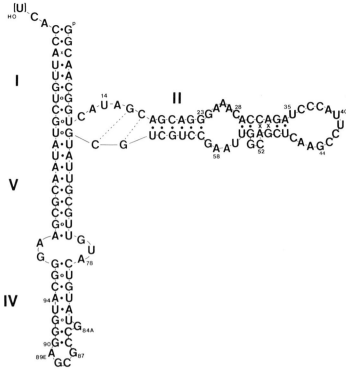

FIG. 9. *Thermoplasma acidophilum* 5 S rRNA. This 5 S rRNA is the only one known in which helix III is not well established due to two consecutive mispairs. There is in fact a third mispair in helix I as well. The 5 S rRNA tree of Fig. 6 places this organism in group 3 of the Division I archaebacteria, whereas the 16 S rRNA catalog data and hybridization data suggest it is properly considered as the sole representative of a fourth Division I group. The many unusual features seen in this 5 S rRNA would seem to support this notion even though the actual primary sequence data apparently do not. Helix IV is identical to that seen in other group 2 or 3 organisms. Helix V is in contrast quite different. A strongly paired helix is certain, and it resembles the eukaryotic case by allowing coaxial stacking between either helices I and V or II and V. There is, however, considerable uncertainty about what is happening at the junction of helices I, II, and V, and there is no comparative evidence available to resolve the situation. For example, one can extend helix I by four pairs into helix V by looping out the uracil at position 11. Likewise, one can eliminate the extension to helix II by extending helix V into helix I by two more pairs. Although the eight-pair helix II seems the most likely structure, the existence of these alternatives is responsible for the doubt implied by the dotted lines on the figure. Again, the numbers only pertain to those regions of the sequence that can be readily related to the eubacterial or eukaryotic sequences and should not be interpreted as indicative of an end-to-end numbering. From Fox *et al.* (1982).

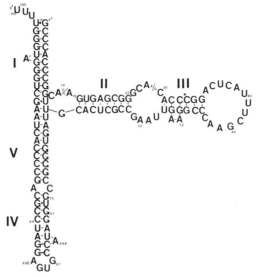

FIG. 10. *Sulfolobus acidocaldarius* 5 S rRNA sequence. This Division II archaebacterium has an unusual 5 S rRNA. Helix I contains a looped-out nucleotide, but it is not in the same position as that seen in *Methanobrevibacter ruminantium* and its immediate relatives. Helix II is extended to eight pairs without the characteristic loop out at *E. coli* position 66. Helix III contains a modified nucleotide, which is unusual in 5 S rRNA. Helix IV has the characteristic archaebacterial version but is unusual in that the universal sequence segment -UA- is missing. Helix V is long and very strong and resembles the eukaryotic cytoplasmic 5 S rRNAs by being consistent with coaxial stacking of either helix I or II with helix V. The numbers provide a frame of reference with the eubacterial sequences and do not represent an end-to-end numbering. From Fox *et al.* (1982).

each exhibits its own anomalies. The *Thermoplasma* structure lacks a well-paired helix III, which cannot be attributed to experimental errors because the sequence has been independently determined twice (Luehrsen *et al.*, 1981b). Helix I is also unusual in that it has the potential to form an alternative pairing arrangement, which would lengthen helix V at the expense of helices I and II. *Sulfolobus* (Stahl *et al.*, 1981) has the only 5 S rRNA that lacks the universal sequence segment $-U_{78}A-$. In addition, it contains a looped-out residue in helix I (in a different location than that seen in the group 1 archaebacteria) and a post-transcriptionally modified nucleotide in helix III.

The suggestion has been made (C. R. Woese, personal communication) that the thermoacidophiles, especially *Sulfolobus,* have adapted to their low pH–high temperature environment by mimicking with secondary structure a geometry that would normally be best accomplished through tertiary interactions. This idea is the root of a proposal (Stahl *et al.*, 1981) that helices I, V, and IV form a domain that is essentially a continuous helical rod in all 5 S rRNAs. The number and types of pairs that those authors invoke to achieve this seem excessive. Nevertheless, Woese's notion in its pure form may be quite correct. Indeed, experimental evidence to this end may exist (Leontis and Moore, as cited in Kime *et al.*, 1984). If this is so, it would be a most instructive example of evolutionary convergence at the molecular level and might provide an explanation for the anomalous result that *H. cutirubrum* 5 S rRNA is active in eubacterial ribosome reconstitution assays (Wrede and Erdmann, 1973).

C. Large *Halococcus* 5 S rRNA

A surprising result was the discovery of a 231 nucleotide 5 S rRNA homolog in *Halococcus morrhuae* (Luehrsen *et al.*, 1981a). Because the extra size is generated by an internal insertion of approximately 108 nucleotides (Fig. 11), it is reasonable to suppose that this large RNA is representative of different phenomonology than was seen in *Clostridium thermosaccharolyticum,* which has a 5 S rRNA of approximately 150 nucleotides. In *C. thermosaccharolyticum,* a normal-sized 5 S rRNA is also present, and the additional bases are appended at the 3' terminus (Sutton and Woese, 1975; C. R. Woese, personal communication). In this latter case, it is likely that the larger-than-normal 5 S rRNA represents a processing defect.

It is presumed that the *H. morrhuae* 5 S rRNA is functional because no normal-sized counterpart is found upon extraction of 70 S ribosomes. That the insertion is internal is readily established by alignment with the known sequence of *Halobacterium cutirubrum* (Nazar *et al.,* 1978), where one sees 88% homology on both sides of the insertion. The insertion itself appears to be minimally disruptive in that all but one of the known structural features of extremely halophilic 5 S rRNAs remain intact. The one exception is the helix V region, which is presumably destroyed by the emergence of the large inserted segment from its midst (see Fig. 11). The inserted piece itself contains numerous possible helical segments, but none are so compelling as to be given special attention in the absence of either comparative or direct experimental evidence. The inclusion into the ribosome of a 5 S rRNA that is almost twice normal size might well require compensating changes elsewhere, but there exists no evidence addressing this point.

Of obvious interest is the origin of such an insertion. The several possibilities are as follows. (A) It could be the result of a duplication event. This is suggested by the coincidence of the insert size, 108 nucleotides, with its location—108 nucleotides from the 5' terminus. However, this does not seem likely because there is not an obvious extensive homology between the two fragments. In order to save the hypothesis, one would have to argue that the event is of considerable age. Yet, this is untenable, because a recent survey of extreme halophile 5 S rRNAs indicated that the large 5 S rRNA is restricted to *Halococcus* strains, which exhibit

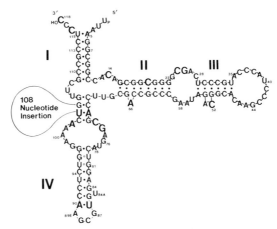

FIG. 11. *Halococcus morrhuae* 5 S rRNA sequence. This large 5 S rRNA homolog contains a 108 nucleotide insertion that is surrounded by normal 5 S rRNA sequences. A comparison with Fig. 7 shows that the usual structural features associated with the halophile 5 S rRNAs are all present though it would seem reasonable to suppose that the insert prevents formation of helix V. The sequence of the insert is known and is published elsewhere (Luehrsen *et al.*, 1981a). From Luehrsen *et al.* (1981a). Reprinted by permission from *Nature*, Vol. 293, pp. 755–756. Copyright © 1981 Macmillan Journals Limited.

little 5 S rRNA sequence variation (Nicholson, 1982). (B) A better hypothesis is that the insertion is an unspliced intervening sequence that does not interfere with the RNA's function. Intervening sequences have not been encountered in either of two published archaebacterial gene sequences that code for proteins (Chang *et al.*, 1981; Dunn *et al.*, 1981) and are so far unknown in 5 S rRNA genes in general. Because intron locations are typically conserved, the 5 S rRNA and the corresponding genes from *Halobacterium volcanii* were sequenced in the hope that positive proof could be had. In that instance, at least one 5 S rRNA gene and the RNA have proven to be colinear (Daniels *et al.*, 1985). These negative observations notwithstanding, the intron hypothesis remains tenable. This is especially true in view of the recent discovery of small introns in what are almost certainly tRNA genes in *Sulfolobus solfataricus* (Kaine *et al.*, 1983). (C) A third possibility is that the insertion is an artifact of a previous transposition event. On the negative side, there is no obvious duplication event, the inserted sequence lacks inverted repeats at its termini, and it is much smaller than a typical transposon. It does, however, have a heptanucleotide 5′ terminus -UGCCUCG-, which is identical to one orientation of the sequenced *H. halobium* transposon, ISH-1 (Simsek *et al.*, 1982). (D) Finally and perhaps most reasonably, the inserted material may be the result of a general recombination event.

Further investigation into the origin of this inserted segment is needed because it may represent an excellent model system for understanding the evolution of increasing size in rRNAs. Although rare in 5 S rRNA, insertion/deletion events of moderate size have been found in tRNA, e.g., in the "extra arm" and in comparisons of the 16 S rRNA of *E. coli,* maize, and various mitochondrial 12 S rRNAs (Glotz *et al.,* 1981). It would not be unreasonable to hope that a general understanding can be had for events of this type. The *Halococcus morrhuae* example might be especially germane for experimental work because its apparent recency suggests that it may be feasible to trace its origins.

D. 5 S rRNA: Summary

By focusing on the subtle differences between the 5 S rRNA types, one can see a molecule that has been subtly reshaped during the evolutionary process. In the case of the archaebacteria as with the other two urkingdoms, the basic character of the molecule remains unchanged. However, sufficient structural variety is encountered in this case to make the comparative approach less straightforward, especially in the helix V region, where one is not sure whether the structure is truly homologous. Despite this difficulty, it is clear that the individual archaebacterial sequences present a mixture of unique features together with some of the specific detailed features we associate with the eubacteria or the eukaryotes. [Two similar cases occur among the eubacteria, where variations in the main theme are well documented. These are the abbreviated helix IV domain of the mycoplasmas (Hori *et al.,* 1981; Walker *et al.,* 1982) and certain other organisms, and the subtly modified helix III region of the chloroplast 5 S rRNAs (Delihas and Andersen, 1982). These variations are unique to the eubacteria and do not correspond to either archaebacterial or eukaryotic 5 S rRNA features.] In view of the occurrence of both eukaryote- and eubacteria-like features in the archaebacterial 5 S rRNAs, it is reasonable to suspect they have a more central position in the evolutionary history of this molecule than the eubacteria do.

VI. 16 S rRNA Primary and Secondary Structure

Relatively little data are available on the large archaebacterial rRNAs. In the case of 16 S rRNA, the primary data base is a collection of 27 16 S rRNA catalogs, many of which are published (Balch *et al.,* 1977, 1979; Fox *et al.,* 1977a; Magrum *et al.,* 1978; Woese *et al.,* 1980b; Stackebrandt *et al.,* 1982; Jones *et al.,* 1983a,b), two complete 16 S rRNA sequences from *Halobacterium volcanii* (Gupta *et al.,* 1983) and *Halobacterium morrhuae* (Leffers and Garrett, 1984), and sequences of several ribosomal protein binding sites (Thurlow and Zimmermann, 1982). In addition, the 3' terminus of *H. halobium* has been sequenced (Kagramanova *et al.,* 1982). In the case of 23 S rRNA, essentially no data is available, though significant progress has been made on the *H. volcanii* primary sequence (C. R. Woese, personal communication).

The availability of an archaebacterial 16 S rRNA sequence makes it possible for the first time to assess the meaning of the large differences seen in comparisons of archaebacterial and eubacterial 16 S rRNA catalogs. In terms of primary structural homology, the *H. volcanii* and *E. coli* comparison gives a value of 58%, which is indeed low compared to the 75% seen in the very distant *Escherichia coli–Zea mays* chloroplast comparison. The 5 S rRNA results would suggest that it will be of more interest to examine the secondary structure.

Given the difficult road that has been followed to widely accepted models for 5 S rRNA secondary structure, the neophyte might view the matter of 16 S rRNA as essentially hopeless. This is not true. The importance of the comparative approach has been immediately recognized, and, as a result, the several models available for 16 S rRNA secondary structure are in good agreement considering the enormity of the problem (Woese *et al.*, 1980a, 1983; Glotz and Brimacombe, 1980; Noller and Woese, 1981; Zweib *et al.*, 1981; Brimacombe *et al.*, 1983; Noller, 1984). It would be counterproductive here to attempt to review the status of each of the proposed helices in *E. coli* 16 S rRNA. Such a detailed analysis can be found elsewhere (Woese *et al.*, 1983).

In this discussion, we will assume that the model of Woese *et al.* (1983) is a reasonable first approximation of a minimal model, and see how *H. volcanii* 16 S rRNA compares to it. This can be accomplished by aligning the 16 S rRNAs of *E. coli* and *H. volcanii* and searching for an equivalent helix in *H. volcanii* for each likely helix in *E. coli*. The alignment issue is again surprisingly easy, as point insertion/deletion events are rare. Once aligned, it is found that the vast majority of helices in *E. coli* are matched by a structure in the *H. volcanii* and *H. morrhuae* sequences of similar length and location. Because of the considerable sequence difference between the *E. coli* and the halophile RNAs, these equivalent structures are supported by a number of evolutionary tests. The net result is that an *H. volcanii* 16 S rRNA secondary structure can be constructed that is extremely similar to that of *E. coli*. This is illustrated by the secondary structure diagram (Fig. 12) (Woese *et al.*, 1983) where the proposed *H. volcanii* structure is compared to reasonable models for *E. coli, Saccharomyces cerevisiae* and a model displaying those features that are common to all 16 S–like rRNAs thus far sequenced. The overall impression given is that the two bacterial structures are very much alike. In fact, they resemble each other more than either resembles the eukaryotic 18 S rRNA. The most obvious changes that occur reflect insertion/deletion events that involve entire local structures rather than individual bases. Thus four hairpin loops, near positions 90, 430, 460, and 1450, that are thought to be present in the *E. coli* case, have no homologous sequence segments in *H. volcanii*. These insertion/deletion events largely account for the smaller size of the *H. volcanii* 16 S rRNA (1472 residues versus 1542) and are typical of the kinds of events that account for the size differences between other homologous ribosomal RNAs. The base composition of the individual helices and their lengths also resemble those of *E. coli*.

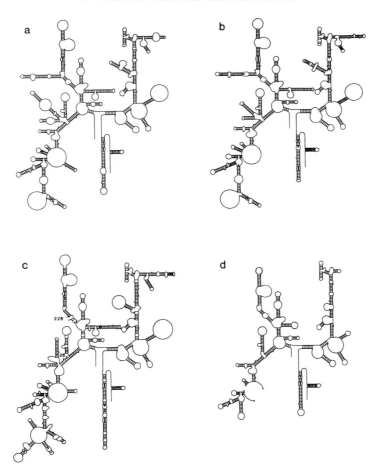

FIG. 12. Schematic comparison of secondary structure models of small ribosomal RNAs. The four RNAs shown are (a) *E. coli* 16 S rRNA; (b) *Halobacterium volcani* 16 S rRNA; (c) *Saccharomyces cerevisiae* 18 S rRNA; and (d) a composite RNA, which contains only those features that are common to all small subunit RNAs that have been sequenced so far, including mitochondria and chloroplasts. This hypothetical construct thus represents the core required structure of a small subunit RNA. More detailed discussion of these structures is presented elsewhere (Woese *et al.*, 1983). These drawings aptly illustrate the extensive structural homology between the *Halobacterium volcanii* 16 S rRNA and all other small subunit RNAs. In particular, it is clear that on the whole the *H. volcanii* is more like *E. coli* than it is like the eukaryotic 18 S rRNA. From Woese *et al.* (1983).

The eukaryotic–eubacterial–archaebacterial comparison was evaluated further (Gupta *et al.*, 1983) in terms of primary structure, by focusing on selected regions of RNA sequence. The idea is to rely only on regions that exhibit both sequence and structural homology. Presumably such regions have similar functional roles so that one has reasonable justification for arguing that the evolutionary constraints involved are comparable. Thirty such regions from 16 S rRNA were examined, and it was found that *E. coli* and *H. volcanii* were significantly closer to one another by this measure than either was to *Xenopus laevis*. The eukaryote, however, was not equidistant from the other two, but instead was significantly closer to the archaebacterium. Examination of the detailed comparisons reveals six regions that suggest a eukaryotic–archaebacterial relationship, whereas the rest imply a eubacterial–archaebacterial relationship. A similar situation actually occurs within the archaebacterial 5 S rRNAs, where the first 60 residues are typically rather eubacteria-like, whereas the next 50 resemble more the eukaryotes. What has not been found in either 5 S or 16 S rRNA are local domains that are suggestive of a specific eukaryote–eubacteria relationship.

Gupta *et al.* (1983) also interpreted their secondary structure findings in a similar fashion. They found that the eukaryotic 18 S rRNA lacks a number of features that are common to the two bacteria. In the one case where a eukaryotic helix is present in one of the bacterial examples but not in the other, it is the archaebacterial sequence that has it. Comparative analysis of ribosomal protein binding sites has allowed study of this phenomenon across phylogenetic lines (Thurlow and Zimmermann, 1982). The S 8–S 15 binding region of several diverse archaebacteria was found to be structurally homologous to the S 8–S 15 regions of the eubacteria *despite* very extensive sequence change. However, in the case of the L1 binding site (in 23 S rRNA), Thurlow and Zimmermann (1982) found that it was the eubacterial version *not* the archaebacterial version (*Thermoplasma*), that had structural homology with the eukaryote. This last fact puts us on notice that the results of further 16 S and 23 S rRNA comparisons are needed before any definite conclusions are reached. It is fortunate that a number of additional large archaebacterial rRNA sequences will soon be forthcoming.

VII. Evolutionary Considerations

With respect to the archaebacteria, there are two separate evolutionary issues. These are the relative phylogenetic position of the individual strains within the archaebacterial urkingdom, and the role of archaebacteria in the origin of other extant biological forms. The data that have primary bearing on the phylogeny issue are the 16 S rRNA catalogs, the 5 S rRNA sequences, and 16 S rRNA–DNA hybridization studies (Tu *et al.*, 1982). The hybridization studies used DNA from 14 strains with 16 S rRNA from seven suitably diverse keys. The results indicate there to be two major divisions of archaebacteria. The first division comprises the five orders: Methanococcales, Methanobacteriales, Methanomicrobiales, Halobacteriales, and Thermoplasmales. The second division includes two orders of thermoacidophiles: Sulfolobales and Thermoproteales. Since the hybridization study was based exclusively on 16 S rRNA, one would expect it to be in good agreement with the 16 S rRNA cataloging data. The earlier dendrograms based on cataloging data did not include any representation from the Thermoproteales and they were in good, but less, agreement with the hybridization results than one would have hoped. The inclusion of additional strains, especially a more representative collection of extreme halophiles, is reflected in Fig. 1. This most recent version of the 16 S rRNA dendrogram is in near-perfect agreement with the hybridization results. The one discrepancy is that cataloging data find the relative branching order of the major methanogen clusters and the halophile cluster to be indeterminate, whereas the hybridization data places the extreme halophiles well outside of both methanogen groups. It should be noted too that although the placement of *Thermoplasma acidophilum* in both cases is the same, the evidence is weak, and it may yet prove to have some other relationship (see Chapter 2 for further discussion).

The 5 S rRNA data (Fig. 5) place the extreme halophiles among the methanogens with *Methanosarcina barkeri* as a specific relative. In other respects, however, the 5 S rRNA tree resembles the 16 S rRNA tree including the detection of a specific relationship between the *Methanococcales* and *Methanobacteriales*. This discrepancy in the position of the halophiles is a nagging one—especially to those who publish papers presenting both views! If the halophiles are descended from the methanogens, as the 5 S data say and the 16 S rRNA cataloging data allow, then one might expect to find an extremely halophilic methanogen in nature. As yet, this hypothetical organism has not been found. A methanogenic ancestry of halophilism is supported by the observation that *Methanobacterium thermoautotrophicum* has an elevated internal salt concentration (Sprott and Jarrell, 1981), but the much smaller information content of the 5 S rRNA in comparison to the 16 S rRNA makes it clear that relationships based on the latter molecule are probably the more reliable. Since the view given by the current version of the 16 S dendrograms is neutral, this issue should be regarded as undecided.

Other data that usefully address phylogenetic relationships among individual strains reinforce this general outline. Immunological mapping of 17 strains of methanogenic bacteria gave excellent agreement with the local relationships seen among the individual strains in Fig. 1 (DeMacario *et al.*, 1981). Examination of the functional analog of elongation factor Tu in 18 archaebacterial strains indicated considerable molecular weight diversity. Separate groupings were found for low (thermoacidophiles), medium (methanogens), and high (extreme halophiles) molecular weight values (Kessel and Klink, 1982). The DNA-dependent RNA polymerases of the archaebacteria contain eight or more components, and two major subtypes can be distinguished, which correspond to the two major archaebacterial clusters (see Chapter 11). By this criterion, *Thermoplasma acidophilum* clusters with the thermoacidophilic–sulfur oxidizing group (Schnabel *et al.*, 1983). If the placement of this organism on Fig. 1 is correct, this would indicate that *T. acidophilum* branched off from the thermoacidophilic–sulfur oxidizing cluster earlier than either the methanogens or extreme halophiles did. An alternative, of course, is that *T. acidophilum* actually belongs within the thermoacidophilic–sulfur oxidizing cluster. The possibility of an extensive coaxial stack of helices I, V, and IV in *T. acidophilum* 5 S rRNA would seem to support this conclusion, though the primary sequence data do not.

Of particular phylogenetic interest is the relative age of the various groups. This might be argued on the basis of diversity. A well-defined phenotype that exhibits extensive diversity by molecular measurements can be presumed to be older than a phenotype that exhibits little molecular diversity. In addition to the usual assumptions about evolutionary rates, one is faced with a problem of definition of a characteristic phenotype and the question of whether or not the full range of phylogenetic diversity of the phenotype has been examined (Fox, 1981). In the present instance such an analysis suggests that the phenotypes represented by the thermoacidophiles and the methanogens are comparably old, whereas those of the extreme halophiles are likely of more recent origin.

Considerable interest has arisen around the role of archaebacteria in the larger picture of evolutionary history. This is primarily a response to the discovery of certain eukaryotic features among the archaebacteria. The bacteriorhodopsin of the halophiles resembles eukaryotic rhodopsin, though the isomers of retinal differ in the two cases (Oesterhelt and Stoeckenius, 1971). As discussed by Bayley and Morton (1964), the pathways utilized for the biosynthesis of β-carotene in the extreme halophiles resemble those of higher plants. *Methanobacterium thermoautotrophicum* may contain repetitive DNA (Mitchell *et al.,* 1979). *Thermoplasma acidophilum* contains a DNA-binding protein that organizes the DNA into globular particles that resemble nucleosomes. The DNA-dependent RNA polymerases of archaebacteria exhibit considerable variety in subunit composition, but they typically resemble RNA polymerase I of eukaryotes (Zillig *et al.,* 1982a,b). Indeed immunological cross-reactivity with yeast polymerase has been demonstrated (see Chapter 11). As has already been noted, the tRNA genes in *Sulfolobus solfataricus* contain intervening sequences (Kaine *et al.,* 1983).

With regard to the translation system, numerous eukaryote-like features have been found. *Halobacterium cutirubrum* lacks the met-tRNA transformylase that is characteristically used in eubacteria to formylate the methionine carried by the initiator tRNA (White and Bayley, 1972). The ribosomal A proteins from *Methanobacterium thermoautotrophicum* and *H. cutirubrum* (Matheson *et al.,* 1980) show significantly more homology with their eukaryotic counterparts than with eubacterial homologs L7/L12. The functional elongation factor G analog from a wide range of archaebacteria has been shown to be ADP-ribosylated by diphtheria toxin (Kessel and Klink, 1980, 1981) and to contain the modified amino acid diphthamide, associated with this reaction (Pappenheimer *et al.,* 1983). (See

Chapter 8). The archaebacterial EF-Tu analog is also immunologically unrelated to eubacterial EF-Tu and is insensitive to two antibiotics, pulvomycin and kirromycin, whose site of action is that factor (Cammarano *et al.*, 1982). Archaebacterial ribosomes also lack binding sites for many 70 S inhibitors like chloramphenicol and streptomycin but are inhibited in at least some cases by the 80 S inhibitor anisomycin (Pecher and Böck, 1981; Schmid *et al.*, 1982). (See Chapter 12.) Finally, as visualized by electron microscopy, the archaebacterial ribosomal subunits bear some resemblance to eukaryotic ribosomal subunits (Lake, 1983a,b).

Lest one forget, it is perhaps useful at this stage to mention that the archaebacteria have a prokaryotic cellular organization and actually resemble the eubacteria in some ways too. Thus, the 2Fe-2S ferredoxin from two strains of *Halobacterium* has been found to be homologous to the ferredoxin of the cyanobacterium, *Nostoc muscorum,* and to various chloroplast-type ferredoxins (Hase *et al.,* 1978). The *Thermoplasma* histone-like protein shows more homology to basic DNA binding proteins of *E. coli* than to the true histones of eukaryotes (Delange *et al.,* 1981). Archaebacterial ribosomes lack a 5.8 S rRNA, and, in the extreme halophiles at least, have subunits of typical eubacterial size.

The point of this discussion is not to deny that the archaebacteria are bacteria. Rather, it is to emphasize that in almost every case in which a trait customarily associated with eukaryotes (but not their organelles, which almost certainly have a eubacterial heritage) is found in the prokaryotic world, it is found among the archaebacteria. This must be interpreted to mean that archaebacteria have a history that is in some way intertwined with that of the eukaryotes. Clarification of this relationship is a central problem for evolutionary biology.

Discussions pertaining to this issue have largely focused on the relative branching order of the three major lines of descent. Specific suggestions have been made (see, for example, Van Valen and Maiorana, 1980; Matheson and Yaguchi, 1982; Fox *et al.,* 1982) and counter arguments have been quickly presented in one case (Woese and Gupta, 1981). Although not without problems, one of the more tenable possibilities (Fox *et al.,* 1982) is that the archaebacterial group split at a very early stage to give rise to the two major archaebacterial divisions. Subsequently, the methanogen/halophile division is seen as giving rise to the eubacterial ribosomes, and the thermoacidophile division leads to the eukaryotic cytoplasmic ribosomes. Electron microscopy has recently shown that each division of the archaebacteria has its own characteristic ribosome morphology (Henderson *et al.,* 1984; Lake *et al.,* 1984). These observations strengthen the view that the archaebacteria separately gave rise to the eubacterial and eukaryotic ribosomes. In Lake's discussion (Lake *et al.,* 1984) the two archaebacterial divisions are each elevated to kingdom status. It is claimed that such a nomenclatural change is necessary as only then would each major kingdom be monophyletic. Without entering the morass surrounding the concept of monophyla, the author finds this nomenclatural change unattractive because the available rRNA sequence data implies that the two archaebacterial divisions are specifically related. Regardless of the outcome of the nomenclatural debate, the new data, if correct, are an important finding pertaining to the history of the earliest evolutionary lineages.

The branching order issue is an obvious one to pursue, but it is not the only one. Evolutionary trees are intuitively very reasonable to most of us, and it is easy to forget that they are only mathematical models of what is in fact a considerably more complex process than is assumed. Such trees ignore the inherent variety that exists in individual species and thus neglect population effects. Modern species are reasonably uniform, and so this approach works well for them. Was the common ancestor of all extant life such a well-defined species? Did the evolutionary process wait for the emergence of the common ancestor of all cellular life before it began? The answer to both of these questions must be no. A relevant issue then is whether or not the character of the evolutionary process might have changed at some point. In particular, the establishment of a genetic apparatus (Woese and Fox, 1977b) could change the nature of evolution (gradually) from an essentially LaMarkian form to the Darwinian form we are so familiar with. If so, it then becomes meaningful to ask whether the relative branching between the major lines of descent occurred before, during, or after this transition (Fox et al., 1982). One possible consequence of a prior branching would be the finding of molecular mosaicism in the major descendent lines (Woese, 1982)—a mosaicism born not of the relatively recent plasmid-mediated gene interchange (that has been so long feared as an obscuring factor in studying molecular evolution) but instead a mosaic that contains a trace of an evolutionary history predating even the earliest populations of entities capable of Darwinian evolution, i.e., true speciation.

Did all the major lines of descent already exist before the dawn of cellular life? We still do not have enough data to be sure of the answer, but the RNA sequence data available to date does not exclude the evolutionary mosaicism this might imply. The 5 S rRNAs of the archaebacteria and eubacteria share an essentially identical structure in their 5' domains, which encompass the first 66 positions. The comparable region in eukaryotic cytoplasmic 5 S rRNAs is subtly different in that a base pair (positions 6 with 114) in helix I and a base at position 41 are deleted. This is borne out by primary sequence comparisons, which suggest that this half of the molecule is more eubacteria-like. The 3' domain of the archaebacteria is not as clear cut. In no case is it completely identical with either eubacterial or eukaryotic 3' domains. In fact, the archaebacterial strains themselves seem to exhibit several different structural themes. That the region is more eukaryote-like is readily seen in sequence comparisons involving this half of the molecule only. The data from the large ribosomal RNAs promises to be even more interesting when they become available. There is already evidence for shared domains of all types, archaebacterial–eukaryotic, archaebacterial–eubacterial and even eubacterial–eukaryotic (Gupta *et al.*, 1983; Woese *et al.*, 1983; Thurlow and Zimmermann, 1982). If this additional data cannot be reconciled with any of the reasonable branching patterns, the case for molecular mosaicism will be strongly enhanced.

The need to consider structure in interpreting the RNA sequence data signals a new age in the field of molecular evolution. No longer can we be content merely to debate the correctness of branching patterns of phylogenetic trees. It is essential to understand how molecular domains are defined and how they can change during evolution. When and where can insertions such as that seen in the *Halococcus morrhuae* 5 S rRNA be accepted? Can point events account for the types of variation seen in the 5S rRNA 3′ domains? Can we usefully extend our trees back further in time if we base our similarity measures on structural ideas rather than on primary sequences? Perhaps we will have to abandon the tree-type approach altogether when we are addressing these early stages in evolution. The goal, after all, is to understand the processes involved in the origin of life. The specific historical record itself is only of consequence to the extent that it helps us achieve this goal.

ACKNOWLEDGMENTS

Financial support was provided by grants from the National Science Foundation (PCM-8215876), the National Aeronautics and Space Administration (NSG-7440), and the Robert A. Welch Foundation (E-784). The author would like to thank Drs. Peter Moore and Charles Manske for helpful discussions.

REFERENCES

Aaronson, R. P., Young, J. F., and Palese, P. (1982). Oligonucleotide mapping: Evaluation of its sensitivity by computer simulation. *Nucleic Acids Res.* **10,** 237–246.
Auron, P. E., Rindon, W. P., Vary, C. P. H., Celentano, J. J., and Vournakis, J. N. (1982). Computer aided prediction of RNA secondary structures. *Nucleic Acids Res.* **10,** 403–419.
Balch, W. E., Magrum, L. J., Fox, G. E., Wolfe, R. S., and Woese, C. R. (1977). An ancient divergence among the bacteria. *J. Mol. Evol.* **9,** 305–311.
Balch, W. E., Fox, G. E., Magrum, L. J., Woese, C. R., and Wolfe, R. S. (1979). Methanogens: Reevaluation of a unique biological group. *Microbiol. Rev.* **43,** 260–296.
Bayley, S. T., and Morton, R. A. (1964). Recent developments in the molecular biology of the extremely halophilic bacterium *Halobacterium cutirubrum. J. Mol. Biol.* **9,** 654–669.
Bell, G. I., DeGennaro, L. J., Gelfland, D. H., Bishop, R. J., Valenzuela, P., and Rutter, W. J. (1977). Ribosomal RNA Genes of *Saccharomyces cerevisiae. J. Biol. Chem.* **252,** 8118–8125.
Bellemare, G., Vigne, R., and Jordan, B. (1973). Interaction between *Eschericia coli* ribosomal

proteins and 5S RNA molecules: Recognition of procaryotic 5S RNAs and rejection of eukaryotic 5S RNA. *Biochimie* **55,** 29–35.

Böhm, S., Fabian, H., and Welfle, H. (1981). Universal sécondary structures of prokaryotic and eukaryotic ribosomal 5S RNA derived from comparative analysis of their sequences. *Acta Biol. Med. Ger.* **40,** K19–K24.

Böhm, S., Fabian, H., and Welfle, H. (1982). Universal structural features of prokaryotic and eukaryotic ribosomal 5S RNA derived from comparative analysis of their sequences. *Acta Biol. Med. Ger.* **41,** 1–16.

Bollon, A. P. (1982). Organization of fungal ribosomal RNA genes. *Cell Nucleus* **10,** 87–125.

Bonen, L., and Doolittle, W. F. (1975). On the prokaryotic nature of red algal chloroplasts. *Proc. Natl. Acad. Sci. U.S.A.* **72,** 2310–2314.

Borsuk, P. A., Nagiec, M. M., Stepien, P. P., and Bartnik, E. (1982). Organization of the ribosomal RNA gene cluster in *Aspergillus nidulans. Gene* **17,** 147–152.

Brimacombe, R., Maly, P., and Zweib, C. (1983). The structure of ribosomal RNA and its organization relative to ribosomal protein. *Prog. Nucleic Acid Res. Mol. Biol.* **28,** 1–48.

Brosius, J., Palmer, M. L., Kennedy, P. S., and Noller, H. F. (1978). Complete nucleotide sequence of a 16S ribosomal RNA gene from *Escherichia coli. Proc. Natl. Acad. Sci. U.S.A.* **75,** 4801–4805.

Brown, D. D., and Gurdon, J. B. (1978). Cloned single repeating units of 5S DNA direct accurate transcription of 5S RNA when injected into *Xenopus* oocytes. *Proc. Natl. Acad. Sci. U.S.A.* **75,** 2849–2853.

Brownlee, G. G., and Sanger, F. (1969). Chromatography of ^{32}P-labelled oligonucleotides on thin layers of DEAE-celluose. *Eur. J. Biochem.* **11,** 395–399.

Brownlee, G. G., Sanger, F., and Barrell, B. G. (1967). Nucleotide sequence of 5S ribosomal RNA from *Escherichia coli. Nature (London)* **215,** 735–736.

Butler, M. H., Wall, S. M., Luehrsen, K. R., Fox, G. E., and Hecht, R. M. (1981). Molecular and evolutionary relationships between two strains of *Caenorhabtis elegans* and the related species *C. briggsae. J. Mol. Evol.* **18,** 18–23.

Cammarano, P., Teichner, A., Chinali, G., Londei, P., De Rosa, M., Gambacorta, A., and Nicolaus, B. (1982). Archaebacterial elongation factor Tu insensitive to pulvomycin and kirromycin. *FEBS Lett.* **148,** 255–259.

Chang, S. H., Majumdar, A., Dunn, R., Makabe, O., RajBhandary, U. L., Khorana, H. G., Ohtsuka, E., Tanaka, T., Taniyama, Y. O., and Ikehara, M. (1981). Bacteriorhodopsin -partial sequence of messenger RNA provides amino acid sequence in the percursor region. *Proc. Natl. Acad. Sci. U.S.A.* **78,** 3398–3402.

Chen, K.-N. B. (1980). Computer simulation and analysis of the 16S rRNA sequence. M.S. Thesis, University of Houston Central Campus.

Cihlar, R. L., and Sypherd, P. S. (1980). The organization of the ribosomal RNA genes in the fungus *Mucor racemosus. Nucleic Acids Res.* **8,** 793–804.

Comay, E., Nussinov, R., and Comay, O. (1984). An accelerated algorithm for calculating the secondary structure of single stranded RNAs. *Nucleic Acids Res.* **12,** 53–65.

Dams, E., Vandenberghe, A., and De Wachter, R. (1983a). Sequences of the 5S rRNAs of *Azotobacter vinelandii, Pseudomonas aeruginosa* and *Pseudomonas fluorescens* with some, notes on 5S RNA secondary structure. *Nucleic Acids Res.* **11,** 1245–1252.

Dams, E., Londei, P., Cammarano, P., Vandenberghe, A., and De Wachter, R. (1983b). Sequences of the 5S rRNAs of the Thermo-Aacidophilic Archaebacterium *Sulfolobus Solfataricus (Calderiella Acidophila)* and the thermophilic eubacteria *Bacillus Acidocaldarius* and *Thermus Aquaticus. Nucleic Acids Res.* **11,** 4467–4676.

Daniels, C., Hoffman, J. D., Leuhrsen, K. R., Woese, C. R., Fox, G. E., and Doolittle, W. F. (1985). Sequence of the 5 S rDNA cistrons of *Halobacterium volcanii. Mol. Gen. Genet.* (in press).

Delange, R. J., Williams, L. C., and Searcy, D. G. (1981). A histone-like protein (HTa) from *Thermoplasma acidophilum*. *J. Biol. Chem.* **256**, 905–911.

Delihas, N., and Andersen, J. (1982). Generalized structures of the 5S ribosomal RNAs. *Nucleic Acids Res.* **10**, 7323–7344.

DeMacario, E. C., Wolin, M. J., and DeMacario, A. J. L. (1981). Immunology of archaebacteria that produce methane gas. *Science* **214**, 74–75.

De Wachter, R., and Fiers, W. (1972). Preparative two-dimensional polyacrylamide gel electrophoresis of ^{32}P-labeled RNA. *Anal. Biochem.* **49**, 184–197.

De Wachter, R., and Fiers, W. (1982). Two dimensional gel electrophoresis of nucleic acids. *In* "Gel Electrophoresis of Nucleic Acids: A Practical Approach" (D. Rickwood and B. D. Hames, eds.), pp. 77–117. IRL Press Ltd., Oxford.

De Wachter, R., Chen, M. W., and Vandenberghe, A. (1982). Conservation of secondary structure in 5S ribosomal RNA: A uniform model for eukaryotic, eubacterial, archaebacterial and organelle sequence is energetically favorable. *Biochimie* **64**, 311–329.

Doering, J., and Brown, D. D. (1978). Sequence of *X. borealis* somatic 5S DNAs. *Year Book—Carnegie Inst. Washington* **77**, 128–130.

Donohue, J., and Trueblood, K. N. (1960). Base pairing in DNA. *J. Mol. Biol.* **2**, 363–371.

Doolittle, W. F., Woese, C. R., Sogin, M. L., Bonen, L., and Stahl, D. (1975). Sequence studies on 16S ribosomal RNA from a bluegreen alga. *J. Mol. Evol.* **4**, 307–315.

Douthwaite, S., and Garrett, R. A. (1981). Secondary structure of prokaryotic 5S ribosomal ribonucleic acids: A study with ribonucleases. *Biochemistry* **20**, 7301–7307.

Douthwaite, S., Garrett, R. A., Wagner, R., and Feunteun, J. (1979). A Ribonuclease-resistant region of 5S RNA and its relation to the RNA binding sites of proteins L18 and L25. *Nucleic Acids Res.* **6**, 2453–2470.

Dunn, R., McCoy, J., Simsek, M., Majumdar, A., Chang, S. H., RajBhandary U. L., and Khorana, H. G. (1981). The bacteriorhodopsin gene. *Proc. Natl. Acad. Sci. U.S.A.* **78**, 6744–6748.

Farber, N. M., and Cantor, C. R. (1981). A slow tritium exchange study of the solution structure of *Escherichia coli* 5S ribosomal RNA. *J. Mol. Biol.* **146**, 223–139.

Fedoroff N. V., and Brown, D. D. (1978). The nucleotide sequence of oocyte 5S DNA in *Xenopus laevis*. I. The A-T rich spacer. *Cell* **13**, 701–716.

Fitch, W. M., and Margoliash, E. (1967). Construction of phylogenetic trees. *Science* **155**, 279–284.

Fox, G. E. (1981). Archaebacteria, ribosomes and the origin of eukaryotic cells. *In* "Evolution Today—Proceedings of the Second International Congress of Systematic and Evolutionary Biology" (G. E. Scudder and J. L. Reveal, eds.), pp. 235–244. Hunt Institute for Botanical Documentation, Carnegie Mellon University, Pittsburgh, Pennsylvania.

Fox, G. E., and Woese, C. R. (1975a). "5S rRNA secondary structure. *Nature (London)* **256**, 505–507.

Fox, G. E., and Woese, C. R. (1975b). The architecture of 5S rRNA and its relation to function. *J. Mol. Evol.* **6**, 61–76.

Fox, G. E., Magrum, L. J., Balch, W. E., Wolfe, R. S., and Woese, C. R. (1977a). Classification of methanogenic bacteria by 16S ribosomal RNA characterization. *Proc. Natl. Acad. Sci. U.S.A.* **74**, 4537–4541.

Fox, G. E., Pechman, K. R., and Woese, C. R. (1977b). Comparative cataloging of 16S ribosomal ribonucleic acid: Molecular approach to procaryotic systematics. *Int. J. Syst. Bacteriol.* **27**, 44–57.

Fox, G. E., Stackerbrandt, E., Hespell, R. B., Gibson, J., Maniloff, J., Dyer, T. A., Wolfe, R. S., Balch, W. E., Tanner, R. S., Mangrum, L. J., Zablen, L. V., Blakemore, R., Gupta, R., Bonen, L., Lewis, B. J., Stahl, D. A., Luehrsen, K. R., Chen, K. N., and Woese, C. R. (1980). The phylogeny of prokaryotes. *Science* **209**, 457–463.

Fox, G. E., Leuhrsen, K. R., and Woese, C. R. (1982). Archaebacterial 5S ribosomal RNA. *Zentralbl. Bakteriol., Mikrobiol. Hyg., Abt. 1, Orig.* C **3**, 330–345.

Garrett, R. A., Douthwaite, S., and Noller, H. F. (1981). Structure and role of 5S RNA-protein complexes in protein biosynthesis. *Trends Biochem. Sci.* **6**, 137–139.

Glotz, C., and Brimacombe, R. (1980). An experimentally derived model for the secondary structure of the 16S ribosomal RNA from *Escherichia coli, Nucleic Acids Res.* **9**, 3287–3306.

Glotz, C., Zweib, C., Brimacombe, R., Edwards, K., and Kössel, H. (1981). Secondary structure of the large subunit ribosomal RNA from *Escherichia coli, Zea mays* chloroplast and human and mouse mitochondrial ribosomes. *Nucleic Acids Res.* **9**, 3287–3306.

Green, M. R., Grimm, M. F., Goewert, R. R., Collins, R. A., Cole, M. D., Lambowitz, A. M., Heckman, J. E., Yin, S., and RajBhandary, U. L. (1981). Transcripts and processing patterns for the ribosomal RNA and transfer RNA region of *Neurospora crassa* mitochondrial DNA. *J. Biol. Chem.* **256**, 2027–2034.

Grimm, M. F., and Lambowitz, A. M. (1979). 32S RNA of *Neurospora crassa* mitochondria is not a precursor of the mitochondrial ribosomal RNAs. *J. Mol. Biol.* **134**, 667–672.

Gupta, R., Lanter, J. M., and Woese, C. R. (1983). Sequence of th 16S ribosomal RNA from *Halobacterium volcanii,* an archaebacterium. *Science* **221**, 656–659.

Hase, T., Wakabayashi, S., Matsubara, H., Kerscher, L., Oesterhelt, D., Rao, K. K., and Hall, D. O. (1978). Complete amino acid sequence of *Halobacterium halobium* ferredoxin containing an N- ε- acetyllysine residue. *J. Biochem. (Tokyo)* **83**, 1657–1670.

Heckman, J. E., and RajBhandary, U. L. (1979). Organization of transfer RNA and ribosomal RNA genes in *Neurospora crassa* mitochondria-intervening sequence in the large ribosomal RNA gene and strand distribution of the RNA genes. *Cell* **17**, 583–595.

Henderson, E., Oakes, M., Clark, M. W., Lake, J. A., Matheson, A. T., and Zillig, W. (1984). A new ribosome structure. *Science* **225**, 510–512.

Hinnebusch, A. G., Klotz, L. C., Blanken, R. I., and Loeblich, A. R. (1981). An evaluation of the phylogenetic position of the dinoflagellate *Crypthecodinium-cohnii* based on 5S ribosomal-RNA characterization. *J. Mol. Evol.* **17**, 334–347.

Hofman, J. D., Lau, R. H., and Doolittle, W. F. (1979). The number, physical organization and transcription of ribosomal RNA cistrons in an archaebacterium: *Halobacterium halobium. Nucleic Acids Res.* **1**, 1321–1333.

Holley, R. W., Apgar, J., Everett, G. A., Madison, J. T., Marquise, M., Merrill, S. H., Penswick, J. R., and Zamir, A. (1965). Structure of a ribonucleic acid. *Science* **147**, 1462–1465.

Hori, H. (1976). Molecular evolution of 5S rRNA. *Mol. Gen. Genet.* **145**, 119–123.

Hori, H., and Osawa, S. (1979). Evolutionary change in 5S RNA secondary structure and a phylogenetic tree of 54 5S rRNA species. *Proc. Natl. Acad. Sci. U.S.A.* **76**, 381–385.

Hori, H., Sawado, M., Osawa, S., Murao, K., and Ishikura, H. (1981). The nucleotide sequence of 5S rRNA from *Mycoplasma capricolum. Nucleic Acids Res.* **9**, 5407–5410.

Hori, H., Itoh, T., and Osawa, S. (1982). The phylogenetic structure of the metabacteria. *Zentralbl. Bakteriol., Mikrobiol. Hyg. Abt. 1, Orig. C* **3**, 18–30.

Jacobson, A. B., Good, L., Simonetti, J., and Zuker, M. (1984). Some simple computational methods to improve the folding of large RNAs. *Nucleic Acids Res.* **12**, 45–52.

Jarsch, M., Altenbuchner, J., and Böch, A. (1983). Physical organization of the genes for ribosomal RNA in *Methanococcus vannielii. Mol. Gen. Genet.* **189**, 41–47.

Jay, E., Bambara, R., Padmanaghan, R., and Wu, R. (1974). DNA sequence analysis: A general simple and rapid procedure for sequencing large oligodeoxyribonucleotide fragments by mapping. *Nucleic Acids Res.* **1**, 331–353.

Jones, W. J., Paynter, M. J. B., and Gupta, R. (1983a). Characterization of *Methanococcus Maripaludis* sp. nov., A new methanogen isolated from salt marsh sediment. *Arch. Microbiol.* **135**, 91–97.

Jones, W. J., Leigh, J. A., Mayer, F., Woese, C. R., and Wolfe, R. S. (1983b). *Methanococcus jannaschii* sp. nov., An extremely thermophilic methanogen from a submarine hydrothermal vent. *Arch. Microbiol.* **136**, 254–261.

Kagramanova, V. K., Mankin, A. S., Baratova, L. A., and Bogdanov, A. A. (1982). The 3'-terminal nucleotide sequence of the *Halobacterium halobium* 16S RNA. *FEBS Lett.* **144**, 177–180.

Kaina, B., Hinz, R., and Schubert, I. (1978). Organization of 5S RNA genes in *Vicia faba*. *FEBS Lett.* **96**, 19–22.

Kaine, B. P., Gupta, R., and Woese, C. R. (1983). Putative introns in tRNA genes of prokaryotes. *Proc. Natl. Acad. Sci. U.S.A.* **80**, 3309–3312.

Kan, L.-S., Chandrasegaran, S., Pulford, S. M., and Miller, P. S. (1983). Detection of a guanine-adenine base pair in a decadeoxyribonucleotide by proton magnetic resonance spectroscopy. *Proc. Natl. Acad. Sci. U.S.A.* **80**, 4263–4265.

Kessel, M., and Klink, F. (1980). Archaebacterial elongation factor is ADP-ribosylated by diphtheria toxin. *Nature (London)* **287**, 250–251.

Kessel, M., and Klink, F. (1981). Two elongation factors from the extremely halophilic archaebacterium *Halobacterium cutirubrum*. *Eur. J. Biochem.* **114**, 481–486.

Kessel, M., and Klink, F. (1982). Identification and comparison of eighteen archaebacteria by means of the diphtheria toxin reaction. *Zentralbl. Bakteriol., Mikrobiol. Hyg., Abt. 1, Orig. C* **3**, 140–148.

Kim, S. H. (1976). Three-dimensional structure of transfer RNA. *Prog. Nucleic Acid Res. Mol. Biol.* **17**, 181–216.

Kime, M. J., and Moore, P. B. (1983a). Physical evidence for a domain structure in *Escherichia coli* 5S RNA. *FEBS Lett.* **153**, 199–203.

Kime, M. J., and Moore, P. B. (1983b). Nuclear Overhauser experiments at 500 MHz on the downfield proton spectrum of a ribonuclease resistant fragment of 5S ribonucleic acid. *Biochemistry* **22**, 2622–2629.

Kime, M. J., Gewirth, D. T., and Moore, P. B. (1984). Assignment of resonances in the downfield proton spectrum of *Escherichia coli* 5S RNA and its nucleoprotein complexes using components of a ribonuclease-resistant fragment. *Biochemistry* **23**, 3559–3568.

Kimmel, A. R., and Gorovsky, M. A. (1978). Organization of 5S RNA genes in macronuclei and micronuclei of *Tetrahymema pyriformis*. *Chromosoma* **67**, 1–20.

Korn, K. J., and Brown, D. D. (1978). Nucleotide sequence of *Xervnpus borealis* oocyte 5S DNA: Comparison of sequences that flank related eucaryotic genes. *Cell* **15**, 1145–1156.

Kramer, R. A., Philippsen, P., and Davis, R. W. (1978). Divergent transcription in the yeast ribosomal RNA coding region shown by hybridization to separated strands and sequence analysis of cloned DNA *J. Mol. Biol.* **123**, 405–416.

Küntzel, H., Piechulla, B., and Hahn, U. (1983). Consensus structure and evolution of 5S rRNA. *Nucleic Acids Res.* **11**, 893–900.

Lake, J. A. (1983a). Evolving ribosome structure: domains in archaebacteria, eubacteria, and eucaryotes. *Cell* **33**, 318–319.

Lake, J. A. (1983b). Ribosome evolution—the structural basis of protein synthesis in archaebacteria, eubacteria and eukaryotes. *Prog. Nucl. Acids Res. Molec. Biol.* **30**, 163–190.

Lake, J. A., Henderson, E., Oakes, M., and Clark, M. W. (1984). Eocytes: A new ribosome structure indicates a kingdom with a close relationship to eukaryotes. *Proc. Natl. Acad. Sci. U.S.A.* **81**, 3786–3790.

Leffers, H., and Garrett, R. A. (1984). The nucleotide sequence of the 16S ribosomal RNA gene of the archaebacterium *Halococcus morrhua*. *EMBO J.* **3**, 1613–1619.

Li, W. H. (1981). Simple method for constructing trees from distance matrices. *Proc.Natl. Acad. Sci. U.S.A.* **78**, 1085–1089.

Lim, B. L., Hori, H., and Osawa, S. (1983). The nucleotide sequences of 5S rRNAs from a multicellular green alga, *Ulva pertura* and two brown algae, *Eisenia bicyclis* and *Sargassum fulvellum*. *Nucleic Acids Res.* **11**, 1909–1910.

Long, E. O., and Dawid, I. B. (1980). Repeated genes in eucaryotes. *Annu. Rev. Biochem.* **49**, 727–764.

Luehrsen, K. R., and Fox, G. E. (1981a). Secondary structure of eukaryotic cytoplasmic 5S ribosomal RNA. *Proc. Natl. Acad. Sci. U.S.A.* **78,** 2150–2154.

Luehrsen, K. R., and Fox, G. E. (1981b). The nucleotide sequence of *Beneckea harveyi* 5S ribosomal ribonucleic acid. *J. Mol. Evol.* **17,** 52–55.

Luehrsen, K. R., Nicholson, D. E., Eubanks, D. C., and Fox, G. E. (1981a). An archaebacterial 5S rRNA contains a long insertion sequence. *Nature (London)* **293,** 755–756.

Luehrsen, K. R., Fox, G. E., Kilpatrick, M. W., Walker, R. T., Domdey, H., Krupp, G., and Gross, H. J. (1981b). The nucleotide sequence of the 5S rRNA from the archaebacterium *Thermoplasma acidophilum*. *Nucleic Acids Res.* **9,** 965–970.

MacKay, R. M., and Doolittle, W. F. (1981). Nucleotide sequences of *Acanthamoeba castellanii* 5S and 5.8S ribosomal ribonucleic acids: Phylogenetic and comparative structural analyses. *Nucleic Acids Res.* **9,** 3321–3333.

MacKay, R. M., Spencer, D. F., Schnare, M. N., Doolittle, W. F., and Gray, M. W. (1982). Comparative sequence analysis as an approach to eualuating structure, function, and evolution of 5S and 5.8S ribosomal ribonucleic acids: Phylogenetic and comparative structural analyses. *Nucleic Acids Res.* **9,** 3321–3333.

MacKay, R. M., Spencer, D. F., Schnare, M. N., Doolittle, W. F., and Gray, M. W. (1982). Comparative sequence analysis as an approach to evaluating structure, function, and evolution of 5S and 5.8S ribosomal RNAs. *Can. J. Biochem.* **60,** 480–489.

McNamara, P. M. (1969). Comparison of catalogues of purine-rich oligonucleotides from *Escherichia coli* and *Bacillus subtilis* ribosomal ribonucleic acids. Ph.D. Dissertation, University of Illinois at Urbana-Champaign.

Magrum, L. J., Luehrsen, K. R., and Woese, C. R. (1978). Are extreme halophiles actually bacteria? *J. Mol. Evol.* **11,** 1–8.

Mankin, A. S., Kagramanova, V. K., Belova, E. N., Teterina, N. L., Baratova, L. A., and Bogdanov, A. A. (1982). Primary and secondary structure of 5S rRNA from *Halobacterium halobium*. *Biochem. Int.* **5,** 719–726.

Manske, C. L. (1983). The application of information theory to: The comparative sequence analysis of 5S rRNA, computer simulations of molecular evolution, and methods of constructing phylogenetic trees from nucleotide sequences. Ph.D. Dissertation, University of California at Los Angeles.

Mao, J., Appel, B., Schaack, J., Sharp, S., Yamada, H., and Söll, D. (1982). The 5S RNA genes of *Schizosaccharomyces pombe*. *Nucleic Acids Res.* **10,** 487–500.

Matheson, A. T., and Yaguchi, M. (1982). The evolution of the Archaebacterial ribosome. *Zentralbl. Bakteriol., Mikrobiol. Hyg. Abt. 1, Orig. C* **3,** 192–199.

Matheson, A. T., Yaguchi, M., Balch, W. E., and Wolfe, R. S. (1980). Sequence homologies in the N-terminal Region of the Ribosomal A protein from *Methanobacterium thermoautotrophicum* and *Halobacterium cutirubrum*. *Biochim. Biophys. Acta* **626,** 162–169.

Miller, J. R., Cartwright, E. M., Brownlee, G. G., Fedoroff, N. V., and Brown, D. D. (1978). The nucleotide sequence of oocyte 5S DNA in *Xenopus laevis*. II. The G-C rich region. *Cell* **13,** 717–725.

Mitchell, R. M., Loeblich, L. A., Klotz, L. C., and Loeblich, A. R., III (1979). DNA organization of *Methanobacterium thermoautotrophicum*. *Science* **204,** 1082–1084.

Moore, R. L., and McCarthy, B. J. (1967). Comparative study of ribosomal ribonucleic acid cistrons in *Enterobacteria* and *Myxobacteria*. *J. Bacteriol.* **94,** 1066–1074.

Morgan, E. A. (1982). "Ribosomal RNA genes in *Escherichia coli*." *Cell Nucleus* **10,** 1–29.

Nath, K., and Bollon, A. P. (1977). Organization of yeast ribosomal RNA gene cluster via cloning and restriction analysis. *J. Biol. Chem.* **252,** 6262–6276.

Nazar, R. N., Matheson, A. T., and Bellemare, G. (1978). Nucleotide sequence of *Halobacterium cutirubrum* ribosomal 5S ribonucleic acid. *J. Biol. Chem.* **253,** 5464–5469.

Nicholson, D. E., Jr. (1982). Structure of 5S Ribosomal RNA from Halobacteriacerae. Ph.D. Dissertation, University of Houston, Houston, Texas.

Nicholson, D. E., Jr., and Fox, G. E. (1983). Molecular evidence for a close phylogenetic relationship among box-shaped halophilic bacteria, *Halobacterium vallismortis*, and *Halobacterium marismortui*. *Can. J. Microbiol.* **29**, 52–59.

Ninio, J. (1979). Prediction of pairing schemes in RNA molecules—loop contributions and energy of wobble and non-wobble pairs. *Biochimie* **61**, 1133–1150.

Nishikawa, K., and Takemura, S. (1974a). Nucleotide sequence of 5S RNA from *Torulopsis utilis*. *FEBS Lett.* **40**, 106–109.

Nishikawa, K., and Takemura, S. (1974b). Structure and function of 5S ribosomal ribonucleic acid from *Torulopsis utilis* II. Partial digestion with ribonucleases and derivation of the complete sequence. *J. Biochem. (Tokyo)* **76**, 935–947.

Noller, H. F. (1984). Structure of ribosomal RNA. *Ann. Rev. Biochem.* **53**, 119–162.

Noller, H. F., and Woese, C. R. (1981). Secondary structure of 16S rRNA. *Science* **212**, 403–411.

Nomura, M., Traub, P., and Bechmann, H. (1968). Hybrid 30S ribosomal particlesreconstituted from components of different bacterial origins. *Nature (London)* **219**, 793–799.

Oesterhelt, D., and Stoeckenius, W. (1971). Rhodopsin like protein from purple membrane of *Halobacterium halobium*. *Nature (London), New Biol.* **233**, 149–152.

Orozco, E. M., Jr., Gray, P. W., and Hallick, R. B. (1980). *Euglena gracilis* chloroplast ribosomal RNA transcription units. I. The location of transfer RNA, 5S, 16S, and 23S ribosomal RNA genes. *J. Biol. Chem.* **255**, 10991–10996.

Pace, B., and Campbell, L. L. (1971). Homology of ribosomal ribonucleic acid of diverse bacterial species with *Escherichia coli* and *Bacillus stearothermophilus*. *J. Bacteriol.* **107**, 543–547.

Papanicolaou, C., Gouy, M., and Nino, J. (1984). An energy model that predicts the correct folding of both the tRNA and the 5S RNA molecules. *Nucleic Acids Res.* **12**, 31–44.

Pappenheimer, A. M., Jr., Dunlop, P. C., Adolph, K. W., and Bodley, J. W. (1983). Occurrence of diphthamide in archaebacteria. *J. Bacteriol.* **153**, 1342–1347.

Patel, D. J., Kozlowski, S. A., Ikuta, S., and Itakura, K. (1984). Deoxyguanosine-deoxyadenosine pairing in the d(CGAGAATTCGCG) duplex: Conformation and dynamics at and adjacent to the dG.dA mismatch site. *Biochemistry* **23**, 3207–3217.

Peattie, D., Noller, H. F., Douthwaite, S., and Garrett, R. A. (1981). A bulged double helix in an RNA protein contact site. *Proc. Natl. Acad. Sci. U.S.A.* **77**, 4170–4174.

Pecher, T., and Böck, A. (1981). In vivo susceptibility of halophilic and methanogenic organisms to protein synthesis inhibitors. *FEMS Lett.* **10**, 295–297.

Pechman, K. J., and Woese, C. R. (1972). Characterization of the primary structural homology between the 16S ribosomal RNAs of *Escherichia coli* and *Bacillus megaterium* by oligomer cataloging. *J. Mol. Evol.* **1**, 230–240.

Philippsen, P., Thomas, M., Kramer, R. A., and Davis, R. W. (1978). Unique arrangement of coding sequences for 5S, 5.8S, 18S and 23S ribosomal RNA in *Saccharomyces cerevisiae* as determined by R-Loop and hybridization analysis. *J. Mol. Biol.* **123**, 387–404.

Pieler, T., and Erdmann, V. A. (1982). Three-dimensional structural model of eubacterial 5S RNA that has functional implications. *Proc. Natl. Acad. Sci. U.S.A.* **79**, 4599–4603.

Procunier, J. D., and Tartof, K. D. (1975). Genetic analysis of 5S rRNA genes in *Drosophila melanogaster*. *Genetics* **81**, 515–523.

Rich, A. (1977). Three dimensional structure and biological function of transfer RNA. *Acc. Chem. Res.* **10**, 388–396.

Rozek, C. E., and Timberlake, W. E. (1979). Restriction endonuclease mapping by crossed contact hybridization: The ribosomal RNA genes of *Achlya ambisexualis*. *Nucleic Acids Res.* **7**, 1567–1578.

Ryan, J. L., and Morowitz, H. J. (1969). Partial purification of mature rRNA and tRNA cistrons from *Mycoplasma sp.* (KID). *Proc. Natl. Acad. Sci. U.S.A.* **63**, 1282–1289.

Salemink, P. J. M., Raue, H. A., Heerschap, A., Planta, R. J., and Hilbers, C. W. (1981). Hydrogen-1 and Phosphorus-31 nuclear magnetic resonance study of the solution structure of *Bacillus licheniformis* 5S ribonucleic acid. *Biochemistry* **20**, 265–272.

Sanger, F., Brownlee, G. G., and Barrell, B. G. (1965). A two dimensional fractionation procedure for radioactive nucleotides. *J. Mol Biol.* **13**, 373–398.

Schmid, G., Pecher, T., and Böch, A. (1982). Properties of the translational apparatus of archaebacteria. *Zentralbl. Bakteriol., Mikrobiol. Hyg., Abt. 1, Orig. C* **3**, 209–217.

Schnabel, R., Thomm, M., Gerardy, Schahn, R., Zillig, W., Stetter, K. O., and Hoet, J. (1983). Structural homology between different archaebacterial DNA dependent RNA polymerases analyzed by immunological comparison of their component. *EMBO J.* **2**, 751–755.

Schwarz, Z., and Kössel, H. (1980). Primary structure of 16S rDNA from *Zea mays* chloroplast is homologous to *Escherichia coli* 16S ribosomal RNA. *Nature (London)* **283**, 739–742.

Selker, E. U., Yanofsky, C., Driftmier, K., Metzenberg, R. L., DeWeerd, B. A., and RajBhandary, U. L. (1981). Dispersed 5S RNA genes in *N. crassa:* Structure, expression and evolution. *Cell* **24**, 819–828.

Silberklang, M., Gillum, A. M., and RajBhandary, U. L. (1977). Studies on the sequence of the 3′ terminal region of Turnip-Yellow-Mosaic-virus RNA. *Nucleic Acids Res* **4**, 4091–4108.

Silberklang, M., Gillum, A. M., and Rajbhandary, U. L. (1979). Use of in vitro ^{32}P labelling in the sequence analysis of nonradioactive tRNAs. *In* "Methods in Enzymology" (K. Moldave and L. Grossman, eds.), Vol. 59, pp. 58–109. Academic Press, New York.

Silberklang, M., RajBhandary, U. L., Luck, A., and Erdmann, V. A. (1983). Chemical reactivity of *E. coli* 5S RNA in situ in the 50S ribosomal subunit. *Nucleic Acids Res.* **11**, 605–617.

Simsek, M., DasSarma, S., RajBhandary, U. L., and Khorana, H. G. (1982). A transposable element from *Halobacterium halobium* which inactivates the bacteriorhodopsin gene. *Proc. Natl. Acad. Sci. U.S.A.* **79**, 7268–7272.

Singhal, R. P., and Shaw, J. K. (1983). Prokaryotic and eukaryotic 5S RNAs: Primary sequences and proposed secondary structures. *Prog. Nucleic Acid Res Mol. Biol.* **28**, 177–209.

Sogin, S. J., Sogin, M. L., and Woese, C. R. (1972). Phylogenetic measurement in procaryotes by primary structural characterization. *J. Mol. Evol.* **1**, 173–184.

Sprott, G. D., and Jarrell, K. F. (1981). K^+, Na^+, and Mg^{2+} content and permeability of *Methanospirillum hungatei* and *Methanobacterium thermoautotrophicum. Can. J. Microbiol.* **27**, 444–451.

Stackebrandt, E., Ludwig, W., Schleifer, K.-H., and Gross, H. J. (1981). Rapid cataloging of Ribonuclease T$_1$ resistant oligonucleotides from ribesomal RNAs for phylogenetic studies. *J. Mol. Evol.* **17**, 227–236.

Stackebrandt, E., Seewaldt, E., Ludwig, W., Schleifer, K.-H., and Huser, B. A. (1982). The phylogenetic position of *Methanothrix soehngenii* elucidated by a modified technique of sequencing oligonucleotides from 16S rRNA. *Zentralbl. Bakteriol., Mikrobiol. Hyg., Abt. 1, Orig. C* **3**, 90–100.

Stahl, D. A., Luehrsen, K. R., Woese, C. R., and Pace, N. R. (1981). An unusual 5S rRNA from *Sulfolobus acidocaldarius* and its implications for a general 5S rRNA structure. *Nucleic Acids Res.* **9**, 6129–6137.

Studnicka, G. M., Eiserling, F. A., and Lake, J. A. (1981). A unique secondary folding pattern for 5S RNA corresponds to the lowest energy homologous secondary structure in 17 different prokaryotes. *Nucleic Acids Res.* **9**, 1885–1904.

Sutton, L. A., and Woese, C. R. (1975). Stable large variant 5S RNA in *Clostridium thermosaccharolyticum. Nature (London)* **256**, 64–66.

Tabata, S. (1980). Structure of the 5S ribosomal RNA gene and its adjacent regions in *Torulopsis utilis. Eur. J. Biochem* **110**, 107–114.

Tabata, S. (1981). Nucleotide sequences of the 5S ribosomal RNA genes and their adjacent regions in *Schizosaccharomyces pombe*. *Nucleic Acids Res.* **9**, 6429–6437.

Thurlow, D. L., and Zimmermann, R. A. (1982). Evolution of protein binding regions of archaebacterial, eubacterial and eukaryotic ribosomal RNAs. *In* "Archaebacteria" (O. Kandler, ed.), p. 347. Fischer, Stuttgart.

Tinoco, J., Jr., Uhlenbeck, O. C., and Levine, M. D. (1971). Estimation of secondary structure in ribonucleic acids. *Nature (London)* **230**, 362–367.

Tinoco, I., Jr., Borer, P. N., Dengler, B., Levine, M. D., Uhlenbeck, O. C., Crothers, D. M., and Gralla, J. (1973). Improved estimation of secondary structure in ribonucleic acids. *Nature (London), New Biol.* **246**, 40–41.

Traub, W., and Elson, D. (1966). RNA composition and base pairing. *Science* **153**, 178–180.

Traub, W., and Sussman, J. L. (1982). Adenine-guanine base pairing in ribosomal RNA. *Nucleic Acids Res.* **10**, 2701–2708.

Troutt, A., Savin, T. J., Curtiss, W. C., Celentan, J., and Vournakis, J. N. (1982). Secondary structure of *Bombyx mori* and *Dictyostelium discoideum* 5S rRNA from S1 nuclease and cobra venom ribonuclease susceptibility, and computer assisted analysis. *Nucleic Acids Res.* **10**, 653–664.

Tu, J., and Zillig, W. (1982). Organization of rRNA structural genes in the archaebacterium *Thermoplasma acidophilum*. *Nucleic Acids Res.* **10**, 7231–7245.

Tu, J., Prangishvilli, P., Huber, H., Wildgruber, G., Zillig, W., and Stetter, K. O. (1982). Taxonomic relations between archaebacteria including 6 novel genera examined by cross hybridization of DNAs and 16S rRNAs. *J. Mol. Evol.* **18**, 109–114.

Uchida, T., Bonen, L., Schaup, H. W., Lewis, B. J., Zablen, L., and Woese, C. R. (1974). The use of ribonuclease U_2 in RNA sequence determination. *J. Mol. Evol.* **3**, 63–77.

Van Valen, L. M., and Maiorana, V. C. (1980). The archaebacteria and eukaryotic origins. *Nature (London)* **287**, 248–250.

Vigne, R., and Jordan, B. R. (1977). Partial enzyme digestion studies on *Escherichia coli, Pseudomonas, Chlorella, Drosophila,* HeLa and yeast 5S RNAs support a general class of 5S RNA models. *J. Mol. Evol.* **10**, 77–86.

Walker, R. T., Chelton, E. T. J., Kilpatrick, M. W., Rogers, M. J., and Simmons, J. (1982). The nucleotide sequence of the 5S rRNA from *Spiroplasma* species BC3 and *Mycoplasma mycoides* sp. *Capri* PG 3. *Nucleic Acids Res.* **10**, 6363–6367.

White, B. N., and Bayley, S. T. (1972). Methionine transfer RNAs from the extreme halophile, *Halobacterium cutirubrum*. *Biochim. Biophys. Acta* **272**, 583–587.

Williamson, S. E., and Doolittle, W. F. (1983). Genes for tRNA Ile and tRNA Ala in the spacer between the 16S and 23S rRNA genes of a blue green alga: strong homology to chloroplast tRNA genes and tRNA genes of the *E. coli rrn D* gene cluster. *Nucleic Acids Res.* **11**, 225–235.

Wimber, D. E., and Steffensen, D. M. (1970). Localization of 5S RNA genes on *Drosophila* chromosomes by RNA-DNA hybridization. *Science* **170**, 639–641.

Woese, C. R. (1982). Archaebacteria and cellular origins: An overview. *Zentralbl. Bakteriol., Mikrobiol, Hyg., Abt. 1, Orig. C* **3**, 1–17.

Woese, C. R., and Fox, G. E. (1977a). Phylogenetic structure of the prokaryotic domains: The primary kingdoms. *Proc. Natl. Acad. Sci. U.S.A.* **74**, 5088–5090.

Woese, C. R., and Fox, G. E. (1977b). The concept of cellular evolution. *J. Mol. Evol.* **10**, 1–6.

Woese, C. R., and Gupta, R. (1981). Are archaebacteria merely derived prokaryotes? *Nature (London)* **289**, 95–96.

Woese, C. R., Fox, G. E., Zablen, L., Uchida, T., Bonen, L., Pechman, K., Lewis, B. J., and Stahl, D. (1975). Conservation of primary structure in 16S ribosomal RNA. *Nature (London)* **254**, 83–86.

Woese, C., Sogin, M., Stahl, D., Lewis, B. J., and Bonen, L. (1976). A comparison of the 16S ribosomal RNAs from mesophilic and thermophilic *Bacilli:* Some modifications in the Sanger method for RNA sequencing. *J. Mol. Evol.* **7,** 197–213.

Woese, C. R., Magrum, L. J., and Fox, G. E. (1978). Archaebacteria. *J. Mol. Evol.* **11,** 245–252.

Woese, C. R., Magrum, L. J., Gupta, R., Siegel, R. B., Stahl, D. A., Kop, J., Crawford, N., Brosius, J., Gutell, R., Hogan, J. J., and Noller, H. F. (1980a). Secondary structure model for bacterial 16S ribosomal RNA-phylogenetic, enzymatic and chemical evidence. *Nucleic Acids Res.* **8,** 2275–2293.

Woese, C. R., Maniloff, J., and Zablen, L. B. (1980b). *Phylogenetic analysis of the mycoplasmas. Proc. Natl. Acad. Sci. U.S.A.* **77,** 494–498.

Woese, C. R., Gutell, R., Gupta, R., and Noller, H. F. (1983). Detailed analysis of the higher-order structure of 16S-like ribosomal ribonucleic acids. *Microbiol. Rev.* **47,** 621–669.

Wrede, P., and Erdmann, V. A. (1973). Activities of *B. stearothermophilus* 50S ribosomes reconstituted with procaryotic and eukaryotic 5S RNA. FEBS Lett. **33,** 315–319.

Young, J. F., Taussig, R., Aaronson, R. P., and Palese, P. (1981). Advantages and limitations of the oligonucleotide mapping technique for the analysis of viral RNAs. *In* "Replication of Negative Strand Viruses" (D. H. L. Bishop and R. W. Compans, eds.), pp. 209–215. Elsevier/North-Holland, New York.

Yu, P.-L., Hohn, B., Falk, H., and Drens, G. (1982). Molecular cloning of the ribosomal RNA genes of the photosynthetic bacterium *Rhodopseudomonas capsulata. Mol. Gen. Genet.* **188,** 392–398.

Zablen, L. and Woese, C. R. (1975). Procaryotic phylogeny. IV. Concerning the phylogenetic status of a photosynthetic bacterium. *J. Mol. Evol.* **5,** 25–34.

Zablen, L., Kissil, M. S., Woese, C. R., and Buetow, D. E. (1975). Phylogenetic origin of the chloroplast and prokaryotic nature of its rRNA. *Proc. Natl. Acad. Sci. U.S.A.* **72,** 2418–2422.

Zillig, W., Stetter, K. O., Schnabel, R., Madon, J., and Gierl, A. (1982a). Transcription in archaebacteria. *Zentralbl. Bakteriol., Mikrobiol. Hyg., Abt. 1, Orig. C* **3,** 218–227.

Zillig, W., Schnabel, R., and Tu, J. (1982b). The phylogeny of archaebacteria, including novel anaerobic thermoacidophiles in the light of RNA polymerase structure. *Naturwissenschaften* **69,** 197–204.

Zingales, B., and Colli, W. (1977). Ribosomal RNA genes in *Bacillus subtilis* evidence for a cotranscription mechanism. *Biochim. Biophys. Acta* **474,** 562–577.

Zuckerkandl, E., and Pauling, L. (1965). Molecules as documents of evolutionary history. *J. Theor. Biol.* **8,** 357–366.

Zweib, C., Glotz, C., and Brimacombe, R. (1981). Secondary structure comparisons between small subunit ribosomal RNA molecules from six different species. *Nucleic Acids Res.* **9,** 3621–3640.

CHAPTER 6

Transfer Ribonucleic Acids of Archaebacteria

RAMESH GUPTA

I. Introduction

The transfer RNAs play a major role in protein biosynthesis. Besides their role in decoding the messenger RNA, tRNAs take part in several other biological processes. These include cell-wall biosynthesis, serving as a reverse transcriptase primer, and various control functions. The aim of this chapter is to discuss the current information available regarding the tRNAs of the archaebacteria. These tRNAs fit the generalized tRNA structure, but show certain peculiarities of their own. They resemble eukaryotic tRNAs in some ways, eubacterial tRNAs in others. For a detailed background in the tRNAs, the reader is referred to some of the more recent comprehensive reviews (Rich and RajBhandary, 1976; Goddard, 1977; Singhal and Fallis, 1979; Ofengand, 1982) and monographs (*Accounts of Chemical Research*, Vol. 10, No. 11, 1977, pp. 385–425; Altman, 1978; Schimmel *et al.*, 1979; Söll *et al.*, 1980). Frequently updated compilations of tRNA sequences are also published, the latest being by Gauss and Sprinzl (1983).

A limited literature is available regarding tRNAs of the archaebacteria. However, the recent report (Kaine *et al.*, 1983) of introns in the tRNA genes of

Copyright © 1985 by Academic Press, Inc.
All rights of reproduction in any form reserved.
ISBN 0-12-307208-5

archaebacteria—the first such example of introns in any prokaryotic genes—
should spur a great deal of interest in this area and in archaebacteria in general.

The next two sections of this chapter are devoted to the structural studies of
tRNAs. The discussion in the first section will be confined essentially to the
eubacterial and the eukaryotic tRNAs and will serve as background for the
subsequent detailed consideration of the archaebacterial tRNAs.

II. Generalized tRNA Structure

All organisms produce 40–50 types of tRNAs, one or more for each of the 20
amino acids. Since the basic translation process is to a first approximation the
same for all organisms, all tRNAs show certain basic similarities in their primary, secondary (cloverleaf form), and tertiary structure (although some mammalian mitochondrial tRNAs are exceptional). The initiator tRNA, whose role in
the protein synthesis is distinct from the other (elongator) tRNAs, has certain
additional special characteristics. Although there are common features of tRNAs
of all organisms, there are some kingdom-specific, phylogenetic characteristics
for tRNAs as well.

A. Secondary Structure

The first sequence of a tRNA, the yeast tRNA[Ala], was published by Holley *et
al.* (1965). One of the possible secondary structures proposed by them was the
now classical cloverleaf form. All of the tRNAs sequenced so far (with the
exception of some of the mitochondrial ones) fit this secondary structure, providing strong comparative support for the existence of base paired stems. X-ray
crystallographic analysis (Kim, 1978, 1979) of yeast tRNA[Phe] further confirmed
this cloverleaf structure and determined its three-dimensional shape. The mitochondrial tRNAs, especially certain mammalian ones, show varying degrees of
exception to this generalized structure. These seem to be peculiar adaptations to
some specialized translation mechanism. These highly idiosyncratic mitochondrial tRNAs will not be discussed in this chapter, nor will the various chloroplast
and phage tRNAs be emphasized.

The tRNA molecule is phosphorylated at its 5′ end, but not at the 3′ end. It is
between 74 and 93 residues long and folds back upon itself to produce four (or
five) double-helical "stems," which define three (or four) single-stranded
"loops," as shown in Fig. 1. A loop–stem combination is referred to as an *arm*.
The stems and loops are variously named by different authors. These names are
sometimes confusing. The nomenclature used here is the one (by Ofengand,

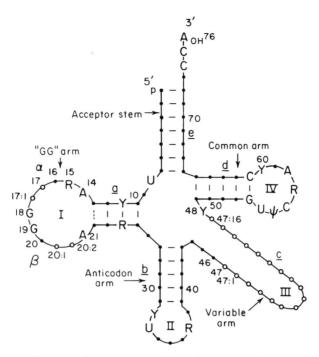

FIG. 1. A generalized secondary structure model showing common features of the tRNAs. The numbering system is the one after yeast tRNA^Phe (Schimmel et al., 1979; Gauss and Sprinzl, 1983). Nucleoside positions are indicated by circles or by letters when these are invariant or semi-invariant residues. Solid circles refer to nucleosides always present; open circles, to those which sometimes may be absent. a through e refer to the stems, in which connecting lines indicate base pairs (dotted lined pair may not always be present). I through IV are the loops. α, β, and stem c with loop III are the variable regions.

1982) shown in Fig. 1, i.e., stem *a* capped by loop I, stem *b* by loop II, stem *c* by loop III, stem *d* by loop IV, and, lastly, stem *e* terminating in the two free ends. The arms are generally recognized by trivial names as well; i.e., stem *a* and loop I are the D-arm, since one or more dihydrouridine (D) residues* are usually present in the loop in eubacteria and eukaryotes; stem *b* and loop II are the

---

*The modified residues are referred to as: m¹A, 1-methyl A; t⁶A, N-[(9-β-D-ribofuranosylpurin-6-yl)carbamoyl]threonine; m⁵C, 5-methyl C; ac⁴C, N⁴-acetyl C; m¹G, 1-methyl G; m²G, N²-methyl G; m²₂G, N²,N²-dimethyl G; m⁷G, 7-methyl G; s⁴U, 4-thio U; mo⁵U, 5-methoxy U, Ů, a specific unidentified modified U; Ψ, pseudouridine; m¹Ψ, 1-methyl Ψ; D, dihydrouridine; T, ribothymidine; s²T, 2-thio T; I, inosine; m¹I, 1-methyl I; R, purine; Y, pyrimidine; X, a specific unidentified modified G; N, unknown nucleoside. 2'O–ribose methylated residues are indicated by ''m'' after the residue. These are Cm, Gm, m²₂Gm, Um, Tm, Ψm. 5'-phosphate or 3'-phosphate is indicated, respectively, by ''p'' preceding or following a residue letter.

anticodon arm, for the obvious reason; stem c and loop III, if present, are the variable or extra arm; stem d and loop IV are the TΨC-, or common arm, due to the nearly invariant occurrence of this sequence in its loop in eubacterial and eukaryotic tRNAs; and stem e, the acceptor, amino acid, or CCA stem, since the amino acid is attached to 3'-terminal A residue of this stem. Because of the lack of the nucleosides T and D in the archaebacterial and some other tRNAs (Sections II,D and III,A, 1), the names D-arm and TΨC-arm are no longer appropriate. Therefore, the "TΨC-arm" will be referred to here by the older designation *common arm* (as the basic sequence of its loop and adjacent first pair is common to most tRNAs). For the "D arm", a new name, *GG arm,* will be used here (because of the conserved adjacent GG residues in its loop).

The following generalizations apply to the tRNA structure. The stems involve almost solely canonical, i.e., Watson–Crick base pairs (G-C or A-U). A few "wobble" (G·U) pairs are encountered, as rarely are other juxtapositions. Occasionally Ψ replaces U in some stems. There are always two bases between the acceptor stem and the GG arm, one between the GG and anticodon arms, none between the common arm and the acceptor stem, while the variable arm separates the common arm from the anticodon arm (Fig. 1). Four unpaired bases constitute the 3'-terminus of the acceptor stem. The GG stem contains either three or four base pairs. [But, by convention it is always considered to contain four, whether or not the final pair is a canonical one (Rich and RajBhandary, 1976; Clark, 1978).] The loop of this arm comprises from 7 to 11 bases. There are two variable regions in this loop—α and β in Fig. 1—which flank the two invariant G residues. The α-region comprises three to five bases and the β-region two to four. The anticodon stem and the common arm stem each comprise five base pairs, and their loops each have seven bases. The acceptor stem contains seven base pairs, except for tRNA^{His}, which forms eight pairs by adding one residue on the 5' end of the stem, thus leaving only three unpaired bases on the 3'-terminus. (In the eubacteria and archaebacteria, this new pair is a canonical one, but in the eukaryotes it is a G·A juxtaposition.) The variable arm is either small (four to five bases) or large (13–21 bases, except for two plant chloroplast leucine tRNAs, each of which have only 11). In the latter case, most of the bases are paired, to form stem c, which is capped by a short loop (loop III). Those tRNAs with small variable arms are called *class I tRNAs*; those with large arms, *class II tRNAs*. All known leucine and serine tRNAs are of class II, as are the tyrosine tRNAs of the eubacteria. A length shorter than four bases in the variable arm may not be possible according to the yeast tRNA^{Phe}-type tertiary structure (Clark, 1978). Initially, the sequences of yeast tRNA^{Gly} and tRNA^{Val} from both brewers' and Torula yeast were reported to have three nucleotides in the variable arms. More accurate sequencing reveals that yeast tRNA^{Gly} actually has four (cited in Clark, 1978) and brewer's yeast tRNA^{Val} has five (see Rich and Raj-Bhandary, 1976) bases in the variable arm. Therefore, it may be desirable to re-examine the sequence of Torula yeast tRNA^{Val}.

B. Invariant and Semi-Invariant Nucleosides

Positions in the tRNA sequence are classified as "invariant" when their basic composition is always the same (with at most a few exceptions), and similarly as "semi-invariant" when they are constrained to be either a purine or a pyrimidine. The invariant and semi-invariant positions in elongator tRNAs are as follows: U always occurs as the first residue between the acceptor and GG stems (position 8).[*] The residue is commonly modified to s^4U in the eubacteria, but remains unmodified in the eukaryotes. In the GG arm, position 10 is usually G, which in the eukaryotes is often modified to m^2G. Its pairing complement, at position 25, is a pyrimidine. [However, a sufficient number of exceptions to this pattern have been reported in both eubacteria and eukaryotes, so that these two residues can no longer be classified as invariant and semi-invariant (Ofengand, 1982; Gauss and Sprinzl, 1983).] Similarly, the pair involving positions 11 and 24, in the GG arm, is a semi-invariant $Y_{11}-R_{24}$ in all elongator tRNAs of eubacteria and eukaryotes. Positions 14 and 21, which define the base of the loop of the GG arm, are always A residues; the first A is modified to m^1A in some eukaryotes, and position 21 is occupied by D in *Escherichia coli* $tRNA^{Cys}$ and by G in several $tRNA^{Leu}$. Position 15 is a purine in nearly all normal tRNAs. Levitt (1969) pointed out that a replacement of G by A at this position was always accompanied by a replacement of C by U at position 48 (at the 3' end of the variable arm). Bases at these two positions are now known to form tertiary hydrogen bonds in the three-dimensional structure of tRNA. The residue at position 15 is unmodified, whereas the one at position 48 in eukaryotes is frequently m^5C. However, there are a few ostensible exceptions to this co-varying pair (Gauss and Sprinzl, 1983). An invariant GG or GmG doublet at positions 18 and 19 separates the α and β variable regions of the loop of the GG arm.[†] The anticodon loop has an invariant U_{33}, and semi-invariant Y_{32} and R_{37}; the nucleosides at the last two positions are frequently modified. C replaces U_{33} in one case—rat $tRNA^{Asp}$ (Gauss and Sprinzl, 1983). The first pair adjacent to the loop of the common arm (53–61) is always G-C. The invariant or semi-invariant residues in the common arm loop are U_{54}, Ψ_{55}, C_{56}, R_{57}, A_{58}, and Y_{60}. Usually, the base at position 54 is modified to T, rarely to Ψ or Tm; in *Bombyx mori* $tRNA^{Ala}$, it is A. A_{58} is frequently modified to m^1A in eukaryotes. Y_{60} is replaced by A in *B. mori* $tRNA^{Ala}$, bovine liver $tRNA^{Asp}$, and mam-

---

[*] The numbering system used here is the one proposed at the 1978 Cold Spring Harbor tRNA meeting and is based on the yeast $tRNA^{Phe}$ numbering (see Appendix I of Schimmel *et al.*, 1979; Gauss and Sprinzl, 1983).

[†] Cell-wall biosynthesis tRNAs of *Staphylococcus* sp., which are glycine tRNAs not involved in ribosome-dependent protein synthesis, have UU instead of GG at positions 18 and 19 (Singhal and Fallis, 1979; Gauss and Sprinzl, 1983) and also have exceptions at several other invariant positions (e.g., occurrence of Y_{37}, UGY replacing TΨC).

malian $tRNA^{Val}$. The terminal CCA_{OH} sequence (positions 74–76) is constant for all tRNAs. This sequence is not encoded in some tRNA genes, but is post-transcriptionally added to the tRNA by a specific enzymatic process.

Besides these invariant and semi-invariant bases, Kim (1979) recognizes the *conserved presence* (with more than 90% occurrence) or *conserved absence* (less than 10% occurrence) of particular residues at certain positions. (Conserved presence is basically the same as the term *invariant* mentioned above.)

C. INITIATOR tRNAS

The initiator tRNAs do not conform completely to the pattern of elongator tRNAs. The special role these tRNAs play in the protein biosynthesis may account for certain of their unique features. The initiator methionine tRNA is not related to elongator methionine tRNA. In the eubacteria, methionine is formylated after aminoacylation; in eukaryotes, it is not.

In all initiator tRNAs so far sequenced (with the exception of certain organelle ones), the anticodon stem possesses three contiguous G-C pairs adjacent to the loop. The three Gs (position 29–31) are on the 5′ side of the stem. Nearly all elongator tRNAs lack this feature.

An initiator-specific feature found only in eubacteria is the occurrence of a C·A juxtaposition replacing the first base pair (1–72); the eukaryotic initiators have a normal $(A_1–U_{72})$ pair. Another feature of initiators limited to eubacteria is the occurrence of the $A_{11}–U_{24}$ pair, which replaces the usual $Y_{11}–R_{24}$ pair in the GG arm stem; the eukaryotic initiators remain normal in this regard.

Other differences between the eukaryotic and eubacterial initiators are the following: In the common arm loop of eukaryotic initiator tRNAs, A_{54} replaces the normal T_{54}, and U_{55} often replaces Ψ_{55}. The purine at position 57 is A in eubacterial initiators (except *Mycoplasma* sp., which has G), while in eukaryotes, it is G. A_{60} is present in the eukaryotic initiators, instead of the normal Y_{60}. The purine at position 37 (on the 3′-side of the anticodon) is a normal A in the eubacterial initiators, while it is t^6A in the eukaryotic ones. Initiators of the higher eukaryotes have C_{33}, instead of constant U_{33} (on the 5′-side of the anticodon).

D. MODIFIED NUCLEOSIDES

Transfer RNAs are the most highly modified RNA species in the cell, and the eukaryotic tRNAs are more so than the eubacterial ones. Some of the modifications are common to nearly all tRNAs of all organisms. Others are restricted to one or a few specific tRNAs of nearly all organisms. Still others are present in nearly all tRNAs of some groups of organisms only. Certain modified nucleotides are found in only one particular position in the tRNA molecule, while others occur at any of the several positions. Various recent reviews treat the modified residues in tRNAs in detail (McCloskey and Nishimura, 1977; Feldman, 1977; Nishimura, 1978, 1979; Singer and Kröger, 1979; Dirheimer *et al.*, 1979; Dirheimer, 1983). The discussion of modified residues here is restricted to those that have relevance to the archaebacterial tRNAs discussed below (Sections III,A,1 and III,C).

The most common modified residues found both in the eubacteria and in the eukaryotes are Ψ, D, T, and m^7G. T is absent from all tRNAs of certain eubacteria, e.g., *Mycobacterium smegmatis* (Vani *et al.*, 1979), *Micrococcus luteus* (Delk *et al.*, 1976), and *Thermus thermophilus* (where it is replaced by s^2T; Watanabe *et al.*, 1974). However, it is never absent from all species in a major phylogenetic grouping of eubacteria.

Table I lists the positions in the tRNAs of the three kingdoms occupied by various modified residues. (Only those residues that have some bearing on the archaebacterial tRNAs are listed in Table I.) Some modified residues at a given position are found only in one kingdom, some in all, whereas others are found in two of the three.

III. Archaebacterial tRNAs

Archaebacteria are an extremely diverse group of organisms, both phylogenetically and ecologically. Therefore, one must be cautious in interpreting any characteristic of the tRNAs of a particular subgroup of archaebacteria, as being characteristic of archaebacteria in general. Group-specific features may be due to specialized internal cell environment, e.g., high salt in the extreme halophiles.

To date, a total of 48 archaebacterial tRNAs have been sequenced. A number of these are initiator tRNAs from a variety of different species (*Halobacterium volcanii*, Gupta, 1981; *Sulfolobus acidocaldarius, Thermoplasma acidophilum*, and *Halococcus morrhuae*, Kuchino *et al.*, 1982). The bulk of them, however, i.e., 41 (including one of the initiators), belong to a nearly complete set of *H. volcanii* tRNAs, which include at least one tRNA for each of the 20 amino acids (Gupta, 1981, and unpublished). The other sequences are those of tRNA^{Met} of *T. acidophilum* (Kilpatrick and Walker, 1981) and tRNA^{Ala} and two tRNA^{Val} of *Halobacterium cutirubrum* (Gu *et al.*, 1983). These, together with the large collection of eubacterial and eukaryotic tRNA sequences (especially nearly complete sets of tRNAs from *E. coli,* yeasts, and mammals), permit a meaningful phylogenetic comparison of the three kingdoms. The sequences of the known 48

TABLE I

Positions of Some Modified Residues in tRNAs of the Three Kingdoms

Modified residue	Positions[a,b]		
	Eubacteria	Eukaryotes	Archaebacteria
m^{1}A	22,58*	14,58,59	58
t^{6}A	37	37	37
ac^{4}C	34	12**	34
m^{5}C	48†	34,38,40,48,49,50,72	39,40,48,49
Cm	32	4,13,32,34	32,34,56
D	16,17,20,20:1,21,47	16,17,20,20:1,20:2,47	
m^{1}G	37	9,37	37
m^{2}G		6,7,9,10,26	10,26
m$^2_2$G		26	10,26
m^{7}G	46	46	
Gm	18,34	18,34,39,64	29
I	34	34	57
m^{1}I		37	57
T	54	54	
s^{4}U	8,9		8
Um	32	4,32,44	39,54
Ψ	13,32,38,39,40,55,65	1,13,20:2,25,26,27,28, 31,32,35,38,39,40,45, 46,47:1,50,54,55,67,68	13,22,28,38,39,52, 54,55
m^{1}Ψ			54

[a] Double underline, common to all three kingdoms; single underline, present at these positions only in archaebacterial and eukaryotic tRNAs; ~, observed in these positions in archaebacteria and eubacteria; ········, seen at these positions only in eubacteria and eukaryotes.

[b] *, only in *Thermus thermophilus* initiator; **, in all class II tRNAs only; †, only in blue-green algae.

archaebacterial tRNAs are tabulated in a linear form in the Appendix of this chapter.

A. MODIFIED NUCLEOSIDES

1. MODIFICATION PATTERN

Overall, the modification pattern of archaebacterial tRNAs is neither typically eubacterial nor typically eukaryotic (Gupta and Woese, 1980). The total tRNAs of ten archaebacterial species (*Halobacterium volcanii, Halococcus morrhuae, Methanobacterium bryantii, Methanobrevibacter smithii, Methanococcus vannielii, Methanococcus voltae, Methanomicrobium mobile, Methanosarcina barkeri, Thermoplasma acidophilum,* and *Sulfolobus acidocaldarius*), which span the three recognized divisions of the archaebacteria, have been analyzed for their modified nucleoside content (Best, 1978; Gupta and Woese, 1980).

Of the nucleosides T, m^7G, and D, so characteristic of eubacterial and eukaryotic tRNAs, only D is found in archaebacteria and that only in one (*M. barkeri*) of the ten species examined (Fox *et al.*, 1980; Gupta and Woese, 1980). Ψ, Cm, m^1G, and m_2^2G are present in all of them. All, except *H. volcanii* and *H. morrhuae,* have m^1A, whereas m^5C is restricted to these two halophiles and *S. acidocaldarius* (Gupta and Woese, 1980). Thiolated nucleosides are also observed in some of the archaebacteria (Best, 1978; Kilpatrick and Walker, 1981). These posttranscriptional modification patterns of the archaebacteria are distinct from those of typical eubacteria (*E. coli*) and typical eukaryotes (*Saccharomyces cerevisiae*), although they resemble the latter somewhat more than the former.

Position 54, which in nearly all tRNAs of almost all eubacteria and eukaryotes contains T, is occupied in archaebacteria by Ψ, which in some groups becomes modified to $m^1\Psi$ (Gupta and Woese, 1980; Pang *et al.*, 1982). The modified version seems to be confined to the orders Methanococcales and Methanomicrobiales and to the extreme halophiles (Gupta and Woese, 1980). As Fig. 2 shows, $m^1\Psi$ and T have similar molecular profiles and base-pairing properties. The methyl groups in the two cases (N-1 in the former and C-5 in the latter) have a similar orientation, relative to ribose and polynucleotide chain (Gupta and Woese, 1980; Pang *et al.*, 1982). This may be an example of an evolutionary convergence of structures, at position 54 in the tRNAs.

The presence of m^1I at position 57 (in common arm loop) has been established for *H. volcanii* and *S. acidocaldarius* (Yamaizumi *et al.*, 1982). A nucleotide with the same enzymatic, electrophoretic, and chromatographic characteristics is also present at the same position in *H. morrhuae, M. bryantii, M. smithii, M. vannielii, M. voltae,* and *M. mobile* tRNAs (Gupta and Woese, 1980;

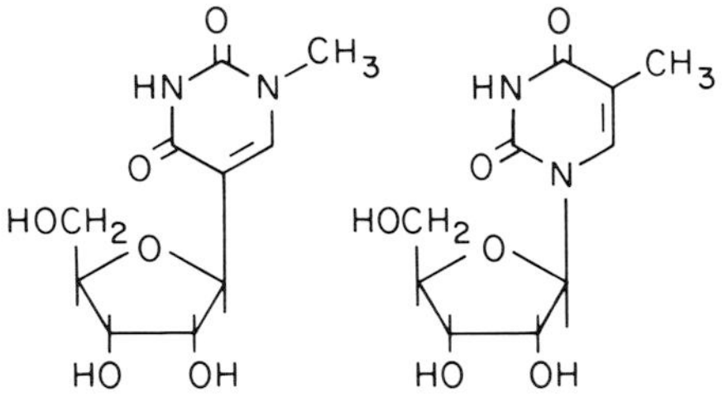

I-Methylpseudouridine   Ribothymidine

Fig. 2.   Structures of 1-methylpseudouridine and ribothymidine.

Yamaizumi *et al.*, 1982). m^{1}I has been found at position 57 *only* in archaebacterial tRNAs. So far, it has not been found in archaebacteria at position 37,
where it is sometimes observed in the eukaryotic tRNAs (Table I). In the initiator
tRNA of *T. acidophilum,* an unmodified I seems to be present at position 57
(Kuchino *et al.,* 1982).

The constant (unmodified) C at position 56 in the common arm loop of all
tRNAs is modified to Cm in *all* archaebacterial tRNAs examined, which can be
seen not only in the 48 sequences mentioned in the Appendix, but also in the
oligonucleotide patterns from ribonuclease A and T$_1$ digests of bulk tRNA of several species (R. Gupta and C. R. Woese, unpublished results). (Cm is also present
occasionally at a few other positions in some specific archaebacterial tRNAs.)

An interesting modified nucleoside (unidentified, and referred to as ''X'' in
the Appendix) is present at position 15 (in the GG arm loop) in *H. volcanii* and in
some methanogens (Gupta, 1981). It seems also to be present in *T. acidophilum*
(Kilpatrick and Walker, 1981) and may be present in *Sulfolobus* species as well
(R. Gupta, unpublished observation). The nucleoside at position 48, which
forms a tertiary pair with the one at position 15 (Section II, B), is C or m^{5}C in all
cases when modified nucleoside ''X'' is present at position 15 (see Appendix).
Preliminary data indicate this nucleoside to be a modified G whose pK_a is
considerably higher than that of the parent nucleoside (Gupta and Woese, 1980).
(It may carry a positive charge under physiological conditions.) So far, no
modifications have been found at position 15 in the tRNAs of eubacteria or
eukaryotes.

The positions in the tRNA molecule of archaebacteria-specific modified residues are listed in Table I, along with those for the other two kingdoms. Some
other archaebacteria position-specific modified residues are m$_2^2$G$_{10}$, m^{5}C$_{39}$, Ψ_{22},
Ψ_{52}, etc. (see Table I).

2. Modifying Enzymes

Little is known about the process of posttranscriptional modification of the
archaebacterial tRNAs or the enzymes involved in modification. Of these pre-

sumed enzymes, probably pseudouridine methylase (though not identified yet) would be unique to the archaebacteria (Pang *et al., 1982).

Best (1978) studied *in vitro* methylation of tRNAs by *Methanococcus vannielii* extracts, using *S*-adenosylmethionine as the methyl donor and *E. coli* under-methylated tRNAs as the acceptor. She observed active tRNA methyltransferases for the formation of those methylated residues found in the native *M. vannielii* tRNA, that is, m^1A, Cm, m^1G, m^2G, and m_2^2G. No m^7G or T was formed in these experiments, suggesting the absence of *S*-adenosylmethionine-dependent methyltransferases for these residues, which is consistent with the absence of these modified residues in archaebacterial tRNAs. Interestingly, *E. coli* extracts did produce m^7G with native *M. vannielii* tRNAs as acceptor, but no T was formed in such cases. Lack of T formation in these experiments is probably due to lack of the substrate, for *M. vannielii* tRNAs have Ψ or $m^1\Psi$ (not the required U residue) at position 54 (Gupta and Woese, 1980). 5,10-Methylenetetrahydrofolate is known to be a methyl donor for T formation in *Streptococcus faecalis* and *Bacillus subtilis* (Delk *et al., 1976), but this compound was not tested as the methyl donor for methylation of tRNAs in the archaebacteria.

B. INITIATOR tRNAs

Initiation of protein synthesis in *Halobacterium cutirubrum* occurs via a methionyl-tRNA (Bayley and Morton, 1978), as in the eukaryotes; not by an *N*-formylmethionyl-tRNA, as in the eubacteria (Section II,C). Met-tRNA^{Met} of *H. cutirubrum* resolves into two peaks by chromatography on BD-cellulose columns, one of which contains the initiator tRNA, which can be formylated by the *E. coli* methionyl-tRNA formyltransferase (White and Bayley, 1972a). This tRNA accounts for about 65% of the total methionine acceptance activity. It alone can be aminoacylated by the *E. coli* methionyl-tRNA synthetase. [It is also charged in a system from baker's yeast and is capable of initiating the synthesis of coat protein in an *E. coli* cell-free system directed by f2 RNA (Bayley and Morton, 1978).] *Halobacterium volcanii* also has two tRNA^{Met} species (Gupta, 1981), one of which (tRNA$_i^{Met}$) has higher methionine acceptance activity than the other. It is extremely similar in sequence to *Halococcus morrhuae* initiator tRNA (Kuchino *et al., 1982).

Heterologous charging of archaebacterial methionine tRNAs by *E. coli* aminoacyl-tRNA synthetase was used by Kuchino *et al. (1982) to identify initiators of *H. morrhuae, S. acidocaldarius,* and *T. acidophilum.* The specific tRNA fraction of *T. acidophilum* that is isolated after this heterologous charging of methionine further separates into two species by BD-cellulose column chromatography. However, only the initiator can be formylated by *E. coli* enzymes.

The sequences of these four initiator tRNAs are presented in the Appendix. All of them have the characteristic three contiguous G-C pairs (one is Gm-C in

S. acidocaldarius) in the anticodon stem adjacent to the loop (Section II,C). The first base pair of the acceptor arm in all of the archaebacterial initiators, as in eukaryotes, is a Watson–Crick pair (A-U). The archaebacterial initiators resemble eubacterial initiators in a number of ways, i.e., in having a U_{33}, an unmodified A_{37}, some form of U at position 54, and a U_{60}. The residue at position 57 (in the common arm loop) is basically A, as it is in the eubacteria, but in this case an A that has been modified to I or m^1I. In the *S. acidocaldarius* initiator, the residue at position 15 is C, not a purine, while its co-varying residue at position 48 is m^5C, so that a recognized tertiary pairing, 15·48, may not be possible in this case. The 11–24 pair in the stem of the GG arm is G-C, which is unique to the archaebacteria (see Section III,C) and their initiators. In all other initiators, it is either Y-R or A-U (Section II,C). In the standard cloverleaf form, the two halophile initiators can form the typical three pairs in the GG stem; however, an alternate five-pair stem is possible using bases at positions 9–13 to pair with those at positions 21–25 (see also Section III,C).

Remarkably, the *Holabacterium volcanii* initiator has a triphosphate at its 5′ end (Gupta, 1981) the only known such case. This indicates that the gene for this tRNA is located at the 5′ extreme of its transcriptional unit, and the first base of this tRNA is the first base of the transcript. The occurrence of 5′-triphosphates in other archaebacterial initiator tRNAs may have been overlooked because most of these were sequenced using *in vitro* labeling techniques (which would remove it).

C. General tRNAs

All but one of the 44 elongator tRNAs listed in the Appendix are from the extreme halophiles. All are accommodated by the generalized tRNA structure derived from eubacterial and eukaryotic tRNAs. This indicates that the basic reactions in which tRNA is involved are essentially similar in all three major kingdoms.

The archaebacterial tRNAs, as those of the other kingdoms, can be classified into class I (short variable arm) and class II (large variable arm). As is the case with eukaryotes, archaebacterial tRNAs for leucine and serine but not for tyrosine are class II. Interestingly, halobacterial cysteine tRNA has six bases in the variable arm (see Appendix); no other case of any tRNA with a six-base variable loop is so far reported (Gauss and Sprinzl, 1983).

The largest loop size of the GG arm in the tRNAs of eubacteria and eukaryotes is 10 residues (Gauss and Sprinzl, 1983). However, seven *H. volcanii* tRNAs (for Cys, Glu_1, Glu_2, Leu_1, Leu_2, Leu_5, and Ser_1) have 11 residues in this loop! In these cases, both the α and β variable regions of this loop (Fig. 1) have their maximum sizes, i.e., 5 and 4 residues, respectively.

There are nine cases (other than the four initiators discussed in Section III,B) where an R_{11}–Y_{24} pair (six G-C and three A-U) is present, instead of the usual Y_{11}–R_{24} pairing seen in eubacterial and eukaryotic elongators. In five of these tRNAs which have the G_{11}–C_{24} pair (for Arg_1, Gln, His, Pro_1, and Pro_2) an alternate (nonstandard) pairing in the GG stem is possible—G_{11} pairing to C_{23} instead of C_{24} and so on, as is also possible for the halophile initiators (Section III,B). This novel pairing would seem to be energetically superior to the standard pairing.

In the eukaryotes and eubacteria, the anticodon stem of five base pairs can, in rare cases, be extended (distal to the loop) by an additional pair (26–44). However, in the archaebacterial examples, this extension is frequent. Out of 44 elongator tRNAs, there are 12 Watson–Crick and 10 G·U (or modified G·U) pairs possible at this position (26–44). All of the G·U pairs have the G at position 26, i.e., on the 5′ end of the stem. [This G·U arrangement is more stable than the U·G alternative because of the increased stacking it creates (Mizuno and Sundarlingam, 1978).] In this regard, it is noteworthy that in the crystal structure of yeast phenylalanine tRNA, there are two hydrogen bonds between $m_2^2G_{26}$ and A_{44} (tertiary pair), though the two bases are not co-planar. A number of such G·A juxtapositions (both modified and unmodified G's) are also found among the archaebacterial examples, as are $m_2^2G_{26}$·C_{44} juxtapositions.

The anticodon stem can also rarely be extended by one pair towards the antidocon loop side in eubacteria and eukaryotes (32–38). The incidence of this is very high in the archaebacterial cases, i.e., 12 cases of Watson–Crick pairs and another three of U·G pairs. Here again, G in all U·G pairs is at position 38, which is the more stable arrangement (Mizuno and Sundarlingam, 1978). A "tertiary" hydrogen bond between C_{32} and A_{38} is also reported in the crystal structure of yeast phenylalanine tRNA (Rich and RajBhandary, 1976; Kim, 1978).

One must be cautious in interpreting these extra pairs or aforementioned alternate pairs (in the GG stem) as being general archaebacterial characters. There are insufficient data from nonhalophilic archaebacteria, so these features could reflect the very high internal salt concentration (Bayley and Morton, 1978) of the halophiles.

Residues at certain positions are nearly invariant in the archaebacterial tRNAs. All 48 tRNAs listed in the Appendix have G or a modified G at position 10 in the GG stem, as do most (but not all) tRNAs of eubacteria and eukaryotes. The residue that pairs with it (at position 25) is always U whenever m_2^2G is present at position 10 ($m_2^2G_{10}$ so far is observed only in the archaebacteria; see Table I). Position 15 is occupied by G or its unidentified modified form—X (see Section III,A,1) in 46 tRNAs. The base at position 48, which forms the tertiary pair with the base at position 15, is always C or m^5C. The most frequent sequences in the common arm of the extreme halophile tRNAs are G-$m^1\Psi$-Ψ-Cm-m^1I-A-A-U-C, and G-m$^1\Psi$-Ψ-Cm-G-A-A-U-C (positions 53–61) and are the same in other archaebacteria (though some of the modifications may vary).

Nineteen of the class I archaebacterial tRNAs do not show normal pairing between positions 13 and 22 in the stem of the GG arm. Seventeen of these show $\Psi \cdot$U juxtapositions here, and one a $\Psi \cdot \Psi$ juxtaposition. The incidence of Ψ_{13} and U_{22} is extremely high, when compared to other tRNAs. The remaining tRNA (for cysteine) forms only *two* pairs in this stem, i.e., 10–25 and 11–24. (This tRNA is also unusual in having six bases in the variable arm.)

There are some exceptions in archaebacteria to the invariant or semi-invariant residues found in the generalized archaebacterial tRNA structure. At position 21, *H. volcanii* tRNA$_1^{\text{Ile}}$ and *T. acidophilum* tRNA^{Met} have U and G, respectively, in place of the invariant A. In the former case, an extra pair can be formed between A_{14} and U_{21}, which extends the GG stem by one pair. There are very few exceptions to this invariant A in eubacteria and eukaryotes. Some other exceptions seen in *H. volcanii* tRNAs are occurrence of A_{32} (not Y_{32}) in the anticodon loop of tRNA$_2^{\text{Pro}}$ and G_{58} and G_{60} (not A_{58} and Y_{60}) in the common arm loop of tRNA^{Met}.

D. General Structural Features and Possible Tertiary Interactions

The folding and backbone structure of a tRNA was first determined for yeast phenylalanine tRNA by X-ray crystallographic studies (Kim *et al.*, 1973). Higher resolution details were provided by several different research groups (Kim *et al.*, 1974; Robertus *et al.*, 1974; Quigley *et al.*, 1975; Ladner *et al.*, 1975; Stout *et al.*, 1976; Sussman and Kim, 1976). Kim (1978, 1979) has reviewed these and other studies, and has suggested some general structural features of the tRNAs. He has also correlated the crystal structure with the structure in solution. Crystal structures for the yeast (Schevitz *et al.*, 1979) and the *E. coli* (Woo *et al.*, 1980) initiator tRNAs have also been determined, though at a coarser resolution than that for the yeast phenylalanine tRNA. The overall shape of these two initiators is basically similar to that of yeast phenylalanine tRNA, with some differences in the anticodon loop region.

Most of the invariant and semi-invariant residues are involved in tertiary interactions. As mentioned before, Kim (1979) has noted the conserved presence (more than 90% occurrence) and the conserved absence (less than 10% occurrence) of certain bases in tRNA sequences. He explains conserved absence as due to the impossibility of forming a necessary equivalent tertiary pair (or triple) with the (absent) base. In other cases, the absence may be due to tight spatial constraints required to form correct tertiary structure. Conserved presence or absence of residues at various positions in the archaebacterial tRNAs are listed in Table II, which, for comparison, also includes Kim's (1979) analysis. Since

TABLE II

CONSERVED PRESENCE OR ABSENCE OF RESIDUES IN THE tRNAs[a]

Position no.	Occurrence in eubacterial and eukaryotic tRNAs[b] y-Phe-like		Occurrence in archaebacterial tRNAs			
			y-Phe-like[c]		Class I[d]	
	>90%	<10%	>90%	<10%	>90%	<10%
1		A,C		Y		Y
2		A		A,U		A,U
3	—	—		A,U		A,U
4		A		A,U		A
5	—	—		A		A
6	—	—		A		A,U
7		C		C		C
8	U		U		U	
9		Y		Y		Y
10	G		G		G	
11		R		U		A
12		A		A,C		A
13		A		G		R
14	A		A		A	
15		Y	G		G	
16	—	—		R		R
17						R
17:1						G
18	G		G		G	
19	G		G		G	
20				G		G
20:1						A
20:2						
21	A		A		A	
22		Y		G,C,U		C
23		U		G,C,U		U
24		Y		A		U
25		R		R		R
26		Y		Y		Y
27		A		G		R
28		G		R		A
29	—	—		A	—	—
30		A,U		A,U		A,U
31		U		U		U
32		R		R		R
33	U		U		U	
34		A		A		A

(*continued*)

TABLE II (*Continued*)

Position no.	Occurrence in eubacterial and eukaryotic tRNAs[b] y-Phe-like		Occurrence in archaebacterial tRNAs			
			y-Phe-like[c]		Class I[d]	
	>90%	<10%	>90%	<10%	>90%	<10%
35	—	—	—	—	—	—
36	—	—	—	—	—	—
37		Y		Y		Y
38		G	A			G,C
39		A		A		A
40		A,U		A,U		A,U
41	—	—		U	—	—
42		C		U		U
43		C		C		Y
44	—	—		G	—	—
45	—	—		A,C		C
46	—	—		Y		Y
47	—	—		R		R
48		R	C		C	
49		U		U		A,U
50	—	—	—	—	—	—
51	—	—		A		A
52		Y		A,C,U		Y
53	G		G		G	
54	U		U		U	
55	U		U		U	
56	C		C		C	
57		Y		Y		Y
58	A		A		A	
59		C		Y		G,C
60		R		R	U	
61	C		C		C	
62		R		A,U		R
63		A	—	—		U
64	—	—	—	—	—	—
65		A		A		A
66	—	—		G		G
67	—	—		A,U		A,U
68	—	—		U		U
69	—	—		A,U	—	—
70	—	—		A		A
71	—	—		A,U		A,U
72		G,U		R		R

(*continued*)

TABLE II (*Continued*)

Position no.	Occurrence in eubacterial and eukaryotic tRNAs[b] y-Phe-like		Occurrence in archaebacterial tRNAs			
			y-Phe-like[c]		Class I[d]	
	>90%	<10%	>90%	<10%	>90%	<10%
73	—	—		C		Y
74	C		C		C	
75	C		C		C	
76	A		A		A	

[a] Modified bases are listed as unmodified parent base. —, no particular pattern.

[b] Data of Fig. 8 of Kim (1979) based on analysis of a total of the 51 eubacterial and eukaryotic tRNAs having four base pairs in stem of the GG arm and five bases in the variable arm (yeast tRNA^Phe-like, y-Phe-like).

[c] Nineteen archaebacterial tRNAs, which are yeast tRNA^Phe-like (as in [a]) are examined.

[d] Analysis of all class I tRNAs of archaebacteria (40 tRNAs).

most of the archaebacterial tRNAs are from a single organism (*H. volcanii*), probably not all of the conserved (presence or absence) residue designations will ultimately hold up. However, the overall archaebacterial pattern is not remarkably different from the others, suggesting that the tertiary structure of the archaebacterial tRNAs is similar to the other tRNAs.

Each stem in eubacterial and eukaryotic tRNAs is relatively rich in G + C. Kim's (1979) analysis shows the G + C richness to be concentrated at specific regions or even specific base pairs within a stem. The pairs with the highest average G + C content are: 1–72 at the terminus of the acceptor stem; 53–61 at the loop end of the common arm stem; 10–25 at the end of the GG stem distal from the loop; and 30–40, the penultimate pair from the loop end of the anticodon stem. The long helix formed in the tertiary structure of a tRNA by the coaxial association of the acceptor and the common arm stems, therefore, has its highest G + C at the extremes, but average G + C composition elsewhere. The stems of archaebacterial tRNAs are also rich in G + C. However, the base pairs at *both* ends of the anticodon and common arm stems, as well as the acceptor stem, are high in G + C content, when compared to their internal pairs. Even so, the distal ends of these three stems (i.e., near their loops, or the terminal pair in the acceptor stem) in general show slightly more G + C than do the other ends. In the fourth stem, i.e., of the GG arm, the G + C richness is variable among different archaebacterial tRNAs. Although this distribution of G + C is distinct from all other tRNAs, it may merely reflect the higher internal salt concentration of the extreme halophiles.

Tertiary hydrogen bonds in the tRNA molecule involve not only the bases, but the ribose and phosphate moieties as well. The crystal structure of yeast phenylalanine tRNA seems representative of class I tRNAs, in that most of its tertiary base pairs can be replaced by equivalent tertiary pairs in other class I tRNAs (Kim, 1979). Kim lists the replaceable tertiary base pairs found in the tRNAs that have four pairs in the GG stem and five bases in the variable arm. Most of these pairs can form approximately equivalent hydrogen bonds. These pairs, along with their homologous counterparts in archaebacterial tRNAs, are presented in Table III. It is evident that most of the comparable tertiary pairs in the archaebacteria are the same as those in the other two kingdoms.

TABLE III

REPLACEABLE TERTIARY BASE PAIRS IN tRNAs

Position of pair	Base pair[d]	Eubacterial and eukaryotic tRNAs y-Phe-like[a]	Archaebacterial tRNAs	
			y-Phe-like[b]	Remaining class I[c]
8·14	U·A	51	19	20
9·23	A·A	33	15	—
	G·C	10	—	8
	G·G	3	1	5
	A·C	3	—	3
	A·G	1	—	—
	G·A	1	1	2
	G·U	—	1	2
	N·C	—	1	—
10·45	G·G	38	13	14
	G·U	2	6	1
	G·A	1	—	4
	G·C	—	—	1
15·48	G·C	42	6	13
	A·U	7	—	1
	G·A	1	—	—
	A·C	1	—	—
	X·C	—	12	6
	C·C	—	1	—
18·55	G·U	51	19	20
19·56	G·C	51	19	20
22·46	G·G	46	15	1
	A·A	3	2	1
	C·A	1	—	—
	G·U	1	—	—
	U·A	—	1	14

(*continued*)

TABLE III (*Continued*)

Position of pair	Base pair[d]	Eubacterial and eukaryotic tRNAs y-Phe-like[a]	Archaebacterial tRNAs	
			y-Phe-like[b]	Remaining class I[c]
	A·C	—	—	1
	U·G	—	1	3
26·44	G·A	29	9	4
	A·G	13	—	1
	A·C	3	1	—
	A·U	2	2	3
	C·U	1	1	—
	G·G	1	—	—
	A·A	1	1	3
	U·G	1	—	—
	G·C	—	3	3
	C·G	—	—	3
	G·U	—	1	—
	U·A	—	—	1
	C·A	—	1	—
	U·U	—	—	2
54·58	U·A	51	19	19
	U·G	—	—	1

[a] Data of Table 2 of Kim (1979), based on analysis of a total of 51 eubacterial and eukaryotic tRNAs with four bases in stem of the GG arm and five bases in the variable arm (yeast tRNA[Phe]-like, y-Phe-like).

[b] Based on 19 archaebacterial tRNAs, which are yeast tRNA[Phe]-like (as in [a]).

[c] Analysis of remaining archaebacterial class I tRNAs (total of 21).

[d] Modified bases are given as the unmodified parent base (except for N, unidentified base, and X, a specific unidentified modification of G).

IV. Aminoacyl-tRNA Synthetases and Aminoacylation

Studies of aminoacylation of the archaebacterial tRNAs and the synthetases involved have been done mainly with the halobacteria. The aminoacyl-tRNA synthetases of *Halobacterium cutirubrum* require nearly saturated KCl (3.8 M) for both stability and activity; further addition of NaCl stimulates aminoacylation with most amino acids (Griffiths and Bayley, 1969; White and Bayley, 1972b). Griffiths and Bayley (1969) tested aminoacylation of *H. cutirubrum* tRNAs for 16 amino acids (all except Cys, His, Trp, and Tyr). *Halobacterium cutirubrum* synthetases charged some *E. coli* tRNAs with amino acids in the presence of high salt, and charging of several *H. cutirubrum* tRNAs was observed with *E. coli* synthetases in low salt (White and Bayley, 1972b). These studies did not determine whether the heterologous charging aminoacylated only the cognate tRNAs in all cases.

Halobacterium volcanii total tRNAs were charged by homologous crude synthetase preparations in the presence of a single added amino acid, and following periodate oxidation and deacylation of these tRNAs, [5'-^{32}P]pCp was ligated to them (R. Gupta, unpublished). These labeled tRNAs after separation by gel electrophoresis were sequenced. In these studies, it was possible to charge *H. volcanii* tRNAs with 18 of the 20 amino acids (only glutamine and asparagine were not accepted), and the anticodons in each case agreed with the accepted genetic code.

Charging by glutamic acid in the *H. volcanii* system yielded one species of tRNA, which had the anticodon for glutamine in addition to those specific for glutamic acid. One tRNA having the anticodon for asparagine was sequenced, but this tRNA could not be charged by any of the 20 common amino acids. White and Bayley (1972c) found that in *H. cutirubrum,* glutaminyl-tRNA^{Gln} was made via an intermediate glutamyl-tRNA^{Gln}. In this respect, the tRNA^{Gln} of the extreme halophiles resemble those of some gram-positive eubacteria (Wilcox and Nirenberg, 1968; Wilcox, 1969). White and Bayley (1972c) did not find any evidence for an analogous route for asparaginyl-tRNA^{Asn} in *H. cutirubrum.* The alternatives here are that the extreme halophiles form their asparaginyl-tRNA^{Asn} by a yet unidentified route or that their asparaginyl-tRNA synthetase is exceptionally labile.

Kwok and Wong (1980) reported that the tRNAs of *H. cutirubrum,* unlike eubacterial tRNAs, are preferentially aminoacylated by synthetases from yeast, as compared to synthetases from *E. coli* or *Rhodopseudomonas spheroides.* However, their conclusion does not seem warranted. In that report, some eubacterial tRNAs appear to be charged better by yeast synthetases than are *H. cutirubrum* tRNAs. Also, there appears as much variation among different eubacterial tRNAs with regard to heterologous charging for a given amino acid as there is between halobacterial and some of the eubacterial or eukaryotic tRNAs.

V. Codon and Anticodon Usage

A number of codon assignments in *H. cutirubrum* were determined by Bayley and co-workers (Bayley and Griffiths, 1968; White and Bayley, 1972b,c), using amino acid incorporation and ribosome binding assays in the presence of random heteropolyribonucleotides. No disagreements were found with the established triplet code (Bayley and Morton, 1978). A proof of the use of the normal genetic code in this organism came only with the comparison of the nucleotide sequence of a structural gene with its corresponding amino acid sequence, as was done with bacteriorhodopsin in *Halobacterium halobium* (Dunn *et al.,* 1981), although not all of the codons are used in this gene. The correspondence between aminoacylation patterns and anticodon sequences of the tRNAs of *H. volcanii* is also consistent with the use of a normal genetic code in the halobacteria.

All codon usage data in the archaebacteria come from the aforementioned bacteriorhodopsin gene sequence (Dunn *et al.,* 1981). The protein contains no His or Cys, so the corresponding codons could not be measured. Among the other codons, those ending in U and A are used either relatively sparingly or not at all (compared to their C and G ending counterparts). AGR codons are not present in the gene. Interestingly, tRNAs (arginine) reading AGA and AGG codons, if present in *H. volcanii,* are in extremely low amounts, for they were not detected in our sequencing studies.

The examination of the "wobble" position (34) in archaebacterial anticodons shows some special features. At this position in the tRNAs of eubacteria and eukaryotes, the A residue is always modified to I, G sometimes is modified to Gm or some other derivative, C occurs in both modified or unmodified form, and U is nearly always modified. In *H. volcanii* tRNAs, A_{34} has never been observed (in any form), G_{34} is *never* modified, and both modified and unmodified pyrimidines are found. If C is modified to Cm in a given tRNA, then it is completely so, but when it is modified to ac^4C, the replacement is only partial.

These observations indicate that archaebacterial codon–anticodon interactions are not exactly the same as those involved in the eubacteria and the eukaryotes. It appears that the so-called "wobble" interactions in extreme halophiles are to some extent unique, which may reflect the high internal salt concentration in these organisms or may, alternatively, be a general characteristic of archaebacterial tRNAs.

VI. Archaebacterial tRNA Genes

A limited amount of information is available regarding archaebacterial tRNA genes and their organization. Recently, it has been found that at least two presumed tRNA genes in *Sulfolobus solfataricus* (for leucine and serine) contain intervening sequences (Kaine *et al.,* 1983). Before this discovery, introns had been observed only in the genes of eukaryotes and their organelles. The sequences of these intron-containing genes are shown in Fig. 3. The positions of the introns in both cases are the same, and very similar to those seen for eukaryotic tRNA genes. The possible secondary structure in the intron locale is similar in the two examples, but differs somewhat from the structure possible in the eukaryotic cases. Each secondary structure has two symmetrically placed, 3-base bulge loops (see Fig. 3), which might represent symmetric cleavage sites for the excision of the introns. If so, a simultaneous cleavage could be made by a symmetric (dimeric) enzyme. These *S. solfataricus* tRNA genes do not encode the CCA 3'-termini of the tRNAs.

A *Methanospirillum hungatei* tRNA gene for glutamic acid has been sequenced (B. P. Kaine and C. R. Woese, personal communication). It lacks the CCA 3'-terminus. Its sequence differs from *H. volcanii* tRNA$_2^{Glu}$ (same anticodon) at 17 positions only, 11 of which involve double-stranded regions. Sig-

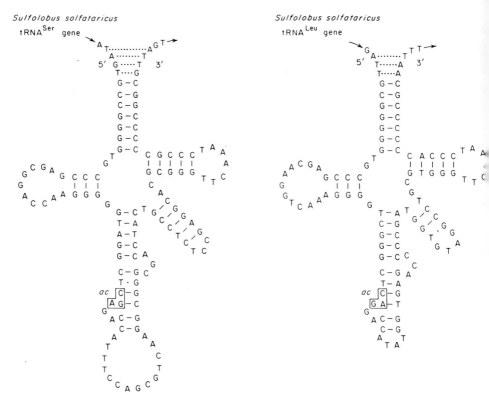

FIG. 3. Serine and leucine tRNA genes of *Sulfolobus solfataricus*, folded into typical cloverleaf structures. Possible paired extensions of the acceptor stem are indicated by dotted lines. *ac* indicates the anticodon sequence (boxed). (After Kaine *et al.*, 1983.)

nificantly, the last four pairs in the acceptor stem (4–7 versus 69–66) are different in the two cases. In addition, two bases, G_{26} and C_{44}, of the *H. volcanii* tRNA are replaced by C_{26} and G_{44} in the *M. hungatei* tRNA. These two transversions lend additional support to the existence of a sixth pair (26–44) at the top of the anticodon stem in the archaebacteria.

All of the seven ribosomal RNA operons in *E. coli* have tRNA genes in the spacer region between the 16 S and 23 S rRNA genes; four spacers contain tRNA^{Glu} genes and the remaining three contain two genes each, for tRNA^{Ile} and tRNA^{Ala} (Morgan *et al.*, 1977). Some of these genes have been sequenced (Sekiya and Nishimura, 1979; Young *et al.*, 1979; Morgan *et al.*, 1980). Some spacer tRNA genes of *Bacillus subtilis* (Loughney *et al.*, 1982), *Anacystis nidulans* (Williamson and Doolittle, 1983), and chloroplasts of *Zea mays* (Koch *et al.*, 1981), *Nicotiana tabacum* (Takaiwa and Sugiura, 1982), and *Euglena gracilis* (Graf *et al.*, 1980; Orozco *et al.*, 1980) have also been sequenced. In all these cases, both tRNA^Ile and tRNA^Ala genes occur in the same spacer. In *H. volcanii* spacer, there occurs a single tRNA gene (Gupta *et al.,* 1983, and unpublished). The gene corresponds exactly to the sequence of tRNA$_3^{Ala}$ (except for the absence of the 3'-terminal CCA). As with the other spacer tRNA genes, this one also has the same sense strand as do the rRNA genes. In all known spacer tRNA^Ala genes, be they eubacterial, chloroplast, or archaebacterial, the anticodon sequence is UGC.

VII. Concluding Remarks

Compared to eukaryotic and eubacterial tRNAs, we know very little about archaebacterial tRNAs. Moreover, most of this information is confined to the tRNAs of the extreme halophiles. However, based on this limited information, it can be said that in their basic organization, archaebacterial tRNAs are very similar to those of eubacteria and eukaryotes. Yet in specific details, the archaebacterial tRNAs are as different from the other two groups as these two were known to be from one another. Based on their general characteristics, tRNAs should be broadly grouped into three major phylogenetic categories (eubacterial, archaebacterial, and eukaryotic). Although there is insufficient evidence to make a definite statement at present, the archaebacterial tRNAs, if they show any phylogenetic bias at all, are closer to their eukaryotic counterparts than to eubacterial tRNAs.

Appendix: Sequences of Archaebacterial tRNAs

The numbering system followed here is that of Gauss and Sprinzl (1983). Positions 47:11 through 47:16 are not shown, as they are not needed.

The base-paired regions in the "cloverleaf" form are underlined. The dots under G and U (or F) indicate G·U (or G·F) pairs. Anticodon sequences are doubly underlined. A normal residue, when it partially replaces a modified one, is indicated in parentheses under the modified form.

The organisms and reference are H.v., *Halobacterium volcanii* (Gupta, 1981, and unpublished results); H.c., *Halobacterium cutirubrum* (Gu *et al.,* 1983); methionine initiators of H.m., *Halococcus morrhuae;* T.a., *Thermoplasma acidophilum;* and S.a., *Sulfolobus acidocaldarius* (Kuchino *et al.,* 1982); and tRNA^Met of T.a. (Kilpatrick and Walker, 1981).

For technical reasons some of the abbreviations used in the text (see footnote on p. 313 for the abbreviations) are changed here. These are: A1, m^{1}A; A6, t^{6}A;

	Acceptor Arm / e-stem	a-stem	"GG" Arm loop I	a-stem	b-stem	Anticodon Arm loop II	b-stem

Position numbers: 1 2 3 4 5 6 7 8 9 10 11 12 13 14 15 16 17 :1 18 19 20 :1 :2 21 22 23 24 25 26 27 28 29 30 31 32 33 34 35 36 37 38 39 40 41 42 4...

Alanine

H.v. 1   G G G C U C C U A G A U C A G G   G G U   A G A U C A C U C C C U U G G C A U G G G A G

H.v. 2   G G G C U C C U A G A U C A G U   G G C   A G A U C G4 C U U C C U U C G C A A G G A A C

H.v. 3   G G G C C C A U A G C U C A G U   G G U   A G A G U G2 C C U C C U U U G C A A G G A G C

H.c.   G G G C U C C U A G A U C A G C   G G U   A G A U C G4 C U U C C U U C G C A A G G A A C

Arginine

H.v. 1   G U C C U G A U A G4 G G F A G U   G G A C U A U C C U C C U G G C U U G C G GI A G C C A C

H.v. 2   G G G C C C G U A G C U C A X U   G G A C   A G A G U G4 C U U G G U U C C G GI A C C A A C

H.v. 3   G G G C G C U U A G C U C A X U C U G G A C   A G A G U G2 C U U G G C U U C G GI A C C A A (G)

Asparagine

H.v.   G C C G C C C U A G C U C A X U U   G G U   A G A G C A C C U C G C U G U U A6 A C G A G (G)

Aspartic Acid

H.v.   G C C C G G G U G G4 U G F A G U   G G C C C A U C A U A C G A C C C U G U C A C G G U C

Cysteine

H.v.   G C C A A G G U G G C A G A X U U C G G C C C A A C G C A U C C G C C U G C A GI A G C G G

Glutamic Acid

H.v. 1   G C U C U G U U U G G4 U G F A G U C C G G C C A A U C A U A U C A C C C U C4 U C A C G G U G   (C)

H.v. 2   G C U C G G U U U G G4 U G F A G U C C G G C C A A U C A U C U U U G G C C U U U U C GI A G C C G

Glutamine

H.v.   A G U C C C A U G G4 G G F A G U   G G C C A A U C C U G U U G C C U U C4 U G GI G G G C A   (C)

Glycine

H.v. 1A   G C G C U G G U A G2 U G F A G U   G G U   A U C A C G U G A C C U U G G C C A U G G U C

H.v. 1B   G C G U C G G U A G2 U G F A G U   G G U   A U C A C G U G A C C U U G G C C A U G G U C

H.v. 2   G C A C C G G U G G2 U C U A A U   G G U   A A G A C A U U G G C C U U U C C A A G C C A

H.v. 3   G C G C C G A U G G2 U C C A G U   G G U   A G G A C A C G A G C U U C C C A A G C U C

Histidine

H.v.   G U C C G G G U U G G4 G G F A G U   G G A C U A U C C U U C A G C C U U G U G GI A G G C U

Isoleucine

H.v. 1   G G G C C A A U A G C U C A G U C A G G U   U G A G C G4 C F C G G C U G A U A6 A C C5 G G

H.v. 2   G G G C C C C U A G C U C A X U C U G G U C   A G A G C G4 C U C G G C U N A U Ab A C C G G (G)

	Variable Arm		Common Arm			Acceptor Arm
		d-stem	loop IV		d-stem	e-stem

Positions: 44 45 46 47 1 :2 :3 :4 :5 :6 :7 :8 :9 :10 48 49 50 51 52 53 54 55 56 57 58 59 60 61 62 63 64 65 66 67 68 69 70 71 72 73 74 75 76

Alanine

```
H.v. 1   A G G C      C5  C5 C G G G  Fl F  C3 Il A A U  C C C G G  C G A G  U  C C A C C A
                          (C)
H.v. 2   A G G C      C5  C5 G G G G  Fl F  C3 Il A A U  C C C C G  C G A G  U  C C A C C A
                          (C)
H.v. 3   A U G C      C5  C5 A G G G  Fl F  C3 G A A U   C C C U G  U G G G  U  C C A C C A
                          (C)
H.c.     A G G C      C   C5 U G G G  Fl F  C3 Il A A U  C C C A G  C G A G  U  C C A C C A
```

Arginine

```
H.v. 1   G G A        C   C5 G G A G  Fl F  C3 Il A A U  C U C C G  U C A G G A C  G C C A
                          (U)
H.v. 2   A U G C      C5  G  C G G G  Fl F  C3 Il A A U  C C C G U  C G G G  U  C C G C C A
                             (A)
H.v. 3   U U G C      C5  A C  G G G  Fl F  C3 Il A A U  C C U G  U G U  A G C G U C C A C C A
```

Asparagine

```
H.v.     U U G U      C5  C5 C A G G  Fl F  C3 G A G U  C C U G G · C G G  U  G G C  G C C A
                          (C)
```

Aspartic Acid

```
H.v.     U G A        C   G C G G G  Fl F  C3 Il A A U  C C C G C C  U  C G G G G C  G C C A
                             (F)
```

Cysteine

```
H.v.     A C C C      C5  G C C G G  Fl F  C3 Il A A U  C C G G C  C C U U G G G  U C C A
```

Glutamic Acid

```
H.v. 1   U G A        C   C5 A G G G  Fl F  C3 G A A U  C C C U G A C  G G A G C  A C C A
                          (F)
H.v. 2   G G A        C   C5 A G G G  Fl F  C3 Il A A U  C C C U G  A C C G A G C  A C C A
                          (C)
```

Glutamine

```
H.v.     C G A        C   C C A G G  F F  C3 G A A U  C C U G G  U G G G A C U  A C C A
```

Glycine

```
H.v. 1A  C A A        C   C5 U G G G  Fl F  C3 Il A A U  C C C A G  C C A G C G C  A C C A
H.v. 1B  C A A        C   C5 U G G G  Fl F  C3 Il A A U  C C C A G  C C G A C G C  A C C A
H.v. 2   U U A        U   C5 U G G G  Fl F  C3 G A U U   C C C A G  C C G G U G C  A C C A
H.v. 3   G A G        C   C5 C G G G  Fl F  C3 Il A U U  C C C G G  U C G G C G C  A C C A
```

Histidine

```
H.v.     A G A        C   G C G G G  F F  C3 A A U U  C U C G C  A C C U G G A C  C C A
```

Isoleucine

```
H.v. 1   A G G C      C   C5 G C G G  Fl F  C3 Il A A U  C C G C C G  U U G G C C C  A C C A
H.v. 2   U G G U      C5  A U G G G  Fl F  C3 G A A C  C C C A U  G G G G C C C  A C C A
```

	Acceptor Arm e-stem		a-stem		"GG" Arm loop I		a-stem		b-stem		Anticodon Arm loop II		b-stem	

```
              1  2 '3  4  5  6  7  8  9   10 11 12 13 14 15 16 17 :1 18 19 20 :1 :2 21 22 23 24 25 26 27 28 29 30 31 32 33 34 35 36 37 38 39 40 41 42 43

Leucine

H.v. 1   G C G U G G G  U A  G C C  A A X  C C A G G  G C C A A C  G G C  G4 C A G C G  U  U  G A G  G1 G  C5 G  C U G

H.v. 2   G C A G G G A  U A  G C C  A A X  U C U G G  G C C A A C  G G C  G4 C A G C G  U  U  C A G  G1 G  C G C U G

H.v. 3   G C G A G G G  U A  G C U  A A X  U C A G G  A A       A A A  A G C  G4 G C G G A  C  U  C A A  G1 A  F C C G C

H.v. 4   G C G C G G G  U A  G C C  A A X  U        G G C C A A A  G G C  G4 C A G C G  C  U  Ů A G  G1 A  C G C U G

H.v. 5   G C G G G G G  U G  G C U  G A X  C C A G G  G C C A A A  A G C  G2 G C G G A  C  U  U A A  G1 A  F C C G C

Lysine

H.v. 1   G G G C C G G  U A  G C U C  A X U U A G G C        A G A G C  G4 U C U G A  C3 U  C4 U U  A6 A  F C A G A
                                                                                                  (C)        (U)

H.v. 2   G G G C U G G  U A  G C U C  A X U U A G G C        A G A G C  G4 U C U G G  C3 U  Ů U U  A6 A  C C A G A

Methionine

H.v.     G C C C G G G  U G  G C U  F A X C U    G G A C     A F A G C G  C C C C A C  U  C3 A U  A6 A  F G C G G

T.a.     G C C G G G G  U4 G  G C U C  A N* C U    G G A     G G A G C  G4 C C G G A  C3 U  C A U  A6 A  U C C G G

Methionine-initiator

H.v.     A* G C G G G A  U G  G G G A  F A X C C A G G A G    A U U C C G  C C G G G C  U C A U  A A  C C C G G

H.m.     A G C G G G A  U G  G G G A  F A G C C A G G A G    A U U C C  G2 G C G G G C  U C A U  A A  C C C G C

S.a.     A G C G G C G  U N  G2 G G A  A C U G    G G A G U  A U C C C C  N* C A G3 G G  C3 U  C A U  A A  C C C U G

T.a.     A G C G G G G  C U G  G G G F  A G U C A G G A      A A U C C  G2 A U G G G C  U C A U  A A  C C C G U
                                                                                                  (C3)

Phenylalanine

H.v.     G C C G C C U  U A  G C U C  A X A C U G G G       A G A G C A  C U C G A  C U G A A  G1 A  F C G A G

Proline

H.v. 1   G G G C C G  C U G  G4 G G  F A X C U U G G U      A U C C U U  C G G C C U U  C4 G G  G1 F  G G C C G
                                                                                       (C)        (U)

H.v. 2   G G G A C C  G U G  G4 G G  F A G U    G G U        A U C C U C U G  C C G A U  G G G  G1 U  C G G U A

h.v. 3   G G G A C C  G U G  G4 G U  F A X C C U G G U      A U A C U U  C G G G C C U U  U G G  G1 U  G C C C G
                            (G)
```

	Variable Arm			Common Arm			Acceptor Arm
			d-stem	loop IV		d-stem	e-stem

44 45 46 47 1 :2 :3 :4 :5 :6 :7 :8 :9 :10 48 49 50 51 52 53 54 55 56 57 58 59 60 61 62 63 64 65 66 67 68 69 70 71 72 73 74 75 76

Leucine

H.v. 1 U C C U ; U A G A G G U C   C5 G C C G G Fl F C3 Il A A U C C G G U C C C A C G C A C C A

H.v. 2 U C U C ، U A G G A G U C   C5 G C A G G Fl F C3 Il A A U C C U G C U C C C U G C A C C A
                                                        (A)

H.v. 3 U C C C ·, U A G G G G U C   C5 G U G G G Fl F C3 Il A A U C C C U C C C C U C G C A C C A

H.v. 4 U G G U ., U A G A C C U U   C5 G C A G G Fl F C3 G A A C C C U G U C C C G C G C A C C A

H.v. 5 U C C C ·, U A G G G G U U   C G C G A G Fl F C3 G A A U C U C G U C C C C C G C A C C A

Lysine

H.v. 1 C G G U   C5 G C G F G Fl F C3 Il A A U C G C G U C C G G C C C A C C A·
                                                       (U)

H.v. 2 C G G U   C5 G G G G G Fl F C3 Il A G U C C C U C C C A G C C C G C C A

Methionine

H.v.   A G A U   C5 G U G G G Fl F C3 G G A G C C C A C C C C G G G C A C C A

T.a.   A G G U   C U C G G G F F C3 G A U C C C C G A U C C C G G C A C · ·-

Methionine-initiator

H.v.   A G A U   C G G U A G Fl F C3 Il A A U C U A C C U C C C G C U A C C A

H.m.   A G A U   C A G U A G Fl F C3 Il A A U C U A C U U C C C G C U A C C A

S.a.   A G G U   C5 C C U G G U3 U C3 Il Al A U C C A G G C G C C G C U A C C A

T.a.   A G A U   C G A U G G F F C3 N* Al A U C C A U C C C C C G C U A C C A

Phenylalanine

H.v.   C U G U   C5 C C C G G Fl F C3 Il A A U C C G G G A G G C G G C A C C A

Proline

H.v. 1 U A A   C5 C5 U C A G Fl F C3 G A A U C U G A G C C G G C C C A C C A
        (C)   (U)

H.v. 2 G G A   C C5 U G A G Fl F C3 G A C U C U C A G C G G U C C C A C C A
                                                             (U)

H.v. 3 U G A   C5 C5 C C G G Fl F C3 Il A A U C C G G G C G G U C C C A C C A
        (C)(C)

| | Acceptor Arm e-stem | a-stem | "GG" Arm loop I | a-stem | b-stem | Anticodon Arm loop II | b-stem |

```
              1  2  3  4  5  6  7  8  9  10 11 12 13 14 15 16 17 :1 18 19 20 :1 :2 21 22 23 24 25 26 27 28 29 30 31 32 33 34 35 36 37 38 39 40 41 42 43

Serine

H.v. 1    G  U  U  G  C  G  G  U  A  G  C  C  A  A  X  C  C  U  G  G  C  C  C  A  A  G  G  C  G4 C  U  G  G  G  U  U  G  C  U  A6 A  C  U  C  A  G

H.v. 2    G  C  C  G  A  G  G  U  A  G  C  C  F  A  X  C  C  C  G  G  C  C     A  A  G  G  C  G4 G  U  A  G  A  U  U  C4 G  A  A  A  F  C  U  A  C
                                                                                                              (C)

H.v. 3    G  C  C  A  G  G  A  U  G  G  C  C  G  A  X  C        G  G  U     A  A  G  G  C  G4 C  A  C  G  C  C  U  G  G  A  A  A  G  C  G  U  G


Threonine

H.v. 1    G  C  C  U  G  G  G  U  A  G  C  U  F  A  X  C        G  G  U     A  A  A  G  C  G4 C  G  U  C  C  U  U  G  G  U  A6 A  G  G  A  C  G
                                               (G)

H.v. 2    G  C  C  G  G  U  G  U  A  G  C  U  C  A  X  U  U     G  G  C     A  G  A  G  C  G4 A  U  U  C  C  U  U  C  G  U  A6 A  G  G  A  A  U


Tryptophan

H.v.      G  G  G  G  C  U  G  U  G  G  G  C  C  A  A  X  C  C  C  G  G  C     A  U  G  G  C  G4 A  C  U  G  A  C3 U  C3 C  A  G1 A  U3 C  A  G  U


Tyrosine

H.v.      C  C  G  C  U  C  U  U  A  G  C  U  C  A  X  C  C  U  G  G  C     A  G  A  G  C  A  G  C  C  G  A  C3 U  G  U  A  G1 A  F  C  G  G  C
                                                                                                        (C)

Valine

H.v. 1    G  G  G  U  U  G  G  U  G  G  U  C  F  A  G  U  C  U  G  G  U  U     A  U  G  A  C  A  C  C  U  C  C  U  U  G  A  C  A  U  G  G  A  G  G

H.v. 2    G  G  G  U  U  G  G  U  G  G  U  C  F  A  X  C  C  A  G  G  U  U     A  U  G  A  C  G  G  C  U  C  C  U  U  C  A  C  A  C  G  G  A  G  C

H.c. 1    G  G  G  U  U  G  G  U  G  G  U  C  F  A  G  U  C  A  G  G  C  U     A  U  G  A  C  A  C  C  U  C  C  U  U  G  A  C  A  U  G  G  A  G  G

H.c. 2    G  G  G  U  U  G  G  U  G  G1 U  C  F  A  G  U  C  A  G  G  C  U     A  U  G  A  C  A  C  C  U  C  C  U  U  C  A  C  A  U  G  G  A  G  G
```

	Variable Arm	Common Arm			Acceptor Arm
		d-stem	loop IV	d-stem	e-stem

```
            44 45 46 47 :1 :2 :3 :4 :5 :6 :7 :8 :9 :10 48 49 50 51 52 53 54 55 56 57 58 59 60 61 62 63 64 65 66 67 68 69 70 71 72 73 74 75 76

Serine

H.v. 1   U  G  G  C  ;  U  C  A  A  G  C  C  C          C5 C5 G  G  G  G  Fl F  C3 G  A  A  U  C  C  C  C  G  C  C  G  C  A  A  C  G  C  C  A

H.v. 2   U  G  U  C  :  A  U  U  C  G  G  A  C  A        C5 G  U  G  A  G  Fl F  C3 Il A  A  U  C  U  C  A  C  C  C  U  C  G  G  C  G  C  C  A

H.v. 3   U  U  C  C  :  U  C  U  G  G  G  A  U           C5 G  G  G  G  G  Fl F  C3 Il A  A  U  C  C  C  U  C  U  C  C  U  G  G  C  G  C  C  A


   Threonine

H.v. 1   A  G  A  C                                      C5 C5 C  G  G  G  Fl F  C3 Il A  A  U  C  C  C  G  G  C  C  U  A  G  G  C  U  C  C  A

H.v. 2   A  G  G  C                                      C5 G  A  G  G  G  Fl F  C3 Il A  A  U  C  C  C  U  C  C  A  C  C  G  G  C  U  C  C  A


   Tryptophan

H.v.     C  G  A  U                                      C5 G  G  G  G  G  Fl F  C3 Il A  A  U  C  C  C  U  C  C  G  G  C  C  C  C  A  C  C  A


   Tyrosine

H.v.     U  U  G  U                                      C5 C  C  C  U  G  Fl F  C3 Il A  A  U  C  G  G  G  G  A  G  A  G  C  G  G  A  C  C  A


   Valine

H.v. 1   A  G  G  C                                      C5 G  G  C  A  G  Fl F  C3 Il A  A  U  C  U  G  U  C  C  C  A  A  C  C  C  A  C  C  A
                                                                  (U)

H.v. 2   A  G  G  C                                      C5 G  G  C  G  G  Fl F  C3 G  A  A  U  C  C  G  C  C  C  C  A  A  C  C  C  A  C  C  A

H.c. 1   A  G  G  U                                      C  G  G  C  G  G  Fl F  C3 Il A  A  U  C  C  G  C  C  C  C  A  A  C  C  C  A  C  C  A

H.c. 2   A  G  G  U                                      C  G  G  C  G  G  Fl F  C3 Il A  A  U  C  C  G  C  C  C  C  A  A  C  C  C  A  C  C  A
```

C3, Cm; C4, ac^{4}C; C5, m^{5}C; G1, m^{1}G; G2, m^{2}G; G3, Gm; G4, m$_2^2$G; U3, Um; U4, s^{4}U; F, Ψ; F1, m^{1}Ψ; I1, m^{1}I. The asterisks indicate the following: H.v. Met$_i$ A$_1$, 5′-triphosphate is present in this tRNA; T.a. MetN$_{15}$, probably X; T.a. Met$_i$ N$_{57}$, probably I; S.a. Met$_i$ N$_{26}$, probably m$_2^2$Gm.

NOTE ADDED IN PROOF

Most of the *Halobacterium volcanii* tRNA sequences listed in the Appendix have now been published [Gupta, R., (1984). *Halobacterium volcanii* tRNAs: Identification of 41 tRNAs covering all amino acids, and the sequences of 33 class I tRNAs. *J. Biol. Chem.* **259,** 9461–9471].

Recently a 105 base intron has been detected in a tRNA gene of *H. volcanii* [Daniels, C. J., Gupta, R., and Doolittle, W. F. (1985). Transcription and excision of a large intron in the tRNA^{Trp} gene of an archaebacterium, *Halobacterium volcanii*. *J. Biol. Chem.* (in press)].

There has been a recent report of purification and characterization of an archaebacterial aminoacyl-tRNA synthetase [Rauhut, R., Gabius, H.-J., Kühn, W., and Cramer, F. (1984). Phenylalanyl-tRNA synthetase from the archaebacterium *Methanosarcina barkeri. J. Biol. Chem.* **259,** 6340–6345].

ACKNOWLEDGMENTS

I am grateful to Professors Carl R. Woese and Norman R. Pace for helpful suggestions and critical reading of the manuscript. I thank Lorie Hatfield for typing the manuscript. The author is supported by NASA grant NSG–7044 to Professor C. R. Woese.

REFERENCES

Altman, S., ed. (1978). "Transfer RNA." MIT Press, Cambridge, Massachusetts.

Bayley, S. T., and Griffiths, E. (1968). Codon assignments and fidelity of translation in a cell-free protein synthesizing system from an extremely halophilic bacterium. *Can. J. Biochem.* **46,** 937–944.

Bayley, S. T., and Morton, R. A. (1978). Recent developments in the molecular biology of extremely halophilic bacteria. *CRC Crit. Rev. Microbiol.* **6,** 151–205.

Best, A. N. (1978). Composition and characterization of tRNA from *Methanococcus vannielii. J. Bacteriol.* **133,** 240–250.

Clark, B. F. C. (1978). General features and implications of primary, secondary, and tertiary structure. *In* "Transfer RNA" (S. Altman, ed.), pp. 14–47. MIT Press, Cambridge, Massachusetts.

Delk, A. S., Romeo, J. M., Nagle, D. P., Jr., and Rabinowitz, J. C. (1976). Biosynthesis of ribothymidine in the transfer RNA of *Streptococcus faecalis* and *Bacillus subtilis;* a methylation of RNA involving 5,10-methylenetetrahydrofolate. *J. Biol. Chem.* **251,** 7649–7656.

Dirheimer, G. (1983). Chemical nature, properties, location, and physiological and pathological variations of modified nucleosides in tRNAs. *Recent Results Cancer Res.* **84,** 15–46.

Dirheimer, G., Keith, G., Sibler, A.-P., and Martin, R. P. (1979). The primary structure of tRNAs and their rare nucleosides. *Cold Spring Harbor Monogr. Ser.* **9A,** 19–41.

Dunn, R., McCoy, J., Simsek, M., Majumdar, A., Chang, S. H., RajBhandary, U. L., and Khorana, H. G. (1981). The bacteriorhodopsin gene. *Proc. Natl. Acad. Sci. U.S.A.* **78,** 6744–6748.

Feldman, M. Y. (1977). Minor components in transfer RNA: The location-function relationship. *Prog. Biophys. Mol. Biol.* **32,** 83–102.

Fox, G. E., Stackebrandt, E., Hespell, R. B., Gibson, J., Maniloff, J., Dyer, T. A., Wolfe, R. S., Balch, W. E., Tanner, R. S., Magrum, L. J., Zablen, L. B., Blakemore, R., Gupta, R., Bonen, L., Lewis, B. J., Stahl, D. A., Luehrsen, K. R., Chen, K. N., and Woese, C. R. (1980). The phylogeny of prokaryotes. *Science* **209,** 457–463.

Gauss, D. H., and Sprinzl, M. (1983). Compilation of tRNA sequences. *Nucleic Acids Res.* **11,** r1–r53.

Goddard, J. P. (1977). The structures and functions of transfer RNA. *Prog. Biophys. Mol. Biol.* **32,** 233–308.

Graf, L., Kössel, H., and Stutz, E. (1980). Sequencing of 16S–23S spacer in a ribosomal RNA operon of *Euglena gracilis* chloroplast DNA reveals two tRNA genes. *Nature (London)* **286,** 908–910.

Griffiths, E., and Bayley, S. T. (1969). Properties of transfer ribonucleic acid and aminoacyl transfer ribonucleic acid synthetases from an extremely halophilic bacterium. *Biochemistry* **8,** 541–551.

Gu, X.-R., Nicoghosian, K., Cedergren, R. J. and Wong, J. T.-F. (1983). Sequences of halobacterial tRNAs and the paucity of U in the first position of their anticodons. *Nucleic Acids Res.* **11,** 5433–5442.

Gupta, R. (1981). Structural characterization of the transfer ribonucleic acids from *Halobacterium volcanii* and other archaebacteria. Ph.D. Thesis, University of Illinois, Urbana-Champaign.

Gupta, R., and Woese, C. R. (1980). Unusual modification patterns in the transfer ribonucleic acids of archaebacteria. *Curr. Microbiol.* **4,** 245–249.

Gupta, R., Lanter, J. M., and Woese, C. R. (1983). Sequence of the 16S ribosomal RNA from *Halobacterium volcanii*, an archaebacterium. *Science* **221,** 656–659.

Holley, R. W., Apgar, J., Everett, G. A., Madison, J. T., Marquisee, M., Merrill, S. H., Penswick, J. R., and Zamir, A. (1965). Structure of a ribonucleic acid. *Science* **147,** 1462–1465.

Kaine, B. P., Gupta, R., and Woese, C. R. (1983). Putative introns in tRNA genes of procaryotes. *Proc. Natl. Acad. Sci. U.S.A.* **80,** 3309–3312.

Kilpatrick, M. W., and Walker, R. T. (1981). The nucleotide sequence of the $tRNA_m^{Met}$ from the archaebacterium *Thermoplasma acidophilum*. *Nucleic Acids Res.* **9,** 4387–4390.

Kim, S.-H. (1978). Crystal structure of yeast $tRNA^{Phe}$: Its correlation to the solution structure and functional implications. *In* "Transfer RNA" (S. Altman, ed.), pp. 248–293. MIT Press, Cambridge, Massachusetts.

Kim, S.-H. (1979). Crystal structure of yeast $tRNA^{Phe}$ and general structural features of other tRNAs. *Cold Spring Harbor Monogr. Ser.* **9A,** 83–100.

Kim, S.-H., Quigley, G. J., Suddath, F. L., McPherson, A., Sneden, D., Kim, J. J., Weinzierl, J., and Rich, A. (1973). Three-dimensional structure of yeast phenylalanine transfer RNA: Folding of the polynucleotide chain. *Science* **179,** 285–288.

Kim, S.-H., Suddath, F. L., Quigley, G. J., McPherson, A., Sussman, J. L., Wang, A. H. J., Seeman, N. C., and Rich, A. (1974). Three-dimensional tertiary structure of yeast phenylalanine transfer RNA. *Science* **185,** 435–440.

Koch, W., Edwards, K., and Kössel, H. (1981). Sequencing of the 16S–23S spacer in a ribosomal RNA operon of *Zea mays* chloroplast DNA reveals two split tRNA genes. *Cell* **25,** 203–213.

Kuchino, Y., Ihara, M., Yabusaki, Y., and Nishimura, S. (1982). Initiator tRNAs from archaebacteria show common unique sequence characteristics. *Nature (London)* **298,** 684–685.

Kwok, Y., and Wong, J. T.-F. (1980). Evolutionary relationship between *Halobacterium cutirubrum* and eukaryotes determined by use of aminoacyl-tRNA synthetases as phylogenetic probes. *Can. J. Biochem.* **58,** 213–218.

Ladner, J. E., Jack, A., Robertus, J. D., Brown, R. S., Rhodes, D., Clark, B. F. C., and Klug, A. (1975). Atomic coordinates for yeast phenylalanine tRNA. *Nucleic Acids Res.* **2**, 1629–1637.

Levitt, M. (1969). Detailed molecular model for transfer ribonucleic acid. *Nature (London)* **224**, 759–763.

Loughney, K., Lund, E., and Dahlberg, J. E. (1982). tRNA genes are found between the 16S and 23S rRNA genes in *Bacillus subtilis*. *Nucleic Acids Res.* **10**, 1607–1624.

McCloskey, J. A., and Nishimura, S. (1977). Modified nucleosides in transfer RNA. *Acc. Chem. Res.* **10**, 403–410.

Mizuno, H., and Sundarlingam, M. (1978). Stacking of Crick Wobble pair and Watson-Crick pair: Stability rules of G-U pairs at ends of helical stems in tRNAs and the relation to codon-anticodon Wobble interaction. *Nucleic Acids Res.* **5**, 4451–4461.

Morgan, E. A., Ikemura, T., and Nomura, M. (1977). Identification of spacer tRNA genes in individual ribosomal RNA transcription units of *Escherichia coli*. *Proc. Natl. Acad. Sci. U S.A.* **74**, 2710–2714.

Morgan, E. A., Ikemura, T., Post, L. E., and Nomura, M. (1980). tRNA genes in rRNA operons of *Escherichia coli*. *Cold Spring Harbor Monogr. Ser.* **9B**, 259–266.

Nishimura, S. (1978). Modified nucleosides and isoaccepting tRNA. *In* "Transfer RNA" (S. Altman, ed.), pp. 168–195. MIT Press, Cambridge, Massachusetts.

Nishimura, S. (1979). Modified nucleosides in tRNA. *Cold Spring Harbor Monogr. Ser.* **9A**, 59–79.

Ofengand, J. (1982). Structure and function of tRNA and aminoacyl tRNA synthetases in eukaryotes. *In* "Protein Biosynthesis in Eukaryotes" (R. Pérez-Bercoff, ed.), pp. 1–67. Plenum, New York.

Orozco, E. M., Jr., Rushlow, K. E., Dodd, J. R., and Hallick, R. B. (1980). *Euglena gracilis* chloroplast ribosomal RNA transcriptional units. II. Nucleotide sequence homology between the 16S–23S ribosomal RNA spacer and the 16S ribosomal RNA. *J. Biol. Chem.* **255**, 10997–11003.

Pang, H., Ihara, M., Kuchino, Y., Nishimura, S., Gupta, R., Woese, C. R., and McCloskey, J. A. (1982). Structure of a modified nucleoside in archaebacterial tRNA which replaces ribosylthymine; 1-methylpseudouridine. *J. Biol. Chem.* **257**, 3589–3592.

Quigley, G. J., Seeman, N. C., Wang, A. H.-J., Suddath, F. L., and Rich, A. (1975). Yeast phenylalanine transfer RNA: Atomic coordinatres and torson angles. *Nucleic Acids Res.* **2**, 2329–2341.

Rich, A., and RajBhandary, U. L. (1976). Transfer RNA: Molecular structure, sequence, and properties. *Annu. Rev. Biochem.* **45**, 805–860.

Robertus, J. D., Ladner, J. E., Finch, J. T., Rhodes, D., Brown, R. S., Clark, B. F. C., and Klug, A. (1974). Structure of yeast phenylalanine tRNA at 3 Å resolution. *Nature (London)* **250**, 546–551.

Schevitz, R. W., Podjarny, A. D., Krishnamachari, N., Hughes, J. J., Sigler, P. B., and Sussman, J. L. (1979). Crystal structure of a eukaryotic initiator tRNA. *Nature (London)* **278**, 188–190.

Schimmel, P. R., Söll, D., and Abelson, J. N., eds. (1979). "Transfer RNA: Structure, Properties, and Recognition," Cold Spring Harbor Monogr. Ser., Vol. 9A. Cold Spring Harbor Lab., Cold Spring Harbor, New York.

Sekiya, T., and Nishimura, S. (1979). Sequence of the gene for isoleucine $tRNA_1$ and the surrounding region in a ribosomal RNA operon of *Escherichia coli*. *Nucleic Acids Res.* **6**, 575–592.

Singer, B., and Kröger, M. (1979). Participation of modified nucleosides in translation and transcription. *Prog. Nucleic Acid Res. Mol. Biol.* **23**, 151–194.

Singhal, R. P., and Fallis, P. A. M. (1979). Structure, function, and evolution of transfer RNAs (with appendix giving complete sequences of 178 tRNAs). *Prog. Nucleic Acid Res. Mol. Biol.* **23**, 227–290.

Söll, D., Abelson, J. N., and Schimmel, P. R., eds. (1980). "Transfer RNA: Biological Aspects," Cold Spring Harbor Monogr. Ser., Vol. 9B. Cold Spring Harbor Lab., Cold Spring Harbor, New York.

Stout, C. D., Mizuno, H., Rubin, J., Brennan, T., Rao, S. T., and Sundarlingam, M. (1976). Atomic coordinates and molecular conformation of yeast phenylalanine tRNA. An independent investigation. *Nucleic Acids Res.* **3**, 1111–1123.

Sussman, J. L., and Kim, S.-H. (1976). Idealized atomic coordinates of yeast phenylalanine transfer RNA. *Biochem. Biophys. Res. Commun.* **68**, 89–96.

Takaiwa, F., and Sugiura, M. (1982). Nucleotide sequence of the 16S–23S spacer region in an rRNA gene cluster from tobacco chloroplast DNA. *Nucleic Acids Res.* **10**, 2665–2676.

Vani, B. R., Ramakrishnan, T., Taya, Y., Noguchi, S., Yamaizumi, Z., and Nishimura, S. (1979). Occurrence of 1-methyladenosine and absence of ribothymidine in transfer ribonucleic acid of *Mycobacterium smegmatis*. *J. Bacteriol.* **137**, 1084–1087.

Watanabe, K., Oshima, T., Saneyoshi, M., and Nishimura, S. (1974). Replacement of ribothymidine by 5-methyl-2-thiouridine in sequence GTΨC in tRNA of an extreme thermophile. *FEBS Lett.* **43**, 59–63.

White, B. N., and Bayley, S. T. (1972a). Methionine transfer RNAs from the extreme halophile, *Halobacterium cutirubrum*. *Biochim. Biophys. Acta* **272**, 583–587.

White, B. N., and Bayley, S. T. (1972b). Functional aspects of tRNA from the extreme halophile, *Halobacterium cutirubrum*. *Biochim. Biophys. Acta* **272**, 588–595.

White, B. N., and Bayley, S. T. (1972c). Further codon assignments in an extremely halophilic bacterium using a cell-free protein-synthesizing system and a ribosomal binding assay. *Can. J. Biochem.* **50**, 600–609.

Wilcox, M. (1969). γ-Phosphoryl ester of Glu-tRNA[Gln] as an intermediate in *Bacillus subtilis* glutaminyl-tRNA synthesis. *Cold Spring Harbor Symp. Quant. Biol.* **34**, 521–528.

Wilcox, M., and Nirenberg, M. (1968). Transfer RNA as a cofactor coupling amino acid synthesis with that of protein. *Proc. Natl. Acad. Sci. U.S.A.* **61**, 229–236.

Williamson, S. E., and Doolittle, W. F. (1983). Genes for tRNA[Ile] and tRNA[Ala] in the spacer between the 16S and 23S rRNA genes of a blue-green alga: Strong homology to chloroplast tRNA genes and tRNA genes of the *E. coli rrnD* gene cluster. *Nucleic Acids Res.* **11**, 225–235.

Woo, N. H., Roe, B. A., and Rich, A. (1980). Three-dimensional structure of *Escherichia coli* initiator tRNA[Met]. *Nature (London)* **286**, 346–351.

Yamaizumi, Z., Ihara, M., Kuchino, Y., Gupta, R., Woese, C. R., and Nishimura, S. (1982). Archaebacterial tRNA contains 1-methylinosine at residue 57 in TΨC-loop. *Nucleic Acids Res. Symp. Ser.* **11**, 209–213.

Young, R. A., Macklis, R., and Steitz, J. A. (1979). Sequence of 16S–23S spacer region in two ribosomal RNA operons of *Escherichia coli*. *J. Biol. Chem.* **254**, 3264–3271.

CHAPTER 7

Ribosomes of Archaebacteria

ALASTAIR T. MATHESON

I. Introduction

When Bayley and Kushner (1964) published their initial paper on the properties of the ribosomes from the extreme halophile *Halobacterium cutirubrum*, it was evident that these ribosomes showed many unusual properties. For example, the ribosomal proteins (r-proteins) were mainly acidic, whereas all other organisms studied up to that time had ribosomes with predominantly basic r-proteins. Initial studies on the sequence of these halophilic r-proteins indicated their structures were also unusual (Oda *et al.*, 1974; Matheson *et al.*, 1978). While sequence studies on r-proteins from *Escherichia coli* (Wittmann *et al.*, 1980), *Bacillus stearothermophilus* (Hori *et al.*, 1977; Yaguchi *et al.*, 1978), and *Bacillus subtilis* (Higo *et al.*, 1980) revealed a large amount of sequence homology among equivalent proteins from these prokaryotes, little or no homology was evident when these r-proteins were compared to the r-proteins from *H. cutirubrum*. What was even more intriguing were the observations that certain r-proteins from the extreme halophile showed substantial sequence homology with eukaryotic r-proteins (Amons *et al.*, 1977; Nazar *et al.*, 1979b). It was not until the publication of the classic paper by Woese and Fox (1977), which postulated three lines of evolutionary descent (the eubacteria, the urkaryotes, and a new kingdom, the archaebacteria), and the experiments of Magrum *et al.* (1978), indicating that the extreme halophiles were members of this archaebacterial kingdom, that the earlier observations on the unusual properties of their ribosomes became more meaningful.

Copyright © 1985 by Academic Press, Inc.
All rights of reproduction in any form reserved.
ISBN 0-12-307208-5

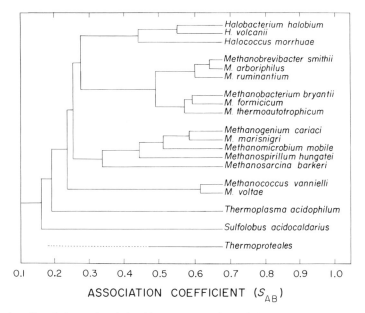

FIG. 1. The phylogenetic relationship among members of the archaebacteria (modified from Woese, 1981), and showing the new order Thermoproteales reported by Tu *et al.* (1982).

The archaebacteria are composed of the methanogens, the extreme halophiles, and the thermoacidophiles (Fox *et al.,* 1980) as outlined in the dendrogram in Fig. 1. Recent studies by Tu *et al.* (1982), based on DNA–rRNA hybridization experiments, suggest that the archaebacteria can be divided into two main subgroups. One group consists of the methanogens, the extreme halophiles, and possibly *Thermoplasma,* while the other group contains the thermoacidophiles, *Sulfolobus,* and a new order, Thermoproteales (Zillig *et al.,* 1981).

In this review, the properties, structure, and evolution of the archaebacterial ribosomes and their protein components will be discussed. The archaebacterial rRNAs are reviewed by Fox in Chapter 5.

II. Archaebacterial Ribosomes

A. GENERAL PROPERTIES

The most definitive study on the properties of an archaebacterial ribosome is the early work by Bayley and Kushner (1964; Bayley, 1966) on the *H. cu-*

tirubrum ribosomes. The ribosome is a 70 S structure composed of two subunits with sedimentation values of 52 S and 31 S. Studies by Douglas *et al.* (1980) on ribosomes from methanogens give similar S values, indicating that the archaebacterial ribosomes are similar in size to the eubacterial ribosomes and smaller than the 80 S eukaryotic ribosome.

The rRNA components of the archaebacterial ribosome are similar to those found in eubacterial ribosomes. Each 70 S ribosome contains one molecule of 5 S, 16 S, and 23 S rRNA (Visentin *et al.*, 1972). There is no evidence for a separate 5.8 S rRNA molecule as is found in eukaryotic ribosomes.

Initially, it was thought that the number of r-proteins in the archaebacterial ribosome (*H. cutirubrum*) was similar to that found in eubacteria, approximately 53–54 proteins (Strøm and Visentin, 1973). However, recent studies by Schmid and Böck (1982a) indicate that some archaebacteria such as *Methanococcus vannielii* and *Sulfolobus acidocaldarius* may contain as many as 60–64 r-proteins. These extra proteins appear to be located mainly in the 30 S subunit. Other methanogens such as *Methanobacterium thermoautotrophicum, M. bryantii, Methanobrevibacter arboriphilus,* and *Methanospirillum hungatei* have the normal, eubacterial-like number of r-proteins (53–54). As these authors point out, proteolysis, enzymatic modification, or coisolation of non-r-proteins could lead to an elevated number of r-proteins. They are now in the process of purifying the r-proteins from the 30 S subunit of *Sulfolobus* to resolve this question (Schmid and Böck, 1982b).

Similar results have been reported by Londei *et al.* (1983) for the ribosomal subunits of the thermoacidophile *Caldariella acidophila* (i.e., *Sulfolobus* sp.). The 30 S ribosomal subunit from this archaebacterium contains 28 proteins with a combined mass of 0.6×10^6 daltons compared with 0.35×10^6 daltons for the eubacterial 30 S subunit. Although no increase in the number of r-proteins was detected in the 50 S subunit the number-average molecular weight of these proteins was greater than those from the 50 S subunit of eubacteria. These ribosomes, therefore, are intermediate in size between those of eubacteria and eukaryotes.

An additional property of the extreme halophilic ribosomes is that they require at least $3.4 \ M \ K^+$ and $0.1 \ M \ Mg^{2+}$ to stabilize their structure (Bayley and Kushner, 1964). Under ionic conditions that are normally used to study eubacterial and eukaryotic ribosomes, the halophilic ribosomes dissociate and lose most of their proteins along with their 5 S rRNA. By controlling the ion concentration, select groups of protein (Strøm *et al.*, 1975a) and certain ribosomal domains can be released from the ribosome (see Section III). It should be noted that Bayley and Griffiths (1968) reported that *in vitro* protein synthesis in *H. cutirubrum* also requires high concentrations of salts ($3.4 \ M$ KCl, $1.0 \ M$ NaCl, $0.4 \ M$ NH$_4$Cl, and 40 mM Mg^{2+}).

B. Electron Microscopy Studies

1. 30 S Ribosomal Subunit

Although the archaebacterial and eubacterial ribosomes show similar sedimentation coefficients (70 S) in the ultracentrifuge, recent EM studies by Lake *et al.* (1982) indicate the structure of the 30 S ribosomal subunits from archaebacteria show unique features and are significantly different from the 30 S subunits of eubacteria and the 40 S subunits of eukaryotes (Fig. 2). The archaebacterial 30 S subunit contains an additional structure, absent in the eubacterial 30 S subunit, called the archaebacterial "bill." The eukaryotic 40 S subunit (which also shows such a "bill") in turn contains a further feature, the eukaryotic "lobe" (see diagrams in Fig. 3A). Within the three evolutionary lines of descent, the structure of the ribosomal subunit is remarkably constant. For example, the methanogens, the extreme halophiles, and the thermoacidophiles all show the archaebacterial bill. This conservation of structure within each lineage implies that the ribosome structure has great evolutionary stability, and it provides morphological data in support of the theory of Woese and Fox (1977) for three evolutionary lines of descent.

The nature of the archaebacterial bill is still unknown. Indirect evidence suggests it may be involved in the factor-dependent steps of protein synthesis. In the 70 S ribosome, the region corresponding to the bill is adjacent to the L7/L12 stalk in the 50 S subunit (see Fig. 3B). In the eukaryotic 40 S subunit, this bill has been shown by immune electron microscopy (Bommer *et al.*, 1980) to contain several proteins and this may be the location of the additional proteins that appear to be present in certain archaebacterial ribosomes. The data from Schmid and Böck (1982a) indicate these extra proteins are mainly in the 30 S subunit.

2. 50 S Ribosomal Subunit

EM studies on the archaebacterial 50 S subunit indicate that it contains many of the features present in the eubacterial subunit (Henderson *et al.*, 1984). For example, it contains the stalk that in eubacteria has been shown to contain the ribosomal "A" protein domain (Strychrz *et al.*, 1978). However, as was the case with the 30 S subunits, the 50 S subunits of archaebacteria also contain unique features.

III. Archaebacterial Ribosomal Domains

In recent years, much effort has gone into locating the r-proteins and specific areas of the rRNA within the ribosome and into determining the function of the various operationally defined domains (see Chambliss *et al.*, 1980, for general

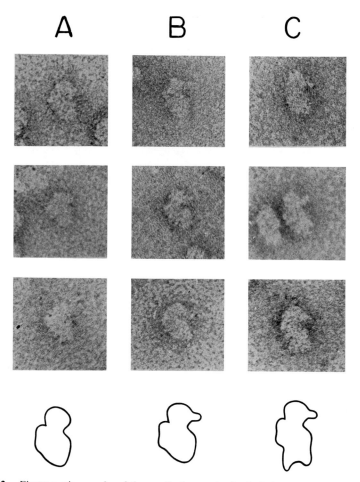

FIG. 2. Electron micrographs of the small ribosomal subunit (Lake *et al.*, 1982). Column A contains (in decreasing order) the eubacterial 30 S subunit from *Thermus aquaticus* (gram-negative thermophile), *Bacillus stearothermophilus* (gram-positive thermophile), *Spinachia oleracea* chloroplasts and a generalized profile of the eubacterial 30 S subunit. Column B contains (in decreasing order) the archaebacterial 30 S subunit from *Methanobacterium thermoautotrophicum, Halobacterium cutirubrum, Sulfolobus acidocaldarius,* and a generalized profile of the archaebacterial 30 S subunit. Column C contains (in decreasing order) the eukaryotic 40 S subunit from *Saccharomyces cerevisiae* (yeast), *Triticum aestivum* (wheat), *Rattus rattus* (rat), and a generalized profile of the eukaryotic 40 S subunit.

discussion). One approach to this problem has been to isolate the various domains and to study the nature of the interactions between the various r-proteins and rRNA components in this region of the ribosome. Two such domains have been studied in some detail in the archaebacterial ribosome. These are the ribosomal "A" protein domain and the 5 S rRNA–protein domain.

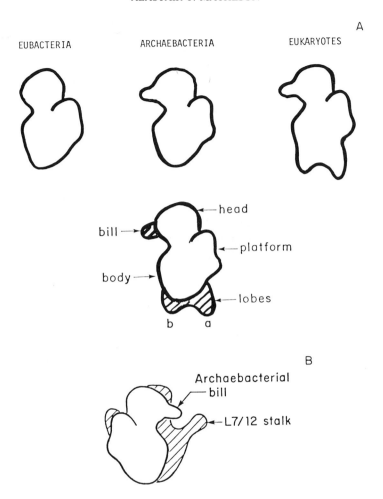

Fig. 3. (A) Schematic representation of the small ribosomal subunits showing the regions present in the three kingdoms. The archaebacterial bill and the eukaryotic lobes (a and b) are shown in diagonal stripes. (From Lake *et al.*, 1982.) (B) Schematic representation of the 70 S archaebacterial ribosome. (From Lake *et al.*, 1982.)

A. Ribosomal "A" Protein Domain

1. Properties of the Eubacterial "A" Protein Domain

A great deal is known of the structure and properties of the eubacterial "A" protein domain. This domain has been isolated from *E. coli* (Pettersson *et al.*,

1976) and from *B. stearothermophilus* (Marquis and Fahnestock, 1980). The complex consists of four copies of protein L7/L12 and one copy of protein L10. In *E. coli*, the complex binds to the 5' region of 23 S rRNA via protein L10, and this binding is stimulated by another r-protein L11 (Dijk *et al.*, 1979). Although the binding site on 23 S rRNA for r-protein L11 has been isolated and sequenced (Schmidt *et al.*, 1981), no data have been reported for the sequence of the L10 binding site.

The C-terminal portion of *E. coli* protein L7/L12 has been crystallized and the three-dimensional structure of this region determined (Lenjonmarck *et al.*, 1980). Recently the L7/L12–L10 complex from *B. stearothermophilus* has been crystallized (A. Liljas, personal communication), and its three-dimensional structure should soon be available.

2. PROPERTIES OF THE ARCHAEBACTERIAL "A" PROTEIN DOMAIN

The ribosomal "A" protein domain has been isolated from the 50 S ribosomal subunit of *H. cutirubrum* (Liljas *et al.*, in preparation). Like its counterpart in eubacterial cells, the complex contains four copies of r-protein HL20 and one copy of HL11. The complex is released from the 50 S subunit by decreasing the Mg^{2+} concentration to 0.3 mM but retaining the high K^+ (3.4 M). When the K^+ concentration is lowered to 100 mM, the 4:1 complex dissociates into a 2:1 complex of HL20 and HL11 with the release of two molecules of HL20. The 2:1 complex is extremely stable, and efforts to dissociate it further using denaturing agents, such as sodium dodecyl sulfate (SDS), have not been successful.

From amino acid sequence data on HL20 and HL11, it is evident that HL20 is equivalent to the eubacterial r-protein EL12 (Oda *et al.*, 1974), whereas HL11 is equivalent to EL11 rather than EL10, the other component in the eubacterial complex (Liljas *et al.*, in preparation). This would indicate that although a 4:1 complex is present in both the eubacterial and archaebacterial 50 S subunits and both ribosomes show the putative L7/L12 stalk, some changes in the topography of this domain have occurred in that the complex is bound via r-protein L10 to the 23 S rRNA in *E. coli* but via L11 in *H. cutirubrum*. We have recently isolated the putative components of the "A" protein domain from *Sulfolobus* (W. J. Garland and A. T. Matheson, unpublished data), and sequence data indicate the complex consists of L12 and L11 in this archaebacterium as well. The structure of these proteins are discussed in Section IV.

Although the stalk in eubacterial 50 S subunits has been identified by immunoelectron microscopy, similar studies have not been carried out on the archaebacterial ribosome. We are now preparing the appropriate antibodies for determining the location of the archaebacterial L12 and L11. In addition, studies are underway to determine the binding site on 23 S rRNA of the "A" domain in archaebacteria.

B. 5 S rRNA–PROTEIN DOMAIN

Both the 5 S rRNA and the 5 S rRNA–binding proteins have proven to be valuable phylogenetic probes (Matheson *et al.*, 1980b). The structure of the 5 S rRNA and the binding sites for the 5 S rRNA–binding proteins will not be discussed in this chapter; they are reported elsewhere in this volume (see Chapter 5 by Fox). In this section I will restrict my remarks to the general properties of the 5 S rRNA protein complex and their 5 S rRNA–binding proteins. The structure of these proteins will be discussed in Section IV.

1. PROPERTIES OF THE EUBACTERIAL AND EUKARYOTIC 5 S rRNA–PROTEIN COMPLEXES

The 5 S rRNA–protein complex from *E. coli* has been extensively studied, and the complete structure of the three proteins that specifically bind to the 5 S rRNA—EL5, EL18 and EL25—has been determined. The complex in *B. stearothermophilus* contains only two proteins—BL5, which is equivalent to EL5, and BL22, which is equivalent to EL18 (Erdmann, 1976).

In eukaryotes, such as rat liver (Blobel, 1971) or yeast (Nazar *et al.*, 1979b), the 5 S rRNA–protein complex contains only a single protein with a molecular weight of 38,000, close to the sum of the molecular weights of the eubacterial 5 S rRNA-binding proteins.

2. PROPERTIES OF THE ARCHAEBACTERIAL 5 S rRNA–PROTEIN COMPLEX

Although the primary structure of a large number of archaebacterial 5 S rRNA molecules is now known (see Fox, Chapter 5), the only archaebacterial 5 S rRNA–protein complex that has been studied in any detail is that from *H. cutirubrum* (Matheson *et al.*, 1980b). The 5 S rRNA–protein domain, like the ribosomal "A" protein domain, can be released from the *H. cutirubrum* 50 S subunit by lowering the Mg^{2+} content of the extraction buffer, while maintaining the K^+ concentration at 3.4 M or greater (Smith *et al.*, 1978). The purified 5 S rRNA–protein complex contains a small but significant amount of the r-proteins present in the "A" protein domain, suggesting these two domains may interact *in situ*.

If the K^+ concentration is also lowered, the 5 S rRNA–protein complex is destabilized and dissociates. It is likely that this is due, at least in part, to the conformational changes that take place in the 5 S rRNA–binding proteins under conditions of lower K$^+$ concentrations (Willick *et al.*, 1978b). The two proteins present in the 5 S rRNA–protein complex, HL13 and HL19, appear, from sequence and other chemical data, to be equivalent to the eubacterial 5 S rRNA–binding proteins, EL18 and EL5, respectively (Nazar *et al.*, 1979a; Willick *et al.*, 1979). The structure of these halophilic proteins will be discussed in Section IV.

IV. Archaebacterial Ribosomal Proteins

A. General Properties

1. Acidic Nature of the Ribosomal Proteins in Many of the Archaebacterial Ribosomes

As mentioned above, one of the most unusual properties noted by Bayley and Kushner (1964) in their study of the ribosomes from the extreme halophile was the preponderance of acidic r-proteins. Whereas most of the r-proteins in eubacterial and eukaryotic ribosomes are basic when fractionated on a Kaltschmidt–Wittmann (1970) two-dimensional polyacrylamide gel, almost all the *H. cutirubrum* r-proteins are acidic under these conditions (Fig. 4). It was thought at the time that this was a unique property of the extreme halophilic ribosome and that it reflected the high internal salt concentrations (over 5 *M*) present in these cells. As Bayley (1979) observed, the halophilic r-proteins are much more polar than the equivalent proteins from other sources. Because they have fewer hydrophobic groups, it was assumed that they require the higher K$^+$ concentration for stability. As will be discussed in Section IV,A,3, the r-proteins from the extreme halophiles are denatured at lower K$^+$ concentrations.

It is now evident that the ribosomes from certain methanogens also contain a large number of acidic proteins (Douglas *et al.*, 1980; Matheson and Yaguchi, 1981). For example, in the *Methanobrevibacter arboriphilus* ribosome, 65% of the r-proteins are acidic, whereas in *Methanococcus vannielii*, this value drops to 30% (Douglas *et al.*, 1980). These authors also noted that there appeared to be a correlation between the acidity of the methanogenic r-proteins and the phylogenetic relationship of these organisms. Our studies on the thermoacidophiles indicate that these ribosomes contain the least number of acidic r-proteins of the archaebacterial ribosomes studied, and the charges on these r-proteins are very similar to those found in eubacterial ribosomes. The slight rearrangement of the phylogenetic "tree" of Fox *et al.* (1980), as shown in Fig. 1, lists the archaebacteria in order of decreasing amounts of acidic r-proteins in their ribosome.

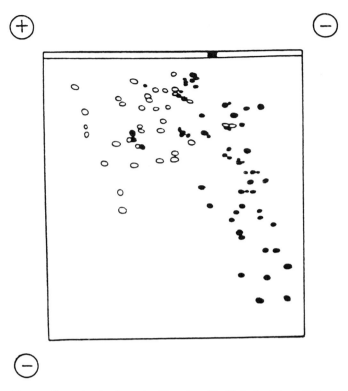

FIG. 4. The two-dimensional polyacrylamide gel profile (Kaltschmidt and Wittmann, 1970) of the 70 S ribosomal proteins from the eubacterium *E. coli* (●) and the archaebacterium *H. cutirubrum* (○). (From Matheson *et al.*, 1978.)

The presence of large numbers of acidic r-proteins in certain methanogens indicates the extreme halophiles are not unique in this respect. These results indicate that the presence of a large number of acidic r-proteins might not be correlated to the high internal K^+ concentration that is present in the extreme halophile, as suggested above. However, internal salt measurements on some methanogens, carried out by Sprott and Jarrell (1981), indeed suggest such a relationship does hold. *Methanobacterium thermoautotrophicum*, which had about 50% of its r-proteins as acidic proteins (Douglas *et al.*, 1980), had an internal cation level 4- to 5-fold higher than that found in *Methanospirillum hungatei*, a methanogen with a much lower amount of acidic r-proteins. Further studies (Jarrell *et al.*, 1984) indicate there is a direct relationship between the internal salt concentration and the number of acidic r-proteins found in the archaebacterial ribosome. For example, *Methanobrevibacter arboriphilus*, the methanogen with the largest number of acidic r-proteins, also contains the highest internal K^+ concentration (860 mM). The dendrogram, shown in Fig. 1, lists the archaebacteria in order of decreasing amounts of acidic r-protein *and* decreasing concentrations of internal K^+.

These internal salt data suggest that the ancestral cell for the extreme halophiles and the group of methanogens, which include *Methanobacterium thermoautotrophicum* and *Methanobrevibacter arboriphilus*, was also halophilic (Matheson and Yaguchi, 1982; Jarrell *et al.*, 1984). One line evolved into the extreme halophiles, whereas the line that evolved into the *Methanobacterium* and *Methanobrevibacter* strains still contains some of these halophilic properties— elevated internal concentrations of K^+ and larger number of acidic r-proteins than found in other methanogens (Fig. 1).

2. IMMUNOLOGICAL STUDIES

Although considerable immunological data are available on the cell wall (Conway de Macario *et al.*, 1982) and RNA polymerases (see Zillig, Chapter 11) of archaebacteria only limited immunological data is currently available on the r-proteins from this group of organisms. Visentin *et al.* (1979) reported that antisera to the ribosomal "A" protein from *H. cutirubrum* (HL20) cross-reacts with the eukaryotic "A" protein from yeast or rat liver but not with the equivalent protein from eubacterial sources.

Using antibodies directed against ribosomes from various methanogens Schmid and Böck (1981) found that antisera against 70 S ribosomes from *Methanobacterium bryantii* gave a strong cross-reaction with the total 70 S r-protein fraction from *Methanobacterium thermoautotrophicum* and *Methanobrevibacter arboriphilus* and a weak reaction from *Methanosarcina barkeri*, *Methanococcus vannielii* and the extreme halophile *Halobacterium halobium*. No cross-reaction was observed between this antisera and total r-proteins from *Sulfolobus acidocaldarius*, *Saccharomyces cerevisiae*, or eubacteria.

Immunoprecipitation experiments gave similar results and also showed a strong reaction between the *Methanobacterium bryantii* antiserum and the 70 S ribosome of *Methanobacterium formicicum*. Schmid and Böck also report that antisera against *Methanosarcina barkeri* ribosomes did not cross-react with ribosomes from the methanogens listed above. A dendrogram, based on the quantitative immunoprecipitation data, was similar to that reported by Fox *et al.* (1980) for the 16 S rRNA data. These results indicate *Methanosarcina barkeri* and *Methanococcus vannielii* are only distantly related to other methanogens and that *Sulfolobus* forms a separate subgroup. Similar results were obtained by Stetter *et al.* (1981) on immunological studies on RNA polymerases from archaebacteria where antisera to the polymerase from *Methanobacterium thermoautotrophicum* did not cross-react with the enzyme from *Methanosarcina barkeri* or *Methanococcus vannielii*. It also did not cross-react with the enzyme from the extreme halophile or thermoacidophiles.

3. STABILITY OF RIBOSOMAL PROTEINS FROM *Halobacterium cutirubrum:* K+ REQUIREMENTS

The r-proteins from the extreme halophile are sensitive to changes in salt concentration. Circular dichroism spectral analysis indicates the ribosomal "A" protein from *H. cutirubrum* undergoes a pronounced change in its secondary structure when the K^+ concentration is decreased from 3 to 1 M or lower as shown in Fig. 5 (Willick *et al.*, 1978a). The α-helix decreases from about 40 to about 18% as the protein is denatured. There is a well-defined transition at about 1.5 M KCl. It is of interest that the equivalent protein in *Methanobacterium thermoautotrophicum* shows a secondary structure expected of a moderate halophile (Gudkov *et al.*, 1984).

Similar structural changes were also observed with HL13 and HL19, the two 5 S rRNA-binding proteins (Willick *et al.*, 1978b). Again the midpoint of the

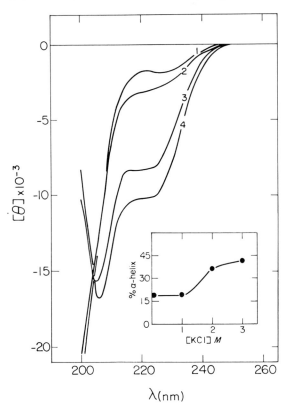

FIG. 5. CD spectrum of the ribosomal "A" protein (HL20) from *H. cutirubrum* in 2 m*M* potassium phosphate, pH 7.6 containing 0.1 M (1), 1 M (2), 2 M (3), and 4 M (4) KCl. The insert shows the effect of KCl concentration on the α-helical content of the protein. (From Willick *et al.*, 1978a.)

transition is at 1.5 M KCl. It was also noted that the 5 S rRNA–protein complex shows a similar dissociation as the K^+ concentration is lowered. The data suggest the destabilization of the complex is correlated with the effect of the K^+ on the secondary structure of the proteins. It should be noted that the denaturation of these proteins appears to be reversible, and the 5 S rRNA–protein complex can reassociated when the denatured proteins (in 6 M urea and 60 mM K^+) are dialyzed into 3.4 M KCl in the presence of the 5 S rRNA.

B. PRIMARY STRUCTURE

1. RIBOSOMAL PROTEINS FROM *Halobacterium cutirubrum*

a. N-terminal Amino Acid Sequences. A large number of the r-proteins from the 30 S (Yaguchi *et al.*, 1982) and 50 S (Matheson *et al.*, 1984) ribosomal subunits of the extreme halophile *H. cutirubrum* have been purified and their N-terminal amino acid sequence determined. This includes the proteins present in the ribosomal "A" protein domain and the 5 S rRNA protein domain. These specific proteins will be discussed later. The amino acid composition of nine of these proteins is shown in Table I. The acidic nature of most of these proteins (except for two basic proteins HL-B3 and HL-B4) is evident from the large number of aspartate and glutamate residues present in these proteins.

The N-terminal sequence of 10 r-proteins has been determined from the 30 S subunit of *H. cutirubrum* (Yaguchi *et al.*, 1982). Four of these proteins have an N-terminal alanyl group, an amino acid that earlier was shown to be the predominant N-terminal residue among the 30 S r-proteins from *H. cutirubrum* (Matheson *et al.*, 1975). Twelve r-proteins have so far been partially sequenced from the 50 S subunit (Matheson *et al.*, 1984) and five of these proteins also have an alanyl N-terminal residue.

Several of the acidic r-proteins show an unusual number of basic amino acids, especially arginyl residues, in the N-terminal region (Fig. 6). For example, the first 23 residues of the 5 S rRNA-binding protein, HL13, contains seven arginine residues, over half the arginine content of this protein. This N-terminal region also includes a tetraalanyl sequence which is conserved in the equivalent protein from eukaryotes (see below). Similarly, the basic regions in the N-terminal portion of HS11 appear to be homologous to a similar region in protein S11 from *E. coli* and *B. stearothermophilus* (Matheson *et al.*, 1978).

Proteins HL-B3 and HL-B4 (Table I) belong to a small group of basic r-proteins present in *H. cutirubrum*. The structure and function of these proteins is currently under investigation.

b. Comparison with Eukaryotic Ribosomal Protein Sequences. A limited amount of amino acid sequence data is now available on yeast (Otaka *et al.*,

TABLE I

AMINO ACID COMPOSITION OF CERTAIN RIBOSOMAL PROTEINS FROM THE **30 S**[a] AND **50 S**[b]
SUBUNITS OF *H. cutrirubrum* (MOL/100 MOL)

Amino acid	HS5	HS11	HS13	HS14	HS20	HL13	HL16	HL-B3	HL-B4
Asx	16.3	12.1	11.1	12.5	11.4	11.0	13.7	8.2	9.2
Thr	6.9	4.8	6.6	5.2	4.2	6.7	4.2	4.9	5.7
Ser	4.6	5.6	5.6	4.2	6.2	5.5	5.2	4.2	5.0
Glx	12.0	18.9	13.5	17.8	11.4	10.0	15.8	13.6	9.3
Pro	3.4	5.2	4.3	3.2	3.7	5.8	1.7	5.9	7.0
Gly	10.5	6.0	8.4	6.9	11.7	8.7	13.3	8.2	14.4
Ala	9.2	8.8	11.3	14.4	9.8	12.7	9.0	10.3	9.9
Val	11.5	6.1	7.9	12.9	9.8	5.9	9.6	9.1	9.7
Met	0.2	0.7	0.4	0.7	1.5	1.1	0.2	2.4	0.7
Ile	4.2	3.6	4.1	4.7	4.6	3.0	1.6	5.0	3.6
Leu	6.1	7.9	8.7	7.0	6.0	9.0	5.1	5.7	3.8
Tyr	2.7	2.0	2.7	1.2	3.6	3.5	2.0	3.5	0.4
Phe	3.3	1.4	1.5	1.8	3.1	1.6	2.1	2.1	2.2
His	2.2	3.7	1.7	0.9	2.4	3.0	2.3	2.0	3.5
Lys	3.0	5.2	4.2	4.1	3.7	3.2	5.4	6.5	5.4
Arg	4.0	8.9	8.0	2.7	6.7	7.9	8.9	8.4	10.2
Cys	ND[c]	0.0	ND	0.0	ND	0.0	ND	ND	ND

[a] Yaguchi *et al.*, 1982.
[b] Matheson *et al.*, 1984. HL-B3 and HL-B4 are basic proteins purified on CM-cellulose.
[c] ND, not determined.

1982) and on rat liver (Wittmann-Liebold *et al.*, 1979) r-proteins. A computer analysis of this sequence data, using the method of Sankoff (1972) as previously described (Yaguchi *et al.*, 1980), did not reveal any strong pattern of matching between the archaebacterial and eukaryotic small subunit r-proteins (Yaguchi *et al.*, 1982). However, several regions of sequence homologies were observed that were definitely above random expectation. Figure 7 shows two examples of such homology (HS9/YS11 and HS20/YS22).

When the 50 S r-protein sequence data from *H. cutirubrum* was compared to the published sequences from the eukaryotic large subunit r-proteins little homology was evident. It should be emphasized, however, that only a few, mostly partial sequences are available on the r-proteins from eukaryotes; three rat liver and six yeast r-proteins from the 40 S subunit and nine rat liver and four yeast r-proteins from the 60 S subunit. Until more sequence data are available on the eukaryotic r-proteins, it is not possible to make a definitive statement on the amount of homology between archaebacterial and eukaryotic r-proteins. In two cases, where the identity of the protein is known from both sources, significant homology was observed. These two examples, the 5 S rRNA–binding protein HL13 and the ribosomal "A" protein HL20, are discussed later in this section.

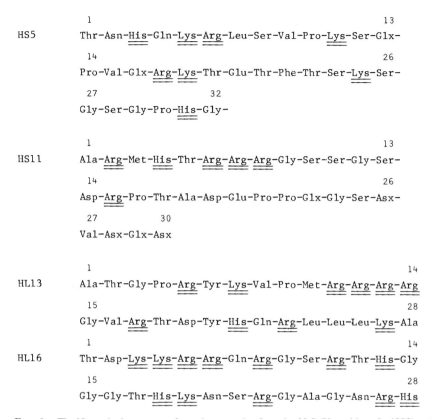

1   13
HS5   Thr-Asn-His-Gln-Lys-Arg-Leu-Ser-Val-Pro-Lys-Ser-Glx-

14   26
Pro-Val-Glx-Arg-Lys-Thr-Glu-Thr-Phe-Thr-Ser-Lys-Ser-

27   32
Gly-Ser-Gly-Pro-His-Gly-

1   13
HS11   Ala-Arg-Met-His-Thr-Arg-Arg-Arg-Gly-Ser-Ser-Gly-Ser-

14   26
Asp-Arg-Pro-Thr-Ala-Asp-Glu-Pro-Pro-Glx-Gly-Ser-Asx-

27   30
Val-Asx-Glx-Asx

1   14
HL13   Ala-Thr-Gly-Pro-Arg-Tyr-Lys-Val-Pro-Met-Arg-Arg-Arg-Arg

15   28
Gly-Val-Arg-Thr-Asp-Tyr-His-Gln-Arg-Leu-Leu-Leu-Lys-Ala

1   14
HL16   Thr-Asp-Lys-Lys-Arg-Arg-Gln-Arg-Gly-Ser-Arg-Thr-His-Gly

15   28
Gly-Gly-Thr-His-Lys-Asn-Ser-Arg-Gly-Ala-Gly-Asn-Arg-His

FIG. 6.   The N-terminal sequence of certain r-proteins from the 30 S (Yaguchi *et al.*, 1982) and 50 S (Matheson *et al.*, 1984) ribosomal subunits of *H. cutirubrum* showing the clustering of basic residues (underlined) in the N-terminal region of these proteins.

c. Comparison with Eubacterial Ribosomal Protein Sequences. A considerable amount of sequence data is available on the eubacterial r-proteins. The extent of homology within the eubacterial kingdom is such that it is possible to establish a one-to-one correspondence between the 30 S subunit r-proteins between *E. coli, B. stearothermophilus,* and *B. subtilis* (Hori *et al.*, 1977; Yaguchi *et al.*, 1978; Higo *et al.*, 1980). An attempt to use sequence homology to establish a one-to-one correspondence between 10 N-terminal sequences of *H. cutirubrum* 30 S r-proteins and the 20 N-terminal sequences available from the equivalent proteins in *E. coli* and *B. stearothermophilus* failed. Apparent sequence homologies could be observed between some archaebacterial and eubacterial proteins (Fig. 8), but any one-to-one correspondence between *H. cutirubrum* and *E. coli* r-proteins was inconsistent with a similar correspondence calculated between *H. cutirubrum* and *B. stearothermophilus* r-proteins (Yaguchi *et al.*, 1982). It would appear, therefore, that *H. cutirubrum* r-proteins do

FIG. 7. Comparison of the amino acid sequences of r-proteins HS9 and HS20 from *H. cutirubrum* (Yaguchi *et al.*, 1982) with the eukaryotic r-proteins YS11 and YS22 from yeast (Otaka *et al.*, 1982).

not show significant homology to eubacterial r-proteins. This observation provides additional support for the concept that the archaebacteria belong to a different kingdom than do the eubacteria.

2. RIBOSOMAL "A" PROTEIN (L12)

a. Archaebacterial Sequence Data. Archaebacterial "A" proteins from an extreme halophile, *H. cutirubrum* (Oda *et al.*, 1974; Duggleby *et al.*, 1975), a methanogen, *Methanobacterium thermoautotrophicum* (Matheson *et al.*, 1980c),

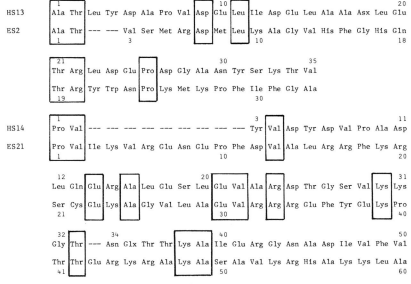

FIG. 8. Comparison of the N-terminal amino acid sequence of r-proteins HS13 and HS14 from *H. cutirubrum* (Yaguchi *et al.*, 1982) with eubacterial r-proteins ES2 (Wittmann-Liebold and Bosserhoff, 1981) and ES21 (Vandekerckhove *et al.*, 1975) from *E. coli*.

and a thermoacidophile, *Sulfolobus* (W. J. Garland, unpublished data), have been characterized and partially sequenced. The amino acid composition of these three "A" proteins is shown in Table II and is compared to the composition of the equivalent proteins from eubacteria and eukaryotes. The archaebacterial protein shows several "eukaryotic-like" features that are absent in eubacterial "A" proteins. For example, these proteins contain considerably less valine than the eubacterial protein and have two tyrosine residues, an amino acid absent in all "A" proteins isolated thus far from eubacteria.

The partial sequences of two archaebacterial "A" proteins are shown in Fig. 9. There is a 56% sequence homology observed between the *H. cutirubrum* and *Methanobacterium thermoautotrophicum* sequences and all the remaining amino acid changes, except one, can be explained by a single point mutation. Recent data on the sequence of the "A" protein from *Sulfolobus* indicate substantial sequence homology between this protein and the "A" proteins from the extreme halophile and methanogen. It is evident, therefore, that the three major subgroups within the archaebacterial kingdom contain a single class of ribosomal "A" protein (archaebacterial type). As will be discussed below, the archaebacterial "A" protein shows different structural features than either the eukaryotic or eubacterial classes of "A" protein.

b. Comparison with Eukaryotic Ribosomal "A" Protein. Sequence data have been published on eukaryotic "A" proteins from four sources: yeast (Itoh, 1980), *Artemia salina* (Amons *et al.,* 1979), rat liver (Lin *et al.,* 1982), and wheat germ (Visentin *et al.,* 1979). The eukaryotic "A" proteins form a well-defined group of "A" proteins (eukaryotic type) but also show a surprising amount of homology to the archaebacterial type (Matheson *et al.,* 1980c). For example, when the N-terminal half of the "A" proteins from yeast and *Methanobacterium thermoautotrophicum* (Fig. 10) are compared, the amount of sequence homology present (42%) is about as great as that found between the two eukaryotes yeast and *A. salina* (43%). This is in marked contrast to the small amount of homology evident when the archaebacterial "A" proteins are compared to the eubacterial type (see Section IV,B,2,c).

The regions of sequence homology between the archaebacterial and eukaryotic "A" proteins is schematically represented in Fig. 11. The two classes of proteins are of similar size (109–112 amino acids), significantly smaller than the eubacterial class (118–125 amino acids). Since the proteins are of similar size and since both the N-terminal (region D, Fig. 11) and the C-terminal portions (region G) have been conserved, the data strongly suggest that the archaebacterial and eukaryotic "A" proteins evolved from a common ancestral molecule (Matheson and Yaguchi, 1982). Other regions that are conserved include an acidic region E, which contains the conserved arginine residue, and an alanine-rich region A, which is followed by a highly charged region H.

TABLE II

Amino Acid Composition of Ribosomal "A" Proteins from Eubacteria, Archaebacteria, and Eukaryotes[a]

Amino acid	Eubacteria			Archaebacteria			Eukaryotes		
	E. coli	B. subtilis	M. lysodeikticus	H. cutirubrum	M. thermoautotrophicum[b]	Sulfolobus[b]	S. cerevisiae	A. salina	Rat liver
Asx	7	6	7	23	8	6	5	9	12
Thr	3	4	4	2	4	4	6	3	0
Ser	6	3	3	4	2	6	9	7	9
Glx	17	23	21	21	28	21	13	18	12
Pro	2	3	2	3	2	3	7	5	5
Gly	8	11	8	9	6	7	14	10	12
Ala	28	21	21	26	29	18	21	22	22
Val	16	14	16	7	7	11	6	3	9
Met	3	0	2	1	3	2	2	4	2
Ile	6	9	7	3	5	7	4	4	5
Leu	8	12	9	8	8	10	9	10	8
Tyr	0	0	0	2	2	2	2	2	2
Phe	2	2	2	1	1	1	3	2	2
His	0	0	0	0	1	1	0	0	0
Lys	13	13	15	1	4	12	7	10	9
Arg	1	1	1	1	2	1	1	1	2
Cys	0	0	0	0	0	ND[c]	0	1	0
Total	120	122	118	112	112	112	109	111	111

[a] Escherichia coli (Terhorst et al., 1973), Bacillus subtilis (Itoh and Wittmann-Liebold, 1978), Micrococcus lysodeikticus (Itoh, 1981), Halobacterium cutirubrum (Strøm et al., 1975b), Sulfolobus (W. J. Garland et al., unpublished data), Saccharomyces cerevisiae (Itoh, 1980), Artemia salina (Amons et al., 1979), rat liver (Lin et al., 1982).

[b] Assume that total residues are 112.

[c] ND, not determined.

	1		5		10			15	
H. cutirubrum	Met Glu Tyr	Val	Tyr Ala Ala	Leu Ile	Leu	Asn	Glu	Ala Asp Glu	Glu
M. thermoautotrophicum	Met Glu Tyr	Ile	Tyr Ala Ala	Met Leu	Leu	His	Thr	Thr Gly Lys	Glu

	17		20		25		30	
H. cutirubrum	Leu Thr	Glu	Asp	Asn	Ile Thr Gly	Val Leu Glu Ala Ala Gly	Val Asp	
M. thermoautotrophicum	Ile Asn	Glu	Glu	Asn	Val Lys Ser	Val Leu Glu Ala Ala Gly	Ala Glu	

	33	35	37	40	45	48
H. cutirubrum	Val	Glu Glu Ser	Arg	Ala Lys Ala Leu	Val	Ala Ala Leu Glu Asp Val
M. thermoautotrophicum	Val	Asp Asp Ala	Arg	Val Lys Ala Leu	Ile	Ala Ala Leu Glu Asp Val

FIG. 9. The sequence homology present in the N-terminal region of the ribosomal "A" protein from the archaebacteria *Halobacterium cutirubrum* and *Methanobacterium thermoautotrophicum* (Yaguchi *et al.*, 1980).

Although an arginine residue (position 38) is present in region E of the "A" protein from yeast and rat liver, lysine has replaced the arginine at this location in wheat germ, whereas in *A. salina,* the basic residue has been replaced by glutamine, and the single arginine residue is found in position 2. The amino acid sequence in positions 34–38 of the *A. salina* protein (Amons *et al.*, 1979) is completely different from the other eukaryotic "A" proteins in this region. The *A. salina* protein has a cysteine residue in position 34, an amino acid absent from all the other "A" proteins sequenced. It is also of interest that two different "A" proteins are present in the *A. salina* ribosome (Amons *et al.*, 1982). These proteins (eL12 and eL12′) show little sequence homology in the N-terminal half of the molecule, whereas the C-terminal portion is highly conserved in the two proteins, suggesting that the N-terminal domain has undergone significant changes in the eukaryote.

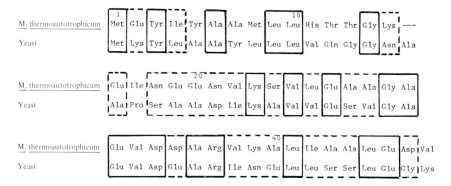

FIG. 10. Sequence homology (first 48 residues) between the ribosomal "A" protein from the archaebacterium *Methanobacterium thermoautotrophicum* (Matheson *et al.*, 1980c) and the eukaryote *Saccharomyces cerevisiae* (Itoh, 1980). The solid lines indicate regions of sequence homology while the dotted lines indicate changes in the amino acid due to a point mutation.

ARCHAEBACTERIA

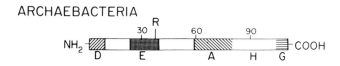

EUKARYOTES

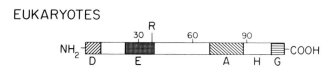

FIG. 11. Schematic outline of the ribosomal "A" protein structure from archaebacteria and eukaryotes indicating regions (D,E,A,H,G) of sequence similarities (see text for details). R indicates location of conserved arginine residue.

c. Comparison with Eubacterial "A" Protein. The complete amino acid sequence of the "A" protein from six eubacteria and partial sequence from three other eubacteria have been determined. These proteins show extensive sequence homology and form an easily recognizable eubacterial class (Matheson *et al.,* 1980a). Recently, the "A" protein from spinach chloroplasts has been sequenced and it is clearly a member of the eubacterial class of "A" proteins (Bartsch *et al.,* 1982).

When the archaebacterial and eubacterial "A" proteins are compared, very little sequence homology is evident (Oda *et al.,* 1974). However by computer analysis, significant homology has been determined. Yaguchi *et al.* (1980) compared the *H. cutirubrum* and *E. coli* "A" proteins by using the single conserved arginine (Arg-73 in EL12 and Arg 37 in HL20) as an equivalent site. The two proteins were then compared in a linear fashion from the N-terminal to C-terminal end using the computer program of Sankoff (1972). With three deletions in HL20, the optimum homology obtained had a standard deviation 3.3 away from the mean, indicating significant homology (Fig. 12). The diagrammatic representation of these results is shown in Fig. 13A. It is obvious that certain areas of similarity such as region A could not be detected by this analysis.

Lin *et al.* (1982) have also carried out a computer analysis on the ribosomal "A" proteins to determine if there is homology between the eubacterial and the archaebacterial–eukaryotic molecules. These authors used two computer programs: ALIGN, which compares the two proteins starting, in each case, at the N-terminal end, and RELATE, which compares each fragment of a given length from one protein with every fragment of a similar length in the second protein. Both programs indicated that the archaebacterial "A" protein was homologous with the eukaryotic protein but not with the eubacterial protein, although the RELATE program did indicate some homology between the eukaryotic and eubacterial proteins. However, Lin *et al.* (1982) found that by transposing the

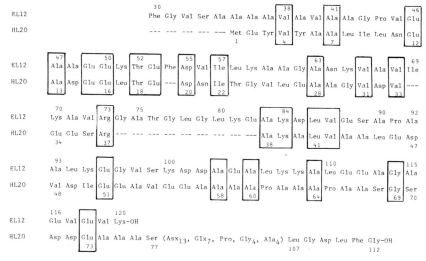

Fig. 12. Sequence homology between the archaebacterial ribosomal "A" protein HL20 (*H. cutirubrum*) and the eubacterial "A" protein EL12 (*E. coli*) as obtained by computer analysis (from Yaguchi *et al.*, 1980). In this analysis, the computer compares each protein in a linear fashion from N-terminal to the C-terminal (Sankoff, 1972), using the conserved arginine (residue 73 in EL12) as the focal point. Some areas of homology, such as the alanine-rich region, would not be detected by this approach.

first 30 residues of the eukaryotic and archaebacterial protein classes to the C-terminal end of the molecule, significant homology was detected between the eubacterial and the other two classes of "A" protein (Fig. 13B).

It is evident from looking at Figs. 13A and B that although both approaches show significant homology between the *H. cutirubrum* and *E. coli* "A" proteins, the two models are not compatible. One possible compromise is shown in Fig. 13C. The "A" protein can be divided into three main segments: I, which contains the alanine-rich region A; II, which contains region E and the conserved arginine residues; and III, which contains the highly charged acidic regions in the C-terminal portion. If the eubacterial protein were split near residues 48 and 84, segment II could be transposed to the N-terminal side of segment I and the C-terminal side of I would join with the N-terminal portion of III to give the archaebacterial–eukaryotic model.

In summary, a large amount of sequence homology is found between the archaebacterial and eukaryotic "A" proteins, and there is good evidence to suggest that the two proteins evolved from a common ancestral protein. However, much less homology is evident when these proteins are compared to the eubacterial "A" protein. Computer analysis indicated significant homology is present, but rearrangement of the protein molecule appears to be required.

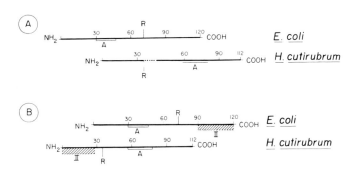

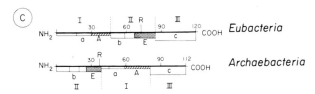

Fig. 13. A schematic comparison of the eubacterial and archaebacterial ribosomal "A" proteins. (A) Alignment of HL20 (*H. cutirubrum*) and EL12 (*E. coli*) by Yaguchi *et al.* (1982) using the computer program of Sankoff (1972) (see Fig. 12). (B) Alignment of HL20 and EL12 by Lin *et al.* (1982) using the RELATE and ALIGN computer programs (the transposition model). (C) Modified transposition model with the ribosomal "A" protein divided into three sections (I, II and III) and rearranged to give maximum homology. Area A indicates the alanine-rich region and E indicates the region containing the conserved arginine residue (R). The regions marked a, b, and c are all highly charged, mainly acidic regions in this protein.

3. Ribosomal Protein HL11

As was discussed above in Section III,A,2, the ribosomal "A" protein domain in the archaebacteria *H. cutirubrum* consists of four molecules of HL20 and one molecule of HL11 (Liljas *et al.*, in preparation). Equivalent proteins have recently been isolated from *S. acidocaldarius* (W. J. Garland, K. A. Louie, and A. T. Matheson, unpublished data). The N-terminal sequence (first 35 residues) of HL11 and the equivalent protein in *Sulfolobus* is shown in Fig. 14. About 60% homology in primary structure is present between these archaebacterial proteins.

In *E. coli*, the ribosomal "A" domain consists of four molecules of EL12 and one molecule of EL10. If we compare the primary structure of the N-terminal portion of the archaebacterial proteins to EL10, no homology is evident. However, when the sequence of these proteins is compared to EL11, the protein located close to the ribosomal "A" domain but not part of the domain in the *E. coli* 50 S subunit, a large amount of homology (about 45%) is observed. In addition, the large number of proline residues clustered in the N-terminal regions of these proteins (positions 15, 17, 18, and 21 in HL11) are also conserved (Fig. 14). This large amount of homology between an archaebacterial and eubacterial protein is surprisingly high, because minimal sequence homology is usually detected between proteins from these two kingdoms (see above). As indicated in Section II,A,2, these data would indicate that the topographies of the "A" domains are different in eubacteria and archaebacteria in that L10 binds the four molecules of L12 to the 50 S subunit in eubacteria while a protein equivalent to L11 in *E. coli* plays a similar role in archaebacteria.

An interesting question, currently under investigation, is the nature of the 23 S rRNA-binding site for the ribosomal "A" protein domain in archaebacteria. Experiments by Thurlow and Zimmermann (1982) indicate there has been a conservation of rRNA–binding sites between archaebacteria and eubacteria for specific r-proteins. Proteins ES8 and ES15 from *E. coli* will bind to the 16 S rRNA from archaebacteria and the archaebacterial RNA binding sites shows common secondary structures to those found on the *E. coli* 16 S rRNA. Since HL11 is the putative protein for the binding of the archaebacterial "A" protein domain to 23 S rRNA and since the N-terminal region of HL11 shows sequence homology to EL11 rather than to EL10, one might predict the binding site on the archaebacterial 23 S rRNA for HL11 will show common structural features to the 23 S rRNA binding site for EL11 previously determined by Schmidt *et al.* (1981). This is currently under investigation.

4. THE 5 S rRNA–BINDING PROTEINS

The 5 S rRNA–binding proteins from *H. cutirubrum* have been partially sequenced (Smith *et al.*, 1978). As shown in Fig. 15, some sequence homology

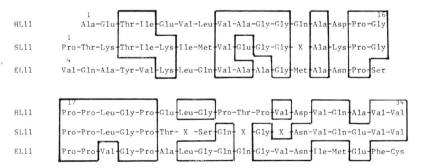

FIG. 14. Homology between the archaebacterial proteins HL11 from *H. cutirubrum* (Liljas *et al.*, 1984) and SL11 from *S. acidocaldarius* (W. J. Garland, K. A. Louie, and A. T. Matheson, unpublished data). The proteins are also compared to the eubacterial protein EL11 from *E. coli* (Dognin and Wittmann-Liebold, 1977).

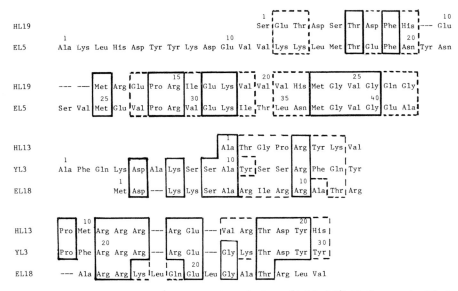

FIG. 15. Sequence homologies between the archaebacterial 5 S rRNA-binding proteins HL13 and HL19 from *H. cutirubrum* (Smith *et al.*, 1978) and the equivalent proteins EL 18 (Brosius *et al.*, 1975) and EL5 (Chen and Ehrke, 1976) from *E. coli* and the single 5 S rRNA-binding protein YL3 (Nazar *et al.*, 1979b) from yeast. Solid lines indicate regions of sequence homology while dotted lines indicate amino acid changes that could result from a point mutation.

is shown between the equivalent eubacterial and archaebacterial 5 S rRNA–binding proteins (Matheson *et al.*, 1980b). The N-terminal portions of *H. cutirubrum* protein HL19 and the equivalent *E. coli* protein EL5 show about 40% homology, whereas HL13 and EL18 show about 30% homology. When HL13, however, is compared to the YL3, the yeast 5 S rRNA–binding protein (Nazar *et al.*, 1979b), about 50% homology is detected.

It would appear, therefore, that there is considerably more homology between the archaebacterial and eukaryotic 5 S rRNA–binding proteins than is the case between the equivalent eubacterial and archaebacterial proteins.

V. Evolutionary Considerations

It is evident from the structural studies on the archaebacterial ribosomes and on their rRNA and r-protein components that these ribosomes show many unique properties, and these data provide additional support for the concept of Woese and Fox (1977) that the archaebacteria are a unique group of organisms, a third evolutionary line of descent. However, these ribosomes also share some of the characteristics of their eubacterial and/or eukaryotic counterparts.

Although the archaebacterial 30 S subunit has a different shape in the electron microscope than that found in eubacteria (Lake *et al.*, 1982) and although there is evidence to indicate that in some archaebacteria the ribosome contains more protein than is present in eubacterial ribosomes (Schmid and Böck, 1982a), nevertheless the archaebacterial and eubacterial ribosomes share many properties. The sedimentation values are identical (70 S), as are the number and approximate size of the rRNA molecules. The organization of their rRNA genes is similar (Hofman *et al.*, 1979) in these organisms, but unlike that seen in eukaryotes. (The latter also have larger rRNAs and an extra rRNA, i.e., the 5.8 S rRNA.) The protein binding studies of Thurlow and Zimmermann (1982) indicate that the rRNA molecules in archaebacteria and eubacteria have similar binding sites although their data also suggest that the eukaryotic ribosome may share this property. Kagramanova *et al.* (1982) have shown the 3' end proximal region of *H. halobium* 16 S rRNA contains a putative Shine–Dalgarno sequence and similar secondary structure in this region to eubacterial 16 S rRNA. Gupta *et al.* (1983) have determined the complete sequence of the 16 S rRNA gene for another extreme halophile *Halobacterium volcanii*. Its secondary structure resembles more closely the secondary structure of eubacterial 16 S rRNA than it does the 18 S rRNA of eukaryotes. It does, however, show some eukaryotic properties. Finally, there is no evidence for phosphorylation of r-proteins (Schmid and Böck, 1982b), a property shown by several eukaryotic r-proteins.

The ribosomal domains discussed in this chapter also show similar features in the archaebacteria and eubacteria. The ribosomal "A" protein domain consists as a 4:1 complex, which is evident in the EM as a stalk in the 50 S subunit (Strycharz *et al.*, 1978). This stalk is also present in the eukaryotic 60 S subunit (Henderson *et al.*, 1984), although no evidence for a 4:1 "A" protein complex in eukaryotes has so far been reported. It should be noted that the "A" protein domains in archaebacteria and eubacteria are not identical in that the "A" protein tetramer is bound to the ribosome via protein L10 in eubacteria and via protein L11 in archaebacteria. The 5 S rRNA protein complex is also similar in that 2–3 proteins are bound to the 5 S rRNA in the archaebacterial–eubacterial complexes, whereas only one, much larger protein, is present in the eukaryotic complex (Nazar *et al.*, 1979b).

However, the archaebacterial protein-synthesizing system shows many eukaryotic features. Archaebacterial tRNAs from *H. cutirubrum* show a slight preference for amino acylation by eukaryotic synthetases than by eubacterial synthetases (Kwok and Wong, 1980). Protein synthesis is initiated by methionine rather than by N-formylmethionine tRNA (White and Bayley, 1972). Although the initiator tRNA from archaebacteria shows unique structural features, *Sulfolobus* tRNA$_i$, shows remarkable resemblance to yeast tRNA$_i$ (Kuchino *et al.*, 1982). Elhardt and Böck (1982) have shown, using an *in vitro* polypeptide synthesis system from methanogens, that the archaebacterial ribosomes lack binding sites for a number of eubacterial 70 S antibiotic inhibitors such as chloramphenicol and streptomycin but possess sites for other 70 S inhibitors. Similarly, they have binding sites for certain eukaryotic (80 S) inhibitors such as anisomycin but lack sites for others such as cycloheximide. Kessell and Klink (1981) have shown that archaebacteria contain an elongation factor, which is ADP-ribosylated by diphtheria toxin, a reaction that was thought to be unique to eukaryotic systems. Cammarano et al. (1982) have shown the EF-Tu-equivalent factor in the archaebacterium *Caldariella acidophilia* (genus *Sulfolobus*) does not cross-react with antibodies against *E. coli* EF-Tu, and the factor is totally insensitive to pulvomycin and kinomycin, antibiotics that inhibit eubacterial EF-Tu.

In addition, when the structure of the individual ribosomal components is considered, these molecules frequently show structural features significantly closer to their eukaryotic counterparts than to the equivalent structures in eubacteria. (Hori and Osawa, 1979; Matheson et al., 1980b; also see Chapter 5, by Fox). For example, for all archaebacterial proteins that have been studied, the sequence is significantly closer to the equivalent eukaryotic r-protein than to the eubacterial r-protein. In the case of the ribosomal "A" protein, the sequence homology between eukaryotic and eubacterial "A" proteins is greater than between the archaebacterial and eubacterial proteins (Lin et al., 1982). The data also indicate that the archaebacterial and eukaryotic "A" proteins, which show a substantial amount of sequence homology (Fig. 11), have evolved from a common ancestral protein (Matheson et al., 1980c), suggesting this portion of the eukaryotic cell has evolved from an archaebacterial origin (Fig. 16).

What do these data tell us about the evolution of the ribosome? One scheme that has been proposed (Matheson and Yaguchi, 1982) is shown in Fig. 16, which is based on a modification of an earlier proposal by Fox et al. (1980). It is proposed that two cell types evolved from the "common ancestral state" or progenote, rather than from three main lines as proposed by Fox et al. (1980); an ancestral archaebacterial line and an ancestral eubacterial line. Since the 70 S ribosomes in present day archaebacteria and eubacteria share many common features, it is likely that the ancestral cell types also contained 70 S–like ribosomes. The ancestral eubacterial line evolved into the present-day eubacteria, which contains a 70 S ribosome of unique structure (Fig. 3) containing 5 S, 16 S, and 23 S rRNA and 53 r-proteins. The available data on chloroplast ribosomes, such as shape of the ribosome (Lake et al., 1982), and eubacterial nature of the chloroplast ribosomal "A" protein (Bartsch et al., 1982), clearly indicate these ribosomes arose from an eubacterial source. The mitochondrial ribosomes may also have evolved from a eubacterial source, although the evidence is much less certain. [The bovine mitochondrial ribosome, for example, has much smaller rRNA molecules and many more proteins (85–90) than the eubacterial ribosome.

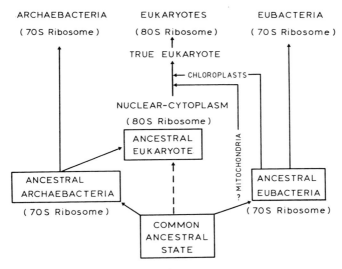

FIG. 16. Proposed scheme, based on a modification of a general scheme proposed by Fox *et al.*, (1980), for the evolution of the ribosomal components of archaebacteria, eubacteria, and eukaryotes. (From Matheson and Yaguchi, 1982.)

In addition, the individual proteins appear to be larger and less basic than eubacterial r-proteins (Matthews *et al.*, 1982).]

The ancestral archaebacterial ribosome evolved into the current 70 S structure containing a 30 S subunit with a unique shape (Fig. 3) and, in certain archaebacteria, more r-protein molecules (Schmid and Böck, 1982b). The 5 S rRNA and the r-protein components of this ribosome also show unique properties but share other properties with the RNA and protein components of the eukaryotic cytoplasmic ribosomes. These shared properties suggest that some of the components of the cytoplasmic ribosomes in eukaryotes may have evolved from the ancestral archaebacterial cell types as outlined in Fig. 16. Significant changes have taken place during the evolution of the modern 80 S ribosome. These changes include the addition of the eukaryotic lobes to the small ribosome subunit of the archaebacteria (Fig. 3). These changes are also reflected in the increased size and number of the r-RNA molecules and the substantial increase in the number of protein molecules. There would also be a change in the organization and structure of the ribosomal RNA genes in the eukaryote.

The translational apparatus may not be the only nuclear–cytoplasmic component that supposedly has arisen from an archaebacterial source. Other cellular components in the archaebacteria, such as the RNA polymerases (Prangishvilli *et al.*, 1982) and the glycoprotein in the cell wall (Mescher and Strominger, 1976) also show eukaryotic-like properties, suggesting that a considerable portion of the nuclear–cytoplasm in the ancestral eukaryote may have evolved from an archaebacterial source.

ACKNOWLEDGMENTS

I would like to thank Drs. A. Böck, S. Osawa, D. Sprott, A. Subramanian, I. Wool, M. Yaguchi, and their colleagues for communication of data prior to publication.

REFERENCES

Amons, R., Van Agthoven, A., Pluijms, W., Möller, W., Higo, K., Itoh, T., and Osawa, S. (1977). A comparison of the amino-terminal sequence of the L7/L12-type proteins of *Artemia salina* and *Saccharomyces cerevisiae*. *FEBS Lett.* **81**, 308–310.

Amons, R., Pluijms, W., and Möller, W. (1979). The primary structure of ribosomal protein eL12/eL12-P from *Artemia salina* 80S ribosomes. *FEBS Lett.* **104**, 85–89.

Amons, R., Pluijms, W., Kriek, J., and Möller, W. (1982). The primary structure of protein eL12'/eL12'-P from the large subunit of *Artemia salina* ribosomes. *FEBS Lett.* **146**, 143–147.

Bartsch, M., Kimura, M., and Subramanian, A. R. (1982). Purification, primary structure, and homology relationship of a chloroplast ribosomal protein. *Proc. Natl. Acad. Sci. U.S.A.* **79**, 6871–6875.

Bayley, S. T. (1966). Composition of ribosomes of an extremely halophilic bacterium. *J. Mol. Biol.* **15**, 420–427.

Bayley, S. T. (1979). Halobacteria—a problem in biochemical evolution. *Trends Biochem. Sci.* **4**, 223–225.

Bayley, S. T., and Griffiths, E. (1968). A cell-free amino acid incorporating system from an extremely halophilic bacterium. *Biochemistry* **7**, 2249–2256.

Bayley, S. T., and Kushner, D. J. (1964). The ribosomes of the extremely halophilic bacterium, *Halobacterium cutirubrum*. *J. Mol. Biol.* **9**, 654–669.

Blobel, G. (1971). Isolation of a 5S RNA-protein complex from mammalian ribosomes. *Proc. Natl. Acad. Sci. U.S.A.* **68**, 1881–1885.

Bommer, U.-A., Noll, F., Lutsch, G., and Bielka, H. (1980). Immunochemical detection of proteins in the small subunit of rat liver ribosomes involved in binding of the ternary initiation complex. *FEBS Lett.* **111**, 171–174.

Brosius, J., Schiltz, E., and Chen, R. (1975). The primary structure of the 5S RNA binding protein L18 from *Escherichia coli*. *FEBS Lett.* **56**, 359–361.

Cammarano, P., Teichner, A., Chinali, G., Londei, P., De Rosa, M., Gambacorta, A., and Nicolaus, B. (1982). Archaebacterial elongation factor insensitive to pulvomycin and kirromycin. *FEBS Lett.* **148**, 255–259.

Chambliss, G., Craven, G. R., Davies, J., Davis, K., Kahan, L., and Nomura, M., eds. (1980). ''Ribosomes: Structure, Function and Genetics.'' University Park Press, Baltimore, Maryland.

Chen, R., and Ehrke, G. (1976). The primary structure of the 5S RNA binding protein L5 of *Escherichia coli* ribosomes. *FEBS Lett.* **69**, 240–245.

Conway de Macario, E., Macario, A. J. L., and Wolin, M. J. (1982). Specific antisera and immunological procedures for characterization of methanogenic bacteria. *J. Bacteriol.* **149**, 320–328.

Dijk, J., Garrett, R. A., and Muller, R. (1979). Studies on the binding of the ribosomal protein complex L7/12-L10 and protein L11 to the 5' one third of the 23S RNA: A functional centre of the 50S subunit. *Nucleic Acids Res.* **6**, 2717–2729.

Dognin, M. J., and Wittmann-Liebold, B. (1977). The primary structure of L11, the most heavily methylated protein from *Escherichia coli* ribosomes. *FEBS Lett.* **84**, 342–346.

Douglas, C., Achatz, F., and Böck, A. (1980). Electrophoretic characterization of ribosomal proteins from methanogenic bacteria. *Zentralbl. Bakteriol., Abt. 1: Orig. [Reihe] C* **1**, 1–11.

Duggleby, R. G., Kaplan, H., and Visentin, L. P. (1975). Carboxyl-terminal sequences of procaryotic ribosomal proteins from *Escherichia coli, Bacillus stearothermophilus,* and *Halobacterium cutirubrum. Can. J. Biochem.* **53**, 827–833.

Elhardt, D., and Böck, A. (1982). An *in vitro* polypeptide synthesizing system from methanogenic bacteria: Sensitivity to antibiotics. *Mol. Gen. Genet.* **188**, 128–134.

Erdmann, V. A. (1976). Structure and function of 5S and 5.8S RNA. *Prog. Nucleic Acid Res. Mol. Biol.* **18**, 45–90.

Fox, G. E., Stackebrandt, E., Hespell, R. B., Gibson, J., Maniloff, J., Dyer, T. A., Wolfe, R. S., Balch, W. E., Tanner, R. S., Magrum, L. J., Zablen, L. B., Blakemore, R., Gupta, R., Bonen, L., Lewis, B. J., Stahl, D. A., Luehrsen, K. R., Chen, K. N., and Woese, C. R. (1980). The phylogeny of prokaryotes. *Science* **209**, 457–463.

Gudkov, A. T., Venyaminov, S. Y., and Matheson, A. T. (1984). Physical studies on the ribosomal "A" protein from two archaebacteria, *Halobacterium cutirubrum* and *Methanobacterium thermoautotrophicum. Can. J. Biochem. Cell. Biol.* **62**, 44–48.

Gupta, R., Lanter, J. M., and Woese, C. R. (1983). Sequence of the 16 S ribosomal RNA from *Halobacterium volcanii,* an archaebacterium. *Science* **221**, 656–659.

Henderson, E., Oakes, M., Clark, M. W., Lake, J. A., Matheson, A. T., and Zillig, W. (1984). A new ribosome structure. *Science* **225**, 510–512.

Higo, K., Itoh, T., Kumanzaki, T., and Osawa, S. (1980). Ribosomal proteins from mesophilic and thermophilic bacteria. *In* "Genetics and Evolution of RNA Polymerase, t-RNA and Ribosomes" (S. Osawa, H. Ozeki, H. Uchida, and T. Yura, eds.), pp. 655–666. Univ. of Tokyo Press, Tokyo.

Hofman, J. D., Lau, R. H., and Dolittle, W. F. (1979). The number, physical organization and transcription of ribosomal RNA cistrons in an archaebacterium: *Halobacterium halobium. Nucleic Acids Res.* **7**, 1321–1333.

Hori, H., and Osawa, S. (1979). Evolutionary change in 5S RNA secondary structure and a phylogenic tree of 54 5S RNA species. *Proc. Natl. Sci. U.S.A.* **76**, 381–385.

Hori, H., Higo, K., and Osawa, S. (1977). Molecular evolution of ribosomal components. *In* "Molecular Evolution and Polymorphism. Proceedings of the Second Taniguchi International Symposium on Biophysics" (M. Kimura, ed.), pp. 240–260.

Itoh, T. (1980). Primary structure of yeast acidic ribosomal protein YPA1. *FEBS Lett.* **114**, 119–123.

Itoh, T. (1981). Primary structure of an acidic ribosomal protein from *Micrococcus lysodeikticus. FEBS Lett.* **127**, 67–70.

Itoh, T., and Wittmann-Liebold, B. (1978). The primary structure of *Bacillus subtilis* acidic ribosomal protein BL9 and its comparison with *Escherichia coli* protein L7/L12. *FEBS Lett.* **96**, 392–394.

Jarrell, K. E., Sprott, G. D., and Matheson, A. T. (1984). *Can. J. Microbiol.* **30**, 663–668.

Kagramanova, V. K., Mankin, A. S., Baratova, L. A., and Bogdanov, A. A. (1982). The 3' terminal nucleotide sequence of the *Halobacterium halobium* 16S rRNA. *FEBS Lett.* **144**, 177–180.

Kaltschmidt, E., and Wittmann, H. G. (1970). Ribosomal proteins. VII. Two-dimensional polyacrylamide gel electrophoresis for fingerprinting of ribosomal proteins. *Anal. Biochem.* **36**, 401–412.

Kessel, M., and Klink, F. (1981). Two elongation factors from the extremely halophilic archaebacterium *Halobacterium cutirubrum*. Assay systems and purification at high salt concentration. *Eur. J. Biochem.* **114**, 481–486.

Kuchino, Y., Ihara, M., Yabusaki, Y., and Nishimura, S. (1982). Initiator tRNAs from archaebacteria show common unique sequence characteristics. *Nature (London)* **298**, 684–685.

Kwok, Y., and Wong, J. T.-F. (1980). Evolutionary relationship between *Halobacterium cutirubrum* and eukaryotes determined by use of aminoacyl-tRNA synthetases as phylogenetic probes. *Can. J. Biochem.* **58**, 213–218.

Lake, J. A., Henderson, E., Clark, M. W., and Matheson, A. T. (1982). Mapping evolution with ribosome structure: Intralineage constancy and interlineage variation. *Proc. Natl. Acad. Sci. U.S.A.* **79**, 5948–5952.

Lenjonmarck, M., Eriksson, S., and Liljas, A. (1980). Crystal structure of a ribosomal component at 2.6Å resolution. *Nature (London)* **286**, 824–826.

Liljas, A., Matheson, A. T., Yaguchi, M., and Christensen, P. (1984). In preparation.

Lin, A., Wittmann-Liebold, B., McNally, J., and Wool, I. G. (1982). The primary structure of the acidic phosphoprotein P2 from rat liver 60S ribosomal subunits: Comparison with ribosomal 'A' proteins from other species. *J. Biol. Chem.* **257**, 9189–9197.

Londei, P., Teichner, A., Cammarano, P., De Rosa, M., and Gambacorta, A. (1983). Particle weights and protein composition of the ribosomal subunits of the extremely thermoacidophilic archaebacterium *Caldariella acidophila*. *Biochem. J.* **209**, 461–470.

Magrum, L. J., Leuhrsen, K. R., and Woese, C. R. (1978). Are extreme halopholes actually "bacteria"? *J. Mol. Evol.* **11**, 1–8.

Marquis, D. M., and Fahnestock, S. R. (1980). Stoichiometry and structure of a complex of acidic ribosomal proteins. *J. Mol. Biol.* **142**, 161–179.

Matheson, A. T., and Yaguchi, M. (1981). The ribosome as a phylogenetic probe. *In* "International Cell Biology 1980–1981" (H. G. Schweiger, ed.), pp. 103–110. Springer-Verlag, Berlin and New York.

Matheson, A. T., and Yaguchi, M. (1982). Evolution of archaebacterial ribosomes. *Zentralbl. Bakteriol., Abt. 1: Orig. C (Reihe)* **3**, 192–199.

Matheson, A. T., Yaguchi, M., and Visentin, L. P. (1975). The conservation of amino acids in the N-terminal position of ribosomal and cytosol proteins from *Escherichia coli, Bacillus stearothermophilus,* and *Halobacterium cutirubrum*. *Can. J. Biochem.* **53**, 1323–1327.

Matheson, A. T., Yaguchi, M., Nazar, R. N., Visentin, L. P., and Willick, G. E. (1978). The structure of ribosomes from moderate and extreme halophilic bacteria. *In* "Energetics and Structure of Halophilic Microorganisms" (S. R. Caplan and M. Ginzburg, eds.), pp. 481–501. Elsevier/North-Holland Biomedical Press, Amsterdam.

Matheson, A. T., Möller, W., Amons, R., and Yaguchi, M. (1980a). Comparative studies on the structure of ribosomal proteins, with emphasis on the alanine-rich, acidic ribosomal 'A' protein. *In* "Ribosomes: Structure, Function, and Genetics" (G. Chambliss, G. R. Craven, J. Davies, K. Davis, L. Kahan, and M. Nomura, eds.), pp. 297–332. University Park Press, Baltimore, Maryland.

Matheson, A. T., Nazar, R. N., Willick, G. E., and Yaguchi, M. (1980b). The evolution of the 5S RNA-protein complex. *In* "Genetics and Evolution of RNA Polymerase, t-RNA and Ribosomes" (S. Osawa, H. Ozeki, H. Uchida, and T. Yura, eds.), pp. 625–637. Univ. of Tokyo Press, Tokyo.

Matheson, A. T., Yaguchi, M., Balch, W. E., and Wolfe, R. S. (1980c). Sequence homologies in the N-terminal region of the ribosomal 'A' protein from *Methanobacterium thermoautotrophicum* and *Halobacterium cutirubrum*. *Biochim. Biophys. Acta* **626**, 162–169.

Matheson, A. T., Yaguchi, M., Christensen, P., Rollin, C. F., and Hasnain, S. (1984). The purification, properties and N-terminal amino acid sequence of certain 50S ribosomal subunit proteins from the archaebacterium, *Halobacterium cutirubrum*. *Can. J. Biochem. Cell Biol.* **62**, 426–433.

Matthews, D. E., Hessler, R. A., Denslow, N. D., Edward, J. S., and O'Brien, T. W. (1982). Protein composition of the bovine mitochondrial ribosome. *J. Biol. Chem.* **257**, 8788–8794.

Mescher, M. F., and Strominger, J. L. (1976). Purification and characterization of a prokaryotic glycoprotein from the cell envelope of *Halobacterium salinarium*. *J. Biol. Chem.* **251**, 2005–2014.

Nazar, R. N., Willick, G. E., and Matheson, A. T. (1979a). The 5S RNA protein complex from an extreme halophile, *Halobacterium cutirubrum:* Studies on the RNA-protein interaction. *J. Biol. Chem.* **254**, 1506–1512.

Nazar, R. N., Yaguchi, M., Willick, G. E., Rollin, C. F., and Roy, C. (1979b). The 5-S RNA binding protein from yeast (*Saccharamyces cervisiae*) ribosomes. Evolution of the eukaryotic 5S RNA binding protein. *Eur. J. Biochem.* **102**, 573–582.

Oda, G., Strøm, A. R., Visentin, L. P., and Yaguchi, M. (1974). An acidic, alanine-rich 50S ribosomal protein from *Halobacterium cutirubrum:* Amino acid sequence homology with *Escherichia coli* proteins L7 and L12. *FEBS Lett.* **43**, 127–130.

Otaka, E., Higo, K., and Osawa, S. (1982). Isolation of seventeen proteins, and amino-terminal amino acid sequences of eight proteins from cytoplasmic ribosomes of yeast. *Biochemistry* **21**, 4545–4550.

Pettersson, I., Hardy, S. J. S., and Liljas, A. (1976). The ribosomal protein L8 is a complex of L7/L12 and L10 *FEBS Lett.* **64**, 135–138.

Prangishvilli, D., Zillig, W., Gierl, A., Biesert, L., and Holz, I. (1982). DNA-dependent RNA polymerases of thermoacidophilic archaebacteria. *Eur. J. Biochem.* **122**, 471–477.

Sankoff, D. (1972). Matching sequences under deletion/insertion constraints. *Proc. Natl. Acad. Sci. U.S.A.* **69**, 4–6.

Schmid, G., and Böck, A. (1981). Immunological comparison of ribosomal proteins from archaebacteria. *J. Bacteriol.* **147**, 282–288.

Schmid, G., and Böck, A. (1982a). The ribosomal protein composition of five methanogenic bacteria. *Zentralbl. Bakteriol., Mikrobiol. Hyg., Abt. 1, Orig. C* **3**, 347–353.

Schmid, G., and Böck, A. (1982b). The ribosomal protein composition of the archaebacteria *Sulfolobus*. *Mol. Gen. Genet.* **185**, 498–501.

Schmidt, F. J., Thompson, J., Lee, K., Dijk, J., and Cundliffe, E. (1981). The binding site for ribosomal protein L11 within 23S ribosomal RNA of *Escherichia coli*. *J. Biol. Chem.* **256**, 12301–12305.

Smith, N., Matheson, A. T., Yaguchi, M., Willick, G. E., and Nazar, R. N. (1978). The 5-S RNA-protein complex from an extreme halophile, *Halobacterium cutirubrum:* Purification and characterization. *Eur. J. Biochem.* **89**, 501–509.

Sprott, G. D., and Jarrell, K. E. (1981). K^+, Na^+, and Mg^{2+} content and permeability of *Methanospirillum hungatei* and *Methanobacterium thermoautotrophicum*. *Can. J. Microbiol.* **27**, 444–451.

Stetter, K. O., Thomm, M., Winter, J., Wildgruber, G., Huber, H., Zillig, W., Jane-Covic, D., König, H., Palm, P., and Wunderl, S. (1981). *Methanothermus fervidus, sp. nov.*, a novel extremely thermophilic methanogen isolated from an Icelandic hot spring. *Zentralbl. Bakteriol., Mikrobiol. Hyg., Abt. 1, Orig. C* **2**, 166–178.

Strøm, A. R., and Visentin, L. P. (1973). Acidic ribosomal proteins from the extreme halophile, *Halobacterium cutirubrum:* The simultaneous separation, identification and molecular weight determination. *FEBS Lett.* **37**, 274–280.

Strøm, A. R., Hasnain, S., Smith, N., Matheson, A. T., and Visentin, L. P. (1975a). Ion effect on protein-nucleic acid interactions: The disassembly of the 50S ribosomal subunit from the halophilic bacterium, *Halobacterium cutirubrum*. *Biochim. Biophys. Acta* **383**, 325–337.

Strøm, A. R., Oda, G., Hasnain, S., Yaguchi, M., and Visentin, L. P. (1975b). Temperature related alterations in the acidic alanine-rich 'A' protein from the 50S ribosomal particle of the extreme halophile, *Halobacterium cutirubrum. Mol. Gen. Genet.* **140**, 15–27.

Strycharz, W. A., Nomura, M., and Lake, J. A. (1978). Ribosomal proteins L7/L12 localized at a single region of a large subunit by immune electron microscopy. *J. Mol. Biol.* **126**, 123–140.

Terhorst, C., Möller, W., Laursen, R., and Wittmann-Liebold, B. (1973). The primary structure of an acidic protein from 50S ribosomes of *Escherichia coli* which is involved in GTP hydrolysis dependent on elongation factors G and T. *Eur. J. Biochem.* **34**, 138–152.

Thurlow, D. L., and Zimmermann, R. A. (1982). Evolution of protein-binding regions of archaebacterial, eubacterial and eukaryotic ribosomal RNAs. *In* "Archaebacteria" (O. Kandler, ed.), p. 347. Fischer, Stuttgart.

Tu, J., Prangishvilli, D., Huber, H., Wildgruber, G., Zillig, W., and Stetter, K. O. (1982). Taxonomic relations between archaebacteria including 6 novel genera examined by cross hybridization of DNAs and 16S rRNAs. *J. Mol. Evol.* **18**, 109–114.

Vandekerckhove, J., Rombauts, W., Peeters, B., and Wittmann-Liebold, B. (1975). Determination of the complete amino acid sequence of protein S21 from *Escherichia coli* ribosomes. *Hoppe-Seyler's Z. Physiol. Chem.* **356**, 1955–1976.

Visentin, L. P., Chow, C., Matheson, A. T., Yaguchi, M., and Rollin, F. (1972). *Halobacterium cutirubrum* ribosomes. Properties of the ribosomal proteins and ribonucleic acid. *Biochem. J.* **130**, 103–110.

Visentin, L. P., Yaguchi, M., and Matheson, A. T. (1979). Structural homologies in alanine-rich acidic ribosomal proteins from procaryotes and eukaryotes. *Can. J. Biochem.* **57**, 719–726.

White, B. N., and Bayley, S. T. (1972). Methionine transfer RNAs from the extreme halophile, *Halobacterium cutirubrum. Biochim. Biophys. Acta* **272**, 583–587.

Willick, G. E., Williams, R. E., and Matheson, A. T. (1978a). Salt-induced conformational changes in, and secondary structure analysis of, halophilic ribosomal 'A' proteins equivalent to EL7/EL12. *FEBS Lett.* **85**, 279–282.

Willick, G. F., Williams, R. E., Matheson, A. T., and Sendecki, W. (1978b). Salt stabilization of a 5S RNA-protein complex from an extreme halophile, *Halobacterium cutirubrum. FEBS Lett.* **92**, 187–189.

Willick, G. E., Nazar, R. N., and Matheson, A. T. (1979). 5S RNA-protein complex from an extreme halophile, *Halobacterium cutirubrum:* Comparative studies on reconstituted complexes. *Biochemistry* **13**, 2855–2859.

Wittmann, H. G., Littlechild, J. A., and Wittmann-Liebold, B. (1980). Structure of ribosomal proteins. *In* "Ribosomes: Structure, Function, and Genetics" (G. Chambliss, G. R. Craven, J. Davies, K. Davis, L. Kahan, and M. Nomura, eds.), pp. 51–58. University Park Press, Baltimore, Maryland.

Wittmann-Liebold, B., and Bosserhoff, A. (1981). Primary structure of protein S2 from the *Escherichia coli* ribosome. *FEBS Lett.* **129**, 10–16.

Wittmann-Liebold, B., Geissler, A. W., Lin, A., and Wool, I. G. (1979). Sequence of the amino-terminal region of rat liver ribosomal proteins S4, S6, S8, L6, L7a, L18, L27, L30, L37, L37a and L39. *J. Supramol. Struct.* **12**, 425–433.

Woese, C. R. (1981). The archaebacteria. *Sci. Am.* **244**, 98–122.

Woese, C. R., and Fox, G. E. (1977). Phylogenetic structure of the prokaryotic domain: The primary kingdoms. *Proc. Natl. Acad. Sci. U.S.A.* **74**, 5088–5090.

Yaguchi, M., Visentin, L. P., Nazar, R. N., and Matheson, A. T. (1978). Structure and thermal stability of ribosomal components from thermophilic bacteria. *In* "Biochemistry of Thermophily" (S. M. Friedman, ed.), pp. 169–190. Academic Press, New York.

Yaguchi, M., Matheson, A. T., Visentin, L. P., and Zuker, M. (1980). Molecular evolution of the alanine-rich, acidic ribosomal A protein. *In* "Genetics and Evolution of RNA Polymerase, t-

RNA and Ribosomes'' (S. Osawa, H. Ozeki, H. Uchida, and T. Yura, eds.), pp. 585–599. Univ. of Tokyo Press, Tokyo.

Yaguchi, M., Visentin, L. P., Zuker, M., Matheson, A. T., Roy, C., and Strøm, A. R. (1982). Amino-terminal sequences of ribosomal proteins from the 30S subunit of the archaebacterium *Halobacterium cutirubrum*. *Zentralbl. Bakteriol., Mikrobiol. Hyg., Abt. 1, Orig. C* **3**, 200–208.

Zillig, W., Tu, J., and Holz, I. (1981). Thermoproteales—a third order of thermoacidophilic archaebacteria. *Nature (London)*, **293**, 85–86.

CHAPTER 8

Elongation Factors

FRIEDRICH KLINK

I. Introduction

The concept of archaebacteria as a third kingdom of organisms has been derived from analyses of ribosomal 16 S RNA (Woese and Fox, 1977; Fox *et al.*, 1980). Beyond that, the protein synthesis machinery appears to be a promising object for phylogenetic comparisons among the three kingdoms. The subunit structure of the ribosome, 5 S rRNA, and ribosomal proteins have been investigated in a variety of archaebacteria (Bayley and Kushner, 1964; Matheson and Yaguchi, 1982; Matheson *et al.*, 1978; Yaguchi *et al.*, 1980; Schmid and Böck, 1981; Visentin *et al.*, 1979). Of the numerous ''soluble'' nonribosomal proteins that cooperate with ribosomes in various stages of peptide synthesis, only the elongation factors of the archaebacteria have been studied (Kessel and Klink, 1980; 1981; Klink *et al.*, 1983; Uschtrin and Klink, 1982). Isolation of these factors makes it possible to compare them among the three kingdoms as has been done for ribosomal structures (Lake *et al.*, 1982; Matheson and Yaguchi, 1981;

Copyright © 1985 by Academic Press, Inc.
All rights of reproduction in any form reserved.
ISBN 0-12-307208-5

Schmid and Böck, 1982; Schmid *et al.*, 1982). Furthermore, functionally active elongation factors being available will allow the study of archaebacterial ribosomal mechanisms in detail.

II. Elongation

A. GENERAL FEATURES

Protein synthesis on ribosomes comprises three distinct stages: chain initiation, elongation, and termination (Weissbach and Pestka, 1977). Elongation adds all amino acids but the first one to a peptide chain, which had been started by the rather complicated initiation reaction. Elongation is a cyclic process (Miller and Weissbach, 1977), each cycle of which attaches one amino acid to the carboxy-terminal end of the growing chain. Elongation is thought to comprise three successive steps (Fig. 1). In the first (decoding) step, codon-directed binding of aminoacyl-tRNA to the ribosomal binding site A takes place. In the second, (synthesis) step, the peptide bond is formed by peptidyl transfer from a peptidyl-tRNA attached to the ribosomal P-site to the amino group of the suceeding aminoacyl-tRNA in the A-site. Finally, in the translocation step a series of

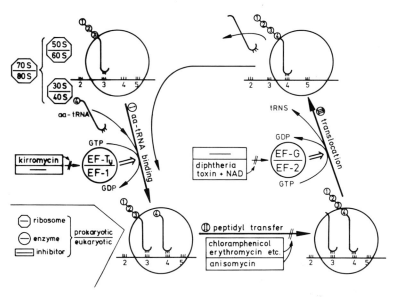

FIG. 1. Schematic diagram of the elongation cycle signifying differences between the two classical kingdoms.

rearrangements on the ribosome follow, which comprise the removal of the now deacylated tRNA, a shift of the newly formed peptidyl-tRNA from the A-site to the P-site, and the concomitant movement of mRNA by one codon triplet (in 5′ direction).

Although peptidyltransferase activity resides in the structure of the large ribosomal subunit (Moore *et al.*, 1975), the decoding and translocation steps involve the catalytic function of cytosolic enzymes, the elongation factors (EF) (Bermek, 1978; Kaziro, 1978). One of these factors catalyzes aminoacyl-tRNA binding by forming a ternary complex with aminoacyl-tRNA and GTP; this complex is subsequently attached to the ribosome. The aminoacyl-tRNA binding factor has been designated as EF-Tu in prokaryotes, as EF-1α in eukaryotes. After the peptidyltransfer has been completed, another elongation factor is bound to the ribosome (again as a complex with GTP) where it is somehow involved in the translocation step. This translocating factor is called EF-G in prokaryotes and EF-2 in eukaryotes. Both of the GTP molecules are split into GDP and P_i in the course of catalytic action of the respective factors. A third elongation factor is required for recycling the tRNA binding factor, by replacement of the tightly bound GDP for GTP. The recycling factor has been designated as EF-Ts in prokaryotes, as EF-1β in eukaryotes.

B. ELONGATION FACTORS FROM THE TWO CLASSICAL KINGDOMS

In all cells studied till now, including chloroplasts and mitochondria having autonomous protein synthesis apparatus, elongation follows the basic scheme outlined above. However, with respect to many structural and functional details of the macromolecules involved in elongation, very clear-cut differences exist between the two groups of translation machineries, the eubacterial and organellar on the one hand, and the (cytoplasmic) eukaryotic on the other. Chapter 7 by Matheson in this volume deals with the differences involving ribosomes per se. The differences among factors of the two groups are also substantial. These differences are most striking in the case of the two translocation factors, EF-G and EF-2: Only the latter one can be ADP-ribosylated and inactivated by reaction with NAD through the catalytic effect of fragment A of diphtheria toxin. In yeast, a novel amino acid "diphthamide," 2-[3-carboxamido-3-(trimethylammonio)propyl]histidine (Van Ness *et al.*, 1980) which lies within a highly conserved sequence, has been shown to serve as acceptor of the ADP-ribose moiety (Brown and Bodley, 1979; Oppenheimer and Bodley, 1981). Elongation factor-G (EF-G), which lacks this domain, is no substrate for the toxin (Bermek, 1978; Collier, 1975; Kessel and Klink, 1982).

The two translocation factor types also differ in ribosome specificity: the eubacterial type will not work in an otherwise eukaryotic system and vice versa.

It cannot be determined whether the specificity has its root in evolutionary structural changes of ribosomes, of factors, or of both.

Eukaryotic EF-2s possess somewhat higher molecular weights, about M_r 96,000 according to sodium dodecyl sulfate–polyacrylamide gel electrophoresis (SDS–PAGE) (Laemmli, 1970; Bermek, 1978; Kaziro, 1978; Kessel and Klink, 1980), than eubacterial EF-G factors (roughly M_r 80,000). Other properties of both factor types have been found to be quite similar.

Differences between aminoacyl-tRNA binding factors are also substantial but less clearly defined than between translocating factors. Eukaryotic EF-1s possess roughly equal affinities for GDP and GTP, whereas their eubacterial analog EF-Tu binds GDP between 50 and 75 times stronger than GTP (Arai *et al.*, 1978; Kaziro, 1978).

Ribosome specificity of these factors is for the most part group specific, but this apparently is not absolute (Graves *et al.*, 1980; Krisko *et al.*, 1969; Slobin, 1981). Elongation factor (EF)-1 from a lower eukaryote (Beck and Spremulli, 1982) will interact with eubacterial ribosomes.

EF-Tu from all eubacteria studied in this respect is inhibited by the antibiotics kirromycin and pulvomycin (Chinali *et al.*, 1977; Wolf *et al.*, 1978), whereas eukaryotic EF-1α is not.

The molecular weights of EF-Tu and EF-1 from various sources lie in the same range with M_r 44,000–54,000 (Bermek, 1978). As far as we know, no elongation factor has a subunit structure. But at least in higher eukaryotes EF-1α is found partially in the form of molecular aggregates with EF-1β, the EF-1 recycling factor, and with a third protein EF-1γ of unknown function (Iwasaki and Kaziro, 1979; Kaziro, 1978). EF-1α, being as labile as eubacterial EF-Tu, is stabilized in the oligomeric organization. Furthermore, the complex EF-1αβγ is able to form higher aggregates described as EF-1H, which possibly represent a storage form of EF-1 (Slobin and Möller, 1976).

EF-Tu and EF-1α are major proteins of the cells. In *Escherichia coli*, EF-Tu accounts for about 5.5% of the total protein (Wittinghofer *et al.*, 1979). The amounts of EF-Ts and EF-G are several times smaller.

III. Archaebacterial Elongation Factors

A. INTRODUCTION

Since elongation factors of eubacteria and eukaryotes show clear-cut differences, the question arises as to whether archaebacteria, a distinct third kingdom, might form a third group with respect to their elongation machinery or whether they could be classed with one of the other groups.

In 1979, M. Kessel in my laboratory observed the inhibition of halobacterial *in vitro* protein synthesis by diphtheria toxin, and we subsequently found ADP-ribosylation of elongation factors to be a common feature of all archaebacteria (Kessel and Klink, 1982). Our attention focused on the ADP-ribosylated domain of archaebacterial EF-2 as well as the ribosome specificity of both archaebacterial EF-1 and EF-2. For direct functional comparisons between factors and ribosomes from different species, the extremely halophilic system was unsuitable because of its very high salt requirements. Therefore, nonhalophilic archaebacterial systems had to be developed, and factors had to be separated and characterized. Our research program is still in its early stages. Therefore, the following survey of archaebacterial elongation factors is necessarily incomplete and often fragmentary.

B. Assay Methods

1. General Remarks

Three methods were chosen from a considerable number of assays that had been developed for eubacterial or eukaryotic elongation factors (Collier, 1975; Iwasaki and Kaziro, 1979; Kaziro *et al.*, 1972; Nirenberg and Leder, 1964): (i) Cell-free polyphenylalanine synthesis, (ii) GDP binding, and (iii) ADP-ribosylation. Some other assay methods measuring only partial functions of the factors will be adapted to our problems in the further course of investigations. For extremely halophilic factors, a number of standard methods cannot be used due to the ionic milieu required.

2. Polyuridylic Acid-Directed Polyphenylalanine Synthesis Assay

A system with artificial mRNA was chosen because the endogenous archaebacterial mRNA, the translation of which could be demonstrated in crude incorporation systems (Uschtrin and Klink, 1982), was lost during ribosome purification. In order to get comparable results from all systems used, poly(U) also was employed for the liver system, though it contained endogenous mRNA in amounts sufficient for elongation assays.

Composition of standard polyphenylalanine synthesis assays and isolation of labeled peptides was performed as described for the systems from *Halobacterium cutirubrum* (Kessel and Klink, 1981) and *Thermoplasma acidophilum* (Klink *et al.*, 1983). For incubation conditions, see legends to the figures.

3. GDP Binding Assay

GDP binding has been known for a long time as simple method for assaying EF-Tu and EF-1α but EF-G and EF-2 from several sources also can be detected in this way (Miller and Weissbach, 1974; Nirenberg and Leder, 1964). Under the conditions used here (Kessel and Klink, 1981; Klink *et al.*, 1983), archaebacterial EF-2 did not bind significant amounts of GDP.

4. ADP-Ribosylation Assay

ADP-ribosylation by incubation with radioactive NAD and diphtheria toxin and subsequent isolation of labeled protein is the simplest assay for eukaryotic EF-2 but naturally unfit for use with EF-G (Hardesty and McKeehan, 1971). Archaebacterial factors react under the same conditions as eukaryotic ones (Kessel and Klink, 1980, 1982), but demand far higher concentrations of toxin (see Fig. 11; Pappenheimer, Jr. *et al.*, 1983). The requirement for high amounts of toxin and longer incubation times is especially pronounced for extremely halophilic factors, there partially due to the concentrated salt milieu unfavorable for enzymatic activity of diphtheria toxin (Kessel and Klink, 1981).

C. Isolation of Archaebacterial Elongation Factors

1. General Remarks

For the present, the research program outlined in Section III,A required the separation of two factors from each organism to be studied, factors that complemented one another in polyphenylalanine synthesis activity. Because we have not applied a special method for detecting Ts activity, we do not know whether partially purified fractions additionally may contain a Ts-like factor if such a protein exists in archaebacteria (see also Section III,D,3).

2. Preparation of Two Factors from *Halobacterium cutirubrum*

Preparation of extremely halophilic factors is complicated by the fact that their stability and activity absolutely depend on high salt concentrations, as is the case for most proteins from these organisms. The purification procedures therefore are restricted to methods which could be adapted to these conditions (Kessel and Klink, 1981). Briefly, we started with Sepharose chromatography of a 150,000 *g* supernatant in the presence of 2.0 *M* ammonium sulfate; the enzymatically active protein eluted by a decreasing gradient of $(NH_4)_2SO_4$ was separated into GDP binding and ADP-ribosylated fractions by gel filtration (see Fig. 2A); the two crude fractions were purified by hydroxyapatite chromatography (Figs 2 B and C). Figure 2 D gives a survey of purification by means of SDS–PAGE according to Laemmli (1970).

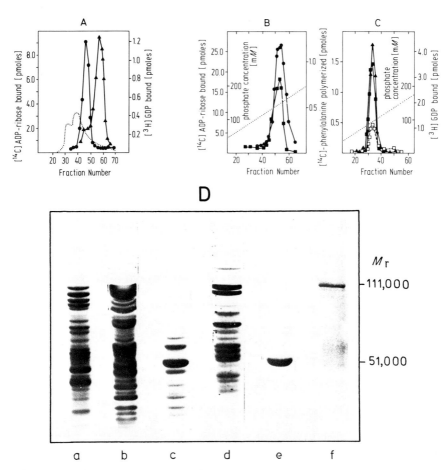

FIG. 2. Purification of two elongation factors from *Halobacterium cutirubrum*. (A) Separation of GDP binding (▲) and ADP-ribosylation (●) activities on Sephadex G-100. Dotted line: Absorbance at 280 nm. (B) Purification of EF-2 [fractions 40–50 from (A)] on hydroxyapatite. (C) Purification of EF-1 [fractions 52–62 from (A)] on hydroxyapatite. Polyphenylalanine synthesis in the presence of the complementary factor (■), and by eluate fractions alone (○). Dotted line: Phosphate concentration. (D) Purification steps documented by SDS–PAGE. a, 150,000 g supernatant; b, ammonium sulfate-Sepharose peak; c, fraction 46 from Sephadex G-100; d, fraction 57 from Sephadex G-100; e, peak fraction from (C); f, peak fraction from (B). (Adapted from Kessel and Klink, 1981.)

The ribosomes required for synthesis assays were isolated in a rather simple way as described by Kessel and Klink (1981) and were not completely factor-free as revealed by the blanks (see legend to Fig. 2 B).

3. Preparation of Two Factors from *Thermoplasma acidophilum* DSM 1728 and from *Methanococcus vannielii* DSM 1224

From these nonhalophilic archaebacteria GDP binding and ADP-ribosylating activities could be largely separated by ion exchange chromatography of the 150,000 g supernatants on DEAE cellulose (Klink *et al.*, 1983; Uschtrin and Klink, 1982). Elution diagrams are shown in Figs. 3A and B. SDS–PAGE analyses revealed that the active fractions do not consist of pure elongation factors (Fig. 4). However, according to the results of GDP binding and ADP-ribosylation assays and to polyphenylalanine synthesis under standard conditions (see legend to Fig. 5), the two factors are not significantly contaminated with one another. But at low concentrations of monovalent cations, synthesis assays showed that crude EF-2 contained small amounts of the complementary factor (see Section III,D,3). It is remarkable that application of the same chromatographic method to cell supernatant from rat liver leads to a completely analogous picture (Fig. 3C), whereas eubacterial factors under comparable conditions are eluted at considerably higher salt concentrations (Gordon *et al.*, 1971).

Again, further purification of factors was achieved with hydroxyapatite, as documented in Fig. 4 for *T. acidophilum*. Ribosomes from *T. acidophilum* and *M. vannielii* were isolated and purified by a method that combined high-salt washing and density gradient centrifugation. Ribosomes isolated in this way were apparently factor-free.

4. Factors from *Sulfolobus* Strains

DEAE-cellulose chromatography under the conditions successfully applied with *Thermoplasma* and *Methanococcus* supernatants did not satisfactorily separate GDP binding and ADP-ribosylation activities from 150,000 g supernatants of *S. solfataricus* or *S. acidocaldarius*. In both cases, the two activities were eluted with the break-through fractions and overlapped one another. At higher pH values (8.0) EF-2 from both strains was absorbed on the column (F. Klink, unpublished results).

Fig. 3.   Separation of GDP binding activity (●) and ADP-ribosylation activity (▲) from three 150,000 g supernatants on DEAE-cellulose (Whatman DE 23). Dotted line: Absorbance at 280 nm; dashed line: KCl gradient. Supernatant sources: *T. acidophilum* (A), *M. vannielii* (B), and rat liver (C).

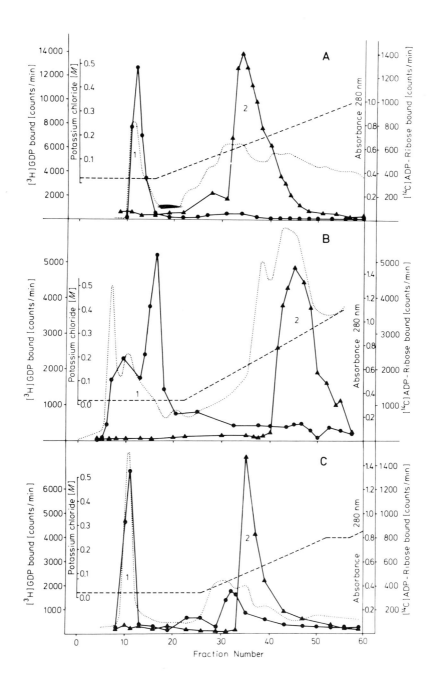

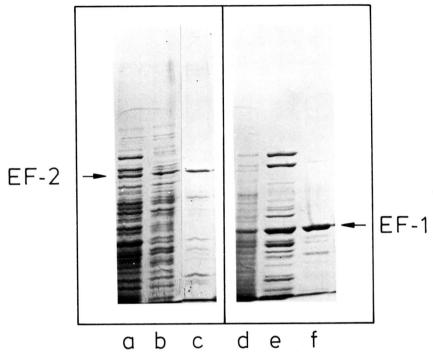

Fig. 4. Purification of elongation factors from *T. acidophilum* documented by SDS–PAGE. a, 150,000 *g* supernatant; b, ADP-ribosylated peak from Fig. 3A; c, protein eluated at 150 m*M* phosphate after hydroxyapatite chromatography of this peak; d, 150,000 *g* supernatant; e, GDP binding peak from Fig. 3A; f, protein eluated at 90 m*M* phosphate after hydroxyapatite chromatography of e.

D. Characterization of Cell-Free Polyphenylalanine Synthesizing Systems from Archaebacteria

1. General Remarks

Protein synthesis is a basic process in growing cells. Therefore, it is of great interest to see to what extent optimal conditions for mRNA translation fit the internal milieu of the living cell, especially during logarithmic growth. Investigations in this respect have been made with only a few archaebacteria, in such cases where extreme growth conditions suggested dramatic differences between external and cell-internal milieu. Although it must be taken into account that *in vitro* polyphenylalanine synthesis is not fully comparable to naturally initiated translation, estimation of requirements for optimal protein synthesis in a poly(U) system may help in understanding the physiological behavior of archaebacteria. The

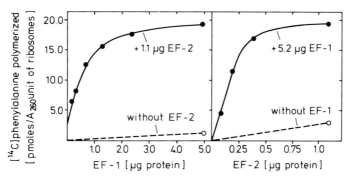

FIG. 5. Dependence of phenylalanine polymerization with *Thermoplasma* ribosomes on elongation factor amounts. Incubation at pH 6.5 and 56°C for 10 min in buffer containing 50 mM hepes, 90 mM KCl, 10 mM magnesium acetate, 0.69 mM spermine, 5 mM mercaptoethanol, 1 mM GTP, 1 mM ATP, 320 μg/ml of polyuridylic acid, and 28.8 nM [^{14}C]phenylalanyl-tRNA (specific activity 500 Ci/mol). The pooled peaks from DEAE (see Fig. 3A) had been used as EF-1 and EF-2, respectively.

optimization of cell-free systems is not only of interest but is essential for further research: Experiments with hybrid systems composed of parts of different translation machineries require detailed knowledge as to the conditions under which the individual translation systems are active.

2. POLYPHENYLALANINE SYNTHESIS SYSTEMS FROM *Halobacterium cutirubrum*

Bayley and Griffith (1968) first described a system from *H. cutirubrum* using phenylalanine as the substrate and a 150,000 *g* supernatant as the source of elongation factors and tRNA charging system. In the system developed by Kessel and Klink (1981), synthesis starts with phenylalanyl-tRNA; its properties reflect the special requirements for elongation. Using a 150,000 *g* supernatant as factor source, these authors found an optimal KCl concentration of 3.4 *M,* a broad magnesium optimum between 48 and 72 m*M,* and only very little influence of sodium ions. Ammonium ion could be substituted for potassium to at least 80%. A rather narrow temperature optimum lay at 37°C; a convenient pH was 7.6, but the tolerable pH range was not defined. Synthesis was completely inhibited by 0.5 m*M* anisomycin, i.e., by the same concentration range as used with eukaryotes (Vazquez, 1979). Kirromycin did not impair halobacterial elongation even in a concentration as high as 2 m*M.* Synthesis activity was dependent on GTP and polyuridylic acid; spermine was not necessary.

Under optimal conditions, 1.1 to 1.3 pmoles of phenylalanine were incorporated into trichloroacetic acid insoluble material by the *H. cutirubrum* system per A_{260} unit of ribosomes.

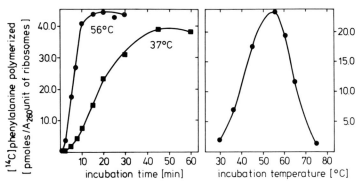

FIG. 6.   Dependence of phenylalanine polymerization with *Thermoplasma* ribosomes and factors on incubation time (left) and temperature (right). Assay mixture and other conditions as in Fig. 5, except that the time course assays contained 46 n*M* [14C]phenylalanyl-tRNA.

3. A POLYPHENYLALANINE SYNTHESIS SYSTEM FROM
Thermoplasma acidophilum

A cell-free system from *T. acidophilum* was optimized using purified ribosomes and two separated elongation factors (Klink *et al.*, 1983; Klink and Uschtrin, 1982). Plots of factor amounts versus synthesis are shown in Figs. 5A and B.

The time course of synthesis under optimal conditions is shown for two temperatures in Fig. 6A. About 40 pmoles of phenylalanine were polymerized per A_{260} unit of *Thermoplasma* ribosomes within 20 min. A temperature optimum of 56°C is close to the optimal growth temperature of *T. acidophilum* (Fig. 6B). At this temperature a rather narrow pH optimum at 6.4 was found (Fig. 7A), but at 37°C the pH peak was broadened, to higher pH values. The system needed 10–20 m*M* magnesium ions and was at this magnesium concentration completely dependent on spermine (Fig. 7B). Maximal synthesis was reached at somewhat differing spermine values for different incubation temperatures.

The *Thermoplasma* and *Methanococcus* systems differed markedly in their dependence on the monovalent cations potassium and ammonium (see Fig. 8). In both systems, ammonium had a more pronounced stimulation effect than potassium—as was known from eubacterial *in vitro* elongation systems. But in the *Thermoplasma* system (Fig. 8, dashed lines), low concentrations of these ions were sufficient to allow 50% of the maximal synthesis rate. The optimal ammonium ion concentrations were about 120 m*M*, whereas potassium inhibited strongly at values higher than 90 m*M*. If EF-1 was limiting the reaction, inhibition already began beyond 50 m*M* KCl (Fig. 8, dotted line). This suggests a different requirement of the two factors (or the respective elongation steps) for

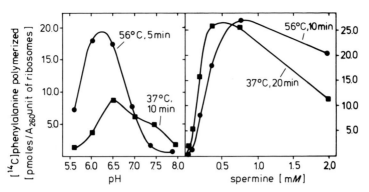

FIG. 7. Dependence of the *Thermoplasma* polyphenylalanine synthesis system on pH (left) and spermine concentration (right). Assay mixture and other conditions as in Fig. 5.

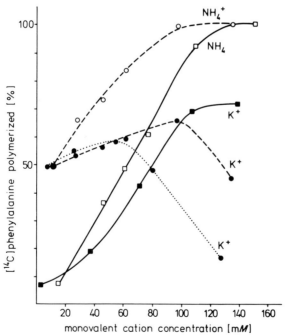

FIG. 8. Dependence on potassium and ammonium concentrations of polyphenylalanine synthesis systems from *T. acidophilum* (dashed lines) and *M. vannielii* (solid lines). Dotted line: *Thermoplasma* system with rate-limiting amounts of EF-1. Incubation 56°C; 10 min; pH 6.4; 0.69 mM spermine (*T. acidophilum* assays) and 37°C; 15 min; pH 7.6; 0.23 mM spermine (*M. vannielii* assays). Other conditions see legend of Fig. 5.

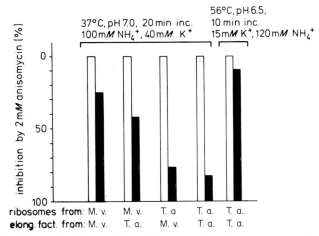

FIG. 9. Inhibition of cell-free polyphenylalanine synthesis by anisomycin in homologous and heterologous (mixed) systems from *Thermoplasma acidophilum*, T. a., and *Methanacoccus vannielii*, M. v. Empty columns: Assays without antibiotic; black columns: assays containing 2 mM anisomycin. For pH 7.0, 30 mM Tris/HCl was used, for pH 6.5, 30 mM "bis-tris" (2-bis[2-hydroxyethyl]-amino-2-[hydroxymethyl]-1,3-propanediol). For other conditions see legend Fig. 5.

potassium; another possible explanation may be that our EF-2 preparation was contaminated by a third EF-1 recycling factor, which works best at low potassium concentration. The ability to synthesize protein at low potassium concentrations with a half-maximal rate may reflect the cell's internal potassium concentration, reported by Searcy (1976) to be as low as 17 mM in *T. acidophilum*.

At 37°C and pH 7.0, *Thermoplasma* ribosomes were sensitive to anisomycin. The extent of inhibition was similar to that observed by Elhardt and Böck (1982) for Methanobacteriales and by Kessel and Klink (1981) for *H. cutirubrum* but was considerably greater than that obtained with *Methanosarcina* and *Methanococcus* ribosomes. Interestingly, inhibition of purified *Methanococcus vannielii* ribosomes could be significantly increased by replacing the homologous elongation factors with factors from *T. acidophilum* (Fig. 9). At optimal conditions of temperature and pH, the *Thermoplasma* system proved to be nearly completely resistant against anisomycin, though the antibiotic was not thermally inactivated at 56°C.

4. POLYPHENYLALANINE SYNTHESIS SYSTEMS FROM *Methanococcus vannielli*

Elhardt and Böck (1982) constructed polyphenylalanine synthesis systems from several methanogenic bacteria, among them *M. vannielii*, with purified ribosomes and unseparated postribosomal supernatants. Provided that homologous tRNA was used, this mesophilic system worked best at 100 mM NH$_4$Cl plus 40 mM KCl, 10 mM Mg^{2+}, and 1 mM spermidine. The sensitivity against various antibiotics was compared. Anisomycin in a concentration of about 2.3 mM caused only 30% inhibition in the *Methanococcus* system, but 90% inhibition with a system from *Methanobacterium formicicum*, which is somewhat more closely related to *H. cutirubrum* than to *M. vannielii* (Fox *et al.*, 1980).

A system from *M. vannielii* utilizing ribosomes purified as described in Section III,C,3 and with separated crude elongation factors (A. Thomsen and F. Klink, unpublished results) was compared to the *Thermoplasma* system (see Section III,D,3). In contrast to the latter, the *M. vannielii* system was not dependent on spermine, though slightly stimulated by this compound; the optimal pH value lay in the range of 7.6–7.8, the temperature optimum at about 37°C. The most striking difference found was in the requirement for monovalent cations (Fig. 8, solid lines). At concentrations of 30 mM ammonium or potassium, only 15% of the maximal activity was observed, whereas 140 mM was needed for optimal function, quite similar to the more complex system of Elhardt and Böck (1982). In contrast to these authors, we used phenylalanyl-tRNA prepared from *E. coli* tRNA^{Phe} as a substrate instead of phenylalanine. The failure to obtain activity with other than homologous tRNA reported by Elhardt and Böck, is therefore probably due to specificity of the phenylalanyl-tRNA synthetase. Under optimal conditions up to 20 pmoles of phenylalanine could be polymerized per A_{260} unit of ribosomes by our system.

All operations with the obligatory anaerobic *M. vannielii* starting with homogenization were performed without precluding oxygen.

5. A CELL-FREE SYSTEM FROM *Caldariella acidophila* MT4

The only cell-free polyphenylalanine synthesizing system from a member of the *Sulfolobus* group working on the ribosomal level was recently described (Cammarano *et al.*, 1982). No supernatant factors have been separated. Cell-free synthesis in this system is highly thermophilic (temperature maximum 75°C), obligatory spermine-dependent, and inhibited by ammonium ions at concentrations exceeding 10 mM.

E. CHARACTERIZATION OF ARCHAEBACTERIAL ELONGATION FACTORS

1. ARCHAEBACTERIAL EF-1

The molecular weights of EF-1 from *H. cutirubrum*, *T. acidophilum*, and *M. vannielii* as determined by SDS–PAGE were M_r 51,000, 49,000, and 47,000,

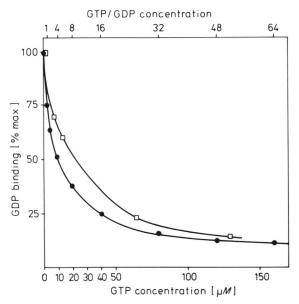

FIG. 10. Exchange of [³H]GDP with cold GTP at different concentration ratios GTP/GDP measured for EF-1 from *H. cutirubrum* (□), and *T. acidophilum* (●).

respectively. With these values archaebacterial factors hold an intermediate position between the M_r values of the eubacterial and eukaryotic proteins; therefore in all probability they are monomeric enzymes like the others.

EF-1 is a dominant protein in all three species. The factors from *H. cutirubrum* and *T. acidophilum* were eluted in nearly pure form from hydroxyapatite at a phosphate concentration of 90 m*M*. EF-1 from *Thermoplasma* and *Methanococcus* was not absorbed by DEAE-cellulose at pH 7.0.

Archaebacterial EF-1 binds GDP as do all analogous factors from the other kingdoms so far studied (Kaziro, 1978). As demonstrated in Fig. 10, 50% of bound GDP can be replaced by GTP at GTP–GDP ratios of 10 for *H. cutirubrum,* 5 for *T. acidophilum,* and about 3 for *M. vannielii* (not shown), values again indicating an intermediate position of the archaebacterial factor between eubacterial and eukaryotic ones.

EF-1 from *T. acidophilum* is rather stable to freezing and thawing even in the absence of glycerol. The methanogenic factor is far less stable and should not be frozen without 50% glycerol; stabilization of EF-1α by glycerol has recently been discussed by Beck and Spremulli (1982).

2. ARCHAEBACTERIAL EF-2

a. ADP-Ribosylation of Archaebacterial EF-2. EF-2 is the only archaebacterial protein that is a substrate of fragment A of diptheria toxin; it shares this property with its eukaryotic analogue but not with the eubacterial one. This then provides a very simple method for identifying those prokaryotes that are archaebacteria, and it enables us to estimate molecular weights of ADP-ribosyl–EF-2 from the whole-cell supernatants of numerous archaebacteria (Kessel and Klink, 1982). ADP-ribosylation of nonhalophilic archaebacterial EF-2 proceeds at much lower rates than that of eukaryotic factors (Fig. 11). Pappenheimer, Jr. *et al.*, (1983) found a crude preparation of supernatant proteins from *T. acidophilum* containing the elongation factors to be ADP-ribosylated roughly three orders of magnitude more slowly than similar preparations from yeast and HeLa cells. Halobacterial factor is very slowly ADP-ribosylated; this fact surely is due to the high salt milieu, which is necessary to keep the factor in the native conformation, but unfavorable for toxin activity. The difference between reaction rates of nonhalophilic archaebacterial factors and eukaryotic ones, however, may indicate that archaebacterial factors are poorer substrates for the ADP-ribosyltransferase than are the eukaryotic ones.

We are now developing a promising new phylogenetic probe for analyzing the ADP-ribosylated domains of archaebacterial EF-2. The initial results obtained

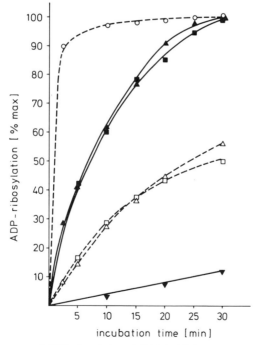

Fig. 11. Time course of ADP-ribosylation of EF-2 by NAD and diphtheria toxin. Sources of EF-2 were rat liver (○), *T. acidophilum* (▲,△), *M. vannielii* (■,□), and *H. cutirubrum* (▼). Amounts of toxin per assay: 10 Lf (flocculation units) for *H. cutirubrum,* 3 Lf (solid lines), or 0.6 Lf (dashed lines). For other conditions see Section III,B,4.

from *H. cutirubrum*, *T. acidophilum*, and *M. vannielii* allow the following conclusions (Gehrmann *et al.*, in press). (i) Archaebacterial EF-2 contains only one site of ADP-ribosylation. (ii) A sequence of nine amino acids containing the ADP-ribosylation site has been nearly completely conserved within the archaebacterial kingdom as well as between archaebacteria and higher eukaryotes. (iii) Just in the two positions in which the sequence from wheat germ differs from the other eukaryotic sequences studied so far (Brown and Bodley 1979), archaebacteria possess their own version.

The ribose-accepting amino acid had not been isolated and analyzed from any archaebacterial EF-2, but Bodley's group has traced diphthamide in *H. cutirubrum* supernatant protein by HPLC after labeling it with radioactive histidine using diphthamide isolated from yeast as a standard (Pappenheimer, Jr. *et al.*, 1983). After that, one may assume that diphthamide is the ADP-ribose acceptor in all archaebacterial factors.

b. Other Properties of Archaebacterial EF-2. The molecular weights of partially or totally purified EF-2 from *H. cutirubrum*, *T. acidophilum*, and *M. vannielii* as determined by SDS–PAGE were M_r 111,000, 81,000, and 83,000. With the same method, M_r values from 31 archaebacterial factors were estimated after incubation of a 150,000 g supernatant with ^{14}C-labelled NAD and toxin. Figures 12A and B show examples; in Table I M_r values are listed. Values from nonhalophilic archaebacteria are located between those of the eubacteria ($M_r \leq 80,000$) and eukaryotes. The considerably higher M_r values of all halophilic factors may at least partially reflect anomalous behavior during electrophoresis due to low hydrophobicities observed for halophilic proteins in general (Werber and Mevarech, 1978).

The factors from *H. cutirubrum* and *T. acidophilum* are eluted from hydroxyapatite at pH 7.0 and a phosphate concentration of 150 mM, the nonhalophilic factors from DEAE-cellulose at KCl concentrations of 140 mM (*T. acidophilum*) or 225 mM (*M. vannielii*).

No significant GDP binding activity could be observed with these three factors under the reaction conditions described in Section III,B,3. Only marginal ribosome-dependent GTP hydrolysis was seen with *H. cutirubrum* components (Kessel and Klink, 1981).

All archaebacterial EF-2 factors proved to be more stable than EF-1 factors

---

FIG. 12. SDS–PAGE of 150,000 g supernatants from six methanogenic (A) and five thermoacidophilic archaebacteria (B) after ADP-ribosylation with diphtheria toxin. Sources: (A) *Methanobacterium thermoautotrophicum*, a; *Methanospirillum hungatei*, b; *Methanobrevibacter arboriphilus*, c; *Methanococcus vannielii*, d; *Methanosarcina barkeri*, e; *Methanothermus fervidus*, f. (B) *Sulfolobus acidocaldarius*, a; *S. solfataricus*, b; *S. brierleyi*, c; *Thermoplasma acidophilum*, d; *Thermoproteus tenax*, e. Multiple bands [most clearly visible in (A,e)] are due to proteolysis during dialysis. Figures are molecular weights. From Kessel and Klink (1982).

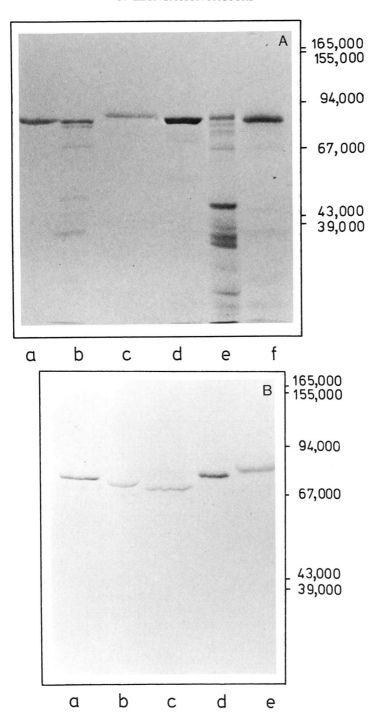

TABLE I

M_r Values of Archaebacterial and Eukaryotic ADP-Ribosyl-EF-2
from SDS–PAGE (Laemmli, 1970) of Labeled 150,000 g Supernatants

Organism	M_r	Organism	M_r
Halobacterium cutirubrum	111,000	Methanococcus vannielii	83,000
H. halobium	111,000	Methanosarcina barkeri	87,000
H. marismortui	111,000		
H. salinarium	111,000	Sulfolobus acidocaldarius	81,000
H. volcanii	101,000	S. solfataricus	76,000
H. saccharovorum	101,000	S. brierleyi	74,000
Halococcus morrhuae	105,000		
		Thermoproteus tenax	83,000
Methanobacterium	83,000		
thermoautotrophicum			
Methanobrevibacter arboriphilus	89,000	Thermoplasma acidophilum	81,000
Methanospirillum hungatei	83,000		
Methanothermus fervidus	83,000	Wheat germ	96,000
		Rat thymocytes	96,000
		Bovine liver	96,000

from the same sources. They could be stored at $-70°C$ for several months and frozen and thawed in the absence of glycerol without serious loss of activity.

IV. Ribosome Specificity of Elongation Factors from Three Primary Kingdoms

A. General Remarks

In Section III,A the question was raised whether the archaebacterial elongation apparatus represents a third type in addition to the eubacterial–organellar and the eukaryotic–cytoplasmatic versions. The results cited above are not sufficient to answer this question with respect to elongation factors. The domain of the molecule common to all archaebacterial EF-2 factors that can be ADP-ribosylated undoubtedly represents a eukaryote-like feature, but nobody knows whether this domain, though highly conserved, is of any importance for the physiological function of EF-2. A hint to a possible regulatory function of diphthamide in eukaryotic EF-2 has recently been given by detecting in beef liver and in transformed BHK cells a cellular ADP-ribosyltransferase activity modifying EF-2 apparently at its diphthamide moiety.

As pointed out in Section II,B, the two established types of elongation apparatus are easily discernible by ribosome specificity of translocating elongation factors. In the following paragraphs experiments are described showing that the archaebacteria represent a separate group with regard to ribosome specificity (Klink *et al.*, 1983).

B. HYBRID POLYPHENYLALANINE SYNTHESIS SYSTEMS INVOLVING COMPONENTS FROM DIFFERENT SPECIES

1. REACTION CONDITIONS IN HYBRID SYSTEMS

When constructing cell-free systems made from different sources, difficulties may arise from dissimilar requirements of the heterologous components for temperature, ionic milieu, pH, and co-factors. In former studies with hybrid systems from eubacteria and eukaryotes, the components have been chosen from among a large number of possibilities, so compatible reactions conditions can be established (Ciferri and Parisi, 1970; Krisko *et al.*, 1969; Slobin, 1981). However, it is difficult to construct hybrid systems using some archaebacterial components because most of them have optimal (internal) reaction conditions far from those that are optimal for eubacteria or eukaryotes (see Section III,D). Components from the extreme halophiles, for example, cannot be used in hybrid experiments because of such an incompatibility.

In most, if not all, such hybrid systems, some kind of compromise has to be made between the differing requirements. The problem is further complicated by the fact that optimal conditions always had been established for the entire homologous systems, not for isolated components of them. Hence, an optimal value for a system may in itself represent a compromise between the requirements of the individual components. We have assumed that reaction conditions lying not too far from optimal values are more crucial for ribosomes than for factors. Therefore, in a series of assays with one kind of ribosomes, we tried to maintain the conditions unchanged, even if factors were exchanged. The various conditions are defined in the legends.

2. EXCHANGE OF TRANSLOCATING FACTORS

a. Exchange between Kingdoms. We started with three homologous systems, one from each kingdom: An archaebacterial system from *T. acidophilum*, a eukaryotic one from rat liver, and a eubacterial one with ribosomes from *E. coli* and factors from the thermophilic eubacterium *Thermus thermophilus*. (The eubacterial system therefore was "homologous" only with respect to kingdom, not to species.) Reaction conditions, though not optimal for any of the systems, allowed satisfactory synthesis in all homologous systems. Maintaining the ho-

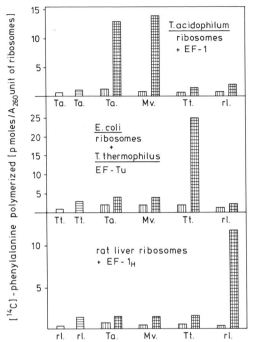

Fig. 13. Exchange of translocating factors between cell-free systems from archaebacteria (*T. acidophilum*, T. a., and *M. vannielii*, M. v.), eubacteria, and a eukaryote. Empty columns: ribosomes without factors; horizontally striped columns: ribosomes plus EF-1 or EF-T$_u$, respectively; vertically striped columns: ribosomes plus EF-2 or EF-G, respectively; cross-hatched columns: ribosomes plus both factors. Sources as indicated. Incubation at 37°C and pH 7.0 for 15 min. Composition of assay mixtures as given in the legend of Fig. 5, except that assays with liver ribosomes contained no spermine. For preparation or source of nonarchaebacterial factors see Klink *et al.* (1983). For relative factor amounts see text. From Klink *et al.*, 1983.

mologous tRNA-binding factor, the translocating factors were exchanged. In addition to *Thermoplasma* EF-2, a second archaebacterial factor, *M. vannielii* EF-2, was employed. The results of this series are shown in Fig. 13. All assays with one system (depicted in one horizontal line) contained equal amounts of ribosomes and of tRNA-binding factors; likewise, the amounts of one individual translocating factor were the same or even larger in heterologous systems compared to the homologous control systems. In this way, each kind of ribosome could be shown to require a translocating factor from its own kingdom for significant polyphenylalanine synthesis (Klink *et al.*, 1983).

In Fig. 14, synthesis effects are plotted against EF-2 amounts for *T. acidophilum*, *M. vannielii*, and rat liver systems. An amount of EF-2 sufficient to catalyze maximal synthesis in its own kingdom had negligible effects in another kingdom even under optimal conditions for all partners, as was the case for rat

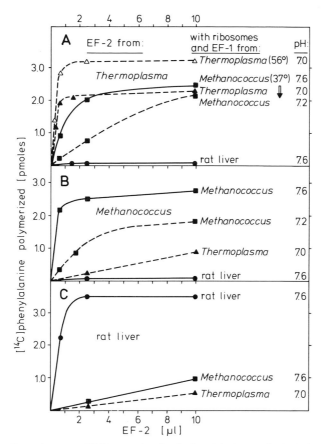

Fig. 14. Dependence on EF-2 amounts of polyphenylalanine synthesis in homologous and heterologous systems. Sources of EF-2: *Thermoplasma* (A), *Methanococcus* (B), or rat liver (C). Sources of EF-1 and ribosomes, conditions of pH, and temperature as indicated. Incubation time 20 min. All assays contained 40 m*M* KCl and 100 m*M* NH$_4$Cl. Assays with ribosomes from *Thermoplasma*, *Methanococcus*, or rat liver ribosomes contained 0.69 m*M*, 0.23 m*M*, or no spermine, respectively. For other constituents see legend of Fig. 5. Protein concentrations in the factor preparations used: EF-2 (*Thermoplasma*) 0.5 mg/ml, EF-2 (*Methanococcus*) 4.1 mg/ml; EF-2 (liver) 0.36 mg/ml. Amounts of ribosomes and EF-1 from individual sources were kept constant throughout the whole series. From Klink *et al.*, 1983.

liver and *Methanococcus*. However, very large amounts of liver EF-2 turned out to give significant synthesis in the archaebacterial systems (Fig. 14c); no analogous effect resulted in the reverse case (Fig. 14a,b).

b. Exchange within the Archaebacterial Kingdom. Figures 13 and 14 include EF-2 exchange experiments between the archaebacterial species *T. acidophilum* and *M. vannielii*, which show the ability of archaebacterial EF-2s to function with ribosomes from different archaebacteria. Figure 14 also illustrates the impairment of this cross-activity by pH conditions unfavorable to ribosomes or factors. Therefore quantitative evaluations are difficult. As our factor preparations were not purified, ADP-ribosylation per milligram of protein may be the best measure of relative molar factor quantities. On this basis, with *Methanococcus* ribosomes Thermoplasma EF-2 was roughly 1.5 times more active than was the homologous EF-2 from *M. vannielii*. When compared in the same way, in the *Methanococcus* system liver EF-2 was at least 25 times less active than the homologous factor.

From this, it may be concluded that EF-2s from the two archaebacteria used here belong to the same class of ribosome specificity although the two organisms are phylogenetically rather distant according to 16 S rRNA analyses (Fox *et al.*, 1980).

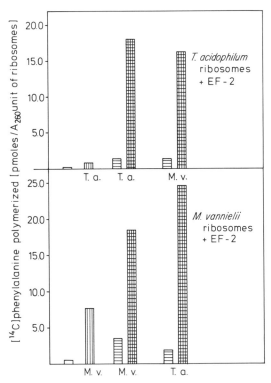

FIG. 15. Exchange of EF-1 between *T. acidophilum*, T. a., and *M. vannielii*, M. v. Hatching of columns as in Fig. 13. Incubations at pH 7.0 and 37°C for 15 min; for assay mixtures see legend of Fig. 14.

3. Exchange of aminoacyl-tRNA Binding Factors

For eubacteria and eukaryotes ribosome specificity for the tRNA-binding factor is not as clear-cut as it is for the translocating factor (see Section II,B,2). Significant activity in heterologous systems (which cross kingdom lines) have been reported only for EF-1 from *Euglena gracilis*, a unicellular alga that was shown to act in polyphenylalanine synthesis in an *E. coli* system (Beck and Spremulli, 1982).

Figure 15 demonstrates that the two archaebacteria studied so far also belong to the same ribosome specificity class with regard to EF-1.

Interchangeability of EF-1 between different kingdoms was investigated in a series of experiments similar to those done with EF-2 (Fig. 16). The experiments show that mammalian and archaebacterial EF-1 factors are rather freely ex-

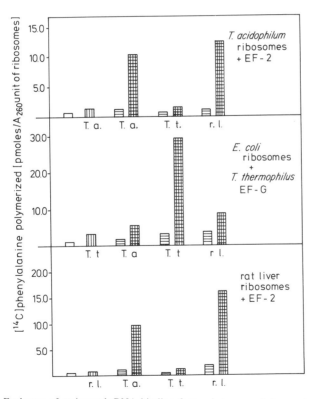

Fig. 16. Exchange of aminoacyl-tRNA binding factors between cell-free systems from archaebacteria, eubacteria, and eukaryotes. For explanations and conditions see legend of Fig. 13. Amounts of ribosomes and factors from individual sources were kept constant throughout the whole series of experiments.

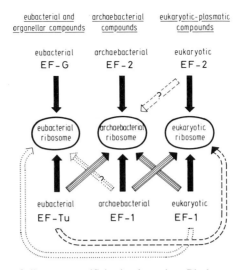

Fig. 17.   Summary of ribosome specificity in elongation. Black arrows: Full cooperation of homologous partners. Striped arrows: Full cooperation of heterologous partners. Dashed arrows: Marginal or insignificant heterologous cooperation. Dotted arrow: The special case of *Euglena gracilis*. Dotted arrow with question mark: no cooperation seen up to now, but experiments with two thermophilic partners or two mesophilic ones are lacking.

changeable. In quantitative evaluation of the results, difficulties arise due to the partially unsuitable reaction conditions and to the lack of fully purified factors, as mentioned in Section IV,B,2.

From the experiments performed at 37°C, eubacterial EF-Tu appears to be without significant effect on archaebacterial ribosomes and conversely. However, an assay under thermophilic conditions showed EF-Tu from *T. thermophilus* to have a considerable effect with ribosomes from *T. acidophilum*. The reverse could not be proven, because no experiments have been conducted as yet with ribosomes from a thermophilic eubacterium; hence, the question of whether cooperation between ribosomes and binding factors from eubacteria and archaebacteria is possible in both directions remains unanswered.

Figure 17 summarizes ribosome specificity of elongation factors in a schematic manner; mammalian mitochondrial ribosomes have been omitted.

V. Conclusions

A. Phylogenetic Relationships

At the present stage of investigation, we feel that caution is necessary when drawing conclusions concerning phylogenetic relationships from results that give an incomplete picture of archaebacterial elongation reactions and reactants. Nevertheless, we will tentatively evaluate some of the findings described above.

Only one fact holds for the entire archaebacterial kingdom, namely the existence of a structural domain in EF-2 rendering it a substrate for diphtheria toxin. The common domain in archaebacterial and eukaryotic EF-2s which can be ADP-ribosylated was either a part of a primordial factor already present at the progenote stage —in which case eubacteria have subsequently lost it—or this sequence was acquired by an ancestor common to archaebacteria and urkaryotes after the eubacterial line had segregated. Hence, this common structural feature does not allow one to discriminate between the conception of Woese (Fox *et al.*, 1980) and that of Van Valen and Maiorana (1980) concerning the early evolution of prokaryotes in connection with eukaryotic origins.

Some other results that were obtained with two or three species belonging to quite different archaebacterial branches may cover the whole kingdom.

1. The inability of archaebacterial EF-2 to work with other than archaebacterial ribosomes for catalyzing polyphenylalanine elongation.

2. The remarkable activity of archaebacterial ribosomes with EF-1 from the other kingdoms, especially with eukaryotic EF-1.

3. The relative affinities of archaebacterial EF-1 for guanine nucleotides which ranges near to the values for eukaryotic factors.

It follows from ribosome specificity that archaebacteria have an elongation apparatus distinct from those of the other kingdoms, but it cannot be decided at the moment in which part(s) of this apparatus—ribosomes and/or translocating factor—the differences have their structural basis.

With respect to tRNA-binding factors, archaebacterial ribosomes apparently have a less strict specificity than eukaryotic or eubacterial ribosomes; further experiments are needed to determine whether this also holds true for the opposite case.

Whether the archaebacterial kingdom itself is homogenous with respect to ribosome specificity must be clarified by involving sulfur-metabolizing archaebacteria in hybrid *in vitro* polyphenylalanine synthesis systems. Initial experiments combining a crude EF-2-containing protein fraction from *S. solfataricus* with ribosomes and EF-1 from *T. acidophilum* suggest such homogeneity (F. Klink, unpublished results).

On the whole, archaebacterial protein synthesis machineries show considerable diversity, ostensibly reflecting extreme growth conditions. Among the results shown in this contribution, this is demonstrated, for instance, by the dependence of cell-free systems on monovalent cations. Concentrations too low for polyphenylalanine synthesis in the *M. vannielii* system were too high for satisfying synthesis in the *Caldariella* (i.e., *Sulfolobus*) system described, whereas ribosomes and factors from *T. acidophilum* tolerated both concentration ranges, which represented denaturing conditions for extremely halophilic factors.

When compared with the other kingdoms, the archaebacterial elongation apparatus holds an "intermediate" position, but some of the structural or functional properties of factors and ribosomes seem to point to a somewhat closer relationship to the eukaryotes. Among these are the ADP-ribosylatable domain in EF-2, the relative affinities for guanine nucleotides of EF-1, and the sensitivity against 80 S-specific antibiotics, such as anisomycin, in the archaebacterial elongation systems. However, much experimental work remains to be done to elucidate the macromolecular structures of archaebacterial translation apparatus before firm phylogenetic conclusions can be drawn.

B. NOMENCLATURE OF ARCHAEBACTERIAL ELONGATION FACTORS

The existence of a unique ribosome specificity class suggests that the archaebacterial factors be given their own nomenclature. Because it would be rather disastrous to turn the present bipartite designation system into a tripartite one, we propose to derive the archaebacterial nomenclature from one of the existing two systems by using the prefix "a" for "archaebacterial." We would prefer the eukaryotic nomenclature as a basis in order to avoid confusion by the eukaryotic feature of aEF-2 and to provide for the case that archaebacteria may possess a protein analogous to EF-1γ. Therefore, we recommend the designations aEF-1α for the archaebacterial aminoacyl-tRNA binding elongation factor and aEF-2 for the archaebacterial translocating elongation factor.

C. SOME FUTURE PROSPECTS

In this chapter, we have described the present state of research on cell-free protein synthesis and elongation factors from archaebacteria. In future investigations, we hope to obtain more information about the structure, catalytic properties, and immunological behavior of purified factors and in particular about the problem of EF-1α recycling in archaebacteria. Continued inclusion of *Sulfolobus* factors into hybrid synthesis assays may further distinguish the archaebacterial kingdom from the other two by ribosome specificity of translocation factors. Sequence comparisons between longer stretches of ADP-ribosylated domains of various archaebacteria may shed light on the phylogenetic structure of the kingdom, and studies with separate elongation steps or partial reactions, which possibly can be brought about by elongation factors on heterologous ribosomes, may further elucidate ribosomal mechanisms.

The most complicated and perhaps most interesting area of translation research is chain initiation, which includes regulation mechanisms and mRNA selection. For this step, fundamental differences between the eukaryotic and prokaryotic machineries are seen, in the number of initiation factors, the mode of mRNA attachment, and the sequence of (partial) reactions. No experimental work with archaebacteria has been reported in this field; a few interesting results concerning mRNA structure and translation with poorly characterized thermoacidophilic strains from Japanese hot springs have been orally presented but have never been published in detail (Ohba and Oshima, 1982). Also an early account of Shine–Dalgarno sequences in 10 methanogenic species (Fox *et al.*, 1977), which was discussed by Steitz (1978), makes it desirable to increase our knowledge of how archaebacteria manage the initiation stage of mRNA translation. One possible approach to this problem may be purification and characterization of archaebacterial initiation factors, now under way in our laboratory.

REFERENCES

Arai, K., Arai, N., Nakamura, S., Oshima, T., and Kaziro, Y. (1978). Studies on polypeptide elongation factors from an extreme thermophile, *Thermus thermophilus* HB8. 2. Catalytic properties. *Eur. J. Biochem.* **92,** 521–531.

Bayley, S. T., and Griffiths, D. (1968). A cell-free amino acid incorporating system from an extremely halophilic bacterium. *Biochemistry* **7,** 2249–2256.

Bayley, S. T., and Kushner, D. J. (1964). The ribosomes of the extremely halophilic bacterium Halobacterium cutirubrum. J. Mol. Biol. **9,** 654–669.

Beck, C. M., and Spremulli, L. L. (1982). Purification of Euglena EF-1: A cytoplasmic factor that also functions on procaryotic and organellar ribosomes. *Arch. Biochem. Biophys.* **215,** 414–424.

Bermek, E. (1978). Mechanisms in polypeptide chain elongation on ribosomes. *Prog. Nucleic Acid Res. Mol. Biol.* **21,** 64–100.

Brown, B. A., and Bodley, J. W. (1979). Primary structure at the site in beef and wheat elongation factor 2 of ADP-ribosylation by diphtheria toxin. *FEBS Lett.* **103,** 253–255.

Cammarano, P., Teichner, A., Chinali, G., Londei, P., De Rosa, M., Gambacorta, A., and Nicolaus, B. (1982). Archaebacterial elongation factor T_u insensitive to pulvomycin and kirromycin. *FEBS Lett.* **148,** 255–259.

Chinali, G., Wolf, H., and Parmeggiani, A. (1977). Effect of kirromycin on elongation factor T_u. *Eur. J. Biochem.* **75,** 55–65.

Ciferri, O., and Parisi, B. (1970). Ribosome specificity of protein synthesis in vitro. *Prog. Nucleic Acid. Res. Mol. Biol.* **10,** 121–144.

Collier, R. J. (1975). Diphtheria toxin: Mode of action and structure. *Bacteriol. Rev.* **39,** 54–85.

Elhardt, D., and Böck, A. (1982). An in vitro polypeptide synthesizing system from methanogenic bacteria: Sensitivity to antibiotics. *Mol. Gen. Genet.* **188,** 128–134.

Fox, G. E., Magrum, L. J., Balch, W. E., Wolfe, R. S., and Woese, C. R. (1977). Classification of methanogenic bacteria by 16S ribosomal RNA characterization. *Proc. Natl. Acad. Sci. U.S.A.* **74,** 4537–4541.

Fox, G. E., Stackebrandt, E., Hespell, R. B., Gibson, I., Maniloff, I., Dyer, T. A., Wolfe, R. S., Balch, W. E., Tanner, R. S., Magrum, L. J., Zablen, L. B., Blakemore, R., Gupta, R., Bonen, L., Lewis, B. J., Stahl, D. A., Luehrsen, K. R., Chen, K. N., and Woese, C. R. (1980). The phylogeny of procaryotes. *Science,* **209,** 457–463.

Gehrmann, R., Henschen, A., and Klink, F. (1984). Primary structure at the site in elongation factor-2 of ADP-ribosylation is highly conserved from archaebacteria to mammals. *FEBS Lett.* (in press).

Gordon, J., Lucas-Lenard J., and Lipmann, F. (1971). Isolation of bacterial chain elongation factors. *In* "Methods in Enzymology" (K. Moldave and L. Grossmann, eds.), Vol. 20, pp. 281–291. Academic Press, New York.

Graves, M. C., Breitenberger, C. A., and Spremulli, C. C. (1980). Euglena gracilis chloroplast ribosomes: Improved isolation procedure and comparison of elongation factor specificity with procaryotic and eucaryotic ribosomes. *Arch. Biochem. Biophys.* **204,** 444–454.

Hardesty, B., and McKeehan, W. (1971). Transfer factor II (T-II) from rabbit reticulocytes. *In* "Methods in Enzymology" (K. Moldave and L. Grossmann, eds.), Vol. 20, pp. 330–337. Academic Press, New York.

Iglewski, W. J., Lee, H., and Muller, P. (1984). ADP-ribosyltransferase from beef liver which ADP-ribosylates elongation factor-2. *FEBS Lett.* **173,** 113–118.

Iwasaki, K., and Kaziro, Y. (1979). Polypeptide chain elongation factors from pig liver. *In* "Methods in Enzymology" (K. Moldave, and L. Grossmann, eds.), Vol. 60, pp. 657–676. Academic Press, New York.

Kaziro, Y. (1978). The role of guanosine-5′-triphosphate in polypeptide chain elongation. *Biochim. Biophys. Acta* **505,** 95–127.

Kaziro, Y., Inoue-Yokosawa, N., and Kawakita, M. (1972). Studies on polypeptide elongation factor from *E. coli* I. Crystalline factor G. *J. Biochem. (Tokyo)* **72,** 853–863.

Kessel, M., and Klink, F. (1980). Archaebacterial elongation factor is ADP-ribosylated by diphtheria toxin. *Nature (London)* **287,** 250–251.

Kessel, M., and Klink, F. (1981). Two elongation factors from the extremely halophilic archaebacterium *Halobacterium cutirubrum. Eur. J. Biochem.* **114,** 481–486.

Kessel, M., and Klink, F. (1982). Identification and comparison of eighteen archaebacteria by means of the diphtheria toxin reaction. *Zentralbl. Bakteriol., Mikrobiol. Hyg., Abt. 1, Orig. C* **3,** 140–148.

Klink, F., Schümann, H., and Thomsen, A. (1983). Ribosome specificity of archaebacterial elongation factor 2. Studies with hybrid polyphenylalanine synthesis systems. *FEBS Lett.* **155,** 173–177.

Krisko, I., Gordon, J., and Lipmann, F. (1969). Studies on the interchangeability of one of the mammalian and bacterial supernatant factors in protein biosynthesis. *J. Biol. Chem.* **244,** 6117–6123.

Laemmli, U. K. (1970). Cleavage of structural proteins during the assembly of the head of bacteriophage T4. *Nature (London)* **227,** 680–685.

Lake, J. A., Henderson, E., Clark, M. W., and Matheson, A. T. (1982). Mapping evolution with ribosome structure: Intralineage constancy and interlineage variation. *Proc. Natl. Acad. Sci. U.S.A.* **79,** 5948–5952.

Lee, H., and Iglewski, W. J. (1984). Cellular ADP-ribosyltransferase with the same mechanism of action as diphtheria toxin and *Pseudomonas* toxin A. *Proc. Natl. Acad. Sci. U.S.A.* **81,** 2703–2707.

Matheson, A. T., and Yaguchi, M. (1981). The ribosome as a phylogenetic probe. *In* "International Cell Biology 1980–1981" (H. G. Schweiger, ed.), pp. 103–110. Springer-Verlag. Berlin and New York.

Matheson, A. T., and Yaguchi, M. (1982). The evolution of the archaebacterial ribosome. *Zentralbl. Bakteriol., Mikrobiol. Hyg., Abt. I, Orig. C* **3,** 192–199.

Matheson, A. T., Yaguchi, M., Nazar, R., Visentin, L. P., and Willick, G. E. (1978). The structure of ribosomes from moderate and extreme halophilic bacteria. *In* "Energetics and Structure of Halophilic Microorganisms" (S. R. Caplan and M. Ginzburg, eds.), pp. 481–501. Elsevier North-Holland Biomedical Press, Amsterdam.

Miller, D. L., and Weissabach, H. (1974). *In* "Methods in Enzymology" (K. Moldave and L. Grossmann, eds.), Vol. 30, pp. 219–232. Academic Press, New York.

Miller, D. L., and Weissbach, H. (1977). Factors involved in the transfer from aminoacyl-tRNA to the ribosome. *In* "Molecular Mechanisms of Protein Biosynthesis" (H. Weissbach and S. Pestka, eds.), pp. 323–373. Academic Press, New York.

Moore, V. G., Atchison, R. E., Thomas, G., Moran, M., and Noller, H. F. (1975). Identification of a ribosomal protein essential for peptidyl transferase activity. *Proc. Natl. Acad. Sci. U.S.A.* **72**, 844–848.

Nirenberg, M., and Leder, P. (1964). RNA codewords and protein synthesis. *Science* **145**, 1399–1407.

Ohba, M., and Oshima, T. (1982). Some biochemical properties of the protein synthesizing machinery of acidothermophilic archaebacteria isolated from Japanese hot springs. *In* "Archaebacteria" (O. Kandler, ed.), p. 353. Fischer, Stuttgart.

Oppenheimer, N. J., and Bodley, J. W. (1981). Diphtheria toxin: Site and configuration of ADP-ribosylation of diphthamide in elongation factor 2. *J. Biol. Chem.* **256**, 8579–8581.

Pappenheimer, A. M., Jr., Dunlop, P. C., Adolph, K. W., and Bodley, J. W. (1983). Occurrence of diphthamide in archaebacteria. *J. Bacteriol.* **153**, 1342–1347.

Schmid, G., and Böck, A. (1981). Immunological comparison of ribosomal proteins from archaebacteria. *J. Bacteriol.* **147**, 282–288.

Schmid, G., and Böck, A. (1982). The ribosomal protein composition of the archaebacterium *Sulfolobus*. *Mol. Gen. Genet.* **185**, 498–501.

Schmid, G., Pecher, T., and Böck, A. (1982). Properties of the translational apparatus of archaebacteria. *Zentralbl. Bakteriol., Mikrobiol. Hyg., Abt. 1, Orig. C* **3**, 209–217.

Searcy, D. (1976). *Thermoplasma acidophilum:* Intracellular pH and potassium concentration. *Biochim. Biophys. Acta* **451**, 278–286.

Slobin, L. I. (1981). The inhibition of eucaryotic protein synthesis by procaryotic elongation factor T_u. *Biochem. Biophys. Res. Commun.* **101**, 1388–1395.

Slobin, L. I., and Möller, W. (1976). Characterization of developmentally regulated forms of elongation factor 1 in *Artemia salina*. *Eur. J. Biochem.* **69**, 351–366.

Steitz, J. A. (1978). Methanogenic bacteria. *Nature (London)* **273**, 10.

Uschtrin, D., and Klink, F. (1982). In vitro polyphenylalanine synthesis with purified ribosomes and two elongation factors from the archaebacterium *Thermoplasma acidophilum*. *Hoppe-Sayler's Z. Physiol. Chem.* **363**, 891.

Van Ness, B. G., Howard, J. B., and Bodley, J. W. (1980). ADP-ribosylation of elongation factor 2 by diphtheria toxin. *J. Biol. Chem.* **255**, 10710–10716.

Van Valen, L. M., and Maiorana, V. C. (1980). The archaebacteria and eucaryotic origins. *Nature (London)* **287**, 248–249.

Vazquez, D. (1979). "Inhibitors of Protein Biosynthesis." Springer-Verlag, Berlin and New York.

Visentin, L. P., Yaguchi, M., and Matheson, A. T. (1979). Structural homologies in alanine-rich acidic ribosomal proteins from procaryotes and eucaryotes. *Can. J. Biochem.* **57**, 719–726.

Weissbach, H., and Pestka, S., eds. (1977). "Molecular Mechanisms of Protein Synthesis." Academic Press, New York.

Werber, M. M., and Mevarech, M. (1978). Purification and characterization of a highly acidic 2Fe-ferredoxin from *Halobacterium* of the Dead Sea. *Arch. Biochem. Biophys.* **187**, 447–456.

Wittinghofer, A., Frank, R., Gast, W. H., and Lberman, R. (1979). Polyphenylalanine synthesis by crystallized trypsin-modified EF-T_u·GDP. *J. Mol. Biol.* **132**, 253–256.

Woese, C. R., and Fox, G. E. (1977). Phylogenetic structure of the procaryotic domain: The primary kingdoms. *Proc. Natl. Acad. Sci. U.S.A.* **74**, 5088–5090.

Wolf, H., Assmann, D., and Fischer, E. (1978). Pulvomycin, an inhibitor of protein biosynthesis preventing ternary complex formation between elongation factor T_u, GTP and aminoacyl-tRNA. *Proc. Natl. Acad. Sci. U.S.A.* **75,** 5324–5328.

Yaguchi, M., Matheson, A. T., Visentin, L.P., and Zucker, M. (1980). Molecular evolution of the alanine-rich acidic ribosomal A protein. *In* ''Genetics and Evolution of RNA Polymerase, t-RNA and Ribosomes'' (S. Osawa, H. Ozeki, H. Uchida, and T. Yura, eds.), pp. 585–599. Univ. of Tokyo Press, Tokyo.

PART III

General Molecular Characteristics
of Archaebacteria

CHAPTER 9

Cell Envelopes of Archaebacteria

OTTO KANDLER AND HELMUT KÖNIG

I. Introduction

Cell wall* chemistry was one of the first phenotypical characteristics to distinguish archaebacteria from eubacteria (Kandler and Hippe, 1977; Kandler and König, 1978; Kandler, 1979, 1982; Fox *et al.,* 1980). Although, morphologically, the cell envelope profiles of the gram-positive members of both kingdoms are very similar, at least at first sight, the presence or absence of the "classical" cell wall components, muramic acid and D-amino acids, in hydrolyzates of whole cells or cell wall preparations allows an unequivocal distinction to be made between archaebacteria and eubacteria, with the exception of the mycoplasmas. No common cell wall component, such as muramic acid in eubacteria, is known within the archaebacteria, which, as a whole, are characterized by a remarkable structural and chemical diversity of their cell envelopes. With the exception of *Thermoplasma* (Darland *et al.,* 1970) and *Methanoplasma* (Rose and Pirt, 1981), which lack any kind of cell envelope, as do the eubacterial

---

*In this paper, the term *cell wall* will be restricted to mean a rigid layer forming a sacculus enclosing an individual cell, which, after isolation, still exhibits the shape of the cell and is insensitive to detergents. Arrays of protein or glycoprotein subunits covering the surface of the cells will be referred to as *surface layers* (S layers) (Sleytr and Glauert, 1982; Sleytr and Messner, 1983). Layers of distinct mechanical stability extending over several cells will be termed *sheaths*. Mucous substances surrounding single cells or groups of cells will be termed *capsules*. Irrespective of the chemical and physical nature of the layers surrounding the cells, their sum will be referred to as the *cell envelope*.

Copyright © 1985 by Academic Press, Inc.
All rights of reproduction in any form reserved.
ISBN 0-12-307208-5

mycoplasmas, the archaebacteria exhibit cell envelope profiles of at least four different morphological types (Fig. 1).

Type 1 represents the most common gram-positive archaebacterial cell envelope profile. Only one more or less homogeneous electron-dense layer, 10–20 nm in width, is seen outside the cytoplasmic membrane. It takes part in septum formation, thus resembling the profile of a gram-positive eubacterium. The electron-dense cell wall can be isolated by the usual techniques for cell wall isolation (Schleifer and Kandler, 1967, 1972). However, the main component of

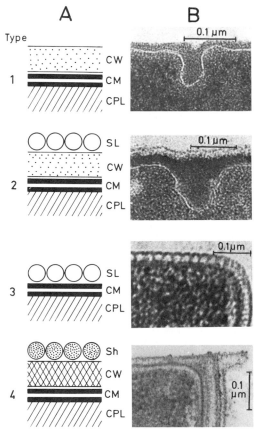

FIG. 1. Cell envelope profiles of archaebacteria (A) Schematic presentation [CW, cell wall; CM, cytoplasmic membrane; CPL, cytoplasm; SL, surface layer (S layer); Sh, fibrillary sheath] (B) Electron micrographs of thin sections of typical representatives of the various types. (1) *Methanobacterium formicicum* with ingrowing septum; (2) *Methanothermus fervidus* with ingrowing septum (K. O. Stetter, unpublished); (3) *Thermoproteus tenax* (W. Zillig, unpublished); (4) *Methanospirillum hungatei.*

the thus obtained rigid cell wall sacculi does not consist of murein, but of either pseudomurein (Methanobacteriales), a nonsulfated (*Methanosarcina*), or a sulfated acidic heteropolysaccharide (*Halococcus*).

Type 2 is only found in the gram-positive *Methanothermus fervidus* (Stetter *et al.*, 1981). Its rigid cell wall sacculus consisting of pseudomurein is covered with protein subunits in hexagonal arrays, thus resembling some eubacteria whose murein sacculi are also covered by a surface layer (S layer) of regularly arranged protein subunits (Sleytr and Glauert, 1982; Sleytr and Messner, 1983).

Type 3 represents the typical gram-negative archaebacterial cell envelope profile. Unlike gram-negative eubacteria, gram-negative archaebacteria exhibit neither a specific sacculus polymer nor an "outer membrane," but only one surface layer consisting of hexagonally or tetragonally arranged protein or glycoprotein subunits. It is found in all thermoacidophiles and many of the halophiles and methanogens.

Type 4 which is found in *Methanospirillum,* is the most complex cell envelope. Here, the individual cell is surrounded by an electron-dense flexible layer, 10 nm in width, of probably proteinaceous nature (Sprott *et al.*, 1979), several cells being held together by a sheath consisting of protein fibrils (Fig. 1). Although much less work has been done on archaebacterial than on eubacterial cell envelopes, at least some data on the structural and chemical features of each of the various genera of archaebacteria are known.

II. Structure and Chemistry of Cell Envelopes

A. CELL ENVELOPES OF GRAM-POSITIVE ARCHAEBACTERIA

1. PSEUDOMUREIN-CONTAINING CELL ENVELOPES OF METHANOBACTERIALES

The cell envelope profile of members of the family Methanobacteriaceae (*Methanobacterium* and *Methanobrevibacter*) corresponds to type 1 (Fig. 1) exhibiting only one electron-dense layer, 15–20 nm in width, closely adjoining the cytoplasmic membrane. This cell wall sacculus appears homogenous in thin sections of organisms of the genus *Methanobacterium,* while it has a distinct triple-layered appearance in thin sections of *Methanobrevibacter ruminantium:* an electron-dense, very thin inner region adjoining the cytoplasmic membrane; an electron-transparent thicker middle region; and a rough, irregular, electron-denser outer region (Zeikus and Bowen, 1975; Zeikus, 1977). The latter is missing in the septum formed during cell division.

The cell envelope of *Methanothermus fervidus,* the only known species of the family Methanothermaceae (Stetter *et al.*, 1981) shows a profile of type 2 (Fig. 1), as already mentioned. The electron-dense layer, 12–15 nm in width, appears homogenous in ultrathin sections and the surface layer, 12–15 nm in width, consists of hexagonally arranged protein subunits of 92,500 daltons (H. König and K. O. Stetter, unpublished).

The cell wall sacculi of all organisms mentioned so far can be isolated by the same methods—mechanical disintegration and digestion with trypsin—as those usually applied to the isolation of cell walls of gram-positive eubacteria (Kandler and König, 1978). Isolated walls still exhibit the shape of the original cell including the septa (Fig. 2), and, morphologically, they cannot be distinguished from isolated cell wall sacculi of gram-positive eubacteria. The isolated rigid sacculi of *Methanothermus fervidus* can be freed from the proteinaceous surface layer by SDS and pronase treatment (Stetter *et al.*, 1981).

a. *Chemistry of Pseudomurein.* Chemical analysis of acid hydrolysates of isolated cell wall sacculi from all known species of the three genera of the Methanobacteriales (Table I; König *et al.*, 1982) has revealed that none of the

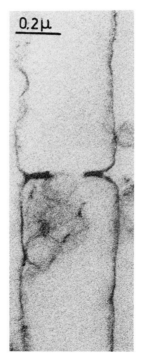

Fig. 2. Thin section of an isolated cell wall of *Methanobacterium formicicum* with ingrowing septum. (From Kandler and König, 1978.)

TABLE I

MOLAR RATIOS OF THE COMPONENTS OF THE CELL WALLS OF METHANOBACTERIALES[a,b]

Species	Amino acids					NH$_3$	N-Acetylamino sugars			Neutral sugars				Phosphate % d.W.
	Ala	Thr	Glu	Lys	Orn		GalNH$_2$	GlcNH$_2$	TalNUA[c]	Gal	Glc	Man	Rha	
Methanobrevibacter														
M. arboriphilus DH1	1.52	—	2.13	1.00	—	1.36	0.67	—	(1.00)	—	O	0.02	0.02	4.54
M. arboriphilus AZ	1.17	—	2.32	1.00	—	3.92	0.91	—	(1.00)	O	O	O	O	O
M. smithii PS	1.42	0.90	2.04	1.00	0.40	0.81	0.44	0.56	(1.00)	0.11	0.14	—	0.09	1.0
M. ruminantium M1[d]	—	0.40	1.85	1.00	—	0.85	0.80	0.60	(1.00)	0.29	0.02	0.02	0.21	4.65
M. ruminantium M1[d]	0.70	0.40	1.80	1.00	—	1.00	0.80	0.50	(1.00)	O	O	O	O	O
Methanobacterium														
M. bryantii M.o.H.	1.39	—	2.39	1.00	—	1.12	0.45	1.25	(1.00)	0.84	0.09	1.18	+	0.20
M. bryantii M.o.H.G.	1.32	—	2.21	1.00	—	0.84	—	0.84	(1.00)	0.53	0.26	0.26	+	O
M. formicicum MF	1.51	—	2.11	1.00	—	1.53	1.00	1.16	(1.00)	0.76	0.27	0.37	0.62	0.64
M. thermoautotrophicum Δ H	1.20	—	2.27	1.00	—	0.61	0.16	1.18	(1.00)	0.02	0.13	—	+	0.28
M. thermoautotrophicum Marburg	1.32	—	2.37	1.00	—	0.77	0.87	0.23	(1.00)	O	O	O	O	O
M. thermoautotrophicum JW 500	1.43	—	2.36	1.00	—	1.64	1.29	1.14	(1.00)	O	O	O	O	O
M. thermoautotrophicum JW 501	1.29	—	2.57	1.00	—	1.07	0.86	1.07	(1.00)	O	O	O	O	O
M. thermoautotrophicum JW 510	1.25	—	2.25	1.00	—	1.25	1.42	1.00	(1.00)	O	O	O	O	O
Methanothermus														
M. fervidus	1.47	—	2.23	1.00	—	1.64	0.35	0.49	(1.00)	—	0.01	—	O	O

[a] From König et al. (1982).

[b] —, not present; +, trace; O, not determined.

[c] The molar ratio of N-acetyltalosaminuronic acid was only estimated from the analysis of partial acid hydrolyzates, since it is completely destroyed in total hydrolyzates.

[d] Two different batches of cells, both obtained from R. Wolfe and W. Balch (Urbana, USA).

species contains muramic acid, meso- or LL-diaminopimelic acid, D-glutamic acid, or D-alanine, typical components of murein. They contain, however, a set of three L-amino acids (Lys, Glu, Ala, or Thr), an *N*-acetylated amino sugar— glucosamine or galactosamine—which are also found in eubacteral cell walls, and *N*-acetyl-L-talosaminuronic acid, which is not known to occur anywhere else in nature. L-Talosaminuronic acid (2-amino-2-deoxy-L-talosaminuronic acid) was identified by ninhydrin degradation to the respective pentose and by several other chemical probes (König and Kandler, 1979b). The configurations of talosaminuronic acid and galactosamine have been elucidated by determining the configuration of serine, obtained by periodate–hypoiodite degradation of the respective compound. In addition, the D-configuration of the two amino sugars has also been shown by phosphorylation with D-specific hexokinase and galactokinase, respectively (König and Kandler, 1980; König et al., 1982).

The primary structure of the polymer responsible for the rigidity of the cell wall sacculus has been elucidated by the analysis of overlapping peptides isolated from partial acid hydrolysates of isolated cell walls (König and Kandler, 1979a) and of soluble breakdown products obtained by treating the cell walls with either hydrofluoric acid, 2 *M* NaOH, or anhydrous hydrazine (König et al., 1982, 1983).

These studies have shown that the new polymer is made up of glycan strands consisting of alternating β(1→3)-bound *N*-acetyl-D-glucosamine and β(1→3)-bound *N*-acetyl-L-talosaminuronic acid residues, which are cross-linked via short peptides attached to the carboxyl group of *N*-acetyl-L-talosaminuronic acid (Fig. 3). Such a polymer must also be considered as a *peptidoglycan,* a very general chemical term. However, it is certainly not merely a further variant of the many well-known murein types (Schleifer and Kandler, 1972; Schleifer and Stackebrandt, 1983), but represents a basically different group of peptidoglycans. In order to indicate the similarity as well as the dissimilarity of the two cell wall peptidoglycans, the new polymer was named *pseudomurein* (Kandler, 1979).

The main differences are that the glycan moiety of pseudomurein is β(1→3)-instead of β(1→4)-linked and contains *N*-acetyl-L-talosaminuronic acid instead of *N*-acetyl-D-muramic acid. Consequently, pseudomurein is resistant to lysozyme. The peptide moiety of pseudomurein possesses an amino acid sequence different from that of murein and lacks D-amino acids. Thus, the biosynthesis of pseudomurein is resistant to antibiotics—such as D-cycloserine, vancomycin, penicillin—which interfere with reactions involving D-alanine (Hammes et al., 1979; Hilpert et al., 1981). Protection against the attack by proteases, which, in the case of murein, is brought about by sequences of alternating D- and L-amino acids, may be caused by the accumulation of the unusual ε- and γ-bonds in pseudomurein. Although the glycan moiety of murein contains only glucosamine but no galactosamine, glucosamine may be partly or completely replaced by galactosamine in the glycan moiety of pseudomurein.

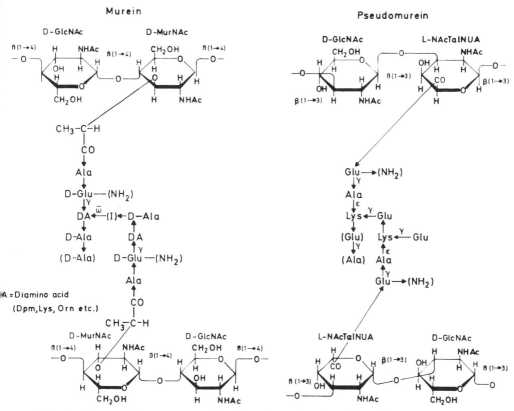

FIG. 3. Proposed primary structure of a dimer of murein and pseudomurein (compounds in parentheses may be missing in some cases). (Modified according to Kandler, 1982.)

However, either of the amino sugars may additionally be present in cell walls of Methanobacteriales in binding forms different from those of the alternating sequences of the glycan strands, i.e., as substituents of the glycan moiety or as components of accessory polysaccharides. As König *et al.* (1982) pointed out, the occurrence of the two amino sugars in different ratios allows four different cases to be distinguished as follows:

1. Cell wall preparations of *Methanobacterium bryantii* M.o. H.-G. and *Methanobacterium thermoautotrophicum ΔH* contain 1 mol glucosamine per mol of lysine, but only traces of galactosamine. Therefore, glucosamine is the main amino sugar of the glycan strand of these pseudomureins.

2. In the cell walls of *Methanobrevibacter arboriphilus* strain AZ and strain DH 1 only galactosamine was found, thus galactosamine replaces glucosamine in the glycan moiety in these two organisms.

3. The glycan strands of *Methanobrevibacter smithii* PS and *Methanobacterium thermoautotrophicum* strain Marburg most likely consist of a mixture of galactosamine and gluco-

samine, since the total of both amino sugars yields a molar ratio of about 1.0 on the basis of lysine.

4. In the other species and in strains of *Methanobacterium thermoautotrophicum* isolated by Wiegel (JW strains), the total of the two amino sugars considerably exceeds a molar ratio of 1.0.

In *Methanobrevibacter ruminantium* M1 glucosamine can be removed by incubating the cell walls with hydrofluoric acid at 0°C indicating that glucosamine is merely a single substituent linked to pseudomurein via a phosphodiester bond (Kandler and König, 1978). This finding is in agreement with the observation that glucosamine is completely destroyed when the cell walls are treated with periodate. It also agrees with the exceptionally high phosphate content of 4.5% found in these cell walls (Table I). Thus *N*-acetyl-D-galactosamine must be assumed to be the main amino sugar of the glycan moiety of the pseudomurein of *Methanobrevibacter ruminantium* M1. In *Methanobacterium bryantii* M.o.H. it was not possible to remove one of the two amino sugars from the isolated cell walls with hydrofluoric acid at 0°C or by extraction with formamide. Thus none of the amino sugars is bound to pseudomurein via a phosphodiester bond nor are they components of the neutral polysaccharides which can be extracted by formamide. At present it cannot be decided which of the two amino sugars is the component of the glycan strand and how the other amino sugar is attached to the cell wall. The frequent mutual replacement of the two amino sugars may be taken as an indirect confirmation of the presence of $(1{\rightarrow}3)$ bonds in pseudomurein, as pointed out by König *et al.* (1982): Since carbons 3 of glucosamine and galactosamine possess the same anomeric configuration a mutual replacement of the two amino sugars does not change the secondary structure of a $(1{\rightarrow}3)$ linked glycan such as pseudomurein. However, in the $(1{\rightarrow}4)$ linked murein a replacement of glucosamine by galactosamine would lead to a change of the secondary structure of the glycan strands since carbons 4 of the two hexosamines possess a different anomeric configuration. Consequently, no replacement of glucosamine by galactosamine is found in murein while this occurs frequently in pseudomurein.

Depending on the species or strain, not only the glycan but also the peptide moiety of pseudomurein exhibits distinct varations. Thus, in the pseudomurein of *Methanobrevibacter ruminantium* Ml, alanine is partly or completely replaced by threonine, and in that of *Methanobrevibacter smithii,* ornithine is attached to the α-carboxylic group of a glutamic acid residue (König *et al.,* 1982) as depicted in Fig. 4. Further studies on yet to be discovered members of Methanobacteriales may well disclose a variety of pseudomurein types comparable to that known in the case of murein. The hydrolysates of isolated sacculi from several species contain varying amounts of monosaccharides but no uronic acid, indicating that in some cases neutral polysaccharides are present in addition to pseudomurein. They can be removed by extracting the sacculi with hot formamide without destroying the integrity of the sacculi, as shown in the case of *Methanobacterium bryantii* M.o.H. (Kandler and König, 1978). Structure, location, and function of these accessory components of the cell wall sacculi still have to be elucidated.

b. Secondary and Tertiary Structure of Pseudomurein. X-Ray diffraction measurements and structure considerations on murein (Formanek *et al.,* 1974; Burge *et al.,* 1977; Labischinski *et al.,* 1979; Formanek, 1982) and pseudo-

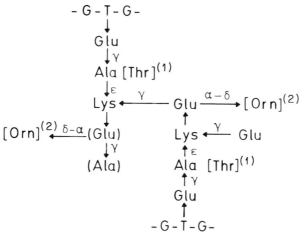

Fig. 4. Proposed amino acid sequence of cross-linked peptide subunits of pseudomurein and its modifications. G, *N*-acetyl-D-glucosamine; T, *N*-acetyl-L-talosaminuronic acid. (1) Modification found in *Methanobrevibacter ruminantium*. (2) Supposed alternative positions of ornithine in *Methanobrevibacter smithii*. (From König *et al.*, 1982.)

murein (Labischinski *et al.*, 1980; Formanek, 1985) have revealed several common structural features in both polymers. The high density in the range of ρ = 1.39–1.46 g/cm^3 of murein (Formanek *et al.*, 1974; Burge *et al.*, 1977; Labischinski *et al.*, 1979) and pseudomurein (Labischinski *et al.*, 1980) indicates a fairly high state of order, since a much lower density, in the range of ρ = 1.24–1.32 g/cm^3, is to be expected for amorphous polymers (Labischinski *et al.*, 1980). X-ray diffraction performed by the same authors yielded similar diffuse Debye–Scherrer rings with Bragg periodicities of about 0.45 and 0.94 nm in the planes of both types of cell walls and reflections corresponding to a periodicity of 4.3–4.5 nm when viewed vertically to the planes. These data have been interpreted in two different ways.

1. The peptide moieties of either murein or pseudomurein are radially oriented around the glycan strands, which form a screw rather than a ribbon. Three to four individual layers are needed to form a 4.5 nm periodicity vertical to the planes of the cell walls (Labischinski *et al.*, 1979, 1980).

2. The glycan strands perform twofold screw axes (ribbons). Consequently, all peptide moieties point in the same direction. Therefore, murein or pseudomurein strands, densely packed in the plane of the cell wall, form a quasi-double layer consisting of a "glycan phase" and a "peptide phase." Two superimposed layers, facing each other with either the glycan or the peptide phase, give rise to a periodicity of 4.5 nm vertical to the planes of the cell walls. In this case the X-ray data obtained may be interpreted as dimensions of an oblique-angled elementary

cell with a height of 2.23 nm and an area of 0.45 $\times$ 1 nm^2, which includes one disaccharide peptide subunit (Formanek 1982, 1983).

The twofold screw axis of the glycan strands of murein has been assumed to be similar to that of chitin (Kelemen and Rogers, 1971; Formanek *et al.,* 1974; Formanek, 1982). In pseudomurein, sterically, a twofold screw axis is also possible, since in the β(1 → 3) linkages of pseudomurein (König *et al.,* 1983) all glycosidically linked oxygen atoms are in equatorial positions (Fig. 5), if the β-*N*-acetyl-D-glucosamine residues are in $^4C_1$ and the β-*N*-acetyl-L-talosaminuronic acid residues in $^1C_4$ conformation* (Formanek, 1985). Such a flat secondary structure of the glycan moiety with all peptides pointing in the same direction allows a very dense packing which is necessary to achieve the experimentally observed high density of the isolated cell walls.

  c. Biosynthesis of Pseudomurein. The overall structural similarity of murein and pseudomurein suggests similar biosynthetic pathways. In fact, a search for possible pseudomurein precursors of structures similar to those involved in murein biosynthesis (cf. Tipper and Wright, 1979) resulted in the isolation of UDP-*N*-acetyl-D-glucosamine, UDP-D-galactosamine and a UDP-activated disaccharide composed of *N*-acetyl-D-glucosamine and *N*-acetyl-L-talosaminuronic acid with UDP bound to the carbonyl group of *N*-acetyl-D-glucosamine. However, neither UDP-*N*-acetyl-L-talosaminuronic acid nor a disaccharide containing a UDP-activated *N*-acetyl-L-talosaminuronyl residue could be detected in cell extracts of *Methanobacterium thermoautotrophicum* (H. König, unpublished). This suggests that L-talosaminuronic acid may be formed by epimerization and/or oxidation at the disaccharide level, subsequent to the transfer of a monomeric precursor to UDP-*N*-acetyl-D-glucosamine. Examples of related mechanisms are known. Thus, epimerization of UDP-*N*-acetyl-D-glucosaminuronic acid to a *N*-acetyl-D-mannosaminuronyl residue of the polymer has been supposed to occur in the course of teichuronic acid biosynthesis in *Micrococcus luteus* (Biely and Jeanloz, 1969), since only UDP-*N*-acetyl-D-glucosaminuronic acid, but not UDP-*N*-acetyl-D-mannosaminuronic acid served as a precursor. The C-5 epimerization of polymannuronic acid to a copolymer of D-mannuronic acid and L-guluronic acid was found to be catalyzed by an enzyme from the culture medium of *Azotobacter vinelandii,* and UDP-D-glucuronic acid was found to be the precursor of L-iduronyl residues in heparin (cf. Sharon, 1975).

  In addition to the UDP-activated disaccharide, UDP-activated peptides exhib-

---

*Designation of the conformation in accordance with: JUPAC-IUB Joint Commission on Biochemical Nocemclature (JCBN). (1980). Conformational nomenclature for five and six-membered ring forms of monosaccharides and their derivatives. *Eur. J. Biochem.* **111,** 295–298. The nomenclature applied in the publication of König *et al.* (1983) was different.

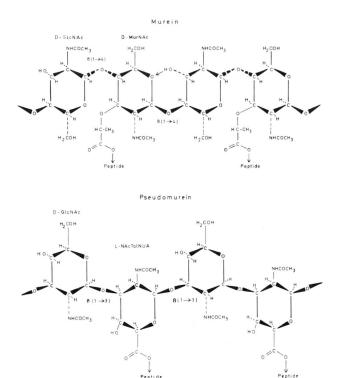

FIG. 5. Proposed secondary structure of glycan strands of murein and pseudomurein. (H. Formanek, unpublished.)

iting amino acid sequences typical of the peptide subunit of pseudomurein, could also be isolated from extracts of *Methanobacterium thermoautotrophicum* (H. König, unpublished). The nature of the UDP linkage and the role of the UDP-linked peptides in the biosynthesis of pseudomurein remains to be elucidated. As recently suggested, the excretion of alanine into the medium, observed with several strains of *Methanobacterium thermoautotrophicum* (Schönheit and Thauer, 1980), may indicate that alanine is liberated during the cross-linking reaction. Thus, the last step in the synthesis of a rigid pseudomurein macromolecule may well follow the same principle as in murein, but L- instead of D-alanine is utilized in pseudomurein.

Lysis of methanogens is often observed in pure cultures of methanobacteria (Jarrell *et al.,* 1982) and pseudomurein is certainly mineralized in nature. However, no particular organism or enzyme preparation that is able to lyse pseudomurein-containing cell wall sacculi is yet known. Thus, much remains to be done to understand the synthesis and breakdown of pseudomurein and of other cell wall components of the methanobacteriales.

2. THE HETEROPOLYSACCHARIDE-CONTAINING CELL ENVELOPES OF *Methanosarcina*

The cell envelopes of those strains of *Methanosarcina barkeri* investigated exhibit a profile of type 1 (Fig. 1). It shows only one extremely thick (up to 200 nm) rigid cell wall layer with a more or less distinct lamination at the outer boundary. The irregular cells of the globoidal packets share a common wall (Fig. 6).

The formation of macro- and microcysts surrounded by a very thick wall was observed in a strain of the so-called *Methanosarcina* biotype 3 by Zhilina and Zavarzin (1973). The cyst skins may not merely be thickened regular cell walls. However, no chemical analysis of such cyst skins exists as yet. Protein-free cell walls (Fig. 7) of several strains of *Methanosarcina barkeri* were isolated by the usual techniques applied to the isolation of cell walls from gram-positive eubacteria. The isolated walls were resistent to cellulolytic enzymes and chondroitin hydrolase. They could be solubilized by incubation in 10% borohydride at 60°C

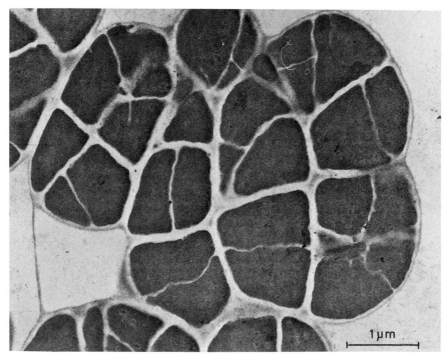

FIG. 6. Thin section of *Methanosarcina barkeri* (type strain). (O. Kandler and G. Wanner, unpublished data.)

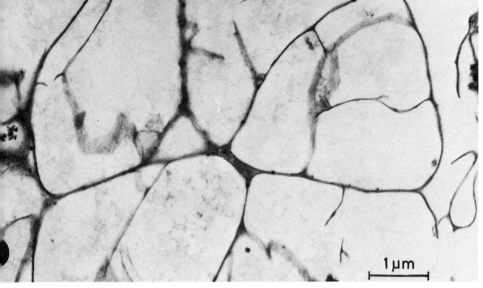

FIG. 7. Thin sections of isolated cell walls of *Methanosarcina barkeri* (type strain). (O. Kandler and G. Wanner, unpublished data.)

for 12 hours. The obtained dialyzed soluble fraction (molecular weight > 10,000) exhibited the same composition as the original cell walls (Kreisl, 1984).

Hydrolysates of cell wall preparations of *Methanosarcina* do not contain muramic acid, talosaminuronic acid, glucosamine, or significant amounts of amino acids, but mainly D-galactosamine, D-glucuronic acid, D-glucose, and acetic acid. The phosphate and sulfate content is neglegible (Kandler and Hippe, 1977).

The data of seven recently analyzed strains are summarized in Table II (P. Kreisl and O. Kandler, unpublished). It shows that the cell wall chemistry is fairly uniform in all strains including *Methanosarcina vacuolata,* whose cell wall was described by Zhilina (1971) as appearing to be triple-layered. Galactosamine and uronic acids are the main components, glucose is a minor component, and mannose is present in traces in all strains. The molar ratio of *N*-acetyl-D-galactosamine to uronic acid is generally about 2:1, but it is 3:1 in *Methanosarcina mazei.* The only distinct variation among the seven strains with respect to the qualitative composition is the partial replacement of glucuronic acid by galacturonic acid in two of the strains.

The main structural element, isolated from partial acid hydrolysates of several strains, is a disaccharide of D-glucuronic acid and *N*-acetyl-D-galactosamine with galactosamine at the reducing end. Its deacetylated derivative is identical with

TABLE II

Composition of Isolated Cell Walls of Seven Strains of *Methanosarcina*[a,b]

Compound	DSM 800[c]		DSM 804		Strain: Bryant G1		DSM 1232[d]		DSM 1825		DSM 2053[e]		DSM 1538	
	% d.w.	mmol/g	% d.w.	mmol/g	% d.w.	mmol/g	% d.w.	mmol/g	% d.w.	mmol/g	% d.w.	mmol/g	% d.w.	mmol/g
D-Glucose	3.75	0.20	5.0	0.27	3.8	0.21	2.2	0.12	0.5	0.027	0.42	0.023	3.5	0.19
D-Galactosamine	27.5	1.54	35	1.95	35	1.95	30.8	1.72	41.45	2.3	40.0	2.23	20.6	1.15
D-Glucuronic acid	16.0	0.82	14.8	0.76	3.5	0.18	18.0	0.92	18.8	0.97	6.9	0.35	11.2	0.58
D-Galacturonic acid	—	—	—	—	14.0	0.72	—	—	—	—	6.9	0.35	—	—
N-Acetyl residues	8.3	1.9	8.7	2.0	8.6	2.0	8.7	2.0	10.5	2.4	9.6	2.23	n.d.	
Ash	20.0		12.2		10.0		2.5		8.35		1.0		n.d.	
H$_2$O	14.0		8.4		14.0		16.0		14.6		16.8		n.d.	

[a] DSM = Deutsche Sammlung von Mikroorganismen. P. Kreisl and O. Kandler, unpublished.

[b] n.d., not determined; —, not present; % d.w., % dry weight.

[c] Type strain of *M. barkeri*.

[d] *M. vacuolata Zavarzin*.

[e] Originally supposed name: "*Methanococcus mazei*."

chondrosin as demonstrated by chemical degradation studies, paper and column chromatography, and ^{1}H NMR spectroscopy (Kreisl, 1984).

Thus, the cell wall polymer of *Methanosarcina* resembles chondroitin common to the connecting tissue of animals. However, it is not sulfated and contains a second mole of *N*-acetyl-D-galactosamine per mole of uronic acid. Most likely these additional amino sugar residues are constituents of the main glycan strands rather than being monomeric or oligomeric side branches, because no decrease of the *N*-acetyl-D-galactosamine content of the polymer was observed upon periodate oxidation of the isolated cell walls. In fact, upon reduction of the carboxyl groups of isolated cell walls, a trimer of the following tentative structure could be isolated:

$$[\rightarrow 4)\text{-}\beta\text{-}D\text{-GlcUA-}(1\rightarrow 3)\text{-}D\text{-GalNAc-}(1\rightarrow 3 \text{ or } 4)\text{-}D\text{-GalNAc-}(1\rightarrow]_x$$

This trimer is supposed to represent the repeating unit of the main cell wall polymer of *Methanosarcina* (Kreisl, 1984).

3. THE CELL ENVELOPE OF *Halococcus* CONTAINING A SULFATED HETEROPOLYSACCHARIDE

The profile of the cell envelope of *Halococcus morrhuae* belongs to type 1 (Fig. 1) and exhibits only one electron-dense layer, 50–60 nm in width, outside the cytoplasmic membrane (Fig. 8; Kocur *et al.,* 1972). The cell wall tends to become laminated as also observed in the case of *Methanosarcina*. The outer lamina is often more electron dense. The cells forming cuboidal packets share a common envelope, which, by lamination, may be derived from the wall of the single cell from which the packet arose by successive cell divisions. Protein-free cell walls were isolated by the usual techniques applied to the isolation of cell walls of gram-positive eubacteria. They are rather stable and could not be lysed by the application of 13 different endo- and exoglycosidases. However, they could partly be solubilized by mild alkaline treatment (0.5 *N* NaOH, 60°C, 12 hours).

The solubilized and the undissolved material exhibited the same complex chemical composition (Schleifer *et al.,* 1982). The cell wall composition of *Halococcus* (Table III) resembles that of *Methanosarcina*. However, it is much more complex in that it contains several neutral and amino sugars, uronic acids, and glycine (Brown and Cho, 1970; Reistad, 1972; Steber and Schleifer, 1975, 1979; Schleifer *et al.,* 1982). Whereas the typical components of murein or pseudomurein are missing, the cell walls contain gulosaminuronic acid as a specific component, as first described by Reistadt (1974). A second peculiarity of the *Halococcus* cell walls is their high sulfate content. Detailed studies have revealed that all neutral and amino sugars and uronic acids are partly sulfated as summarized in Table IV (Schleifer *et al.,* 1982). A further uncommon feature is

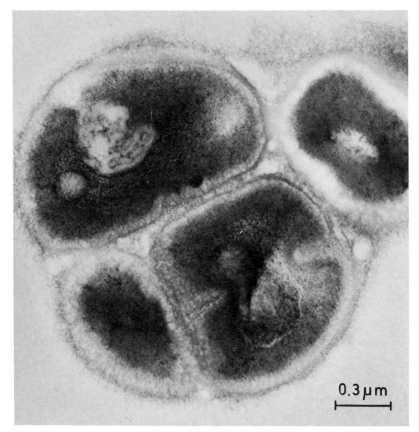

FIG. 8.   Thin sections of *Halococcus morrhuae* (O. Kandler and G. Wanner, unpublished data.)

the partial replacement of *N*-acetyl-D-glucosamine by *N*-glycyl-D-glucosamine. As a result of their studies, Schleifer *et al.* (1982) suggested a scheme of the primary structure of the sulfated heteropolysaccharide shown in Fig. 9 stating:

> Based on extraction and fragmentation studies one can distinguish at least two domains within the heteroglycan. Domain I primarily consists of uronic acid, galactosamine, mannose and glucose. Various oligosaccharides from this domain are found in partial acid hydrolysates of cell walls. Domain II primarily consists of uronic acids, glucosamine, gulosaminuronic acid and galactose and cannot be extracted as easily as components of domain I. The two domains are connected by covalent linkages. Galactose residues of domain II constitute a possible branching point. However, a part of the glycan chains may be covalently linked by glycine residues. At least some of the glucosamine residues are *N*-glycyl substituted. Dinitrophenylation and hydrazinolysis of cell walls indicated that the predominant portion of glycine residues is covalently linked by both functional groups to cell wall components (Steber and Schleifer, 1979). Glycyl peptide bridges may exist between glucosamine and uronosyl residues of the glycan strands. Thus, at least part of the structure of the cell wall of *Halococcus morrhuae* may

TABLE III

CHEMICAL COMPOSITION OF PURIFIED (TRYPSIN-TREATED) CELL WALLS OF THREE
STRAINS OF *Halococcus morrhuae*[a]

Components	μmol/mg cell wall		
	CCM 859	CCM 537	CCM 889
Glucose	0.44	0.39	0.30
Mannose	0.35	0.21	0.30
Galactose	0.27	0.18	0.17
Total neutral sugars	1.06	0.78	0.77
Glucosamine	0.38	0.21	0.35
Galactosamine	0.20	0.08	0.24
Gulosaminuronic acid	0.11	0.09	0.10
Total amino sugars	0.69	0.38	0.69
Uronic acids[b]	0.67	0.24	0.63
Acetate	0.62	n.d.	n.d.
Glycine	0.1	0.13	0.1
Sulfate	1.47	0.8	0.18

[a] From Schleifer *et al.* (1982).

[b] Total uronic acids determined as glucuronic acid. Glucuronic and galacturonic acids were present in cell walls of CCM 859 in a molar ratio of about 2.3:1.

TABLE IV

SULFATE SUBSTITUTION IN SUGARS, URONIC ACIDS, AND AMINO SUGARS AS DETERMINED
BY PERIODATE OXIDATION, SMITH DEGRADATION, AND METHYLATION STUDIES OF INTACT
AND DESULFATED CELL WALL PREPARATIONS OF *Halococcus morrhuae*[a]

Constituent	Hydroxyl groups substituted by ester-bound sulfate
Sugar	
Glucose	—
Mannose	—
Galactose	Major part: C-2
	Minor part: C-3
Amino sugar	
Glucosamine	—
Galactosamine	C-3
Uronic acid	
Gulosaminuronic acid	—
Galacturonic acid	About 20%: C-3
Glucuronic acid	C-3
	(C-4)

[a] Data from Schleifer *et al.*, 1982.

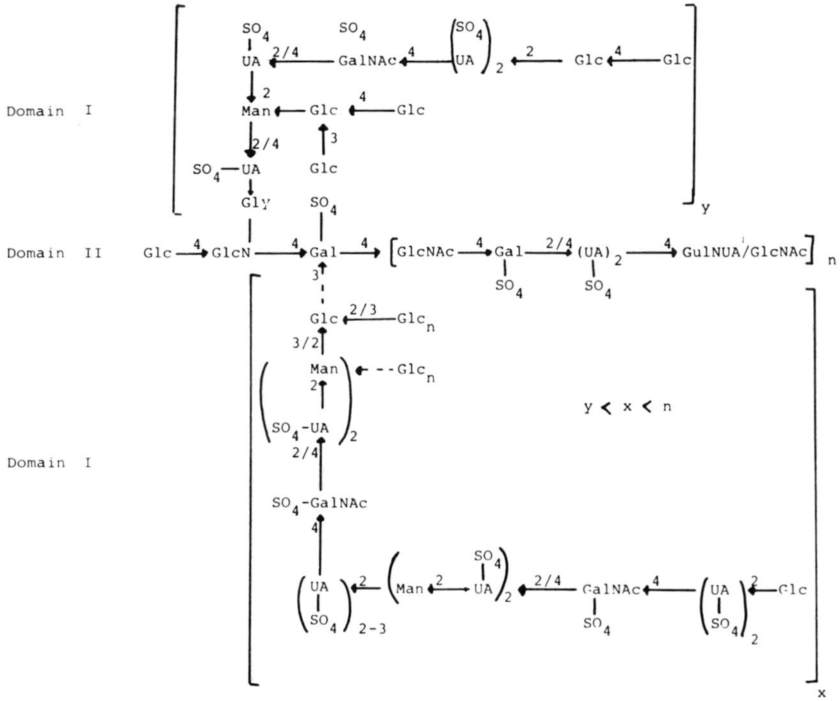

FIG. 9. Proposed linkage of the sugar residues within the cell wall of *H. morrhuae* CCM 859. Gal, galactose; GalNac, *N*-acetylgalactosamine; Glc, glucose; GlcNac, *N*-acetylglucosamine; Gly, glycine; GulNUA, *N*-acetylgulosaminuronic acid; Man, mannose; UA, uronic acid. (From Schleifer *et al.*, 1982.)

be similar to that of bacterial peptidoglycan or pseudomurein of methanobacteria, in that glycan strands are cross-linked through peptide linkages.

B. Cell Envelopes of Gram-Negative Archaebacteria

1. Cell Envelopes Composed of Protein or Glycoprotein Subunits

As pointed out earlier, most gram-negative archaebacteria exhibit only one surface layer (S layer) of regularly arranged protein or glycoprotein subunits outside the cytoplasmic membrane (profile type 3, Fig. 1). The S layers are usually very easily disintegrated by mechanical forces, detergents, change of ionic strength, etc., thus resembling the S layers of eubacteria (Sleytr and Messner, 1983). However, the envelopes of some thermoacidophilic archaebacteria are much more rigid. When they are isolated by mechanical disintegration of the cells, they still show the original shape of the cell even when treated with hot detergents, thus indicating that the subunits may be linked to each other by unusually strong intermolecular forces or even by covalent bonds.

a. S Layers of Thermoacidophiles. The occurrence of detergent-resistant rigid S layer sacculi is typical of the anaerobic, thermophilic, sulfur-respiring organisms belonging to the order Thermoproteales, recently discovered by Zillig *et al.*, (1981, 1982). *Thermoproteus tenax,* a thin, slender rod, often showing true branching, possesses an S layer—17 nm in width—of hexagonally arranged subunits (Zillig *et al.*, 1981). The S layer could be isolated by disrupting the cells by sonification followed by incubation with DNase and RNase, SDS treatment (2% SDS, 30 min, boiling water bath), and centrifugation (H. König and K. O. Stetter, unpublished results). The isolated rod-shaped S layers (Fig. 10) are resistant to proteinase K, trypsin, and pronase (W. Zillig, personal communication). Chemical analysis of acid hydrolysates revealed a spectrum of 18 amino acids and significant amounts (ca. 20% of dry matter) of carbohydrates (Table V) indicating the presence of glycoprotein.

The S-layer of the extremely thin and long rods of *Thermofilum pendens* also shows hexagonal arrays of subunits and possesses the same rigidity as that of *Thermoproteus tenax* (Zillig *et al.*, 1983b).

The spherical cells of *Thermococcus celer* also possess an envelope of hexagonally arranged subunits (Zillig *et al.*, 1983a), whereas the spherical cells of *Desulfurococcus mucosus* and *Desulfurococcus mobilis* exhibit a 25-nm thick

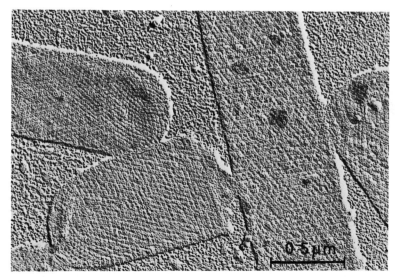

Fig. 10. Platinum-shadowed isolated S layers of *Thermoproteus tenax.* (H. König and K. O. Stetter, unpublished data.)

envelope, which is presumably composed of two layers (Zillig *et al.*, 1982), one of them showing arrays of tetragonally arranged subunits (Fig. 11). *Desulfurococcus mucosus* is surrounded by an additional capsule (Zillig *et al.*, 1982), which is composed mainly of neutral sugars (W. Zillig, personal communication).

Extremely thick S-layers (Fig. 12), 25–40 nm in width, are found in *Pyrodictium occultum* (Stetter *et al.*, 1983), an organism that grows even at 105°C (Stetter, 1982), and in *Thermodiscus maritimus* (König and Stetter, unpublished). The S layers of both organisms were sensitive to SDS treatment (2% SDS, 30 min, boiling water). In the case of *Pyrodictium occultum*, SDS–polyacrylamide gel electrophoresis of the solubilized protein resulted in only one band of about 172,000 daltons. It stained with the periodate–Schiff reagent (PAS) for carbohydrates thus indicating the presence of glycoprotein.

The S layers studied in most detail were those of the coccoid, multilobed

Fig. 11. Platinum-shadowed S-layer patches of *Desulfurococcus mucosus*. (H. König and K. O. Stetter, unpublished data.)

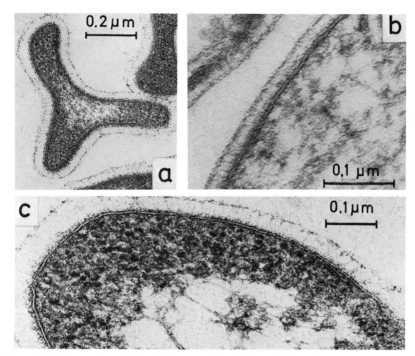

FIG. 12. Thin sections of *Pyrodictium occultum* (a,b) and *Thermodiscus maritimus* (c). (H. König and K. O. Stetter, unpublished data.)

Sulfolobus acidocaldarius (Weiss, 1974). The S layers were isolated by lysing the cells with SDS (0.15%), digestion with DNase, and repeated treatment of the cell-shaped S layers with SDS. The isolated S layers were disintegrated with phosphate buffer pH 9 at 60°C, and the solubilized protein was purified by chromatography on Sepharose (Michel *et al.*, 1980). Chemical analysis revealed that the S layer was composed of a single glycoprotein occurring in two modifications of apparent molecular weights of about 140,000 and 170,000, respectively. Both forms stained with the PAS reagent for carbohydrates and exhibited the same amino acid composition (Table V). High values for serine and threonine and low values for basic amino acids and dicarboxylic amino acids were obtained. Tryptophan and cysteine may be absent. Hexosamine was detected, but the structure of the carbohydrates was not determined (Michel *et al.*, 1980).

A PAS-positive S layer protein with an apparent molecular weight of 140,000 was also found in the second species, *Sulfolobus solfataricus*, while in S layers of *Sulfolobus brierley*, the third member of this genus, no PAS-positive protein was detected (H. König and K. O. Stetter, unpublished).

Under the electron microscope, plane specimens of purified S layers showed a hexagonal lattice, two-sided plane group p6, with a 22-nm unit cell dimension.

TABLE V

Chemical Composition of S Layers or Isolated Surface Glycoproteins of Gram-Negative Archaebacteria[a]

Component	Thermoproteus tenax[b]		Sulfolobus acidocaldarius[c]	Halobacterium salinarium[d]		Methanospirillum hungatei				Methanothrix soehngenii[g]		Thermoplasma acidophilum[h]		Methanococcoides methylutens[i]		Methanolobus tindarius[j]	
						strain JF1[e]		strain GP1[f]									
	I	II	mol %	I	II	I	II	I	II	I	II	I	II	I	II	I	II
Amino acids																	
Ala	0.598	3.70	7.0	0.38	19.00	0.69	23.00	0.47	11.75	0.445	1.92	0.58	5.80	0.147	3.24	0.7033	5.08
Arg	0.149	0.92	0.3	0.13	6.50	0.07	2.33	0.09	2.25	0.155	0.67	0.28	2.80	0.091	2.01	0.3651	2.64
Asp	0.530	3.28	9.3	1.34	67.00	0.88	29.33	0.71	17.75	0.826	3.56	0.80	8.00	0.187	4.14	0.7674	5.54
Cys	0.024	0.15	0.07	n.d.	—	0.04	1.33	0.06	1.50	0.016	0.07	n.d.	—	0.025	0.55	0.0285	0.21
Glu	0.445	2.75	6.6	0.74	37.00	0.52	17.33	0.54	13.50	1.109	4.78	0.54	5.40	0.177	3.92	0.8533	6.16
Gly	0.481	2.98	8.9	0.56	28.00	0.52	17.33	0.36	9.00	0.909	3.92	0.70	7.00	0.152	3.39	0.6992	5.07
His	0.162	1.00	0.3	0.02	1.00	0.03	1.00	0.04	1.00	0.232	1.00	0.10	1.00	0.045	1.00	0.1385	1.00
Ile	0.251	1.55	4.7	0.26	28.00	0.17	5.66	0.26	6.50	0.181	0.78	0.42	4.20	0.113	2.50	0.3677	2.63
Leu	0.516	3.19	9.8	0.31	15.50	0.37	12.33	0.33	8.25	0.342	1.47	0.60	6.00	0.134	2.97	0.5903	4.26
Lys	0.171	1.06	2.2	0.10	5.00	0.22	7.33	0.30	7.50	0.290	1.25	0.31	3.10	0.121	2.67	0.2953	2.14
Met	0.074	0.46	0.9	0.05	2.50	0.07	2.33	0.09	2.25	0.012	0.05	0.04	0.40	0.052	1.15	n.d.	—
Phe	0.224	1.39	4.0	0.11	5.50	0.22	7.33	0.18	4.50	0.135	0.58	0.40	4.00	0.057	1.27	0.2734	1.97
Pro	0.293	1.81	8.7	0.27	13.50	0.15	5.00	0.14	3.50	0.181	0.78	0.08	0.80	0.070	1.56	0.2417	1.66
Ser	0.464	2.87	10.3	0.65	32.50	0.38	12.66	0.39	9.75	0.497	2.14	0.74	7.40	0.100	2.26	0.5216	3.77

Thr	0.588	3.63	12.4	0.58	29.00	0.36	12.00	0.50	12.50	0.626	2.69	0.59	5.90	0.079	1.72	0.4679	3.38
Trp	0.012	0.07	0.06	n.d.	—	0.05	1.66	0.22[k]	5.50	n.f.	—	n.d.	—	n.d.	—	n.f.	—
Tyr	0.388	2.40	6.7	0.16	8.00	0.22	7.33	0.14	3.50	0.219	0.94	0.52	5.20	0.052	1.15	0.7255	1.25
Val	0.419	2.59	7.2	0.47	23.50	0.22	7.33	0.35	8.75	0.439	1.89	0.54	5.40	0.123	2.72	0.4280	3.09
NH₃	0.859	5.31	—	—	—	0.59	19.67	n.d.	—	0.497	2.14	n.d.	—	n.d.	—	1.3263	95.86
Carbohydrates																	
Ara	0.405	2.50	—	—	—	0.19	6.33	n.f.	—	n.f.	—	n.f.	—	—	—	n.f.	n.f.
Gal	n.f.	—	—	—	—	0.13	4.33	0.03	0.75	n.f.	—	0.03	0.30	—	—	n.f.	n.f.
Glc	0.405	2.50	—	0.42[l]	21.00	0.10	3.33	0.02	0.50	0.260	1.11	0.05	0.50	—	—	19.614	141.62
GlcN	0.070	0.43	2.7[m]	0.09[n]	4.50	n.f.	—	n.f.	—	n.f.	—	0.01	0.10	—	—	n.f.	n.f.
Man	0.222	1.37	—	—	—	0.03	1.00	0.02	0.50	0.580	2.50	0.52	5.20	—	—	n.f.	n.f.
Rha	0.182	1.12	—	—	—	0.04	1.33	0.16	4.00	n.f.	—	n.f.	—	—	—	n.f.	n.f.
Rib	n.f.	—	—	—	—	n.f.	—	0.14	3.50	n.f.	—	n.f.	—	—	—	n.f.	n.f.

[a] I, μmol/mg dry weight; II, molar ratio (His = 1); n.d., not determined; n.f., not found.
[b] S layer (H. König and K. O. Stetter, unpublished).
[c] Isolated S layer glycoprotein (Michel et al., 1980).
[d] Isolated S layer glycoprotein (Mescher and Strominger, 1976).
[e] Protein sheath (Kandler and König, 1978; supplemented).
[f] Protein sheath (calculated from Sprott and McKellar, 1980).
[g] Protein sheath (H. König and K. O. Stetter, unpublished).
[h] Isolated cytoplasmic membrane glycoprotein (Yang and Haug, 1979).
[i] S layer (Sowers and Ferry, 1983).
[j] Isolated S layer glycoprotein (F. Fischer, H. König, and K. O. Stetter, unpublished).
[k] Uncertain.
[l] Total hexose content.
[m] Total hexosamine content.
[n] Total amino sugar content.

The three-dimensional structure of the S layer of *Sulfolobus acidocaldarius* (Fig. 13) was elucidated by electron microscopy to 1.7 nm resolution using procedures similar to those applied by Henderson and Unwin (1975) for elucidating the three-dimensional structure of the purple membrane of *Halobacterium halobium* (Taylor *et al.*, 1982; Deatherage *et al.*, 1983). The model shows dimer units arranged to form a series of hexagonal and triangular holes. Although the protein occupies only 30% of the about 10-nm thick S layer, only the hexagonal holes on the sixfold axis pass directly through. These holes show a diameter of about 5 nm at the outside of the S layer and form a large domed cave at the inside. The holes on the threefold axis are 4.5 nm in diameter and split into three channels of a 2.5-nm diameter, which open into the side of the large cave. The external surface of the S layer is fairly smooth, whereas the inner surface is sculptured with large cavities and protruding "pedestals" anchoring the subunits to the surface, rather than the lipid bilayer of the cytoplasmic membrane. Hence there seems to be no room for a periplasmatic space in addition to the caves and channels within the S layer.

The protein substructure consists of three types of globular domains—diad (D), triad (T), ring region (R)—connected by narrow bridges. These may act as "hinges," allowing the S layer to form a curved surface (lobes) and to follow the movements of the cell surface during growth. The very loose network of the *Sulfolobus* S layer glycoprotein with its large channels is certainly not able to act as a molecular sieve as was supposed in other cases (Stewart and Beveridge, 1980; Stewart *et al.*, 1980), but it still maintains its supportive function and provides the cell with as much access to the medium as necessary. It is worthy of note that the larger pores of the S layer are the same size as the pili, which attach the cells to sulfur crystals (Weiss, 1973).

b. S Layers of Halobacteria. The genus *Halobacterium* includes a number of gram-negative, rod-shaped or pleomorphic bacteria (Larsen, 1967; Dundas, 1977). Morphological and chemical studies of cell envelopes of halobacteria

---

FIG. 13. Model of a small area of the S layer of *Sulfolobus acidocaldarius*. (a) The external surface; (b) the surface that lies next to the plasma membrane. The large hole at the center of the model lies on one of the six-fold axes; similar sixfold axes are located at the edges of the portion shown. Smaller channels through the layer lie around the three-fold axes. The darker parts of the model represent the simplest repeating unit from which the S layer could be formed. Even though the S layer consists of a single species of glycoprotein, the repeating unit is very complex and constitutes several different domains. The actual molecular boundaries are unknown but are likely to be even more complex (Deatherage *et al.*, 1983). T, One-third of a triad; D, one-half of a diad; R, one-sixth of a ring domain; H1, H2, H3, H4, "hinges" between domains. (From Taylor *et al.*, 1982. Reprinted by permission from *Nature*, Vol. 299, pp. 840–842. Copyright © 1982 Macmillan Journals Limited.)

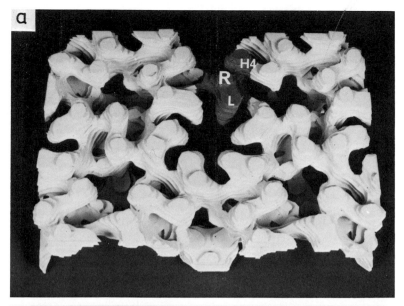

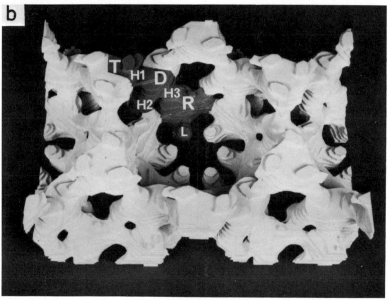

failed to demonstrate a rigid murein sacculus and an outer membrane containing lipopolysaccharide typical of gram-negative eubacteria (Brown and Shorey, 1963; Kushner *et al.*, 1964; Stoeckenius and Rowen, 1967; Steensland and Larsen, 1969). Instead, the cells of *Halobacterium halobium* are surrounded by a single layer of hexagonally arranged subunits, first shown by Houwink (1956). In thin sections, these S layers appear scalloped with rounded projections spaced at periods of 13.0 nm. They show an outer dense band of 3.0 nm, followed by a lighter 5.0 nm wide zone bordering another dense band, 5.0 nm wide, that merges with the outer dense zone of the cytoplasmic membrane (Fig. 14; Robertson *et al.*, 1982). Based on X-ray diffraction studies, Blaurock *et al.* (1976) also assumed the existence of two protein layers of similar width outside the cytoplasmic membrane. Both authors (Robertson *et al.*, 1982; Blaurock *et al.*, 1976) considered the light zone between the two layers as "periplasmatic space." However, the comparison of micrographs of cross-sections of halobacteria and other archaebacteria possessing S layers indicates that the two assumed extracytoplasmic layers may merely represent the inner and outer surface of the S layer, which are more osmophilic and thus more electron-dense than the central portion, as more clearly seen in the thin sections of other archaebacteria (Figs. 1 and 12). In addition, the width of the extracytoplasmic envelope corresponds very well to that of the other S layer bearing archaebacteria. According to this interpretation of the micrographs, the S layer is very tightly joined to the cytoplasmic membrane, as is generally observed in archaebacteria, thus leaving no room for a periplasmatic space. To maintain the rod shape, halobacteria require high concentrations of NaCl (about 25%), which can be replaced by K^+, Mg^{2+}, or Ca^{2+} ions to some extent (Brown, 1964; Abram and Gibson, 1961). At lower NaCl concentration, they form spheres, and lysis occurs. In the case of *Halobacterium halobium,* no cells were left intact at NaCl concentrations of 0.5 M (Stoeckenius and Rowen, 1967). Thus, S layers exhibiting the shape of the original cell can only be obtained by mechanical disintegration of the cells in high salt concentrations (Stoeckenius and Rowen, 1967). Such preparations, still containing material of the cytoplasmic membrane, account for 20% of the cell dry weight and are composed of protein (63%), hexoses (3.6%), amino sugars (0.6%), and lipids (21.0%) (Mescher *et al.*, 1974).

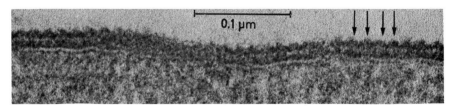

FIG. 14.   Thin section of cytoplasmic membrane and S layer of *Halobacterium halobium*. (From Robertson *et al.*, 1982.)

The isolated S layers are easily solubilized by lowering the salt concentration and treatment with SDS and mercaptoethanol (Mescher and Strominger, 1978). SDS-polyacrylamide gel electrophoresis of solubilized crude S layers yielded 15–20 protein bands. The majority of bands possessed a molecular weight of 130,000 or less and were PAS-negative, whereas one band, accounting for 50–60% of the total protein, exhibited a molecular weight of about 200,000 and was PAS-positive (Mescher and Strominger, 1978). Further purified [extraction with phenol and a mixture of chloroform and methanol (2:1)] and solubilized S layers yielded almost exclusively the high molecular glycoprotein with a carbohydrate content of 10–12% (Table V) and an excess of 12.5 mol% of acidic amino acids over the basic amino acids. The covalent linkage of the carbohydrate moiety to the peptide was demonstrated by isolating glycopeptides subsequent to the digestion of the protein with trypsin and pronase (Mescher and Strominger, 1976). Two types of glycopeptides were found: one carrying di- and trisaccharides of glucose, galactose, and an unidentified uronic acid in O-glycosidic linkages to threonine residues, and one with a larger heterosaccharide N-glycosidically bound to asparagine via N-acetylglucosamine. Recently, Wieland *et al.* (1980, 1981, 1982) found that cell envelopes of *Halobacterium halobium* contain sulfate. A high molecular weight sulfated glycopeptide was isolated after digestion of the glycoprotein with pronase. It has a sulfated repetitive structure similar to animal mucins rather than a heterosaccharide as proposed by Mescher and Strominger (1976). The repeating units of the saccharide moiety are composed of 1 Gal, 1 GalN, 1 GlcN, 2 GalUA, and 2 SO_4^{2-}. In addition, Wieland *et al.* (1982, 1983) found sulfated saccharides of low molecular weight, which contain glucose and glucuronic acid at a molar ratio of 1:1. A new type of N-glycosidic linkage has been found in this latter type of saccharide: Glucose is N-glycosidically linked to the polypeptide chain via the amido nitrogen of an asparagine residue, whereas the only N-glycosidic linkage so far known is Asn-GlcNAc. Biosynthesis of S layer glycoprotein involves lipid (C_{55}) intermediates. The formation of polyisoprenyl derivatives of glucose, mannose, and N-acetyl-D-glucosamine was shown in cell homogenates. N-Acetyl-D-glucosamine transferase activity was demonstrated in cell envelopes and evidence has been obtained for the formation of a lipid containing both glucosamine and mannose. The following pathway for the attachment of the N-linked heterosaccharide to the glycoprotein of *Halobacterium salinarium* was proposed by Mescher and Strominger (1978): Inside the cells:

$$\text{UDP-NAc-GlcN} + \text{lipid-P} \rightarrow \text{lipid-PP-NAcGlcN} + \text{UMP}$$
$$\text{Y-XDP} + \text{lipid-PP-NAc-GlcN} \rightarrow \text{lipid-PP-NAc-GlcN-Y} + \text{XDP}$$

Outside the cells:

$$\text{lipid-PP-NAc-GlcN-Y} + \text{protein} \rightarrow \text{protein-NAc-GlcN-Y} + \text{lipid PP}$$
$$\text{lipid-P} = C_{55-60} \text{ isoprenylphosphate}$$

Y equals additional sugars.

In contrast to the biosynthesis of eukaryotic glycoproteins the saccharides of the high molecular weight glycopeptide is synthesized while bound to the lipid carrier, and the fully sulfated saccharide is then transferred to the protein as found by Wieland *et al.* (1982).

Bacitracin interferes with the biosynthesis of the high molecular weight sulfated heterosaccharide (Wieland *et al.,* 1982) and leads to the formation of spheres (Mescher and Strominger, 1978).

A double-layered cell envelope of a total thickness of 40 nm is found in halophilic, flat, and rectangular bacteria (Walsby, 1980; Stoeckenius, 1981; Kessel and Cohen, 1982). The outer layer, about 20 nm in width, is fairly electron-transparent and shows arrays of globules with a quasi-spacing of 18 nm. Between this S layer and the cytoplasmic membrane, a well-defined electron-dense layer, 14 nm in width, is found. Shadowed replicas revealed the presence of envelopes of two lattice types (Kessel and Cohen, 1982): a hexagonal lattice with a constant of 15.9 or 22 nm and a tetragonal lattice with a constant of 14.6 nm. In addition, cells surrounded by a fibrillar sheath were frequently found. This indicates that the investigated population consisted of different species. No chemical data of envelopes of these organisms are available.

c. S Layers of Gram-Negative Methanogens. The five gram-negative genera *Methanomicrobium, Methanogenium, Methanoplanus, Methanolobus,* and *Methanococcus* possess S layers composed of protein or glycoprotein subunits as the only envelope component (profile type 3, Fig. 1). The cells of *Methanomicrobium mobile,* a rod-shaped methanogen (Paynter and Hungate, 1968; Balch *et al.,* 1979), are osmotically very fragile, and no rigid cell wall sacculi can be isolated. Treatment with SDS (2% SDS, 30 min, 100°C) resulted in complete solubilization of the cells (Kandler and König, 1978). SDS-polyacrylamide gel electrophoresis of disintegrated S layers revealed several protein bands, but none is PAS-positive (H. König and K. O. Stetter, unpublished), thus indicating that the S layer consists of proteins, which do not contain carbohydrates.

Thin sections of *Methanogenium marisnigri* and *Methanogenium cariaci,* irregular peritrichously flagellated cocci (Romesser *et al.,* 1979), revealed an S layer of 9.5–10.5 nm in width, with hexagonal arrays of subunits (Fig. 15; Romesser *et al.,* 1979).

The cells are sensitive to SDS treatment (Kandler and König, 1978). Isolated S layers yielded several protein bands when subjected to SDS-polyacrylamide gel electrophoresis. The main surface protein of *Methanogenium marisnigri* exhibits a molecular weight of 143,000 and is PAS-positive, indicating the presence of a glycoprotein. However, in the S layers of *Methanogenium cariaci,* the heaviest protein band exhibits a molecular weight of 125,000 and does not stain with PAS (H. König and K. O. Stetter, unpublished). Hence, the S layer of *Methanogeni-*

FIG. 15. Platinum-shadowed surface of *Methanogenium marisnigri*. (F. Mayer, Göttingen, unpublished data.)

um cariaci is composed of protein subunits, whereas that of *Methanogenium marisnigri* probably contains glycoprotein.

The two species of *Methanogenium* also differ markedly with respect to their immunological relationship (Conway de Macario *et al.*, 1981) and G + C content (Romesser *et al.*, 1979).

Also the cells of *Methanoplanus limicola*, which form plates with a thickness of only 50–70 nm in the center, exhibit an S layer of hexagonal arrays with a spacing of 14 nm (Wildgruber *et al.*, 1982). The cells lyse completely when suspended in an SDS solution (2 g/100 ml). The main protein band, obtained by SDS-polyacrylamide gel electrophoresis of the disintegrated S layer, possesses a molecular weight of 143,000. It stains with Coomassie blue and is PAS-positive, indicating that it may consist of a glycoprotein (Wildgruber *et al.*, 1982). Because the cells contain no rigid cell wall, it is unclear how the unusual shape of these flat bacteria is maintained.

A glycoprotein of 156,000 daltons is also the main component of the S layer of the methylamine-utilizing *Methanolobus tindarius* (König and Stetter, 1982), as evidenced by SDS gel electrophoresis and a positive PAS reaction of solubilized isolated S layers. Its amino acid composition is given in Table V. Glucose is the only carbohydrate detected. The S layer, 10 nm wide, is the only envelope component outside the cytoplasmic membrane, and it shows hexagonal arrays of subunits with a spacing of 12.4 nm (Fig. 16; F. Fischer, H. König, and K. O. Stetter, unpublished).

The recently described methylamine-utilizing genus *Methanococcoides* (Sowers and Ferry, 1983) may be considered as a synonym of *Methanolobus*, as evidenced by the virtually identical morphological and physiological characters given in the respective descriptions. The differences between the two described

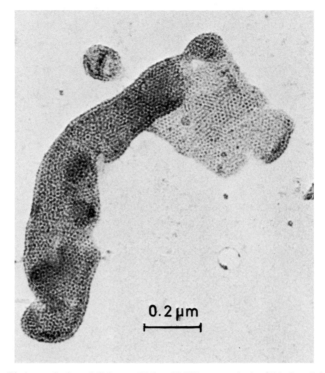

Fig. 16.   Platinum-shadowed S layers (Triton X-100 preparation) of *Methanolobus tindarius.* (F. Fischer, H. König, and K. O. Stetter, unpublished.)

species with respect to flagellation may be species rather than genus specific. The cell envelope consists in both species of an S layer. The overall amino acid composition of the isolated envelope of *Methanococcoides methylutens* is given in Table V. No test for the presence of a glycoprotein is mentioned.

The gram-negative, spherical to irregular cells of *Methanococcus vannielii* exhibit only one layer outside the cytoplasmic membrane. It is 18 nm thick and exhibits a hexagonal pattern with a spacing of 10.8 nm, originally described as possessing a tetragonal pattern (Jones *et al.*, 1977). Cell envelopes obtained after disrupting the cells by sonification showed no arrays of subunits. Regular surface patterns were only visible after treating the envelopes with Triton X-100 (Fig. 17), indicating that material covering the subunits had to be removed.

When the isolated envelopes were subjected to SDS-polyacrylamide gel elec-trophoresis, seven PAS-negative protein bands were obtained (Conway de Mac-ario *et al.*, 1984). The major band possessed a molecular weight of 60,000 and accounts for 20% of the total envelope protein. Similar results were obtained

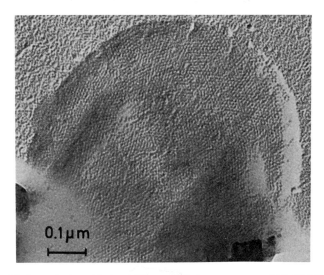

FIG. 17.   Platinum-shadowed isolated S layer of *Methanococcus vannielii*. (H. König and K. O. Stetter, unpublished data.)

with *Methanococcus voltae* and *M. thermolithotrophicus* having main cell wall proteins with molecular weights of 76,000 and 87,000, respectively (H. König and K. O. Stetter, unpublished).

2. SHEATHS MADE UP OF PROTEIN FIBRILS

The most unusual and complex cell envelopes, characterized by a fibrillary sheath, are found in *Methanospirillum hungatei* and *Methanothrix soehngenii*.

a. Methanospirillum hungatei. Several rods are held together by a fibrillary sheath (Fig. 18 a,b) forming filaments, the typical long "spirilla" observed under the light microscope. The single rods are surrounded by an electron-dense layer, 13.6 nm in width (Fig. 19b). Its chemical structure is unknown. This layer does not participate in septum formation during cell division (Fig. 19a), although it forms a continuous layer again shortly after the cell division is completed. It is not ruptured during formation of spheroplasts by dithiothreitol and sticks to the cytoplasmic membrane after Triton X treatment (Sprott *et al.,* 1979). This behavior indicates that it may consist of protein. The sheath (outer wall in the terminology of Zeikus, 1977) does not participate in septum formation (Fig. 19a). However, two "structural elements", which are separated by a "spacer"

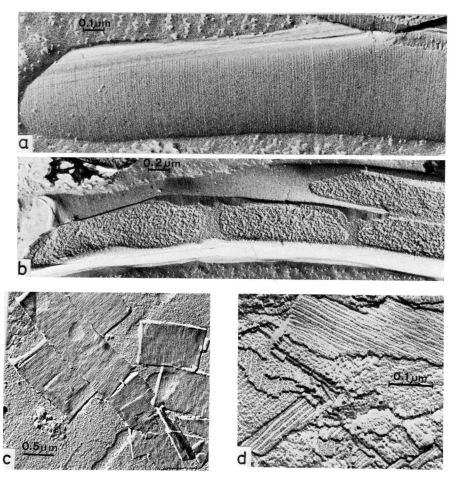

FIG. 18. *Methanospirillum hungatei.* Freeze-etching of the sheath surface (a) and of a longitudinal break through a filament (b); platinum-shadowed isolated sheaths (c), freeze-etching of isolated sheaths (d). (From Kandler, 1979.)

and connected with the sheath, are formed between some of the cells. The filaments break apart at such locations thus leading to the multiplication of filaments.

The usual techniques for cell wall isolation lead to pure preparations of sheath material (Kandler and König, 1978; Kandler, 1979). The cylindrical pieces break very frequently along the fibrils (Fig. 18c), which are oriented perpendicularly to the long axis of the sheath. Higher magnification of a freeze-etched specimen (Fig. 18d) shows each fibril to be composed of two subfibrils.

The isolated sheath material is resistant to detergent and all proteases com-

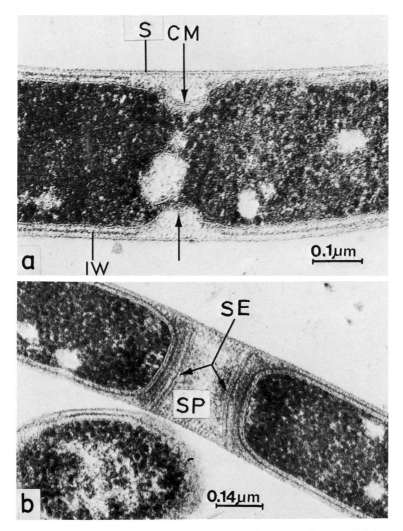

Fig. 19. Thin sections of *Methanospirillum hungatei* during (a) and after (b) cell divisions. IW, Inner wall; S, sheath; SE, structural elements; SP, spacer; CM, cytoplasmic membrane. (Modified according to Zeikus and Bowen, 1975.)

monly used to solubilize or break down cell wall material; it also resists 4 *M* urea. Analysis of hydrolysates of isolated sheaths of strain JF1 (Kandler and König, 1978) showed a complex spectrum of amino acids (65%, Table V), the presence of neutral sugars (8.2%, glucose, galactose, mannose, rhamnose, arabinose), and ash. It is not known whether the carbohydrates are directly bound to the protein as in the case of glycoproteins. The sheaths are dissolved in 1 *N*

NaOH (3 min boiling water bath), resulting in a main polypeptide fragment below 12,000 daltons (E. Conway de Macario and H. Konig, unpublished).

Sprott and McKellar (1980) described a different isolation procedure for sheath material. The lysis of cells of strain GP 1 was carried out by treatment of the bacteria with sodium hydroxide (0.1 N, 25°C, 30 min), followed by incubation with deoxyribonuclease and ribonuclease (30 min, 35°C) and by two more treatments with NaOH. Purification was achieved by treatment with dithiothreitol (50 mM) in sodium carbonate buffer (25 mM, pH 9) for 1.5 hr at 35°C.

By successive incubation with dithiothreitol 23% of the dry weight of the isolated sheaths was solubilized. The molecular weight of the released material exceeded 0.5 million. As assumed by Sprott and McKellar (1980), this material is partly derived from the cell spacer and the adhesive material (see Fig. 19a). The insoluble residue is supposed to be equivalent to the protein sheath. The DDT residue was found to contain 18 amino acids (72.1%) and the neutral sugars (6.03%) of glucose, galactose, mannose, rhamnose, and ribose (Table V). The differences in the amino acid and the neutral sugar composition (ribose instead of arabinose) between the two studies may be due to different isolation procedures as well as to strain dissimilarities.

b. Methanothrix soehngenii. The gram-negative organism forms very long and flexible filaments (Huser, 1981; Zehnder *et al.*, 1980; Huser *et al.*, 1982) often containing several hundred single rods of 1.5–3.3 μm in length and 0.5–0.8 μm in width. Under the scanning electron microscope, the septa between the cells appear as bulges. In negatively stained specimens, the surface of the filaments exhibit a laminar striation of concentric rings, resembling the sheath of *Methanospirillum hungatei.*

In cross-sections, the envelope appears as a double track about 25–30 nm in width, with a very dark inner and a more electron-transparent outer layer. Only the inner layer participates in septum formation during cell division (Fig. 20). Hence, it may be assumed that only the outer, electron-transparent layer represents a striated sheath, embracing many cells, whereas the inner layer represents a rigid cell wall sacculus of individual cells.

However, isolated envelopes, which are obtained by treatment of cells with boiling 2% SDS solution, do not merely show open cylinders as found in respective preparations of *Methanospirillum hungatei,* but also septa. They also show the typical striation not only at the cylindric part of the sacculi, but also at the septa (Fig. 21). This may indicate that both layers seen in cross-sections of whole cells may belong to the same morphological entity, which may not fit the definition of a sheath in the strict sense of the word.

Chemical analysis of the isolated envelopes revealed a complex amino acid pattern and the presence of neutral sugars (Table V), thus resembling the composition of the sheath of *Methanospirillum hungatei.* On the other hand, cytolog-

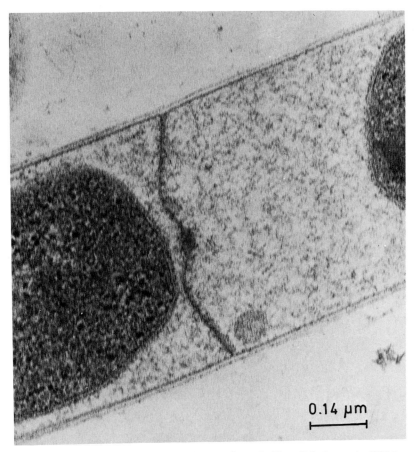

FIG. 20.   Thin sections of *Methanothrix soehngenii*. (From Zehnder *et al.*, 1980.)

ical details (lack of "spacer," "structural elements" and inner cell wall, mode of septum formation) and low S_{AB} values resulting from 16 S rRNA catalogs (Stackebrandt *et al.*, 1982) speak against a close relationship between the two organisms.

III.   Surface Components of Archaebacteria without Cell Envelopes

Two counterparts of the cell envelope-lacking eubacterial mycoplasmas are found among archaebacteria: *Thermoplasma acidophilum*, a thermoacidophilic organism, and *Methanoplasma elizabethii*, a methanogen.

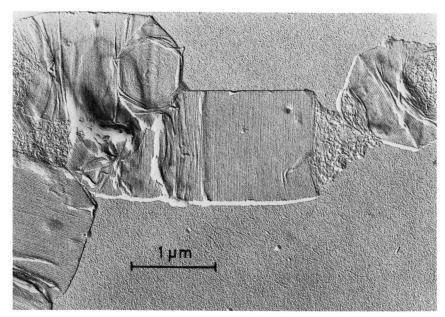

Fig. 21. Platinum-shadowed envelopes of *Methanothrix soehngenii*. (H. König and K. O. Stetter, unpublished data.)

A. *Thermoplasma acidophilum*

The organism grows optimally at about 60°C and at pH 2.0. It forms pleomorphic spheres with 0.3–2.0 μm diameters and possesses a very rigid cytoplasmic membrane but no additional cell envelope (Darland *et al.*, 1970). The cells do not lyse above pH 1.0 and below pH 5.0, nor in deionized water containing 5 mM EDTA (Smith *et al.*, 1973). The unusual stability of the cytoplasmic membrane is probably due to the predominant presence of dibiphytanyldiglycerol tetraethers (cf. Langworthy *et al.*, 1982) and glycoprotein (Yang and Haug, 1979) besides some lipopolysaccharides (Mayberry-Carson *et al.*, 1974).

The cytoplasmic membrane was isolated by lysing the cells in 1 M glycine buffer at pH 9.3, purification by a discontinuous sucrose gradient (25–55%, pH 7.4) at 40,000 rpm and extraction with a solution of 50% phenol at 70°C followed by dialysis. Glycoprotein was isolated from solubilized membranes (1% SDS) by gel filtration on a Sepharose 4B column and affinity chromatography on concanavalin A-Sepharose using 50 mM Tris buffer (pH 8.0) containing 0.2 M α-methylmannoside for elution. The isolated membranes consisted of 60% protein, 25% lipid, and 10% carbohydrate. Twenty-two protein bands were obtained by SDS-polyacrylamide gel electrophoresis. The major band with a molecular weight of 152,000 daltons was PAS-positive. The amino acid composition of this glycoprotein (Table V) showed a high proportion (36%) of hydrophobic amino acid residues. It is worthy to note that the hydrophobic interactions are at their maximum at temperatures optimal for the growth of the organism. The carbohydrate moiety consisted of mannose, glucose, and galactose (molar ratio: 31 : 9 : 4) and was found to be linked to asparagine via a disaccharide composed of N-acetylglucosamine. This unusual mannose-rich glycoprotein is thought to contribute to the rigidity of the membrane by forming a hydroskeleton with its carbohydrate network surrounding the cells. The presence of a mannose-containing carbohydrate on the cell surface was demonstrated by employing peroxidase-labeled concanavalin A (Mayberry-Carson *et al.*, 1978).

B. *Methanoplasma elizabethii*

Methanoplasma elizabethii, a mycoplasma-like penicillin-insensitive methanogen, was isolated from a sewage sludge digester (Rose and Pirt, 1981). It forms branched filaments surrounded only by the cytoplasmic membrane as shown in thin sections. No further data on surface components are as yet known. At present, this organism is not available, and its status is uncertain.

IV. Immunology of Cell Envelopes of Methanogenic Bacteria

As discussed above, methanogenic bacteria have developed a variety of chemically diverse cell envelopes during evolution. Consequently, these bacteria display considerable antigenic diversity as recently demonstrated with antibody probes from rabbit antisera (Conway de Macario *et al.*, 1981, 1982a,b,c). Twenty-one methanogens, representing all the families proposed by Balch *et al.* (1979), were examined by means of calibrated antibody probes derived from the corresponding 21 antisera raised in rabbits. The results shown in Fig. 22 were obtained with the S probe (i.e., the last dilution in the titration curve plateau) applying the standardized indirect immunofluorescence technique (IIF) (Conway de Macario *et al.*, 1982b). Four clusters of antigenically related methanogens were disclosed, as depicted in Fig. 22 (Conway de Macario, 1981, 1982a,c). These clusters coincide with the families defined on the basis of 16 S rRNA nucleotide catalogues (Balch *et al.*, 1979). Intraspecific cross-reactions were intense. Interfamily cross-reactions were not observed, while some weak intergeneric cross-reactions were detected.

While the antigenic determinants of the eubacterial murein have been well

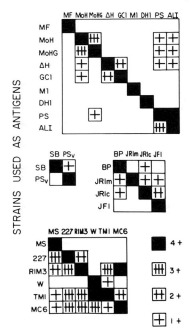

ANTIBODY AGAINST STRAINS

STRAINS USED AS ANTIGENS

FIG. 22. Immunologically related clusters of methanogens. Only the cross-reactions are included in the figure showing increasing intensity from 1 to 4 as revealed by IIF. The following strains were tested: *Methanobacterium formicicum* (MF), *M. bryantii* (MoH and MoHG), *M. thermoautotrophicum* (ΔH and GC1); *Methanobrevibacter ruminantium* (M1), *M. arboriphilus* (DH1), *M. smithii* (PS and ALI); *Methanococcus vannielii* (SB), *M. voltae* (PSv); *Methanomicrobium mobile* (BP); *Methanogenium marisnigri* (JR1m) and *M. cariaci* (JR1c); *Methanospirillum hungatei* (JF1); *Methanosarcina barkeri* (MS, 227, R1M3, W and TM1); *Methanococcus mazei* (= *Methanosarcina mazei*) (MC6). (From Conway de Macario *et al.*, 1981. Copyright 1981 by the American Association for the Advancement of Science.)

investigated (Karakawa *et al.*, 1967; Schleifer and Krause, 1971a,b; Seidl and Schleifer, 1977, 1978), comparable investigations on pseudomurein and other components of the various cell envelopes of methanogens have been started only recently. In a first step, immunochemical characterization of the antigenic determinants of pseudomurein using monoclonal antibodies was undertaken (Conway de Macario *et al.*, 1982d, 1983). Four of these determinants involve a distinctive residue, which is not present in any of the others. Three of these residues are components of the glycan strand, and the other includes the C-terminal sequence of the peptide moiety. The fifth determinant does not contain any of these residues as an immunodominant epitope.

The antigenic mosaic of Methanococcaceae (Conway de Macario *et al.*, 1984)

TABLE VI

Survey on the Cell Envelope Structure in Archaebacteria

Organism	Rigid sacculus	Protein-aceous envelope	Polymer
Methanobacteriaceae	+	−	Pseudomurein
Methanothermaceae	+	+	Pseudomurein, protein subunits
Methanosarcina	+	−	Acidic heteropolysaccharide
Methanococcus	−	+	Protein subunits
Methanomicrobium	−	+	Protein subunits
Methanogenium cariaci	−	+	Protein subunits
Methanogenium marisnigri	−	+	Glycoprotein subunits
Methanoplanus	−	+	Glycoprotein subunits
Methanolobus	−	+	Glycoprotein subunits
Methanothrix	−	Sheath?	(Glyco)protein fibrills
Methanospirillum	−	Sheath	(Glyco)protein fibrills
Halococcus	+	−	Sulfated heteropolysaccharide
Halobacterium[a]	−	+	Glycoprotein subunits
Sulfolobus[a]	−	+	Glycoprotein subunits[a]
Thermoproteus	−	+	Glycoprotein (?) subunits
Thermococcus	−	+	Glycoprotein (?) subunits
Thermofilum	−	+	Glycoprotein (?) subunits
Thermodiscus	−	+	Glycoprotein subunits
Desulfurococcus	−	+	Glycoprotein subunits
Pyrodictium	−	+	Glycoprotein subunits
Thermoplasma	−	−	None[b]

[a] Species with protein subunits also found (H. König and K. O. Stetter, unpublished).
[b] Glycoprotein is found in the cytoplasmic membrane.

and Methanomicrobiaceae using isolated S layers or protein sheaths was also studied with monoclonal antibodies (E. Conway de Macario and H. König, unpublished). The eluted protein bands of the isolated S layers of *Methanococcus vannielii* strain SB and *Methanogenium cariaci* strain JRIc obtained by SDS-polyacrylamide gel electrophoresis were recognized by monoclonal antibodies using a recently developed slide microimmunoenzymatic assay (micro SIA) (Conway de Macario *et al.*, 1982d). The data showed that the antigenic determinants of all isolated antibodies are located within the purified S layer. Most monoclonal antibodies recognize a particular short amino acid sequence rather than more complex tridimensional determinants. The data also indicated that the spectrum of the surface proteins from *Methanogenium cariaci* and *Methanogenium marisnigri* distinguishes each of them despite their assumed taxonomic relatedness. Antisera, and monoclonal antibodies in particular, may become important tools not only for the chemical characterization of the numerous different S layers in archaebacteria but also for facilitating the classification of these organisms.

V. Conclusion

The distinct diversity of the chemistry and structure of cell envelopes in the archaebacteria becomes obvious from the survey given in Table VI. In contrast to the eubacterial kingdom, where the cell wall polymer murein is a common heritage, no common cell wall polymer is found in the archaebacterial kingdom. Obviously, no cell wall polymer was invented by the common ancestor of the archaebacteria. Only after intensive branching and physiological diversification have the various lines of archaebacteria evolved more or less effective polymers to form cell envelopes and even rigid sacculi. The genetic information for cell envelope synthesis could, however, no longer spread across the genetic barriers developed between the then established taxa. Hence, the primary structures and antigenic patterns of the cell envelope components are important phylogenetic and taxonomic markers within the archaebacteria.

REFERENCES

Abram, D., and Gibbons, N. E. (1961). The effect of chlorides and monovalent cations, urea, detergents, and heat on the morphology and the turbidity of suspensions of red halophilic bacteria. *Can J. Microbiol.* **7,** 741–750.

Balch, W. E., Fox, G. E., Magrum, L. J., Woese, C. R., and Wolfe, R. S. (1979). Methanogens: Reevaluation of a unique biological group. *Microbiol. Rev.* **43,** 260–296.

Biely, P., and Jeanloz, R. W. (1969). The isolation and structure identification of uridine 5-(2-acetamido-2-deoxy-α-D-glucopyranosyluronic acid pyrophosphate) from Micrococcus cells. *J. Biol. Chem.* **244,** 4929–4935.

Blaurock, A. E., Stoeckenius, W., Oesterhelt, D., and Scherphof, G. L. (1976). Structure of the cell envelope of Halobacterium halobium. *J. Cell Biol.* **71,** 1–22.

Brown, A. D. (1964). Aspects of bacterial responses to the ionic environment. *Bacteriol. Rev.* **28,** 296–329.

Brown, A. D., and Cho, K. J. (1970). The walls of the extremely halophilic cocci. Gram-positive bacteria lacking muramic acid. *J. gen. Microbiol.* **62,** 267–270.

Brown, A. D., and Shorey, C. D. (1963). The cell envelopes of two extremely halophilic bacteria. *J. Cell Biol.* **18,** 681–689.

Burge, R. E., Fowler, A. G., and Reaveley, D. A. (1977). Structure of the peptidoglycan of bacterial cell walls. I. *J. Mol. Biol.* **117,** 927–953.

Conway de Macario, E., Wolin, M. J., and Macario, A. J. L. (1981). Immunology of the archaebacteria that produce methane gas. *Science* **214,** 74–75.

Conway de Macario, E., Wolin, M. J., and Macario, A. J. L. (1982a). Antibodies analysis of relationships among methanogenic bacteria. *J. Bacteriol.* **149,** 316–319.

Conway de Macario, E., Macario, A. J. L., and Wolin, M. J. (1982b). Specific antisera and immunological procedures for characterization of methanogenic bacteria. *J. Bacteriol.* **149**, 320–328.

Conway de Macario, E., Macario, A. J. L., and Wolin, M. J. (1982c). Antigenic analysis of Methanomicrobiales and Methanobrevibacter arboriphilus. *J. Bacteriol.* **152**, 762–764.

Conway de Macario, E., Macario, A. J. L., and Kandler, O. (1982d). Monoclonal antibodies for immunochemical analysis of methanogenic bacteria. *J. Immunol.* **129**, 1670–1674.

Conway de Macario, E., Macario, A. J. L., Magarinos, M. C., König, H., and Kandler, O. (1983). Dissecting the antigenic mosaic of the archaebacterium Methanobacterium thermoautotrophicum by monoclonal antibodies of defined molecular specificity. *Proc. Natl. Acad. Sci. U.S.A.* **80**, 6346–6350.

Conway de Macario, E., König, H., Macario, A. J. L., and Kandler, O. (1984). Six antigenic determinants in the surface layer of the archaebacterium *Methanococcus vannielii* revealed by monoclonal antibodies. *J. Immunol.* **132**, 883–887.

Darland, G., Brock, T. D., Samsonoff, W., and Conti, S. F. (1970). A thermophilic mycoplasma isolated from a coal refuse pile. *Science* **170**, 1416–1418.

Deatherage, J. F., Taylor, K. A., and Amos, L. A. (1983). Three-dimensional arrangement of the cell wall protein of Sulfolobus acidocaldarius. *J. Mol. Biol.* **167**, 823–852.

Dundas, I. E. D. (1977). Physiology of Halobacteriaceae. *Adv. Microb. Physiol.* **15**, 85–120.

Formanek, H. (1982). Possible models of murein and their Fourier-transforms. *Z. Naturforsch., C:Biosci.* **37C**, 226–235.

Formanek, H. (1985). Three-dimensional models of the carbohydrate moieties of murein and pseudomurein. *Z. Naturforsch.* (in press).

Formanek, H., Formanek, S., and Wawra, H. (1974). A three-dimensional atomic model of the murein layer of bacteria. *Eur. J. Biochem.* **46**, 279–294.

Fox, G. E., Stackebrandt, E., Hespell, R. B., Gibson, J., Maniloff, J., Dyer, T. A., Wolfte, R. S., Balch, W. E., Tanner, R. S., Magrum, L. J., Zablen, L. B., Blakemore, R., Gupta, R., Bonen, L., Lewis, B. J., Stahl, D. A., Luehrsen, K. R., Chen, K. N., and Woese, C. R. (1980). The phylogeny of prokaryotes. *Science* **209**, 457–463.

Hammes, W. P., Winter, J., and Kandler, O. (1979). The sensitivity of the pseudomurein-containing genus Methanobacterium to inhibitors of murein synthesis. *Arch. Microbiol.* **123**, 275–279.

Henderson, R., and Unwin, P. N. T. (1975). Three-dimensional model of purple membrane obtained by electron microscopy. *Nature (London)* **257**, 28–32.

Hilpert, R., Winter, J., Hammes, W., and Kandler, O. (1981). The sensitivity of archaebacteria to antibiotics. *Zentralbl. Bakteriol., Mikrobiol. Abt. 1, Orig. C* **2**, 11–20.

Houwink, A. L. (1956). Flagella, gas vacuoles and cell-wall structure in Halobacterium halobium; an electron microscope study. *J. Gen. Microbiol.* **15**, 146–150.

Huber, H., Thomm, M., König, H., Thies, G., and Stetter, K. O. (1982). Methanococcus thermolithotrophicus, a novel thermophilic lithotrophic methanogen. *Arch. Microbiol.* **132**, 47–50.

Huser, B. A. (1981). Methanbildung aus Acetat: Isolierung eines neuen Archaebakteriums. Thesis No. 6750, Swiss Federal Institute of Technology, Zürich.

Huser, B. A., Wuhrmann, K., and Zehnder, A. J. B. (1982). Methanotrix soehngenii gen. nov. sp. nov., a novel acetotrophic non-hydrogen-oxidizing methane bacterium. *Arch. Microbiol.* **132**, 1–9.

Jarell, K. F., Colvin, J. R., and Sprott, G. D. (1982). Spontaneous protoplast formation in Methanobacterium bryantii. *J. Bacteriol.* **149**, 346–353.

Jones, B. J., Bowers, B., and Stadtman, T. C. (1977). Methanococcus vannielii: Ultrastructure and sensitivity to detergents and antibiotics. *J. Bacteriol.* **130**, 1357–1363.

Kandler, O. (1979). Zellwandstrukturen bei Methanbakterien. Zur Evolution der Prokaryonten. *Naturwissenschaften* **66**, 95–105.

Kandler, O. (1982). Cell wall structures and their phylogenetic implications. *Zentralbl. Bakteriol., Mikrobiol. Hyg., Abt. 1, Orig. C* **3,** 149–160.

Kandler, O., and Hippe, H. (1977). Lack of peptidoglycan in the cell walls of Methanosarcina barkeri. *Arch. Microbiol.* **113,** 57–60.

Kandler, O., and König, H. (1978). Chemical composition of the peptidoglycan-free cell walls of methanogenic bacteria. *Arch. Microbiol.* **118,** 141–152.

Karakawa, W. W., Lackland, H., and Krause, R. M. (1967). Antigenic properties of the hexosamine polymer of streptococcal mucopeptide. *J. Immunol.* **99,** 1179–1182.

Kelemen, M. V., and Rogers, H. J. (1971). Three-dimensional models of bacterial cell wall micopeptides (peptidoglycans). *Proc. Natl. Acad. Sci. U.S.A.* **68,** 992–996.

Kessel, M., and Cohen, Y. (1982). Ultrastructure of square bacteria from a brine pool in southern Sinai. *J. Bacteriol.* **150,** 851–860.

Kocur, M., Smid, B., and Martinee, T. (1972). The fine structure of extreme halophilic cocci. *Microbios* **5,** 101–107.

König, H., and Kandler, O. (1979a). The amino acid sequence of the peptide moiety of the pseudomurein from Methanobacterium thermoautotrophicum. *Arch. Microbiol.* **121,** 271–275.

König, H., and Kandler, O. (1979b). N-Acetyltalosaminuronic acid a constituent of the pseudomurein of the genus Methanobacterium. *Arch. Microbiol.* **123,** 295–299.

König, H., and Kandler, O. (1980). 2-Amino-2-deoxytaluronic acid and 2-amino-2-deoxyglucose from the pseudomurein of Methanobacterium thermoautotrophicum possess the L- and D-configurations, respectively. *Hoppe-Seyler's Z. Physiol. Chem.* **361,** 981–983.

König, H., and Stetter, K. O. (1982). Isolation and characterization of Methanolobus tindarius, sp. nov., a coccoid methanogen growing only on methanol and methylamines. *Zentralbl. Bakteriol., Mikrobiol. Hyg., Abt. 1, Orig. C* **3,** 478–490.

König, H., Kralik, R., and Kandler, O. (1982). Structure and modifications of pseudomurein in Methanobacteriales. *Zentralbl. Bakteriol., Mikrobiol. Hyg., Abt. 1, Orig. C* **3,** 179–191.

König, H., Kandler, O., Jensen, M., and Rietschel, T. (1983). The primary structure of the glycan moiety of pseudomurein from Methanobacterium thermoautotrophicum. *Hoppe-Seyler's Z. Physiol. Chem.* **364,** 627–636.

Kreisl, P. (1984). Chemische Untersuchungen zur Struktur der Zellwandpolymere von *Methanosarcina barkeri.* Ph.D. Thesis, University of Munich.

Kushner, D. J., Bayley, S. T., Boring, J., Kates, M., and Gibbons, N. E. (1964). Morphological and chemical properties of cell envelopes of the extreme halophile, Halobacterium cutirubrum. *Can. J. Microbiol.* **10,** 483–497.

Labischinski, H., Barnickel, G., Bradaczek, H., and Giesbrecht, P. (1979). On the secondary and tertiary structure of murein. *Eur. J. Biochem.* **95,** 147–155.

Labischinski, H., Barnickel, G., Leps, B., Bradaczek, H., and Giesbrecht, P. (1980). Initial data for the comparison of murein and pseudomurein conformations. *Arch. Microbiol.* **127,** 195–201.

Langworthy, T. A., Tornabene, T. G., and Holzer, G. (1982). Lipids of archaebacteria. *Zentralbl. Bakteriol., Mikrobiol. Hyg., Abt. 1, Orig. C* **3,** 228–244.

Larsen, H. (1967). Biochemical aspects of extreme halophilism. *Adv. Microb. Physiol.* **1,** 97–132.

Mayberry-Carson, K. J., Langworthy, T. A., Mayberry, W. R., and Smith, P. F. (1974). A new class of lipopolysaccharide from Thermoplasma acidophilum. *Biochim. Biophys. Acta* **360,** 217–229.

Mayberry-Carson, K. J., Jewell, M. J., and Smith, P. F. (1978). Ultrastructural localisation of Thermoplasma acidophilum surface carbohydrate by using Concanavalin A. *J. Bacteriol.* **133,** 1510–1513.

Mescher, M. F., and Strominger, J. L. (1976). Purification and characterization of a prokaryotic glycoprotein from cell envelope of Halobacterium salinarium. *J. Biol. Chem.* **251,** 2005–2014.

Mescher, M. F., and Strominger, J. L. (1978). The cell surface glycoprotein of Halobacterium salinarium. In "Energetics and Structure of Halophilic Microorganisms" (S. R. Caplan and M. Ginzburg, eds.), pp. 503–513. Elsevier/North-Holland Biomedical Press, Amsterdam.

Mescher, M. F., Strominger, J. L., and Watson, S. W. (1974). Protein and carbohydrate composition of the cell envelope of Halobacterium salinarium. J. Bacteriol. 120, 945–954.

Michel, H., Neugebauer, D.-C., and Oesterhelt, D. (1980). The 2-d crystalline cell wall of Sulfolobus acidocaldarius: Structure, solubilization, and reassembly. In "Electron Microscopy at Molecular Dimensions" (W. Baumeister and W. Vogell, eds.), pp. 27–35. Springer-Verlag, Berlin and New York.

Paynter, M. J. B., and Hungate, R. E. (1968). Characterization of Methanobacterium mobilis, sp. n., isolated from the bovine rumen. J. Bacteriol. 95, 1943–1951.

Reistad, R. (1972). Cell wall of an extremely halophilic coccus. Investigation of ninhydrin-positive compounds. Arch. Mikrobiol. 82, 24–30.

Reistad, R. (1974). 2-Amino-2-deoxyguluronic acid: a constituent of the cell wall of Halococcus sp., strain 24. Carbohydr. Res. 36, 420–423.

Robertson, J. D., Schreil, W., and Reedy, M. (1982). Halobacterium halobium I: A thin-sectioning electron-microscopic study. J. Ultrastruct. Res. 80, 148–162.

Romesser, J. A., Wolfe, R. S., Mayer, F., Spiess, E., and Walther-Mauruschat, A. (1979). Methanogenium, a new genus of marine methanogenic bacteria, and characterization of Methanogenium cariaci sp. nov. and Methanogenium marisnigri sp. nov. Arch. Microbiol. 121, 147–153.

Rose, C. S., and Pirt, S. J. (1981). Conversion of glucose to fatty acids and methane: Roles of two mycoplasmal agents. J. Bacteriol. 147, 248–254.

Schleifer, K. H., and Kandler, O. (1967). Zur chemischen Zusammensetzung der Zellwand der Streptokokken. I. Die Aminosäuresequenz des Mureins von Str. thermophilus und Str. faecalis. Arch. Mikrobiol. 57, 335–364.

Schleifer, K. H., and Kandler, O. (1972). Peptidoglycan types of bacterial cell walls and their taxonomic implications. Bacteriol. Rev. 36, 407–477.

Schleifer, K. H., and Krause, R. M. (1971a). The immunochemistry of peptidoglycan. The immunodominant site of the peptide subunit and the contribution of each of the amino acids to the binding properties of the peptides. J. Biol. Chem. 246, 986–993.

Schleifer, K. H., and Krause, R. M. (1971b). The immunochemistry of peptidoglycan. Separation and characterization of antibodies to the glycan and to the peptide subunit. Eur. J. Biochem. 19, 471–478.

Schleifer, K. H., and Stackebrandt, E. (1983). Molecular systematics of prokaryotes. Annu. Rev. Microbiol. 37, 143–187.

Schleifer, K. H., Steber, J., and Mayer, H. (1982). Chemical composition and structure of the cell wall of Halococcus morrhuae. Zentralbl. Bakteriol., Mikrobiol. Hyg., Abt. 1, Orig. C 3, 171–178.

Schönheit, P., and Thauer, R. K. (1980). L-Alanine, a product of cell wall synthesis in Methanobacterium thermoautotrophicum. FEMS Microbiol. Lett. 9, 77–80.

Seidl, H. P., and Schleifer, K. H. (1977). The immunochemistry of the peptidoglycan. Antibodies against a synthetic immunogen crossreacting with an interpeptide bridge of peptidoglycan. Eur. J. Biochem. 74, 353–363.

Seidl, H. P., and Schleifer, K. H. (1978). Specific antibodies of the N-termini of the interpeptide bridges of peptidoglycan. Arch. Microbiol. 118, 185–192.

Sharon, N. (1975). "Complex Carbohydrates, their Chemistry, Biosynthesis, and Functions," pp. 295–298. Addison-Wesley, Reading, Massachusetts.

Sleytr, U. B., and Glauert, A. M. (1982). Bacterial cell walls and membranes. In "Electron Microscopy of Proteins" (J. R. Harris, ed.), Vol. 3, pp. 41–76. Academic Press, New York.

Sleytr, U. B., and Messner, P. (1983). Crystalline surface layers on bacteria. *Annu. Rev. Microbiol.* **37**, 11–39.

Smith, P. F., Langworthy, T. A., Mayberry, W. R., and Hougland, A. E. (1973). Characterization of the membranes of Thermoplasma acidophilum. *J. Bacteriol.* **116**, 1019–1028.

Sowers, K. R., and Ferry, J. G. (1983). Isolation and characterization of a methylotrophic marine methanogen, Methanococcoides methylutens gen. nov., sp. nov. *Appl. Environ. Microbiol.* **45**, 684–690.

Sprott, G. D., and McKellar, R. C. (1980). Composition and properties of the cell wall of Methanospirillum hungatei. *Can. J. Microbiol.* **26**, 115–120.

Sprott, G. D., Colvin, J. R., and McKellar, R. C. (1979). Spheroplasts from Methanospirillum hungatei formed upon treatment with dithiothreitol. *Can. J. Microbiol.* **25**, 730–738.

Stackebrandt, E., Seewaldt, E., Ludwig, W., Schleifer, K. H., and Huser, B. A. (1982). The phylogenetic position of Methanotrix soehngenii. Elucidated by a modified technique of sequencing oligonucleotides from 16S rRNA. *Zentralbl. Bakteriol., Mikrobiol. Hyg., Abt. 1, Orig. C* **3**, 90–100.

Steber, J., and Schleifer, K. H. (1975). Halococcus morrhuae: A sulfated heteropolysaccharide as the structural component of the bacterial cell wall. *Arch. Microbiol.* **105**, 173–177.

Steber, J., and Schleifer, K. H. (1979). N-Glycyl-glucosamine, a novel constituent in the cell wall of Halococcus morrhuae. *Arch. Microbiol.* **123**, 209–212.

Steensland, H., and Larsen, H. (1969). A study of the cell envelope of the halobacteria. *J. Gen. Microbiol.* **55**, 325–336.

Stetter, K. O. (1982). Ultrathin mycelia-forming organisms from submarine volcanic areas having an optimum growth temperature of 105°C. *Nature (London)* **300**, 258–260.

Stetter, K. O., Thomm, M., Winter, J., Wildgruber, G., Huber, H., Zillig, W., Janékovic, D., König, H., Palm, P., and Wunder, S. (1981). Methanothermus fervidus, sp., nov., a novel extremely thermophilic methanogen isolated from an Icelandic hot spring. *Zentralbl. Bakteriol., Mikrobiol. Hyg., Abt. 1, Orig. C* **2**, 166–178.

Stetter, K. O., König, H., and Stackebrandt, E. (1983). *Pyrodictium occultum* gen. nov. sp. nov., an extremely thermophilic archaebacterium growing at 105°C. *Syst. Appl. Microbiol.* **4**, 535–551.

Stewart, M., and Beveridge, T. J. (1980). Structure of the regular surface layer of Sporosarcina ureae. *J. Bacteriol.* **142**, 302–309.

Stewart, M., Beveridge, R. J., and Murray, R. G. E. (1980). Structure of the regular surface layer of Spirillum putridiconchylium. *J. Mol. Biol.* **137**, 1–8.

Stoeckenius, W. (1981). Walsby's square bacterium: Fine structure of an orthogonal procaryote. *J. Bacteriol.* **148**, 352–360.

Stoeckenius, W., and Rowen, R. (1967). A morphological study of Halobacterium halobium and its lysis in media of low salt concentration. *J. Cell Biol.* **34**, 365–393.

Taylor, K. A., Deatherage, J. F., and Amos, L. A. (1982). Structure of the S-layer of Sulfolobus acidocaldarius. *Nature (London)* **299**, 840–842.

Tipper, K. J., and Wright, A. (1979). Structure and biosynthesis of bacterial cell walls. *In* "The Bacteria: A Treatise on Structure and Function" (I. C. Gunsalus, J. R. Sokatch, and L. N. Ornston, eds.), Vol. 7, pp. 291–426. Academic Press, New York.

Walsby, A. E. (1980). A square bacterium. *Nature (London)* **283**, 69–71.

Weiss, L. R. (1973). Attachment of bacteria to sulphur in extreme environments. *J. Gen. Microbiol.* **77**, 501–507.

Weiss, L. R. (1974). Subunit cell wall of Sulfolobus acidocaldarius. *J. Bacteriol.* **118**, 275–284.

Wieland, F., Dompert, W., Bernhardt, G., and Sumper, M. (1980). Halobacterial glycoprotein saccharides contain covalently linked sulphate. *FEBS Lett.* **120**, 110–114.

Wieland, F., Lechner, J., Bernhardt, G., and Sumper, M. (1981). Sulphation of a repetitive saccharide in halobacterial cell wall glycoprotein. *FEBS Lett.* **132**, 319–323.

Wieland, F., Lechner, J., and Sumper, M. (1982). The cell wall glycoprotein of halobacteria: Structural, functional and biosynthetic aspects. *Zentralbl. Bakteriol., Mikrobiol. Hyg., Abt. 1, Orig. C* **3**, 161–170.

Wieland, F., Heitzer, R., and Schaefer, W. (1983). Asparaginylglucose: A novel type of carbohydrate linkage. *Proc. Natl. Acad. Sci. U.S.A.* **80**, 5470–5474.

Wildgruber, G., Thomm, M., König, H., Ober, K., Ricchiuto, T., and Stetter, K. O. (1982). Methanoplanus limicola, a plate-shaped methanogen representing a novel family, the Methanoplanaceae. *Arch. Microbiol.* **132**, 31–36.

Yang, L. L., and Haug, A. (1979). Purification and partial characterization of a procaryotic glycoprotein from the plasma membrane of Thermoplasma acidophilum. *Biochim. Biophys. Acta* **556**, 277–365.

Zehnder, A. J. B., Huser, B. A., Brock, T. D., and Wuhrman, K. (1980). Characterization of an acetate-decarboxylating, non-hydrogen-oxidizing methane bacterium. *Arch. Microbiol.* **124**, 1–11.

Zeikus, J. G. (1977). The biology of methanogenic bacteria. *Bacteriol. Rev.* **41**, 514–541.

Zeikus, J. G., and Bowen, V. G. (1975). Fine structure of Methanospirillum hungatei. *J. Bacteriol.* **121**, 373–380.

Zhilina, T. (1971). The fine structure of Methanosarcina. *Mikrobiologiya* **40**, 674–680.

Zhilina, T. N., and Zavarzin, G. A. (1973). Trophic relationships between Methanosarcina and its associates. *Mikrobiologiya* **42**, 235–241.

Zillig, W., Stetter, K. O., Schäfer, W., Janékovic, D., Wunderl, S., Holz, I., and Palm, P. (1981). Thermoproteales: A novel type of extremely thermoacidophilic anaerobic archaebacteria isolated from Icelandic solfataras. *Zentralbl. Bakteriol., Mikrobiol. Hyg., Abt. 1, Orig. C* **2**, 205–227.

Zillig, W., Stetter, K. O., Prangishvilli, D., Schäfer, W., Wunderl, S., Janékovic, D., Holz, I., and Palm, P. (1982). Desulfurococcaceae the second family of the extremely thermophilic, anaerobic, sulfur respiring Thermoproteales. *Zentralbl. Bakteriol., Mikrobiol. Hyg., Abt. 1, Orig. C* **3**, 304–317.

Zillig, W., Holz, I., Janékovic, D., Schäfer, W., and Reiter, W. D. (1983a). The archaebacterium Thermococcus celer, a novel genus within the thermophilic branch of the archaebacteria. *Syst. Appl. Microbiol.* **4**, 88–94.

Zillig, W., Gierl, A., Schreiber, G., Wunderl, S., Janékovic, D., Stetter, K. O., and Klenk, H. P. (1983b). The archaebacterium Thermofilum pendens a novel genus of the thermophilic, anaerobic sulfur respiring Thermoproteales. *Syst. Appl. Microbiol.* **4**, 79–81.

CHAPTER 10

Lipids of Archaebacteria

THOMAS A. LANGWORTHY

I. Introduction

Archaebacteria differ from eubacteria and eukaryotic cells in a variety of molecular respects. The nature of the membrane lipids is no exception. In fact, their lipids are perhaps the most invariant and unique aspect of the archaebacteria. Unlike the straight-chain fatty acids and fatty acid ester glycerol lipids that are characteristic of eubacteria and eukaryotes, archaebacterial lipids are distinguished by being isoprenoid and hydroisoprenoid hydrocarbons and isopranyl glycerol ether lipids. Furthermore, the optical activity characteristic of the central carbon of glycerol is in archaebacteria the opposite of what it is in the other two primary kingdoms. These and other facts suggest important differences in the construction of membrane superstructure in archaebacteria, which challenge a number of basic concepts of membrane biochemistry. The uniqueness of the archaebacterial lipids is an important point for defining them as a separate evolutionary path. The lipids, furthermore, serve as a convenient chemical marker for the identification of archaebacteria and provide insight into geochemical processes.

Copyright © 1985 by Academic Press, Inc.
All rights of reproduction in any form reserved.
ISBN 0-12-307208-5

Because the archaebacterial concept is new and new archaebacteria are being isolated at a rapid rate, knowledge of the varieties and function of archaebacterial lipids is necessarily incomplete, and some points concerning their biochemistry are still speculative. The most thoroughly defined lipids are those of the halophilic archaebacteria, but a fairly large number of lipids from representative species of thermoacidophilic, methanogenic, alkalohalophilic, and thermoanaerobic archaebacteria have now been identified. It is the intent of this chapter not to dwell on the experimental details for the identification of lipid structures, but to provide an integrated view of the more important structural aspects and consequences of archaebacterial lipids. For the detailed experimental evidence, the reader is referred to the referenced papers and to the several summaries detailing the methodology employed in archaebacterial lipid chemistry (Kates, 1972a, 1978; Kates *et al.*, 1982; Langworthy, 1982a, 1983; Ross *et al.*, 1981).

II. Glycerolipids

A. Occurrence

Lipids are molecules capable of hydrophobic interactions and are the essence of biological membranes. In thermodynamically and geometrically optimal interactions with proteins and ions, they constitute the cytoplasmic membranes of all living cells. Although the membrane lipids are certainly characterized by great diversity and structural variation (Razin and Rottem, 1982), the principal apolar residues comprising the hydrophobic core of the membrane are the glycerolipids. Generally, glycerolipids consist of a glycerol molecule to which are attached two hydrocarbon chains. These chains may vary in length, monomethyl branching, and unsaturation in response to external environmental parameters such as temperature and pH. When substituted with polar head groups, the glycerolipids also provide the hydrophobic region of the more complex glycolipid and phospholipid structures. Furthermore, the glycerolipids possess an asymmetric center (in the central carbon of glycerol) so that they are optically active.

The glycerolipids of the eubacteria and eukaryotes occur as four principal types predominantly in the double-chain form (Fig. 1): the acylglycerols (glycerides), alk-1-enylglycerols (plasmalogens), alkylglycerols (monoethers), and the dialkylglycerols (diethers). All of these glycerolipids are levorotary having the D- or *sn*-1,2-glycerol stereoconfiguration. (In the *sn* system of nomenclature, if the secondary hydroxyl group in glycerol is oriented to the left of carbon-2 in the Fischer projection, the carbon atom above carbon-2 is designated carbon-1 and the one below is designated carbon-3.) Thus, when polar head groups are involved, they are attached to carbon-3 of glycerol.

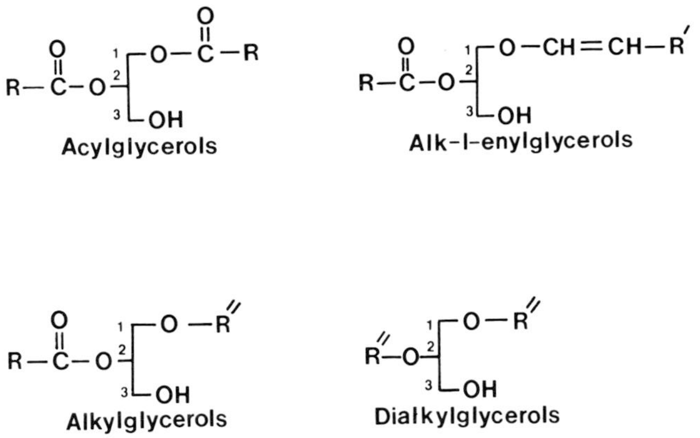

FIG. 1. Glycerolipids of eubacteria and eukaryotes shown as the predominant double chain *sn*-1,2-glycerol derivatives. R, fatty acids (esters); R′, fatty aldehydes (vinyl ethers); R″, fatty alcohols (ethers).

Acylglycerols, composed of fatty acids in ester linkages to glycerol, are the dominant form of glycerolipids in all eubacteria and eukaryotes. Although the 1,2-diacyl-*sn*-glycerols account for the majority, all structural isomers also occur, for example, 1- and 3-monoacyl-*sn*-glycerol, 1,3-diacyl-*sn*-glycerols, and 1,2,3-triacyl-*sn*-glycerols. The alk-1-enylglycerols, containing a fatty aldehyde attached to carbon-1 through an acid-sensitive vinyl ether linkage are fairly uncommon and generally restricted to some anaerobic eubacteria and animal tissues (Snyder, 1972; Goldfine, 1982). The alk-1-enyl-2-acyl-*sn*-glycerols, in which a fatty acid is linked to carbon-2, is the predominant form, although derivatives with either two fatty acids or no fatty acids attached can exist. The alkylglycerols, in which a fatty alcohol is ether-linked to glycerol at carbon-1, is very uncommon, being confined to some animal tissues, fish oils, and some anaerobic eubacteria (Snyder, 1972; Goldfine, 1982). The usual three alkylglycerols to be encountered are 1-*O*-hexadecyl-*sn*-glycerol (chimyl alcohol), 1-*O*-octadecyl-*sn*-glycerol (batyl alcohol), and 1-*O*-octadec-*cis*-enyl-*sn*-glycerol (selachyl alcohol), and these too can occur as the 2-acyl or 2,3-acyl derivatives in the natural state. The dialkylglycerols, containing two ether-linked fatty alcohols at carbon-1 and carbon-2 of glycerol are so rare that they were not known to exist at all in eubacteria or eukaryotes until recently (Langworthy *et al.*, 1983). The only known occurrence is in the thermophilic, anaerobic, sulfate-reducing eubacterium, *Thermodesulfotobacterium commune*, in which 1,2-di-*O*-alkyl-*sn*-glycerols possessing *iso*- and *anteiso*-branched alkyl chains comprise the glycerolipid component.

Thus, with the exception of the very limited occurrence of the fatty aldehyde and fatty alcohol derived glycerolipids, the acylglycerols constitute the basic fundamental glycerolipid component of eubacteria and eukaryotic cells. To be sure, the archaebacteria possess glycerolipids, but, whereas the glycerol ether lipids are only minor to almost nonexistent in other cells, the archaebacteria have evolved glycerol ether lipids as their fundamental glycerolipid component. They differ in the additional respect that they are limited in the variability of the structure of their hydrocarbon chains and are stereochemically the mirror image of the naturally occurring acylglycerols. At least from the human point of view, the archaebacteria have evolved glycerolipids, which in their structure and bio-synthesis are in essence reversed in comparison to other cells.

B. DIETHERS AND TETRAETHERS

1. BASIC STRUCTURE

Although trace quantities of fatty acids have been detected in most archaebac-terial preparations (Kates, 1972a; Langworthy *et al.*, 1974; Ruwart and Haug, 1975; Tornabene and Langworthy, 1979), the glycerolipids of archaebacteria occur as two basic types: glycerol diethers and diglycerol tetraethers (Fig. 2). The diethers contain two C_{20} phytanyl chains in ether linkage to glycerol, where-as tetraethers possess two C_{40}-biphytanyl chains ether-linked simultaneously to two opposite glycerol molecules.

The existence of diethers was first recognized through the extensive and now classical studies of Kates and colleagues (Sehgal *et al.*, 1962; Kates *et al.*, 1965) to constitute the glycerolipids of the halophilic archaebacteria long before the concept of archaebacteria was formulated (see reviews by Kates, 1972a, 1978). The *O*-alkyl chains, invariant in their composition, were shown to be fully saturated isopranoid-branched hydrocarbon phytane ($C_{20}H_{42}$). The absolute ster-eoconfiguration of the three asymmetric -CH_3 groups at carbons 3, 7, and 11 of the phytane skeleton were shown to have the *3R, 7R, 11R* configuration (for example, 3*R*,7*R*,11*R*,15-tetramethylhexadecane). The diether was itself found to be dextarotary ([α]$_D$ = +8.7°, *M* = +55°) demonstrating the diether to have the L- or *sn*-2,3-glycerol stereoconfiguration, opposite to that of the naturally occur-ring diacylglycerols. Thus, polar head groups are found attached to carbon-1 of the glycerol molecule. The diether structure was ultimately confirmed by chem-ical synthesis (Kates, 1978) and fully established as 2,3-di-*O*-phytanyl-*sn*-glycerol ($C_{43}H_{88}O_3$, MW 652). Traces of the glycerol monoethers 2-*O*- and 3-*O*-phytanyl-*sn*-glycerol were also detected, but it is not certain whether they arise from breakdown during chemical workup of the diethers or if they are naturally present. The diether has been found in all archaebacteria examined although in variable quantities (see Section II,B,2).

Tetraethers, as a component glycerolipid of archaebacteria, were a later dis-covery, and their structure is unprecedented (Langworthy, 1977a). Initial studies

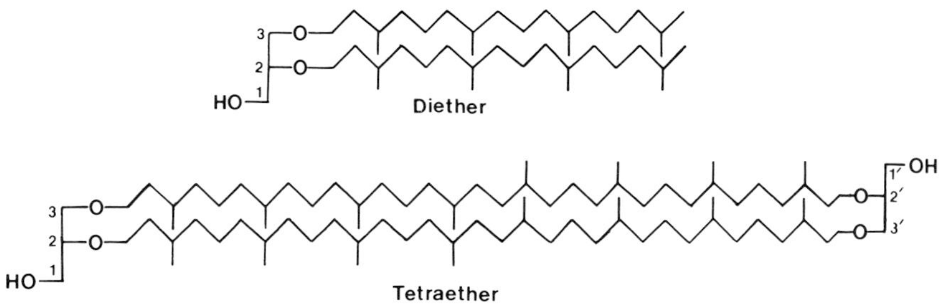

Diether

Tetraether

FIG. 2. Principal glycerolipids of archaebacteria are the phytanylglycerol diether, 2,3-di-*O*-phytanyl-*sn*-glycerol, and the dibiphytanyldiglycerol tetraether 2,3,2′,3′-tetra-*O*-dibiphytanyl-di-*sn*-glycerol.

on *Thermoplasma acidophilum* (Langworthy *et al.*, 1972) and *Sulfolobus acido-caldarius* (Langworthy *et al*, 1974; De Rosa *et al.*, 1974), both of which now constitute members of the thermoacidophilic archaebacteria, indicated the presence of glycerol ether lipids containing fully saturated C_{40}-hydrocarbon chains. But, their structural assembly as tetraethers has been only more recently realized (Langworthy, 1977a) and confirmed (De Rosa *et al.*, 1977a,b, 1980a,e; Yang and Haug, 1979). The C_{40}-alkyl chains are made up of two C_{20}-phytane units joined by head-to-head covalent linkage at the $16,16'$-*gem*-dimethyl ends resulting in the $C_{40}H_{82}$ biphytane skeleton (De Rosa *et al.*, 1977a,b, 1980a,e). The -CH_3 groups, by analogy to those of the phytane, have been assigned the R absolute stereoconfigurate at the $C_{3,3'}, C_{7,7'}, C_{11,11'}$ asymmetric centers, although the stereochemistry of the $C_{15,15'}$ methyl groups is uncertain (De Rosa *et al.*, 1980e). The tetraether structure is then made up of two identical pairs of biphytane molecules, bridged through ether linkages, to two glycerol molecules at the terminal end of the biphytanyl chains (Langworthy, 1977a). Or, viewed differently, the tetraether is the structural equivalent of two phytanyl diether molecules that have been covalently linked through the terminal ends of their phytanyl chains. The tetraethers are dextrarotary ($[\alpha]_D = +8.3°, M = +108°$, De Rosa *et al.*, 1980e; $[\alpha]_D = +8.7°, M = 113°$, Kushwaha *et al.*, 1981a,b) indicating that both glycerols have the L- or *sn*-2,3-glycerol stereoconfiguration and that the primary -OH groups of the two opposed glycerols are in the *trans* configuration. Thus polar head groups may be attached to glycerol at either or both carbon-1 or 1′ when present. The basic tetraether structure has thereby been identified as 2,3,2′3′-tetra-*O*-dibiphytanyl-di-*sn*-glycerol ($C_{86}H_{172}O_6$, MW 1300). The structure of the tetraether has not yet been chemically synthesized, which presents an interesting challenge in itself. No diglycerol diethers (the "monoether" equivalent) containing a single biphytanyl chain have been detected. Unlike the diethers, tetraethers are restricted to the thermoacidophilic, thermoanaerobic, and some methanogenic archaebacteria (see Section II,B,2).

2. STRUCTURAL VARIETY AND DISTRIBUTION

Until recently, the diether possessing two identical phytanyl chains was the only recognized species among archaebacteria. However, several haloalkalophilic isolates from alkaline soda lakes, designated SP1, SP2, MS3 (Tindall *et al.*, 1980; Ross *et al.*, 1981) and *Halobacterium pharaonis* (Soliman and Trüper, 1982) have now been shown to contain two diethers in about equal amounts as the component glycerolipids. One diether was shown to be the usual phytanyl diether, but the second was identified as a C_{20}:C_{25} glycerol diether containing one C_{20}-phytanyl chain attached to carbon-3 of glycerol and one C_{25}-sesterterpane (2,6,10,14,18-pentamethyleicosane) attached to carbon-2 (De Rosa *et al.*, 1982a). This asymmetric diether was dextrorotary ($[\alpha]_D = +7.2°$) and identified (Fig. 3A) as 2-*O*-sesterterpanyl-3-*O*-Phytanyl-*sn*-glycerol ($C_{48}H_{98}O_3$, MW 722). This is so far the only recognized variation in the diether structure, which in effect is the addition of one more C_5-isoprene unit to one of its phytanyl chains.

Unlike the diethers, the variety of tetraether species is greater in terms of structural variation both within the biphytanyl chains and in the nature of the dipolar glycerol backbone. Whereas the dibiphytanyl diglycerol tetraether containing acylic ($C_{40}H_{82}$) biphytanyl chains is the fundamental tetraether structure (Fig. 2), the tetraethers can also differ in the additional feature that the biphytanyl chains may contain from one to four cyclopentyl rings (De Rosa *et al.*, 1977a,b, 1980a,e). Within the tetraether, the identical pairs of C_{40}-pentacyclic biphytanes are in the antiparallel configuration giving rise to the four tetraether species ($C_{86}H_{168\ -156}O_6$) shown in Fig. 3B–E. The pentacyclic biphytanyl tetraethers are thus far known only in the thermoacidophilic archaebacteria *Thermoplasma* (Langworthy, 1977a, 1979a), *Sulfolobus* (De Rosa *et al.*, 1977a,b, 1980e), and *Thermoproteus* (Zillig *et al.*, 1981).

In addition to variation with respect to the alkyl chains, a second class of tetraether assembly, differing in the nature of the polar ends of the tetraethers, occurs in *Sulfolobus* (Langworthy *et al.*, 1974). *Sulfolobus* contains both the basic glycerol-dialkylglycerol tetraether structure and a more polar component now shown by De Rosa *et al.* (1980a) to be a glycerol-alkylnonitol tetraether in which a branched 9-carbon nonitol ($C_9H_{20}O_9$), called by them calditol, substitutes for one of the glycerols in the tetraether structure (Fig. 3B–E). The glycerol-nonitol tetraethers ($C_{92}H_{180-168}O_{12}$, MW 1476–1464) account for about 50% of the total tetraethers in *Sulfolobus* grown heterotrophically (Langworthy *et al.*, 1974) and up to 85% in autotrophically grown cells (Langworthy, 1977b).

In general terms, the structural variety within archebacterial glycerolipids is fairly limited in comparison to the naturally occurring glycerolipids of eubacteria and eukaryotes, although they are certainly unique. Still, the distribution be-

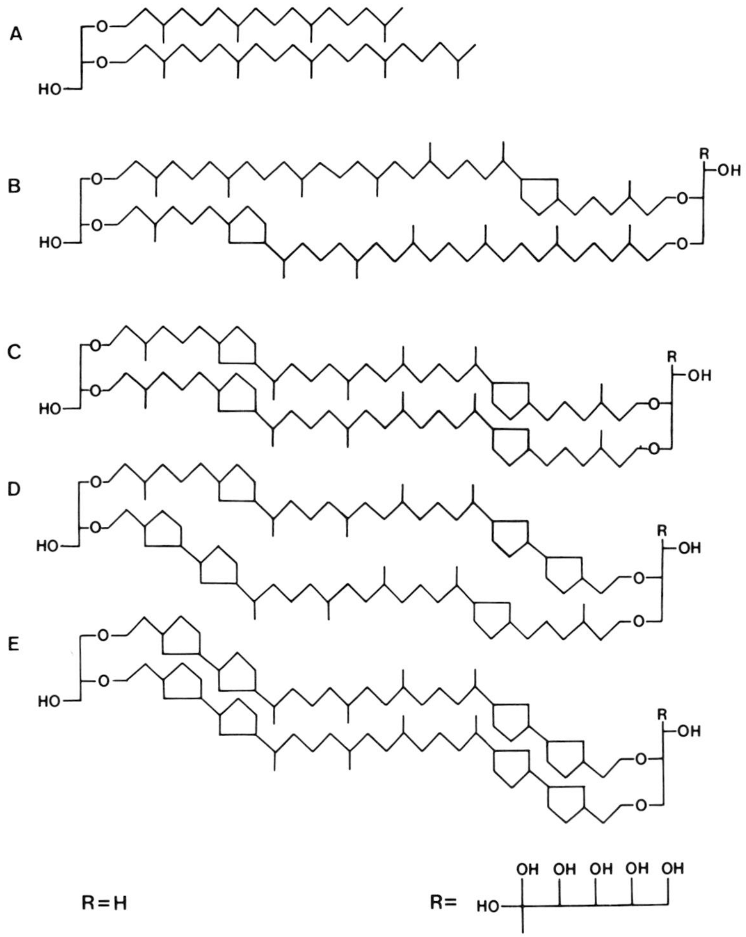

FIG. 3. Variations in diether and tetraether glycerolipid structures of archaebacteria. (A) A C_{20}:C_{25} diether in haloalkalophiles: 2-O-sesterterpanyl-3-O-phytanyl-*sn*-glycerol. (B–E) Glycerol-dialkylglycerol tetraethers (R=H) or glycerol-dialkylnonitol tetraethers (R=$C_6H_{13}O_6$) of thermoacidophiles. These contain either mono- (B), bi- (C), tri- (D), or tetrapentacyclic (E) C_{40} biphytanyl chains.

tween the relative proportions of diethers and tetraethers and the composition of the hydrocarbon chains can be useful, in conjunction with other parameters, in assessing the identification, relationships, and taxonomy of archaebacteria (Table I). Phytanyl diethers, for example, occur in all archaebacteria examined, but their proportions relative to tetraethers vary from 100% diethers in the halophiles (Kates, 1972a, 1978) and methanogens of mostly coccal morphology (Tornabene and Langworthy, 1979) to as low as 5% to trace amounts of diethers

TABLE I
DISTRIBUTION OF DIETHERS AND TETRAETHERS AMONG ARCHAEBACTERIA[a]

Group	Archaebacterium	Diether (%)	Tetraether (%)
Halophiles	*Halobacterium cutirubrum*	100	0
	H. salinarium	100	0
	H. volcanii	100	0
	H. halobium	100	0
	H. saccharovorum	100	0
	H. morrhuae	100	0
	H. simoncinii	100	0
	H. marismortui	100	0
	H. trapanicum	100	0
	H. sp. (Cagliari)	100	0
	H. sp. A-2c	100	0
	H. sp. R-4	100	0
	Halococcus morrhuae	100	0
	Sarcina litoralis	100	0
Haloalkalophiles	*Halobacterium pharaonis*	100[b]	0
	isolate sp1	100[b]	0
	isolate sp2	100[b]	0
	isolate MS3	100[b]	0
Methanogens	*Methanococcus* strain PS	100	0
	M. methylutens	100[c]	0
	Methanothrix söhngenii	100	0
	Methanosarcina barkeri	100	0
	Methanococcus vannielii	99.9	0.1
	Methanobacterium ruminantium M-1	71.8	28.2
	M. ruminantium PS	44.7	55.3
	M. thermoautotrophicum	44.5	55.5
	M. strain M.O.H.	43.5	56.5
	Methanospirillum hungatei	40.5	59.5
	M. strain AZ	37.5	62.4
	Methanothermus fervidus strain V245	13.5	86.5[d]
Thermoacidophiles	*Thermoplasma acidophilum* strain 122-1B2	10	90[e]
	Sulfolobus solfataricus	5	95[f]
	S. brierleyi	t	100[f]
	S. acidocaldarius	t	100[f]
	Thermoproteus tenax	+	+
Thermophilic	*Desulfurococcus mobilis*	+	+
anaerobes	*D. mucosus*	+	+
	Pyrodictium occultum	+	+[d]

[a] Data compiled from Langworthy *et al.*, 1982, also unpublished; Tornabene and Langworthy, 1979; Kate 1978; Ross *et al.*, 1981; Zillig *et al.*, 1981, 1982.

[b] Includes mixture of $C_{20}:C_{20}$ diether and $C_{20}:C_{25}$ diether.

[c] Includes unidentified component (T. A. Langworthy, K. R. Sowers, and J. G. Ferry, unpublished).

[d] Includes unidentified component (T. A. Langworthy and K. O. Stetter, unpublished).

[e] Tetraethers contain acyclic, mono-, and bipentacyclic C_{40} chains.

[f] Mixture of diglycerol and glycerolnonitol tetraethers; contains acyclic to tetrapentacyclic C_{40} chains. Present in trace amounts; +, present but distribution not established.

in the thermoacidophiles (Langworthy, 1979b, also unpublished; Langworthy *et al.*, 1982; De Rosa *et al.*, 1980e; Zillig *et al.*, 1981). Thus, the tetraethers are most pronounced in those archaebacteria that grow at high temperatures. The presence of both phytanyl diethers and the C_{25}-sesterterpanyl chain-containing diethers distinguishes the haloalkalophiles (De Rosa *et al.*, 1982a). Furthermore, the degree of pentacyclic ring formation among the biphytanyl chains discriminates between those methanogens possessing tetraethers and the thermoacidophiles. The tetraethers of all methanogens examined contain only the acyclic biphytanyl chains (Tornabene and Langworthy, 1979; Makula and Singer, 1978). Those of *Thermoplasma* contain acyclic, monocyclic, and bicyclic biphytanes, and *Thermoproteus* has a similar composition, but in *Sulfolobus* the bicyclic to tetracyclic biphytanes predominate. The relative proportions of the pentacyclic biphytanes in thermoacidophiles, however, is influenced by the temperature at which cells are grown (De Rosa *et al.*, 1980d; Langworthy, 1982b; Langworthy *et al.*, 1982). And, to be sure, the glycerolnonitol tetraethers are characteristic of the *Sulfolobus* species. It is of interest that the ultrathin, mycelia-forming organism isolated by Stetter (1982) from submarine volcanic areas grows optimally at 105°, perhaps the highest optimal temperature ever recorded, and its diglycerol tetraethers possess only the acyclic biphytanyl chains (T. A. Langworthy and K. O. Stetter, unpublished).

Although only a relatively limited number of organisms have been examined, the general trend within the glycerolipid components of the archaebacterial subgroups is apparent. Certainly the recognized diversity of glycerolipid structures will expand in the future. Summarized procedures for archaebacterial glycerolipid identification are given in Langworthy (1982a), Kates *et al.* (1982), Ross *et al.* (1981), and Kates (1972b), which are useful guides when using glycerolipids as a marker for establishing new members of the archaebacteria.

III. Polar Lipids

Total extractable lipids (polar plus nonpolar) on a cell dry weight basis, with limited exceptions, comprise about 2–6% of methanogens (Tornabene *et al.*, 1978; Tornabene and Langworthy, 1979) and about 2–4% each of the halophiles (Kates, 1972a, 1978) and the thermoacidophiles *Thermoplasma* and *Sulfolobus* (Langworthy *et al.*, 1972, 1974; Langworthy, 1977b). The polar lipids, including the glycolipids and phospholipids, represent about 80–90% of the total, and the remaining are neutral lipids. With the exception that they contain ether linkages and are stereochemically mirror images, the diether polar lipids of archaebacteria are structurally analogous to their glycosyl or phosphatidyldiacylglycerol counterparts in eubacteria and eukaryotes. Tetraether polar lipids, however, are unique, and the largest percentages occur as phosphoglycolipids in which sugar residues and phosphate radicals are linked to opposite sides of the tetraether molecule. The polar lipids of the halophilic archaebacteria have been well detailed, but polar lipid structures have been established in only one methanogen and those of the thermacidophiles have, for the most part, been only partially characterized.

A. HALOPHILES

The polar lipids of halophilic archaebacteria have been elucidated by Kates and associates (Kates, 1972a, 1978), and most have been proven by chemical synthesis. The four basic polar lipid structures (Fig. 4) are comprised of the diether analogues of phosphatidylglycerol, phosphatidylglycerophosphate, phosphatidylglycerosulfate, and a triglycosyl diether glycolipid containing a terminal sulfate radical.

The major phospholipid present is the phosphatidylglycerophosphate analogue, 2,3-di-O-phytanyl-sn-glycerol-1-phosphoryl-3'-sn-glycerol 1'-phosphate, which approximates 65% of the total polar lipids in *Halobacterium cutirubrum* (Kates *et al.*, 1965; Kates and Kushwaha, 1976; Kates, 1978). The remainder of the phospholipids occur as the diether analogs of phosphatidylglycerol, 2,3-di-O-phytanyl-sn-glycerol-1-phosphoryl-3'-sn-glycerol and phosphatidylglycerosulfate, 2,3-d-O-phytanyl-sn-glycerol-1-phosphoryl-3'-snglycerol 1'-sulfate, representing about 3–4% each of the total polar lipids (Hancock and Kates, 1973; Kates, 1978). These phospholipids have also been shown to occur in *H. halobium, H. salinarium,* and *Sarcina litoralis* (Kates *et al.*, 1966), *H. marismortui* (Evans *et al.*, 1980) and an unclassified strain R-4 (Kushwaha *et al.*, 1982), although relative concentrations of each may vary depending upon species.

Glycolipids, with limited exception, are almost entirely present as their acidic sulfated derivatives (Kates, 1978). The second most prominent lipid of *H. cutirubrum,* which accounts for about 25% of the polar lipids, is the sulfated triglycosyl diether, SO_4^- -3-Gal$p$-(β1→6)-Man$p$-(α1→2)-Glc$p$-(α1→1)-2,3-di-O-phytanyl-sn-glycerol (Kates and Deroo, 1973; Kates, 1978) (Gal$p$, galactopyranose). A minor branched tetraglycosyl glycolipid sulfate in which an additional galactofuranosyl (Gal$f$) residue is linked to the internal mannose has also been detected in *H. cutirubrum* (Smallbone and Kates, 1981). It has the structure SO_4^- -3-Gal$p$-(β1→6)-Man$p$-(3←1αGAlf)(α1→2)-Glc$p$-(α1→1)-2,3-di-O-phytanyl-sn-glycerol. Kushwaha *et al.* (1982) have also identified a novel diglycosyl diether in the unclassified halophile R-4 as, SO_4^- -6-Man$p$-(α1→2)-Glc$p$-(α1→1)-2,3-di-O-phytanyl-sn-glycerol. The same glycolipids as above, but lacking the acidic sulfate radicals, are also found in very small quantities among the halophiles. An exception, however, is seen in the Dead Sea isolate *H. marismortui* (Evans *et al.*, 1980), which lacks any sulfated glycolipids. About 11% of the polar lipids are comprised of the triglycosyl diether, Glc$p$-(β1→6)-

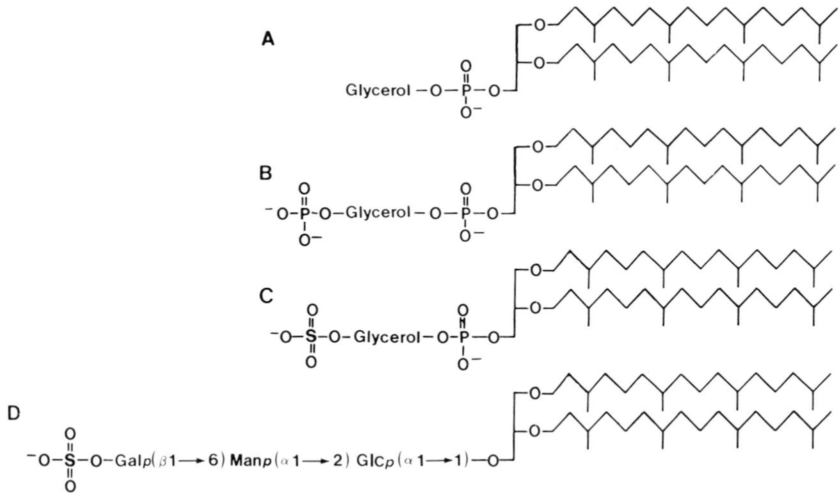

FIG. 4. Major polar lipids of halophilic archaebacteria include the diether analogues of (A) phosphatidylglycerol, (B) phosphatidylglycerophosphate, (C) phosphatidylglycerosulfate, (D) triglycosyl diether glycolipid sulfate.

Man$p$-(α1→2)-Glc$p$-(α1→1)-2,3-di-O-phytanyl-sn-glycerol. The deficiency in sulfate radicals, however, is made up by the high (17%) content of phosphatidylglycerosulfate in this organism.

The sulfated glycolipids and the phospholipids impart a strong acidic character to the surface of the halophilic bacteria. Essentially all of the polar lipids are acidic. The occurrence of glycolipid sulfates in the halophilic archaebacteria might be related to the unusual feature that these organisms possess bacteriorhodopsin (Oesterhelt and Stoekenius, 1973), a photosensitive purple pigment that converts light into chemical energy (Caplan and Ginsburg, 1978). Bacteriorhodopsin is associated with specialized purple membrane regions of the halophilic membrane, and it has been shown (Kushwaha et al., 1975a; Kates and Kushwaha, 1978) that the glycolipid sulfate and phosphatidylglycerosulfate are associated exclusively with the purple membrane fraction. The function of these sulfated polar lipids is not clear, but it has been speculated that they might serve as a proton donor for the functioning of the purple membrane as a light-driven proton pump (Kates and Kushwaha, 1978). The bacteriorhodopsin photosystem is a special feature of the halophilic archaebacteria and so far it, or a facsimile, has not been detected in other archaebacteria.

B. METHANOGENS

Polar lipid structures (Fig. 5) have thus far been established in but one methanogen, *Methanospirillum hungatei* (Kushwaha et al., 1981a,b). The polar lipids

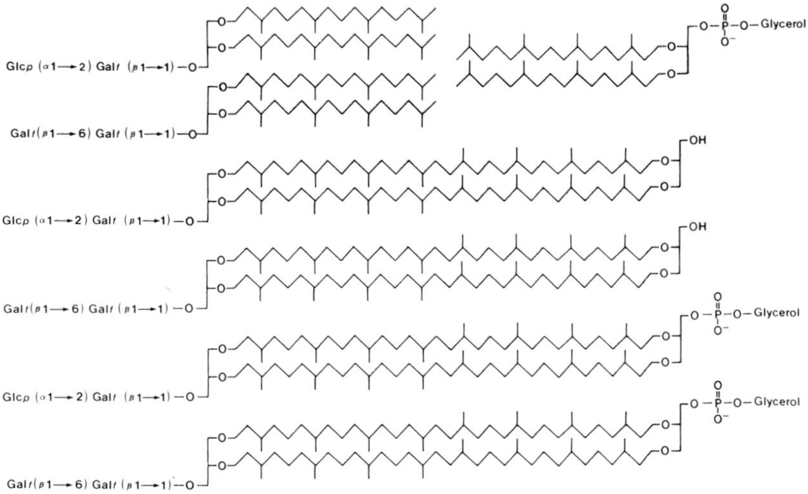

FIG. 5. Polar lipids of the methanogenic archaebacterium. *Methanospirillum hungatei*. Glc*p*, glucopyranose; Gal*f*, galactofuranose.

account for about 94% of the total extractable lipids and consist of both diether and tetraether derivatives. The polar head groups involved are limited to either of the two disaccharides Gal*f*-(β1→6)-βGal*f* or Glc*p*-(α1→2)-βGal*f* and to the acidic phosphate radical glycerophosphate.

Glycolipids are comprised of the two diether derivatives, Gal*f*-(β1→6)-Gal*f*-(β1→1)-2,3-di-*O*-phytanyl-*sn*-glycerol and Glc*p*-(α1→2)-Gal*f*-(β1→1)-2,3-di-*O*-phytanyl-*sn*-glycerol, which account for about 3% and 28%, respectively, of the polar lipids. Two diglycerol tetraether glycolipids are also present in small quantities (0.2–0.3%), which consist of the same two disaccharides glycosidically linked to one side of the tetraether structure as, Gal*f*-(β1→6)-Gal*f*- (β1→1)-*O*-[diglyceryltetraether]-OH and Glc*p*-(α1→2)-Gal*f*-(β1→1)- *O*- [diglyceryltetraether]-OH.

Phospholipids consist of a single component, representing about 10% of the polar lipids, identified as the diether analog of phosphatidylglycerol, 2,3-di-*O*-phytanyl-*sn*-glycerol-1-phosphoryl-1'-*sn*-glycerol. The terminal glycerol in this structure is reportedly diastereoisomeric (*sn*-1') with the phosphatidyl glycerol (*sn*-3') of the halophiles. A trace phospholipid component was noted, however, suggesting that both stereosiomers (*sn*-1-*sn*-1' and *sn*-1-*sn*-3') of phosphatidyl glycerol may be synthesized by *M. hungatei* (Kushwaha *et al.,* 1981b).

Phosphoglycolipids based upon tetraethers are the major polar lipid structures accounting for 58% of the polar lipids and about 64% of the total lipids of the organism. They occur as two components, based upon either of the two tetraether glycolipids, in which a glycerophosphate radical is attached to the free -OH group on the opposite side of the tetraether molecule. They have been identified as Gal*f*- (β1→6) -Gal*f*- (β1→1- *O*- [diglyceryltetraether]- *O*-phosphoryl- 3'-*sn* glycerol and Glc*p*- (α1→2) -Gal*f*- (β1→1) -*O*- [diglyceryltetraether] -*O*- phosphoryl- 3'*sn*- glycerol. The former constitutes 13% of the polar lipids and the latter 45%. Thus, with the possible exception that the glycerophosphate has the *sn*-3' stereoconfiguration, the phosphoglycolipids clearly bear a structural resemblance to one molecule of diether glycolipid plus one molecule of diether phosphatidyl glycerol, which have been covalently condensed through the terminal ends of their phytanyl chains.

C. THERMOACIDOPHILES

Identities of the polar lipid structures among the thermoacidophiles are incomplete. In *Thermoplasma,* polar lipids comprise about 82% of the total lipids. They consist of at least six different glycolipids and the remainder are a mixture of at least seven phosphorus-containing lipids. Structures have not been assigned to any of these lipids. However, the major component comprising 80% of the polar lipids is known to be a phosphoglycolipid. It is a diglycerol tetraether with a *sn*-3-glycerol-phosphate attached to one of the free -OH groups and a single unidentified sugar glycosidically linked to the other (Langworthy *et al.*, 1972, 1982; Langworthy, 1979a, also unpublished). The remaining polar lipids include mostly tetraether derivatives, but also some diether derivatives (Langworthy, 1979b, 1980).

The structure of one tetraether lipid from *Thermoplasma,* which represents about 3% of the cell dry weight, has now been fully established (Mayberry-Carson *et al.*, 1974; Smith, 1980). It is extractable by hot aqueous phenol and can be considered to be either a lipoglycan (MW 5300) or to be a glycolipid with an extended 25 sugar chain (Fig. 6). It contains 24 mannose residues and one glucose and has been shown by Smith (1980) to be [Man*p*-(α1→2)-Man*p*-(α1→4)-Man*p*-(α1→3)]$_8$-Glc*p*-(α1→1)-*O*-[diglyceryltetraether]-OH. The carbohydrate chain is attached to only one side of the tetraether molecule. This lipid is associated with the surface of the *Thermoplasma* membrane (Mayberry-Carson *et al.*, 1978) and has physical properties resembling those of gram-negative lipopolysaccharides (Mayberry-Carson *et al.*, 1975). No lipoglycans have been reported as yet in *Sulfolobus,* nor is any information on them available from methanogens. However, similar types of diacylglycerol-derived lipoglycans, for which the *Thermoplasma* lipoglycan is the prototype, have been shown to occur in the eubacterial *Acholeplasmas* (Smith *et al.*, 1976; Smith, 1981).

Sulfolobus polar lipids (Langworthy *et al.*, 1974; Langworthy, 1977b) have now been somewhat better defined by De Rosa and colleagues (1980c). Polar

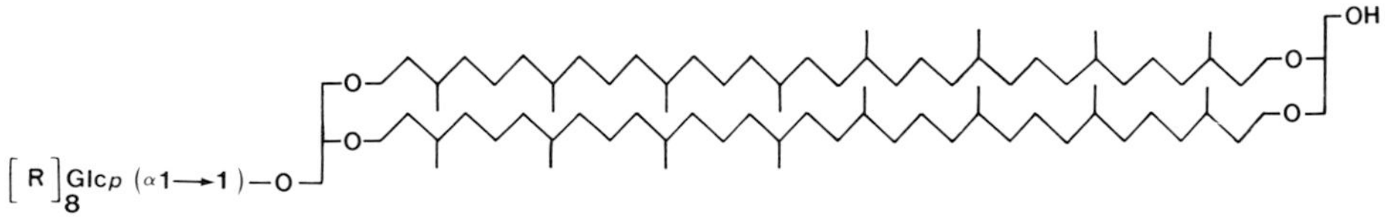

THOMAS A. LANGWORTHY

R = Man$p(\alpha 1 \rightarrow 2)Manp(\alpha 1 \rightarrow 4)Manp(\alpha 1 \rightarrow 3)$

FIG. 6. Structure of the tetraether lipoglycan from the thermoacidophilic archaebacterium *Thermoplasma.*

lipids constitute about 90% of the total lipids. They are almost exclusively based on either of the two diglycerol tetraether or glycerolnonitol tetraether structures. In heterotrophically grown cells, polar lipids of both varieties of tetraethers occur in about equal proportions, but in autotrophically grown cells, the glycerol-nonitol tetraether derivatives predominate. Polar head groups are restricted to either glucose, a glucosylgalactosyl disaccharide, phosphorylinositol, or sulfate.

Glycolipids comprise about 75% of the polar lipids and are made up of about equal amounts of diglycosyldiglycerol tetraether and a glycosylglycerol-nonitol tetraether. The two glycolipids have been partially characterized as βGlc$p$-βGal$p$-O -[diglyceryltetraether] -OH and βGlc$p \rightarrow$ (HO)$_7$-[glycerylnonityltetra-ether]-OH. The disaccharide is glycosidically linked to one side of the diglycerol tetraether structure. The monosaccharide is glycosidically linked to one of the free -OH groups of nonitol in the glycerolnonitol tetraether, but precisely to which one has not been established. A glycolipid sulfate also occurs representing about 1% of the polar lipids. In this instance, a single SO_4^- radical is attached to the glucosyl glycerolnonitol tetraether glycolipid but its location is not known.

Phosphorus-containing lipids are represented by three inositol phosphate de-rivatives, which account for about 24% of the polar lipids. Present in comparable amounts are the diglycerol tetraether analog of phosphatidyl inositol, identified as inositolphosphoryl-O-[diglyceryltetraether]-OH, and two phosphoglycolipids, which are the inositophosphoryl derivatives of the two glycolipids of the orga-nism noted above. Where the inositol phosphate is attached to these two glycoli-pids has not been established. Two unidentified polar lipids also occur when *Sulfolobus* species are grown autotrophically (Langworthy, 1977b).

Although relatively few polar lipid structures of the archaebacteria have been fully established, a general picture is clear. Tetraether polar lipids may contain head groups substituted to either or both sides of the tetraether molecule. And, in those archaebacteria possessing tetraethers, phosphoglycolipids are present such that nearly all of the polar lipid structures (glycolipids plus acidic lipids) contain carbohydrate residues. The unusual structures of the polar lipids of the few archaebacteria studied thus far suggests that many new types of lipids remain to be found.

IV. Nonpolar Lipids

The nonpolar or neutral lipid content, as a percentage of the total lipids ranges from about 7 to 20% in halophiles (Tornabene *et al.*, 1969; Kates, 1978) and thermoacidophiles (Langworthy *et al.*, 1972, 1974) to as high as 30% reported in some methanogens (Tornabene *et al.*, 1978, 1979; Tornabene and Langworthy, 1979). Isoprenoid and hydroisoprenoid hydrocarbons, which can comprise up to 95% of the nonpolar lipid fraction, are a ubiquitous and distinctive feature of archaebacteria. Although acyclic isoprenoid hydrocarbons are universal, generally occurring as minor or trace constituents in other cells (Tornabene, 1981), they are the principal components in archaebacteria. The isoprenoid hydrocarbons display a complete array of C_{15} to C_{30} isoprenoid skeletons, although the C_{30} hexaisoprenoids (squalenes), C_{25} pentaisoprenoids, and C_{20} tetraisoprenoids (phytanes) are the major constituents. Some C_{40} and C_{50} carotenoids are also additional components of the halophilic archaebacteria. Along with a number of minor constituents, it is important to realize that many of these hydrocarbons have previously been detected only within the geosphere in sediments, deposits, shale, and petroleum (Tornabene *et al.*, 1979; Holzer *et al.*, 1979; Langworthy *et al.*, 1982; Hahn, 1982; see Chapter 4 by Hahn and Haug).

A. Acyclic Isoprenoids

Squalenes (C_{30} hexaisoprenoids) are the major neutral lipids of archaebacteria and can occur in a continuous range of hydrosqualenes from dihydrosqualene to squalane (Fig. 7). The squalenes of *Halobacterium cutirubrum* have undergone rigorous proof of structure, being identified as all-*trans*-squalene, dihydrosqualene, and tetrahydrosqualene and a small amount of dehydrosqualene (Tornabene *et al.*, 1969; Kramer *et al.*, 1972; Kates, 1978). Squalene is the major hydrocarbon in the cell. These compounds also occur in *H. halobium, H. salinarium, Sarcina litoralis, S. morrhuae* (Kates, 1978), and *H. marismortui* (Evans *et al.*, 1980). Greater diversity of squalenes is found among the methanogens and thermoacidophiles (Tornabene *et al.*, 1978, 1979; Holzer *et al.*, 1979), in which the full range of more saturated derivatives occur. Squalene is the principal component among most methanogens examined, but exceptions include *Methanococcus voltae*, in which tetrahydrosqualene is the major hydrocarbon, and *Methanosarcina barkeri*, which contains no squalenes. Among thermoacidophiles, *Thermoplasma* contains only squalene, whereas *Sulfolobus* contains the more saturated derivatives including squalane.

In addition to squalenes, the methanogens and thermoacidophiles possess a wide range of C_{15} to C_{30} isoprenoids, including the C_{20} to C_{30} hydrocarbon

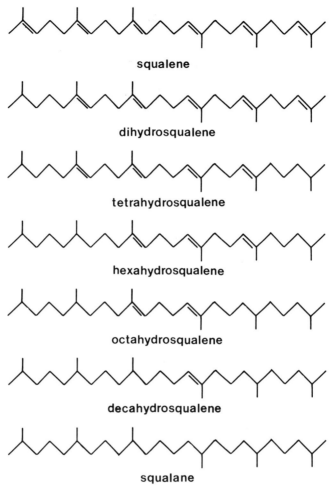

FIG. 7. Structure of the C_{30} isoprenoid hydrocarbon squalene and its hydroisoprenoid derivatives, the predominant nonpolar lipid component among archaebacteria.

skeletons, which are illustrated in Fig. 8 (Tornabene *et al.*, 1979; Holzer *et al.*, 1979). Positions of double bonds have not been established. Although the squalene carbon skeleton present in archaebacteria is that expected from a tail-to-tail (pyrophosphate end-to-pyrophosphate end) condensation of two C_{15} farnesyl units, a second hexaisoprenoid identified in *Sulfolobus* reveals a carbon skeleton consistent with head-to-tail (hydrocarbon end-to-pyrophosphate end) biosynthetic route. The C_{25} pentaisoprenes, with a range of hydropentaisoprene derivatives, are also major hydrocarbons in methanogens and thermoacidophiles. Two derivatives have been identified. One arises from a head-to-tail condensation of a

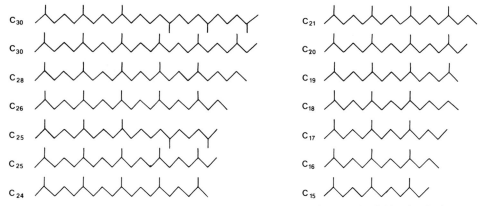

FIG. 8. Isoprenoid hydrocarbon skeletons identified among the nonpolar lipids of archaebacteria.

C_{20} geranylgeranyl unit and a C_5 isopentenyl unit. This 2,6,10,14,18-pentamethyleicosane skeleton is the same as that of the *O*-alkyl diether chain found in the alkalohalophiles (De Rosa *et al.*, 1982a), but its occurrence in the neutral lipid fraction of these organisms has not been determined. The second pentaisoprene skeleton arises from the tail-to-tail condensation of a C_{15}-farnesyl unit and a C_{10}-geranyl unit. Rowland *et al.* (1982) have now confirmed by synthesis the identity of this skeleton as 2,6,10,15,19-pentamethyleicosane and have also confirmed its presence in marine sediments (Brassell *et al.*, 1981). The C_{20} tetraisoprenoids, including phytane and phytenes, derived by a head-to-tail biosynthetic route, have also been identified in thermoacidophiles and some methanogens. Minor isoprenoids occur in the range of C_{15} to C_{28} hydrocarbons, including pristane, along with traces of *iso*-branched C_{16} and C_{17} hydrocarbons and C_{19} to C_{32} *n*-alkanes (Tornabene *et al.*, 1979; Holzer *et al.*, 1979).

The principal acyclic isoprenoid hydrocarbon skeletons of archaebacteria are of the same general nature and structure, but the distribution between the degrees of unsaturation is as varied as the distribution between the carbon ranges of the isoprenoid skeletons themselves (Table II). These differences, in addition to being characteristic of phenotypically diverse species, may also reflect physiological differences within the cells. In *H. cutirubrum*, for example, it has been demonstrated that the squalene-to-hydrosqualene ratios are directly related to whether cells are cultivated aerobically or anaerobically, and proportions are also relative to rates of aeration (Kushwaha *et al.*, 1975b; Tornabene, 1978). Thus, the degree of unsaturation and the relative concentration of hydrocarbons may reflect differences in the growth phases of individual organisms and/or the physiological state of the cells.

A divergence in the general trend of acyclic isoprenoid composition among archaebacteria occurs within the halophilic subgroup. They differ by the lack of

TABLE II

PRINCIPAL NEUTRAL LIPID COMPONENTS OF ARCHAEBACTERIA[a]

Archaebacterium	C_{30} Hexaisoprenoids	C_{25} Pentaisoprenoids	C_{20} Tetraisoprenoids
Halobacterium cutirubrum	$C_{30:4}, C_{30:5}, C_{30:6}, C_{30:8}$	—	—
Methanobacterium thermoautotrophicum	$C_{30:0}, C_{30:1}, C_{30:2}, C_{30:3},$	$C_{25:1}, C_{25:2}, C_{25:3}$	$C_{20:1}, C_{20:2}$
Methanococcus vannielii	$C_{30:4}, C_{30:5}, C_{30:6}$	$C_{25:3}, C_{25:4}, C_{25:5}$	$C_{20:2}, C_{20:3}, C_{20:4}$
Methanobacterium ruminantium PS	$C_{30:3}, C_{30:4}, C_{30:5}, C_{30:6}$	$C_{25:5}, C_{25:3}$	—
Methanococcus strain PS	$C_{30:3}, C_{30:4}, C_{30:5}, C_{30:6}$	$C_{25:3}$	—
Methanosarcina barkeri	—	$C_{25:0}, C_{25:1}, C_{25:2}, C_{25:3}$	—
Methanobacterium strain A2	$C_{30:6}$	—	$C_{20:0}, C_{20:1}, C_{20:4}$
M. strain M.o.H.	$C_{30:4}, C_{30:5}, C_{30:6}$	—	$C_{20:0}, C_{20:1}, C_{20:2};$ $C_{20:3}, C_{20:4}$
M. ruminantium M-1	$C_{30:1}, C_{30:2}, C_{30:3}; C_{30:4}$ $C_{30:5}, C_{30:6}$	—	—
Methanospirillum hungatei	$C_{30:4}, C_{30:5}, C_{30:6}$	—	—
Sulfolobus solfataricus	$C_{30:0}, C_{30:1}, C_{30:2}$	$C_{25:0}, C_{25:1}$	$C_{20:1}, C_{20:2}, C_{20:3}$
S. brierleyi	—	—	$C_{20:0}$
Thermoplasma acidophilum	$C_{30:6}$	$C_{25:5}$	$C_{20:0}, C_{20:4}$

[a] Data from Langworthy *et al.*, 1982; Tornabene *et al.*, 1979; Holzer *et al.*, 1979. The first number represents the chain length and the second number indicates the number of double bonds.

apolar isoprenoids in the C_{20} to C_{28} range, and they are the only group thus far known to possess carotenoid pigments. The only exception is the closely related and pigmented alkalohalophiles (Tindall *et al.*, 1980; Soliman and Trüper, 1982), but the neutral lipid composition of these organisms has not been reported. The additional isoprenoid structures that characterize the halophiles are illustrated in Fig. 9. Together, the C_{40} and C_{50} carotenoids make up about 10% of the nonpolar lipids. The bacterioruberins, which are unusual hydroxylated C_{50} carotenoids, predominate, and they are responsible for the characteristic red color of the halophiles. They include bacterioruberin, monoanhydrobacterioruberin, and bisanhydrobacterioruberin (Kelly and Liaaen-Jensen, 1967; Kelly *et al.*, 1970; Kushwaha *et al.*, 1974, 1975b; Kates, 1978). They are associated with the red membrane fraction of the halophiles (Kushwaha *et al.*, 1975a). The principal C_{40} carotenoids have been identified as neo-α-carotene, α- and neo-β-carotene, lycopersene, *cis*- and *trans*-phytoene, *cis*- and *trans*-phytofluene, and lycopene (Kushwaha *et al.*, 1972; Kushwaha and Kates, 1973). The low concentration of these carotenoids suggests that they may be precursors to either the bacterioruberins or retinal. Colorless strains of halophiles lack both the C_{40} and C_{50} carotenoids. Among the C_{20} isoprenoids, phytane has not been detected per se, but geranylgeraniol together with trace amounts of phytol (Kushwaha *et al.*, 1975a; Kates, 1978) have been identified. Geranylgeraniol has also been detected in *Sulfolobus* (De Rosa *et al.*, 1980b), and since it is an intermediate in isoprenoid biosynthesis, it will likely be found to occur in all archaebacteria. All-*trans*-Retinal, associated with bacteriorhodopsin in the purple membrane, has also been identified (Kushwaha and Kates, 1973; Kushwaha *et al.*, 1974; Kates, 1978). Its presence is dependent on growth conditions, being produced anaerobically in the light.

B. OTHER COMPONENTS

In addition to the principal acyclic isoprenoids, a number of minor nonpolar lipid constituents have been identified which are characteristic of certain of the archaebacterial subgroups. Small quantities of the free alkyl glycerol ethers, which serve as the backbone of the polar lipids, are present, including the phytanyl diether in halophiles (Kushwaha and Kates, 1978b), dibiphytanyldiglycerol tetraethers in thermoacidophiles (Langworthy *et al.*, 1972, 1974), and probably both types in methanogens (Langworthy *et al.*, 1982, also unpublished). De Rosa *et al.* (1976b) have reported the occurrence in *Sulfolobus* of an unusual tri-*O*-phytanylglycerol ether containing partially and fully saturated phytanyl chains. Mevalonic acid has been detected in most halophiles (Kushwaha and Kates, 1978a). Free indol has been found in substantial quantities, up to 7% of the total nonpolar lipid fraction, in a number of halophiles (Kushwaha *et*

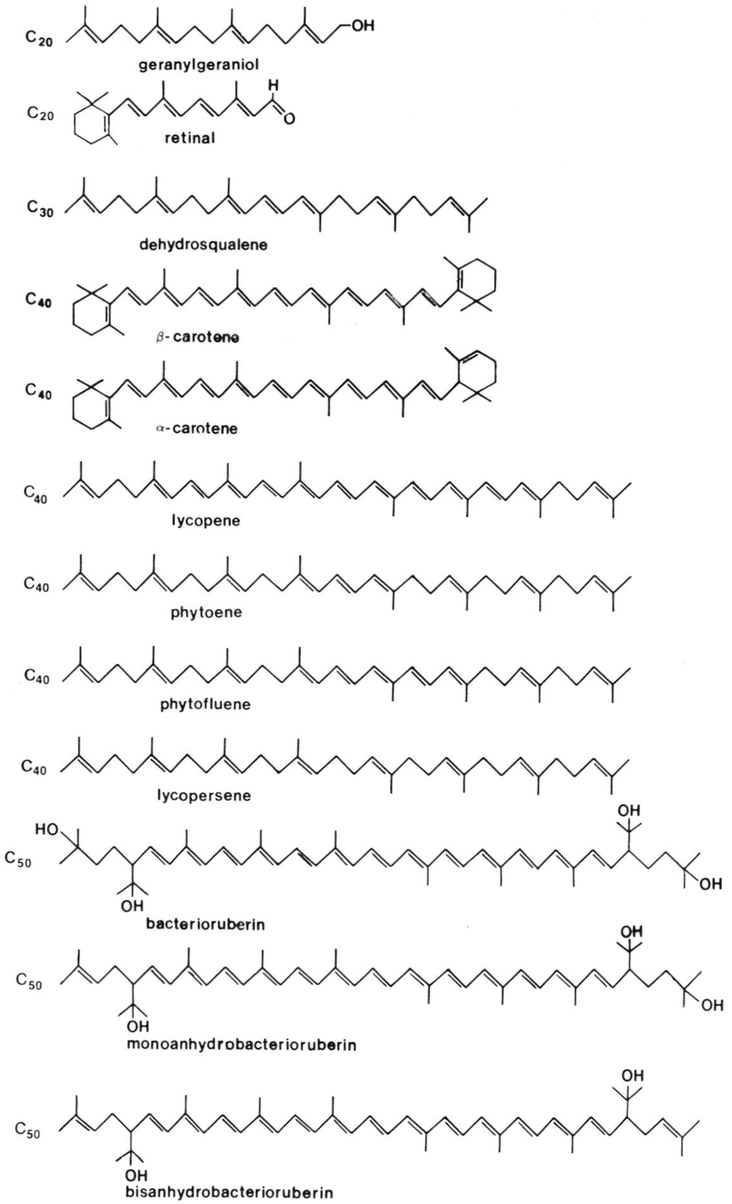

Fig. 9. Minor isoprenoid nonpolar lipid components of halophilic archaebacteria. C_{20}, geranylgeraniol, retinal; C_{30}, dihydrosqualene; C_{40}, carotenoids; C_{50}, bacterioruberins.

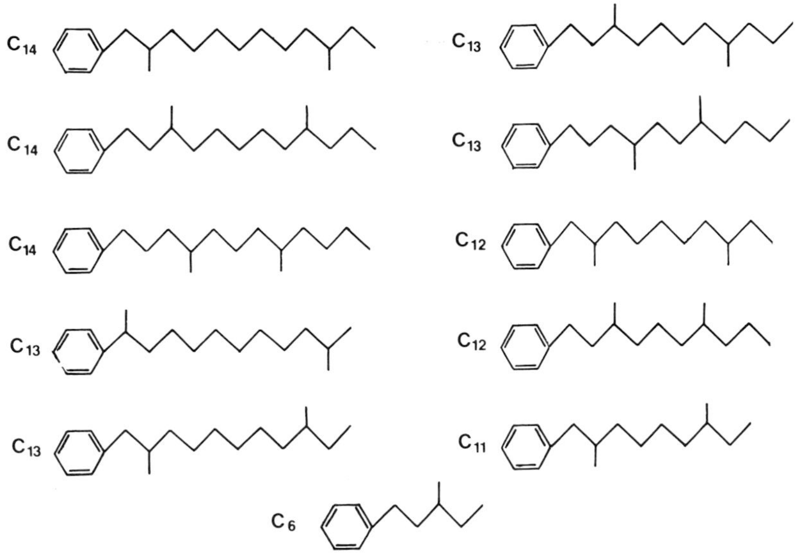

FIG. 10. Alkylbenzenes identified as minor nonpolar lipid constituents of the thermoacidophilic archaebacteria *Thermoplasma* and *Sulfolobus*. Carbon number indicates alkyl chain length.

al., 1977). Though indole derivatives occur in substantial amounts in nature, this is the first known instance of its intracellular accumulation. Its physiological significance and whether intracellular indole is a feature of all archaebacteria are not known. The two thermoacidophiles, *Thermoplasma* and *Sulfolobus*, contain small quantities of a series of branched alkylbenzenes, whose structures, based upon mass spectral interpretation, are shown in Fig. 10 (Langworthy *et al.*, 1982). This is the first instance of such compounds being detected in living organisms, although alkylbenzenes have been isolated from crude oils and sediments.

Lipoquinones are also known to occur, and their structures appear to discriminate between the archaebacterial subgroups in which they have been detected. Halophiles contain menaquinone-8 (vitamin MK-8) [which possesses a C_{40} isoprenoid side chain (Fig. 11)]. It is the major quinone, accounting for 5–10% of the nonpolar lipids (Tornabene *et al.*, 1969; Kushwaha *et al.*, 1974, 1975a; Kates, 1978). In addition, most halophiles have now been found to possess quantities of MK-7, containing a C_{35} isoprenoid side chain, and both menaquinones may occur with one isoprene unit saturated (Collins *et al.*, 1981; Collins and Jones, 1981). Among the thermoacidophiles, *Thermoplasma* possesses MK-7 as the major quinone, but small amounts of MK-6 and MK-5 also occur (Collins and Langworthy, 1983; Langworthy *et al.*, 1972). *Thermoplasma* also contains a new partially characterized quinone structure ($C_{47}H_{66}O_2$), desig-

Menaquinone   **Thianaphthenequinone**

FIG. 11.   Structures of the principal lipoquinones of halophilic archaebacteria (menaquinone-8), *Thermoplasma* (menaquinone-7), and *Sulfolobus* (4,7-thianaphthenequinone, "Caldariellaqui-none").

nated *thermoplasmaquinone,* that makes up 25% of the total quinone fraction (Collins and Langworthy, 1983). *Sulfolobus* species, however, contain thianaphthenequinones (Fig. 11) identified as benzo[b]thiophen-4,7-quinone and called generically *caldariellaquinone* (De Rosa *et al., 1977c*). This quinone contains a saturated C_{30} isoprenoid chain and is the major quinone, although small amounts (4–5% each) of the C_{25} and C_{20} chain derivatives also occur (Collins and Langworthy, 1983). This quinone is the only sulfur-containing quinone known, and *Sulfolobus* provides the only example of its natural occurrence. Whether other archaebacteria contain quinones or analogous type substances is not known, but as pointed out by Collins and Jones (1981), further studies on such structures could provide information of taxonomic value.

Overall, the nonpolar lipid composition and the specific variations in relative concentrations of individual components reflect phenotypic diversity among individual species. However, the generic nature of the isoprenoid hydrocarbons demonstrates the similarities between the increasingly diverse collection of archaebacteria and definitely sets them apart from other organisms. Future detailed studies of nonpolar lipid components should further delineate similarities and specific differences.

V. Biosynthesis

A. ACYCLIC ISOPRENOIDS

The pathways of lipid metabolism in archaebacteria are reasonably clear in general terms, but experimental details are lacking. That the archaebacteria possess the mevalonate pathway for isoprenoid biosynthesis is clearly evidenced by the structure of the hydrocarbon skeletons and by the almost exclusive incorporation of [^{14}C]mevalonate into them (Langworthy *et al.,* 1972, 1974; Langworthy, 1978a; Kates *et al.,* 1968; Kates, 1978; De Rosa *et al.,* 1977a, 1980b). Archaebacteria also possess the malonyl-CoA pathway for fatty acid biosynthesis attested to by the low levels of fatty acids detected in archaebacteria (Langworthy *et al.*, 1974; Ruwart and Haug, 1975; Kates, 1972a; Tornabene and Langworthy, 1979). It has been shown that [^{14}C]acetate incorporates into the fatty acids in *Halobacterium cutirubrum* (Kates *et al.*, 1968) and that the malonyl-CoA system is indeed operative at low levels in cell-free extracts (Pugh *et al.*, 1971).

Archaebacteria therefore possess the same two great pathways for isoprenoid and fatty acyl lipid synthesis present in all cells. Up to a point the pathways are indistinguishable, but a deep divergence occurs in that the major route of lipid synthesis in archaebacteria is that of the mevalonate pathway for isoprenoids, whereas the principal route is that of fatty acids in other organisms. This difference could be ascribed to the extreme environments in which many archaebacteria thrive. This is unlikely, however, since eubacteria possessing conventional fatty acyl lipids can be found in the same habitats (Langworthy, 1978b, 1982a; Langworthy *et al.*, 1976; Kushner, 1978).

The initial steps in isoprenoid synthesis involving mevalonate → (C_5 isopentenyl pyrophosphate ⇆ C_5 dimethylallyl pyrophosphate) → C_{10} geranyl pyrophosphate → C_{15} farnesyl pyrophosphate → C_{20} geranylgeranyl pyrophosphate are those shared by all organisms including archaebacteria. These initial steps also appear to manifest the same stereospecificity in all cases (De Rosa *et al.*, 1980b). In cells other than archaebacteria, these steps can ultimately proceed to the formation of such characteristic components as carotenoids, pigments, higher terpenoids, cyclic terpenoids, sterols, etc. Archaebacteria, however, produce a wide range of C_{20} to C_{30} acyclic isoprenoids and also form the acyclic C_{40} isoprenoid, biphytane (Fig. 12). Perhaps significantly (see Section V,C), the biphytane skeleton is a principal component of archaebacterial lipids, occurring in the tetraether glycerolipids. However, although synthesized via the mevalonate pathway, it has not been detected as a component of the nonpolar lipids (Tornabene *et al.*, 1979; Holzer *et al.*, 1979). Furthermore, archaebacteria differ in the ability to biohydrogenate the isoprenoids. The hydrogenation of double bonds to produce more saturated lipid chains is relatively rare in biological systems. Tornabene (1978) has suggested the possibility that this may reflect the ability of archaebacteria to utilize these compounds as intermediates in the internal regulation of protons. In any case, the mevalonate pathway is the main route of lipid synthesis, but what caused this divergence in direction and the metabolic aspects of its control are unknown.

B. DIETHERS

Diether and tetraether glycerolipid structures distinguish the archaebacteria, yet little is actually known about how these compounds are synthesized. Obviously, the phytanyl and biphytanyl chains arise via the mevalonate pathway,

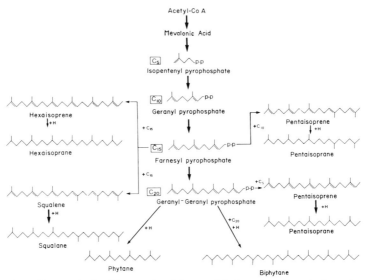

FIG. 12. Synthesis of isoprenyl and isopranyl hydrocarbons in archaebacteria.

but the steps involved in forming the ether linkage to glycerol are unclear. Studies by Kates *et al.* (1970) on the metabolism of phytanyl diethers in *H. cutirubrum* suggested that synthesis may proceed through the stepwise hydrogenation of geranylgeranyl pyrophosphate to phytanylpyrophosphate followed by subsequent condensation via ether linkage to glycerol. Alternatively, synthesis could occur via condensation of geranylgeranyl pyrophosphate, first to glycerol producing a di-*O*-geranylgeranyl glycerol ether, followed by sequential reduction steps producing the saturated phytanyl chains. As pointed out by De Rosa *et al.* (1982b), geranylgeranyl and similar allylic pyrophosphates have the ability to act as alkylating agents in other biosynthetic mechanisms, but such alkylating reactivity would be reduced in a nonallylic molecule such as phytanyl pyrophosphate. It is therefore likely that reduction of the hydrocarbon chains occurs after formation of geranylgeranyl ether linkages to glycerol. That a phosphorylated isoprenoid is involved in the formation of the ether linkages is supported by the report of Basinger and Oliver (1979) in which bacitracin, an inhibitor of phosphorylated lipid carriers, inhibits lipid synthesis in *H. cutirubrum.*

   How the ether linkages to glycerol are actually formed is not known. Obviously the pyrophosphate is lost from the isoprenoid chain, but there are conflicting reports on the involvement of glycerol in the reaction. Kates *et al.* (1970), employing [^{14}C]glycerol plus either [1(3)-^{3}H]glycerol or [2-^{3}H]glycerol as precursors, found that the ^{3}H–^{14}C ratio indicated a dehydrogenation of the glycerol backbone at carbon-2, but not at carbon-1 or carbon-3 of the glycerol molecule. The hydrogen atom lost at carbon-2 was considered as possibly important in the ether-forming step. But, in a similar study of glycerol incorporation into the tetraethers of *Sulfolobus*, De Rosa *et al.* (1982b) found that all of the original hydrogen atoms remained 100% intact in the glycerol backbone. Either the mechanism of ether-linkage formation differs between diether and tetraethers, which seems unlikely, or, as De Rosa *et al.* (1982b) suggested, the observed loss of the hydrogen atom from carbon-2 in halophiles could be due to a discriminating isotopic effect resulting in an efficient interconversion of glycerol. However, all that can be stated now is that diether synthesis appears to be novel and that all steps in the formation must be stereospecific, both in terms of the methyl groups of the phytanyl chains, and in the attachment of these chains to carbon-2 and carbon-3 of the glycerol backbone.

C. Tetraethers

An explanation of how tetraethers are biosynthesized is still within the realm of conjecture. Several routes might be considered to be possible, but one pathway at this time seems highly probable, a conclusion now reached by several laboratories (De Rosa *et al.*, 1980e; Kushwaha *et al.*, 1981b; Langworthy, 1982b,c; Langworthy *et al.*, 1982). It appears that tetraethers are formed through the head-to-head condensation of the phytanyl chains from two diether molecules to yield free tetraethers, or even perhaps by condensation between two diether polar lipid molecules to yield the corresponding tetraether polar lipids.

It has been well established by De Rosa and colleagues (1977a,b, 1980a,e) that the C_{40} biphytanyl skeleton is the structural equivalent of two C_{20} phytane molecules, which are linked head-to-head through the 16,16′ terminal methyl groups. This formation is the reverse of the normal tail-to-tail linkage through the 1,1′ pyrophosphate ends of two C_{20} geranyl-geranyl pyrophosphates in the analagous C_{40} carotenoid skeleton of other organisms (Fig. 13).

One possible pathway involves two C_{20} geranylgeranyl pyrophosphates first condensing head-to-head to form a pool of unsaturated C_{40} biphytanyl diterminal pyrophosphates. These would then further condense with glycerols on either end followed by reduction of the biphytanyl chains as suggested for diether synthesis. Presumably, the formation of pentacyclic rings, when present, could occur during steps leading to the C_{20} geranylgeranyl pyrophosphate unit via a mechanism proposed by Nes and Nes (1980), which will not be dealt with here. But it cannot be ruled out that cyclization could also occur during the final reduction steps of the ether-linked biphytanyl chains. However, this overall route would be expected to yield a wide range of products including all of the possible nonsymmetrical, parallel, and antiparallel combinations of acyclic and pentacyclic bi-

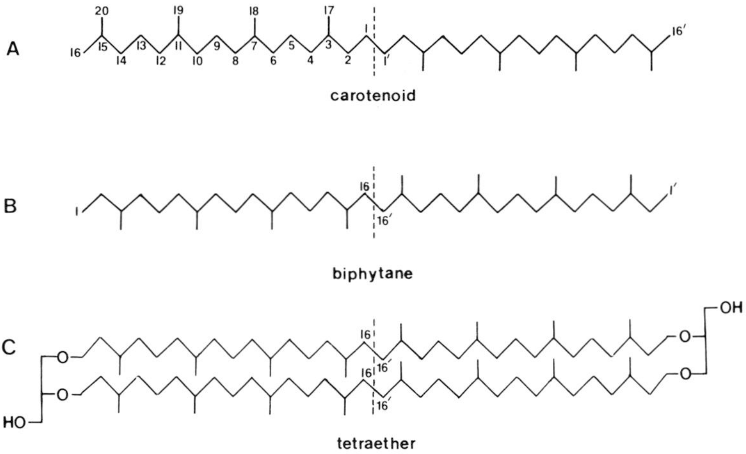

FIG. 13. Comparison of C_{40} isoprenoid structures. (A) An acyclic C_{40} carotenoid skeleton formed by tail-to-tail (C_1 to $C_{1'}$) condensation of two C_{20} units. (B) Archaebacterial C_{40} biphytane skeleton formed by head-to-head (C_{16} to $C_{16'}$) condensation of two C_{20} units. (C) A dibiphytanyl tetraether formed by head-to-head ($C_{16'},C_{16}$ to $C_{16'},C_{16'}$) condensation of two phytanyl diethers.

phytanyl tetraethers. It would be expected that the primary -OH groups of the tetraethers could be found in both the *cis* and *trans* configuration. These organisms should be examined for traces of biphytanyl pyrophosphates, diols or biphytanes among their lipids. Furthermore, diethers containing biphytanyl chains might also occur owing to "mistakes" due to the dimensional biosynthetic problems wherein the long and fluid biphytanyl dipyrophosphate must locate a free glycerol at either end.

By contrast, the actual tetraethers structures found are in fact quite limited (see Fig. 3). They always contain identical pairs of biphytanyl chains. Those chains possessing pentacyclic rings are always arranged symmetrically and stereoselectively in an antiparallel configuration. The primary -OH groups occur only in the *trans* configuration. And, no free biphytanes, diols, or phosphate derivatives have yet been detected among the total lipids. The alternative hypothesis, however, that tetraethers arise by condensation of two diether molecules, explains the limited range and the specificity in which the tetraether structures are found. Furthermore, the tetraethers are in fact the structural analog of a phytanyl diether dimer (Fig. 13).

De Rosa *et al.* (1980e) have nicely refined this concept, which not only explains the structures actually found, but is also compatible with the observed phenotypic variations. Tetraether synthesis could proceed from the same unsaturated geranylgeranyl glycerol diether intermediate proposed in phytanyl diether synthesis. Head-to-head coupling could occur through the terminal 16,16′ double bonds of two of the unsaturated diether intermediates, followed by reduction of the remaining double bonds yielding the dibiphytanyl diglycerol tetraether. Competition between reduction and pentacyclization could yield the variety of specialized tetraethers in the thermoacidophiles. In addition, competition between the complete reduction of the geranylgeranyl glycerol diether to phytanyl diether and head-to-head coupling to another unsaturated intermediate would give rise to the diether–tetraethers found in some methanogens. It can also be added, that a coupling reaction to produce tetraethers would likely occur at the center of the membrane bilayer between two unsaturated intermediates located opposite one another on the inner and outer faces of the membrane (see Section VI).

Only some indirect evidence supports the diether condensation hypothesis. The tri-*O*-phytanylglycerol triether identified in *Sulfolobus* (De Rosa *et al.*, 1976b) contains unsaturated chains including an ether-linked geranylgeranyl chain. This suggests that a diether analog could be possible. An apparently rapid turnover of the small amount of [^{14}C]mevalonate labeled diether component in *Thermoplasma* also lends credence (Langworthy, 1982c). Also, the diethers of *Thermoplasma* occur only in the polar lipid fraction, and as suggested by Kushwaha *et al.* (1981b), the possibility that condensation could occur via an unsaturated diether already substituted with polar head groups cannot be excluded.

The only thing that can be definitely said at this time is that the proposed biosynthetic route for tetraethers is not supposed to exist. But it should be pointed out that tetraethers are not supposed to exist. Obviously, tetraether synthesis requires experimental proof. Especially helpful would be the isolation of the presumed unsaturated diether intermediate. Whatever the mechanism, it is clearly unusual, and its elucidation should provide a new route of hydrocarbon biosynthesis.

VI. Membrane Organization

The glycerol ether lipids are of considerable interest in terms of the consequent molecular structure, function, and biogenesis of the archaebacterial membrane. The generally accepted superstructure of a biologically functioning membrane is based upon a lipid bilayer formed by hydrophobic interactions between separate and opposite fatty acyl chains that are linked to glycerol backbones residing on the opposing inner and outer bilayer faces (Fig. 14). Appropriate fluidity for optimizing physiological functions is basically achieved through alterations in hydrocarbon chain length, methyl branching, or unsaturation (Melchior, 1982). The glycerol diether residues present in archaebacteria would also be expected to be able to provide a lipid bilayer through interactions of opposing phytanyl chains. The only constraint in this case is that the hydrocarbon chains are the same, being inalterably saturated, methyl-branched, and fixed at 20 carbons.

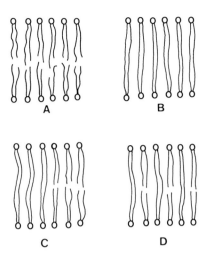

Fig. 14. Schematic comparison between (A) a normal lipid bilayer membrane, (B) a tetraether derived lipid "monolayer" membrane of thermoacidophilic archaebacteria, and (C and D) possible orientations of "monolayer" and bilayer membrane regions of certain methanogenic archaebacteria. Circles, polar head groups; lines, hydrocarbon chains.

Tetraether structures, however, indicate a new type of membrane organization, which cannot be considered to be a lipid bilayer in the strict sense of the word, but is rather an amphiphilic lipid "monolayer" (Langworthy, 1977a, 1979a,b, 1982b; Langworthy *et al.*, 1982; De Rosa *et al.*, 1980e). Such an organization is created by the extension of C_{40} biphytanyl chains across the membrane which are in turn ether-linked to glycerols located on the inner and outer membrane faces (Fig. 14). The tetraethers create what is, in essence, the equivalent of a lipid bilayer in which the two layers are covalently joined together at the center. The membranes of the thermoacidophiles *Thermoplasma* and *Sulfolobus* occur almost entirely as such lipid "monolayers." In some methanogens that contain both diethers and tetraethers, several membrane configurations could be possible ranging from segregated monolayer and bilayer regions at the one extreme to a completely integrated and alternating diether–tetraether arrangement (Fig. 14).

One might consider the alternative possibility that tetraethers could double over forming a U-shape and therefore be capable of forming a true lipid bilayer. However, both direct and indirect evidence for a "monolayer" membrane assembly in archaebacteria exists. The extended length of the tetraethers varies slightly depending upon the number of pentacyclic rings within the biphytanyl chains, but they are such that their dimensions can equal the width of the archaebacterial membrane (Langworthy, 1977a; Ourisson and Rohmer, 1982). The extension of hydrocarbons across a membrane is not unusual. For example, the common C_{40} carotenoids of other organisms can insert through a bilayer adding rigidity to the membrane (Ourisson and Rohmer, 1982). What is unusual is that the membranes of *Thermoplasma* and *Sulfolobus* are composed almost entirely of extended C_{40} biphytanyl chains. Electron-spin resonance studies (Smith *et al.*, 1974) suggest that the *Thermoplasma* membrane is the most rigid one known. Moreover, true bilayer membranes tend to freeze-fracture tangentially, while those of *Thermoplasma, Sulfolobus*, and some methanogens do not (Langworthy, 1979a, 1982b; Doddema *et al.*, 1979); inner and outer membrane faces are not seen. Instead these archaebacterial cells are characteristically observed to cross-fracture perpendicularly through the plane of the membrane (and ultimately the whole cell), exactly as predicted for a "monolayer" membrane assembly. An example of a single *Thermoplasma* cell in typical cross-fracture is shown in Fig. 15.

The concept that tetraethers form a "monolayer" membrane structure helps to explain another phenomenon. The role of pentacyclic rings in the tetraethers of the thermoacidophilic archaebacteria can then be seen as a mechanism for controlling membrane fluidity in response to fluctuations in the high temperatures in the environments in which these archaebacteria thrive. Because the biphytanyl chains are fixed at a total of 40 carbons and ether-linked to glycerols on either edge of the membrane, the introduction or removal of pentacyclic rings would

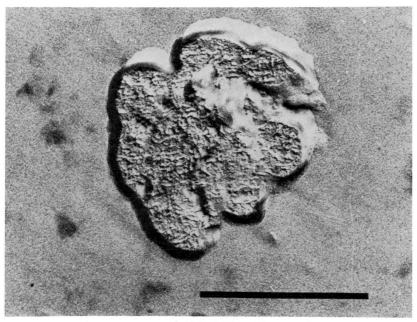

FIG. 15. Freeze–fracture electron micrograph of *Thermoplasma* showing a cell in typical cross-fracture. Bar = 0.5 μm (courtesy of P. H. J. Th. Ververgaert and A. J. Verkleij).

alter chain length (visually apparent in Fig. 3), and so alter the width of the membrane. Ring introduction would also reduce rotational freedom within the biphytanyl chains thereby influencing the viscosity, rigidity, and/or density of the membrane interior. The well-demonstrated correlation that cyclization increases within the biphytanyl chains of *Sulfolobus* in concert with increasingly higher growth temperatures supports this conclusion (De Rosa *et al.*, 1976a, 1980d; Furuya *et al.*, 1980). A similar response can be seen in *Thermoplasma* (Langworthy *et al.*, 1982). The tetraethers in archaebacteria other than the thermoacidophiles, however, contain only acyclic biphytanyl chains. Perhaps in these organisms, the diether–tetraether ratio may be altered to achieve appropriate membrane fluidity, although this has not been tested experimentally. In those archaebacteria possessing only diether based lipid bilayers, there appears to be no way of altering the physical state of the membrane interior, unless the nonpolar lipids would have some effect.

However, considerations such as these may be inconsequential when considering archaebacterial membranes. As pointed out by Lanyi (1976), diether bilayers should be highly cooperative structures with no distinct liquid-to-liquid crystalline phase. Indeed, first-order phase transitions have not been observed (above zero degrees) either for diether lipid dispersions (Plachy *et al.*, 1974; Lanyi, 1976; Ekiel *et al.*, 1981) or for tetraether lipids (P. W. M. van Dijck, personal communication).

The question of how tetraether based "monolayer" membrane asymmetry (Op den Kamp, 1979) or "transbilayer" distribution of lipids (Rottem, 1982) is actually achieved is an interesting one. The tetraethers comprising a "monolayer" membrane are not free entities, but nearly all are substituted with polar head groups. Furthermore, the vast majority are substituted asymmetrically, possessing a carbohydrate moiety on one side and a phosphate radical on the other. How can such a polar lipid structure be incorporated into a "monolayer" membrane and at the same time orient the correctly required head groups on the outer and inner membrane surfaces? One explanation could be that if tetraether biosynthesis proceeds by condensation of two unsaturated diether analogues, as already discussed (see Section V,C), it is entirely possible that a diether polar lipid analog is first inserted with the polar head group oriented toward the outer surface, followed by insertion of the appropriate diether analog with the polar head group oriented toward the inner surface. Condensation of the phytanyl chains of the diethers at the membrane center would subsequently occur, creating both the tetraether polar lipid structure and the "monolayer" membrane with appropriately oriented polar head groups simultaneously. This notion is without any experimental evidence, but correlates with the proposed route of biosynthesis (See Section V,C) and can also explain the variety of polar lipids actually found.

Of considerable interest will be the biophysical properties exhibited by the isolated tetraether lipids, the lipid–protein interactions, interactions with diether lipids or nonpolar lipids, and especially the possible formation of "monolayer" configurations in artificial systems. It is already well established that the diether lipids of archaebacteria can form membrane vesicles (Lanyi, 1976; Caplan and Ginsburg, 1978). However, attempts to assess the physical properties of isolated tetraether lipids and their ability to form "monolayer" structures *in vitro* are few. Gliozzi *et al.* (1982) were able to prepare artificial black membranes from the glycerol-nonitol tetraether lipids of *Sulfolobus*, but not from the diglycerol tetraethers. It was not clear whether the lipids actually formed extended chains and, therefore, the expected "monolayer" or if the polar ends associated, creating a U-shaped structure. Indeed, attempts to produce vesicles or liposomes from the tetraether polar lipids of *Thermoplasma* by a variety of standard procedures have so far proven unsuccessful (T. A. Langworthy, T. Oshima, and A. F. Esser, unpublished observations). As pointed out by Ourisson and Rohmer (1982), it is entirely possible that isolated tetraether lipids may in fact not be able to form "monolayers" in artificial systems, at least by procedures commonly employed. In any case, the study of archaebacterial membranes in natural and artificial systems should ultimately provide unusual phenomena, as well as a useful tool for assessing our perceptions of normal bilayer membranes (Langworthy *et al.*, 1979).

The concept that archaebacterial tetraethers provide a "monolayer" membrane organization appears to be a good one, but it is still in its infancy and obviously awaits considerable study. It should be clear at this point, however, that the archaebacteria have evolved a cytoplasmic membrane structure, which represents an alternative solution to that chosen by other organisms.

VII. Summary

The lipids of archaebacteria are characterized by the almost exclusive presence of nonsaponifiable isopranylglycerol ether lipids and isoprenoid hydrocarbons. This uniqueness in composition implies that the archaebacteria have pursued an evolutionary course that is far removed from either eubacteria or eukaryotic cells. Although all organisms share common pathways for lipid synthesis, the main point of divergence between archaebacteria and the other two groups occurs when the pathway for isoprenoid synthesis becomes the major route for their lipid formation, but the fatty acid pathway becomes the principal route in the other groups. An overall comparison of lipid biosynthesis is summarized in Fig. 16. The initial steps in the pathway employed by archaebacteria are too similar to those of other organisms to be considered coincidence. Perhaps this reflects early common ancestry. What was responsible for the subsequent separation or for when it occurred?

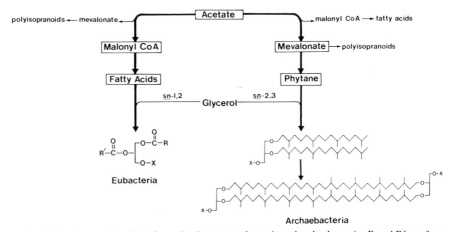

FIG. 16. Comparative lipid formation between eubacteria and archaebacteria. R and R' are fatty acid residues; X indicates polar head groups.

In terms of molecular organization of the lipids, the archaebacteria have also evolved a different approach to the assembly of their cytoplasmic membranes. In addition to reflecting deep evolutionary branching, archaebacterial lipid and membrane structure seem to reflect the extreme and unusual environments in which these organisms are found—environments that could have predominated in the earliest episodes of the earth's history. Ether linkages are resistant to acid and base hydrolysis. The membranes of halophiles and thermoacidophiles are therefore resistant to degradation by exposure of their membranes to their respective hypersaline and alkaline or hot acidic environments (Kates, 1972a; Smith *et al.*, 1973). Added stability is afforded by the tetraether lipids of some archaebacteria in the form of lipid "monolayer" membranes. Methanogens evolve considerable amounts of methane, which at some point is associated with the membrane. Because methane could act as an organic solvent causing disruption of the usual bilayer, in those cases where they occur tetraethers could add stability by effectively holding the membrane together (Kushwaha *et al.*, 1981b). Similarly, the membranes of thermoacidophiles are almost entirely a "monolayer" structure and are stable at high temperatures where the usual bilayer membrane might melt apart. Indeed, the stability of the archaebacterial "monolayer" membrane is such that it may allow archaebacteria to thrive at unprecidented temperatures, well over 100°C, in environments provided by submarine hydrothermal systems (Stetter, 1982).

In more practical terms, the specific nature of the lipids represents a useful chemical marker for the identification of new archaebacteria and for establishing taxonomic relationships. Moreover, the lipids can serve as molecular "fossil" evidence for ascertaining the existence of archaebacteria in the past and their contributions to biogeochemical processes. The same range of isoprenoids including phytane, biphytane, as well as polar lipids as exists in archaebacteria, has been detected in sediments (Anderson *et al.*, 1977; Brassell *et al.*, 1981), kerogen (Michaelis and Albrecht, 1979), shale (Chappe *et al.*, 1979), and petroleum (Moldowan and Seifert, 1979; Chappe *et al.*, 1982). Conversely, some new archaebacterial-type lipids of geochemical origin have been detected suggesting the possible existence of other varieties of yet undiscovered archaebacteria (Chappe *et al.*, 1979, 1982).

At the present time, detailed knowledge of the varieties and diversity of lipids to be found in archaebacteria is necessarily incomplete. Future studies should extend our perceptions of the structural and functional relationships of archaebacterial lipids. Certainly the increasing recognition of new members of the archaebacteria will continue to provide a paradise for the lipid chemist and to reveal new structures that continue to defy conventional nomenclature.

NOTE ADDED IN PROOF

Since completion of this manuscript, and as noted in the concluding paragraph of this chapter, a number of new varieties of archaebacterial lipids and studies on their properties have been reported. These reports include: the isolation of a C_{25},C_{25} diether (2,3-di-O-sesterterpanyl-*sn*-glycerol) from haloalkalophilic archaebacteria (De Rosa *et al.*, 1983a); the identification of asymmetrically assembled tetraethers in *Sulfolobus*, which contain two different cyclized C_{40} chains, as well as a tetraether skeleton containing one C_{40} chain and two C_{20} phytanyl chains (De Rosa *et al.*, 1983b); the effect of isoprenoid cyclization within tetraethers on the transition temperature of *Sulfolobus* lipids (Gliozzi *et al.*, 1983a); studies on monolayer black membranes of tetraether lipids that further support the concept that tetraethers provide a monolayer membrane architecture in archaebacteria (Gliozzi *et al.*, 1983b); studies that indicate that tetraethers can form small unilamellar liposomes when sonicated from lipid mixtures containing at least 25mol% egg phosphatidylcholine (Lelkes *et al.*, 1983); and the isolation of a diether possessing a C_{40} biphytanyl macrocyclic loop as a major lipid of *Methanococcus jannaschii* (Comita and Gagosian, 1983).

ACKNOWLEDGMENTS

The author wishes to thank R. Uecker for illustrations and P. Sivesind and J. Pond for editorial assistance.

REFERENCES

Anderson, R., Kates, M., Baedecker, M. J., Kaplan, I. R., and Ackman, R. G. (1977). The stereoisomeric composition of phytanyl chains in lipids of Dead Sea sediments. *Geochim. Cosmochim. Acta* **41**, 1381–1390.
Basinger, G. W., and Oliver, J. D. (1979). Inhibition of *Halobacterium cutirubrum* lipid biosynthesis by bacitracin. *J. Gen. Microbiol.* **111**, 423–427.

Brassell, S. C., Wardroper, A. M. K., Thomson, I. D., Maxwell, J. R., and Eglinton, G. (1981). Specific acyclic isoprenoids as biological markers of methanogenic bacteria in marine sediments. *Nature (London)* **290**, 693–696.

Caplan, R. S., and Ginzburg, M., (eds.) (1978). "Energetics and Structure of Halophilic Microorganisms." Elsevier, Amsterdam.

Chappe, B., (Chap Sim), Michaelis, W., Albrecht, W., and Ourisson, G. (1979). Fossil evidence for a novel series of arcbaebacterial lipids. *Naturwissenschaften* **66**, 522–523.

Chappe, B., Albrecht, P., and Michaelis, W. (1982). Polar lipids of archaebacteria in sediments and petroleums. *Science* **217**, 65–66.

Collins, M. D., and Jones, D. (1981). Distribution of quinone structural types in bacteria and their taxonomic significance. *Microbiol. Rev.* **45**, 316–354.

Collins, M. D., and Langworthy, T. A. (1983). Respiratory quinone composition of some acidophilic bacteria. *Syst. Appl. Microbiol.* **4**, 295–304.

Collins, M. D., Ross, H. N. M., Tindall, B. J., and Grant, W. D. (1981). Distribution of isoprenoid quinones in halophilic bacteria. *J. Appl. Bacteriol.* **50**, 559–565.

Comita, P. B., and Gagosian, R. B. (1983). Membrane lipid from deep-sea hydrothermal vent methanogen: a new macrocyclic glycerol diether. *Science* **222**, 1329–1331.

De Rosa, M., Gambacorta, A., Millonig, G., and Bu'Lock, J. D. (1974). Convergent characters of extremely thermophilic acidophilic bacteria. *Experientia* **30**, 866–868.

De Rosa, M., Gambacorta, A., and Bu'Lock, J. D. (1976a). The *Caldariella* group of extreme thermoacidophile bacteria: Direct comparison of lipids in *Sulfolobus, Thermoplasma* and the MT strains. *Phytochemistry* **15**, 143–145.

De Rosa, M., De Rosa, S., Gambacorta, A., and Bu'Lock, J. D. (1976b). Isoprenoid triether lipids from *Caldariella*. *Phytochemistry* **15**, 1996–1997.

De Rosa, M., De Rosa, S., and Gambacorta, A. (1977a). ^{13}C-NMR assignments and biosynthetic data for the ether lipids of *Caldariella*. *Phytochemistry* **16**, 1909–1912.

De Rosa, M., De Rosa, S., Gambacorta, A., Minale, L., and Bu'Lock, J. D. (1977b). Chemical structure of the ether lipids of thermophilic acidophilic bacteria of the *Caldariella* group. *Phytochemistry* **16**, 1961–1965.

De Rosa, M., De Rosa, S., Gambacorta, A., Minale, L., Thomson, R. H., and Worthington, R. D. (1977c). Caldariellaquinone, a unique benzo-b-thiopen-4,7 quinone from *Caldariella acidophila,* an extremely thermophilic and acidophilic bacterium. *J. Chem. Soc., Perkin. Trans.* **1** pp. 653–657.

De Rosa, M., De Rosa, S., Gambacorta, A., and Bu'Lock, J. D. (1980a). Structure of calditol, a new branched chain nonitol, and of the derived tetraether lipids in thermoacidophile archaebacteria of the *Caldariella* group. *Phytochemistry* **19**, 249–254.

De Rosa, M., Gambacorta, A., and Nicolaus, B. (1980b). Regularity of isoprenoid biosynthesis in the ether lipids of archaebacteria. *Phytochemistry* **19**, 791–793.

De Rosa, M., Gambacorta, A., Nicolaus, B., and Bu'Lock, J. D. (1980c). Complex lipids of *Caldariella acidophila,* a thermoacidophile archaebacterium. *Phytochemistry* **19**, 821–825.

De Rosa, M., Esposito, E., Gambacorta, A., Nicolaus, B., and Bu'Lock, J. D. (1980d). Effects of temperature on ether lipid composition of *Caldariella acidophila*. *Phytochemistry* **19**, 827–831.

De Rosa, M., Gambacorta, A., Nicolaus, B., Sodano, S., and Bu'Lock, J. D. (1980e). Structural regularities in tetraether lipids of *Caldariella* and their biosynthetic and phyletic implications. *Phytochemistry* **19**, 833–836.

De Rosa, M., Gambacorta, A., Nicolaus, B., Ross, H. N. M., Grant, W. D., and Bu'Lock, J. D. (1982a). An asymmetric archaebacterial diether lipid from alkaliphilic halophiles. *J. Gen. Microbiol.* **128**, 343–348.

De Rosa, M., Gambacorta, A., Nicolaus, B., and Sodano, S. (1982b). Incorporation of labelled glycerols into ether lipids in *Caldariella* acidophila. *Phytochemistry* **21**, 595–599.

De Rosa, M., Gambacorta, A., Nicolaus, B., and Grant, W. D. (1983a). A C_{25}, C_{25} diether core lipid from archaebacterial haloalkalophiles. *J. Gen. Microbiol.* **129**, 2333–2337.

De Rosa, M., Gambacorta, A., Nicolaus, B., Chappe, B., and Albrect, P. (1983b). Isoprenoid ethers: backbone of complex lipids of the archaebacterium *Sulfolobus solfataricus. Biochim. Biophys. Acta* **753**, 249–256.

Doddema, H. J., van der Drift, C., Vogels, G. D., and Veehuis, M. (1979). Chemiosmotic coupling in *Methanobacterium thermoautotrophicum:* hydrogen-dependent adenosine 5'-triphosphate synthesis by subcellular particles. *J. Bacteriol.* **140**, 1081–1089.

Ekiel, I., Marsh, D., Smallbone, B. W., Kates, M., and Smith, I. C. P. (1981). The state of the lipids in the purple membrane of *Halobacterium cutirubrum* as seen by [31]p NMR. *Biochem. Biophys. Res. Commun.* **100**, 105–110.

Evans, R. W., Kushwaha, S. C., and Kates, M. (1980). The lipids of *Halobacterium marismortui,* an extremely halophilic bacterium in the Dead Sea. *Biochim. Biophys. Acta* **619**, 533–544.

Furuya, T., Nagumo, T., Itoh, T., and Kaneko, H. (1980). The effect of growth temperature on the lipids in an extremely thermoacidophilic bacterium, TA-1. *Agric. Biol. Chem.* **44**, 517–521.

Gliozzi, A., Rolandi, R., De Rosa, M., and Gambacorta, A. (1982). Artificial black membranes from bipolar lipids of thermophilic archaebacteria. *Biophys. J.* **37**, 563–566.

Gliozzi, A., Paoli, G., De Rosa, M., and Gambacorta, A. (1983a). Effect of isoprenoid cyclization on the transition temperature of lipids in thermophilic archaebacteria. *Biochim. Biophys. Acta* **735**, 234–242.

Gliozzi, A., Rolandi, R., De Rosa, M., and Gambacorta, A. (1983b). Monolayer black membranes from bipolar lipids of archaebacteria and their temperature-induced structural changes. *J. Membr. Biol.* **75**, 45–56.

Goldfine, H. (1982). Lipids of prokaryotes-structure and distribution. *Curr. Top. Membr. Transp.* **17**, 1–43.

Hahn, J. (1982). Geochemical fossils of a possibly archaebacterial origin in ancient sediments. *Zentralbl. Bakteriol., Mikrobiol. Hyg., Abt. 1, Orig. C* **3**, 40–52.

Hancock, A. J., and Kates, M. (1973). Structure determination of the phosphatidylglycerosulfate (diether analog) from *Halobacterium cutirubrum. J. Lipid Res.* **14**, 422–429.

Holzer, G., Oró, J., and Tornabene, T. G. (1979). Gas chromatographic-mass spectometric analysis of neutral lipids from methanogenic and thermoacidophilic bacteria. *J. Chromatogr.* **186**, 795–809.

Kates, M. (1972a). Ether-linked lipids in extremely halophilic bacteria. *In* "Ether Lipids: Chemistry and Biology" (F. Snyder, ed.), pp. 351–398. Academic Press, New York.

Kates, M. (1972b). Techniques of lipidology. *In* "Laboratory Techniques in Biochemistry and Molecular Biology" (T. S. Work and E. Work, eds.), Vol. 3, pp. 269–610. Am. Elsevier, New York.

Kates, M. (1978). The phytanyl ether-linked polar lipids and isoprenoid neutral lipids of extremely halophilic bacteria. *Prog. Chem. Fats Other Lipids* **15**, 301–342.

Kates, M., and Deroo, P. W. (1973). Structure determination of the glycolipid sulfate from the extreme halophile *Halobacterium cutirubrum. J. Lipid Res.* **14**, 438–445.

Kates, M., and Kushwaha, S. C. (1976). The diphytanyl glycerol ether analogues of phospholipids and glycolipids in membranes of *Halobacterium cutirubrum. In* "Lipids" (R. Paoletti, G. Procellati, and G. Jacini, eds.), Vol. 1, pp. 267–275. Raven, New York.

Kates, M., and Kushwaha, S. C. (1978). Biochemistry of the lipids of extremely halophilic bacteria. *In* "Energetics and Structure of Halophilic Microorganisms" (S. R. Caplan and M. Ginzburg, eds.), pp. 461–480. Elsevier/North-Holland Biomedical Press, Amsterdam.

Kates, M., Yengoyan, L. S., and Sastry, P. S. (1965). A diether analog of phosphatidyl glycerophosprate in *Halobacterium cutirubrum. Biochim. Biophys. Acta* **98**, 252–268.

Kates, M., Palameta, M. B., Joo, C. N., Kushner, D. J., and Gibbons, N. E. (1966). Aliphatic diether analogs of glyceride-derived lipids. IV. The occurrence of di-*O*-dihydrophytylglycerol ether containing lipids in extremely halophilic bacteria. *Biochemistry* **5**, 4092–4099.

Kates, M., Wassef, M. K., and Kushern, D. J. (1968). Radioisotopic studies on the biosynthesis of the glyceryl diether lipids of *Halobacterium cutirubrum*. *Can. J. Biochem.* **46**, 971–977.

Kates, M., Wassef, M. K., and Pugh, E. L. (1970). Origin of the glycerol moieties in the glycerol diether lipids of *Halobacterium cutirubrum*. *Biochim. Biophys. Acta* **202**, 206–208.

Kates, M., Kushwaha, S. C., and Sprott, D. G. (1982). Lipids of purple membrane from extreme halophiles and of methanogenic bacteria. *In* ''Methods in Enzymology'' (L. Packer, ed.), Vol. 88, pp. 98–111. Academic Press, New York.

Kelly, M., and Liaaen-Jensen, S. (1967). Bacterial carotenoids. XXVI. C_{50}-Carotenoids. 2. Bacteriorubrin. *Acta Chem. Scand.* **21**, 2578–2580.

Kelly, M., Norgard, S., and Liaaen-Jensen, S. (1970). Bacterial carotenoids. *Acta Chem. Scand.* **24**, 2169–2182.

Kramer, J. K. G., Kushwaha, S. C., and Kates, M. (1972). Structure determination of squalene, dihydrosqualene and tetrahydrosqualene in *Halobacterium cutirubrum*. *Biochim. Biophys. Acta* **270**, 103–110.

Kushner, D. J., ed. (1978). ''Microbial Life in Extreme Environments.'' Academic Press, London.

Kushwaha, S. C., and Kates, M. (1973). Isolation and identification of ''bacteriorhodopsin'' and minor C_{40}-carotenoids in *Halobacterium cutirubrum*. *Biochim. Biophys. Acta* **361**, 235–243.

Kushwaha, S. C., and Kates, M. (1978a). Mevalonic acid concentrations in halophilic bacteria. *Phytochemistry* **17**, 1793.

Kushwaha, S. C., and Kates, M. (1978b). 2,3-Di-*O*-phytanyl-*sn*-glycerol and prenols from extremely halophilic bacteria. *Phytochemistry* **17**, 2029–2030.

Kushwaha, S. C., Pugh, E. L., Kramer, J. K. G., and Kates, M. (1972). Isolation and identification of dehydrosqualene and C_{40}-carotenoid pigments in *Halobacterium cutirubrum*. *Biochim. Biophys. Acta* **260**, 492–506.

Kushwaha, S. C., Gochnauer, M. B., Kushner, D. J., and Kates, M. (1974). Pigments and isoprenoid compounds in extremely and moderately halophilic bacteria. *Can. J. Microbiol.* **20**, 241–245.

Kushwaha, S. C., Kates, M., and Martin, W. G. (1975a). Characterization and composition of the purple and red membrane from *Halobacterium cutirubrum*. *Can. J. Biochem.* **53**, 284–292.

Kushwaha, S. C., Kramer, J. K. G., and Kates, M. (1975b). Isolation and characterization of C_{50}-carotenoid pigments and other polar isoprenoids from *Halobacterium cutirubrum*. *Biochim. Biophys. Acta* **398**, 303–314.

Kushwaha, S. C., Kates, M., and Kramer, J. K. G. (1977). Occurrence of indole in cells of extremely halophilic bacteria. *Can. J. Microbiol.* **23**, 826–828.

Kushwaha, S. C., Kates, M., Sprott, G. D., and Smith, I. C. P. (1981a). Novel complex polar lipids from the methanogen *Methanospirillum hungatei*. *Science* **211**, 1163–1164.

Kushwaha, S. C., Kates, M., Sprott, G. D., and Smith, I. C. P. (1981b). Novel polar lipids from the methanogen *Methanospirillum hungatei*. *Biochim. Biophys. Acta* **664**, 156–173.

Kushwaha, S. C., Kates, M., Juez, G., Rodriguez-Valera, F., and Kushner, D. J. (1982). Polar lipids of an extremely halophilic bacterial strain (R-4) isolated from salt ponds in Spain. *Biochim. Biophys. Acta* **711**, 19–25.

Langworthy, T. A. (1977a). Long-chain diglycerol tetraethers from *Thermoplasma acidophilum*. *Biochim. Biophys. Acta* **487**, 37–50.

Langworthy, T. A. (1977b). Comparative lipid compositionn of heterotrophically and autotrophically grown *Sulfolobus acidocaldarius*. *J. Bacteriol.* **130**, 1326–1332.

Langworthy, T. A. (1978a). Membranes and lipids of extremely thermoacidophilic microorganisms. *In* ''Biochemistry of Thermophily'' (S. M. Friedman, ed.), pp. 11–30. Academic Press, New York.

Langworthy, T. A. (1978b). Microbial life in extreme pH values. *In* "Microbial Life in Extreme Environments" (D. J. Kushner, ed.), pp. 279–315. Academic Press, London.

Langworthy, T. A. (1979a). Special features of thermoplasmas. *In* "The Mycoplasmas" (M. F. Barile and S. Razin, eds.), Vol. 1, pp. 495–513. Academic Press, New York.

Langworthy, T. A. (1979b). Membrane structure of thermoacidophilic bacteria. *In* "Strategies of Microbial Life in Extreme Environments" (M. Shio, ed.), Dahlem Konf. Life Sci. Res. Rep., Vol. 13, pp. 417–432. Verlag Chemie, Weinheim.

Langworthy, T. A. (1980). Archaebacterial membrane assembly. *In* "Dissipative Structures and Spatiotemporal Organization Studies in Biomedical Research" (W. O. Read and J. M. McMillin, eds.), pp. 82–102. Iowa State Univ. Press, Ames.

Langworthy, T. A. (1982a). Lipids of *Thermoplasma*. *In* "Methods in Enzymology" (L. Packer, ed.), Vol. 88, pp. 396–406. Academic Press, New York.

Langworthy, T. A. (1982b). Lipids of bacteria living in extreme environments. *Curr. Top. Membr. Transp.* **17**, 45–77.

Langworthy, T. A. (1982c). Turnover of di-*O*-phytanylglycerol in *Thermoplasma*. *Rev. Infect. Dis.* **4**, S266.

Langworthy, T. A. (1983). Diglyceryl tetraether lipids. *In* "Ether Lipids: Biomedical Aspects" (H. K. Mangold and F. Paltauf, eds.), pp. 161–175. Academic Press, New York.

Langworthy, T. A., Smith, P. F., and Mayberry, W. R. (1972). Lipids of *Thermoplasma acidophilum*. *J. Bacteriol.* **112**, 1193–1200.

Langworthy, T. A., Smith, P. F., and Mayberry, W. R. (1974). Long chain diether and polyol dialkyl glycerol triether lipids of *Sulfolobus acidocaldarius*. *J. Bacteriol.* **119**, 106–116.

Langworthy, T. A., Mayberry, W. R., and Smith, P. F. (1976). A sulfonolipid and novel glucosamidyl glycolipids from the extreme thermoacidophile *Bacillus acidocaldarius*. *Biochim. Biophys. Acta* **431**, 550–569.

Langworthy, T. A., Brock, T. D., Castenholz, R. W., Esser, A. F., Johnson, E. J., Oshima, T., Tsuboi, M., Zeikus, J. G., and Zuber, H. (1979). Life at high temperatures. *In* "Strategies of Microbial Life in Extreme Environments." (M. Shilo, ed.), Dahlem Konf. Life Sci. Res. Rep., Vol. 13, pp. 489–502. Verlag Chemie, Weinheim.

Langworthy, T. A., Tornabene, T. G., and Holzer, G. (1982). Lipids of Archaebacteria. *Zentralbl. Bakteriol., Mikrobiol. Hyg., Abt. 1, Orig. C* **3**, 228–244.

Langworthy, T. A., Holzer, G., Zeikus, J. G., and Tornabene, T. G. (1983). Iso- and anteiso-branched glycerol diethers of the thermophilic anaerobe *Thermodesulfotobacterium commune*. *Syst. Appl. Microbiol.* **4**, 1–17.

Lanyi, J. K. (1976). Membrane structure and salt dependence in extremely halophilic bacteria. *In* "Extreme Environments: Mechanisms of Microbial Adaptation" (M. R. Heinrich, ed.), pp. 295–303. Academic Press, New York.

Lelkes, P. I., Goldenberg, D., Gliozzi, A., De Rosa, M., Gambacorta, A., and Miller, I. R. (1983). Vesicles from mixtures of bipolar archaebacterial lipids with egg phosphatidylcholine. *Biochim. Biophys. Acta* **732**, 714–718.

Makula, R. A., and Singer, M. D. (1978). Ether-containing lipids of methanogenic bacteria. *Biochem. Biophys. Res. Commun.* **82**, 716–722.

Mayberry-Carson, K. J., Langworthy, T. A., Mayberry, W. R., and Smith, P. F. (1974). A new class of lipopolysaccharide from *Thermoplasma acidophilum*. *Biochim. Biophys. Acta* **360**, 217–229.

Mayberry-Carson, K. J., Roth, I. L., and Smith, P. F. (1975). Ultrastructure of lipopolysaccharide isolated from *Thermoplasma acidophilum*. *J. Bacteriol.* **121**, 700–703.

Mayberry-Carson, K. J., Jewell, M. J., and Smith, P. F. (1978). Ultrastructural location of *Thermoplasma acidophilum* surface carbohydrate by using concanavalin A. *J. Bacteriol.* **133**, 1510–1513.

Melchior, D. L. (1982). Lipid phase transitions and regulation of membrane fluidity in prokaryotes. *Curr. Top. Membr. Transp.* **17,** 263–316.

Michaelis, W., and Albrecht, P. (1979). Molecular fossils of archaebacteria in Kerogen. *Naturwissenshaften* **66,** 420–422.

Moldowan, J. M., and Seifert, W. K. (1979). Head to head linked hydrocarbons in petroleum. *Science* **204,** 169–171.

Nes, W. R., and Nes, W. D. (1980). "Lipids in Evolution." Plenum, New York.

Oesterhelt, D., and Stoeckenius, W. (1973). Functions of new photoreceptor membrane. *Proc. Natl. Acad. Sci. U.S.A.* **70,** 2853–2857.

Op den Kamp, J. A. F. (1979). Lipid asymmetry in membranes. *Annu. Rev. Biochem.* **48,** 47–71.

Ourisson, G., and Rohmer, M. (1982). Prokaryotic polyterpenes: Phylogenetic precursors of sterols. *Curr. Top. Membr. Transp.* **17,** 153–182.

Plachy, W. Z., Lanyi, J. K., and Kates, M. (1974). Lipid interactions in membranes of extremely halophilic bacteria. I. Electron spin resonance and dilatometric studies of bilayer structure. *Biochemistry* **13,** 4906–4913.

Pugh, E. L., Wassef, M. K., and Kates, M. (1971). Inhibition of fatty acid synthetase in *Halobacterium cutirubrum* and *Escherichia coli* by high salt concentrations. *Can. J. Biochem.* **49,** 953–958.

Razin, S., and Rottem, S., eds. (1982). "Current Topics in Membranes and Transport," Vol. 17. Academic Press, New York.

Ross, H. N. M., Collins, M. D., Tindall, B. J., and Grant, W. D. (1981). A rapid procedure for the detection of archaebacterial lipids in halophilic bacteria. *J. Gen. Microbiol.* **123,** 75–80.

Rottem, S. (1982). Transbilayer distribution of lipids in microbial membranes. *Curr. Top. Membr. Transp.* **17,** 235–261.

Rowland, S. J., Lamb, N. A., Wilkinson, C. F., and Maxwell, J. R. (1982). Confirmation of 2,6,10,15,19- pentamethyleicosane in methanogenic bacteria and sediments. *Tetrahedron Lett.* **23,** 101–104.

Ruwart, M. J., and Haug, A. (1975). Membrane properties of *Thermoplasma acidophila*. *Biochemistry* **14,** 860–866.

Sehgal, S. N., Kates, M., and Gibbons, N. E. (1962). Lipids of *Halobacterium cutirubrum*. *Can. J. Biochem. Physiol.* **40,** 69–81.

Smallbone, B. W., and Kates, M. (1981). Structural identification of minor glycolipids in *Halobacterium cutirubrum*. *Biochim. Biophys. Acta* **665,** 551–558.

Smith, G. G., Ruwart, M. J., and Haug, A. (1974). Lipid phase transitions in membrane vesicles from *Thermoplasma acidophila*. *FEBS Lett.* **45,** 96–98.

Smith, P. F. (1980). Sequence and glycosidic bond arrangement of sugars in lipopolysaccharide from *Thermoplasma acidophilum*. *Biochim. Biophys. Acta* **691,** 367–373.

Smith, P. F. (1981). Structure of the oligosaccharide chain of lipoglycan from *Acholeplasma granularum*. *Biochim. Biophys. Acta* **665,** 92–99.

Smith, P. F., Langworthy, T. A., Mayberry, W. R., and Hougland, A. E. (1973). Characterization of the membranes of *Thermoplasma acidophilum*. *J. Bacteriol.* **116,** 1019–1028.

Smith, P. F., Langworthy, T. A., and Mayberry, W. R. (1976). Distribution and composition of lipopolysaccharides from mycoplasmas. *J. Bacteriol.* **125,** 916–922.

Snyder, F., ed. (1972). "Ether Lipids: Chemistry and Biology." Academic Press, New York.

Soliman, G. S. H., and Trüper, H. G. (1982). *Halobacterium pharaonis* sp. nov., a new, extremely haloalkaliphilic archaebacterium with low magnesium requirement. *Zentralbl. Bakteriol., Mikrobiol Hyg., Abt. 1, Orig. C* **3,** 318–329.

Stetter, K. O. (1982). Ultrathin mycelia-forming organisms from submarine volcanic areas having an optimum growth temperature of 105°C. *Nature (London)* **300,** 258–260.

Tindall, B. J., Mills, A. A., and Grant, W. D. (1980). An alkalophilic red halophilic bacterium with a low magnesium requirement from a kenyan soda lake. *J. Gen. Microbiol.* **116**, 257–260.

Tornabene, T. G. (1978). Non-aerated cultivation of *Halobacterium cutirubrum* and its effect on cellular squalenes. *J. Mol. Evol.* **11**, 253–257.

Tornabene, T. G. (1981). Formation of hydrocarbons by bacteria and algae. *In* "Trends in the Biology of Fermentations for Fuels and Chemicals" (A. Hollaender, ed.), pp. 438–521. Plenum, New York.

Tornabene, T. G., and Langworthy, T. A. (1979). Diphytanyl and dibiphytanyl glycerol ether lipids of methanogenic archaebacteria. *Science* **203**, 51–53.

Tornabene, T. G., Kates, M., Gelpi, E., and Oró, J. (1969). Occurrence of squalene, di- and tetrahydrosqualenes and vitamin MK_8 in an extremely halophilic bacterium, *Halobacterium cutirubrum. J. Lipid Res.* **10**, 294–303.

Tornabene, T. G.. Wolfe, R. S., Balch, W. E., Holzer, G., Fox, G. E., and Oró, J. (1978). Phytanyl-glycerol ethers and squalene in the archaebacterium *Methanobacterium thermoautotrophicum. J. Mol. Evol.* **11**, 259–256.

Tornabene, T. G., Langworthy, T. A., Hozer, G., and Oró, J. (1979). Squalenes, phytanes and other isopranoids as major neutral lipids of methanogenic and thermoacidophilic "archaebacteria." *J. Mol. Evol.* **13**, 73–83.

Yang, L. L., and Haug, A. (1979). Structure of membrane lipids and physico-biochemical properties of the plasma membrane from *Thermoplasma acidophilum*, adapted to growth at 37°C. *Biochim. Biophys. Acta* **573**, 308–320.

Zillig, W., Stetter, K. O., Schafer, W., Janekovic, D., Wunderl, S., Holz, I., and Palm, P. (1981). *Thermoproteales*—a novel type of extremely thermoacidophilic anaerobic archaebacteria isolated from Icelandic solfataras. *Zentralbl. Bakteriol., Mikrobiol. Hyg., Abt. 1, Orig. C* **2**, 205–227.

Zillig, W., Stetter, K. O., Prangishvilli, D., Schafer, W., Wunderl, S., Janekovic, D., Holz, I., and Palm, P. (1982). *Desulfurococcaceae*, the second family of the extremely thermophilic, anaerobic, sulfur-respiring *Thermoproteales. Zentralbl. Bakteriol., Mikrobiol. Hyg., Abt. 1, Orig. C* **3**, 304–317.

CHAPTER 11

DNA-Dependent RNA Polymerases of the Archaebacteria

W. ZILLIG, K. O. STETTER, R. SCHNABEL AND M. THOMM

I. Transcription in Eubacteria

DNA-dependent RNA polymerase holoenzymes of eubacteria have the standard composition $\beta'\beta\alpha_2\sigma$ in which $\beta'\beta\alpha_2$ is the core enzyme, involved in elongation, and σ the initiation factor released soon after initiation of transcription (Burgess, 1976; Zillig et al., 1976; Zillig and Stetter, 1980). The role of a normally occurring small component, ω, in enzyme function remains obscure. Depending on the mode of preparation, additional proteins, sometimes bound less tightly to the polymerases than the "normal" components, occur in fractions of them, e.g. δ in the RNA polymerase of *Bacillus subtilis* (Pero et al., 1975), γ and y in that of *Lactobacillus* (Stetter and Zillig, 1974; Stetter, 1977; Gierl et al., 1982). They are engaged in special transcription functions. Variations do not occur in the principal composition but in details, e.g., the relative size of the β' versus the β chain (the former being larger than the latter in *Escherichia coli* and other gram-negative bacteria but smaller in gram-positive bacteria), the large molecular weight of σ in many gram-negative bacteria as opposed to its small size in gram-positive bacteria, and the corresponding occurrence of binding proteins in gram-positive bacteria (δ in *Bacillus* and γ and y in *Lactobacillus*).

Copyright © 1985 by Academic Press, Inc.
All rights of reproduction in any form reserved.
ISBN 0-12-307208-5

RNA polymerases of different eubacteria resemble each other in recognizing the same promoters, e.g., those in coliphage T7 DNA (Wiggs *et al.*, 1979), and thus define, in the limit, "the eubacterial promoter."

Little is known about the function of enzyme components. The β subunit of both gram-negative and gram-positive bacteria forms a complex, $\beta\alpha_2$, with the α subunits. This complex binds rifampicin (Zillig *et al.*, 1976). In gram-negative bacteria, the sites of interaction of RNA polymerase with the antibiotics rifampicin and streptolydigin both map in the *rpoB* gene (Heil and Zillig, 1970). In gram-positive bacteria that for rifampicin is also in *rpoB*, that for streptolydigin, however, is in *rpoC* (Halling *et al.*, 1978).

Antibodies against eubacterial RNA polymerases do not form precipitation lines with RNA polymerases from members of different families of eubacteria in Ouchterlony's in-gel immunodiffusion test (K. O. Stetter, unpublished), though the homology of the components of enzymes from different families was demonstrated by intergeneric reconstitution (Lill *et al.*, 1975) and by means of antisera against single subunits (Leib *et al.*, 1980).

Thus, both the compositions and the functions of all known eubacterial RNA polymerases conform to a prototype, but variations in detail reflect phylogenetic divisions, and can therefore be used for taxonomic purposes.

II. Nuclear Transcription in Eukaryotes

The nuclear DNA-dependent RNA polymerases of eukaryotes are of higher complexity, containing 9 to 12, or more, components, than those of eubacteria. Moreover, three types of polymerase exist that differ in most of their components and transcribe different classes of genes: RNA polymerase I(A) produces rRNA, including 5.8 S rRNA, RNA polymerase II(B) produces mRNA, and RNA polymerase III(C) produces tRNA and 5 S rRNA. RNA polymerase I is insensitive to the fungal toxin α-amanitin; RNA polymerase III shows low and RNA polymerase II shows high sensitivity to the drug. On the other hand, rifampicin and streptolydigin inhibit none of these three enzymes (Roeder, 1976).

The homology between the components of the three functionally different RNA polymerases and also between enzymes from different phyla of eukaryotes has been demonstrated immunochemically by challenging the (spotted) enzymes or Western blots of their components with antibodies against single enzyme components (Huet *et al.*, 1982).

Each of the three functionally different RNA polymerases recognizes corresponding promoters of a distinct type. None of these show -10 and -35 sequences closely resembling those characteristic for eubacterial promoters.

Almost nothing is known about the function of the components of these eukaryotic enzymes. Dissociation and reconstitution approaches, which have helped in elucidating the role of the components of bacterial RNA polymerases (Heil and Zillig, 1970; Palm *et al.*, 1975; Ishihama and Ito, 1972), have not yet been worked out for the eukaryotic RNA polymerases.

In summary, eukaryotic and eubacterial RNA polymerases are clearly distinguished by component patterns and functional characteristics such as promoter recognition sites, which would be expected for homologous molecules from different primary kingdoms.

III. DNA-Dependent RNA Polymerases of Archaebacteria

A. Component Patterns and Insensitivity to Inhibitors

As expected, DNA-dependent RNA polymerases from members of the third primary kingdom, the archaebacteria, differ characteristically from the corresponding enzymes of both eubacteria and eukaryotes (Zillig and Stetter, 1980; Zillig *et al.*, 1982a,b; Prangishvilli *et al.*, 1982) (Fig. 1). In contrast to the polymerases from eubacteria, but like the nuclear enzymes of eukaryotes, the polymerases contain 7 to 12 different components. The two heaviest components exhibit a larger size difference than do the β′ and β subunits of most eubacterial enzymes. The two following types of archaebacterial RNA polymerases have been distinguished on the basis of their component patterns.

1. The RNA polymerases of *Thermoplasma* (Sturm *et al.*, 1980; Zillig *et al.*, 1979, 1980), *Sulfolobus,* and the extremely thermophilic, anaerobic, sulfur-dependent Thermoproteales (Prangishvilli *et al.*, 1982; Zillig *et al.*, 1983a,b) contain three large components, including component C (about 40,000 daltons), in a 1:1:1 ratio. However, in eubacteria, component α, which corresponds in size to C, is present in *two* copies per enzyme monomer. Furthermore the archaebacterial enzymes of this type apparently lack a component corresponding in size to the eubacterial σ component.

2. The RNA polymerases of *Methanobacterium* (Stetter *et al.*, 1980), *Methanococcus, Methanosarcina* (Thomm, 1983), and *Halobacterium* (Zillig *et al.*, 1978) contain four instead of three heavy components, all in one copy per monomer.

All archaebacterial RNA polymerases isolated so far are insensitive to as much as 100 μg/ml of the antibiotics rifampicin and streptolydigin, the former normally inhibiting initiation, the latter inhibiting elongation by eubacterial RNA polymerases (Table I). The rifampicin sensitivity of *Halobacterium halobium* is due to a detergent-like effect of the antibiotic, which leads to cell lysis at

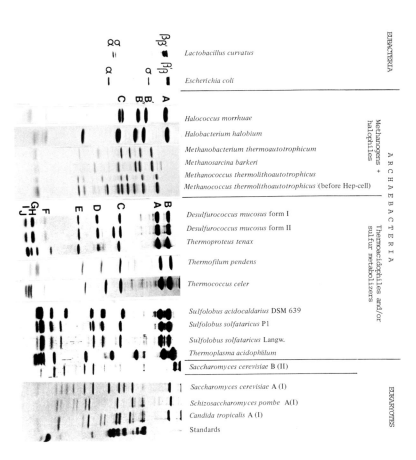

βρ'

Ωq

β'β

Ω   q

C   B'
        B"

Lactobacillus curvatus

Escherichia coli

A

Halococcus morrhuae

Halobacterium halobium

Methanobacterium thermoautotrophicum

Methanosarcina barkeri

Methanococcus thermolithoautotrophicus

Methanococcus thermolithoautotrophicus (before Hep-cell)

GH
I J   F   E   D   C   A
                                                B

Desulfurococcus mucosus form I

Desulfurococcus mucosus form II

Thermoproteus tenax

Thermofilum pendens

Thermococcus celer

Sulfolobus acidocaldarius DSM 639

Sulfolobus solfataricus P.1

Sulfolobus solfataricus Langw.

Thermoplasma acidophilum

Saccharomyces cerevisiae B (II)

Saccharomyces cerevisiae A (I)

Schizosaccharomyces pombe A(I)

Candida tropicalis A (I)

Standards

ARCHAEBACTERIA

Methanogens + halophiles

Thermoacidophiles and/or sulfur metabolizers

EUKARYOTES

concentrations above 10 μg/ml (Zillig *et al.*, 1978). However, archaebacterial transcriptases are also insensitive to the fungal toxin α-amanitin, which inhibits eukaryotic RNA polymerase II(B) and (less efficiently) III(C) (Table I).

Aside from their one major difference (three instead of four heavy components), the component patterns of the RNA polymerases of the thermoacidophiles and the methanogens (including the extreme halophiles) appear similar to each other. Homology between small components of the different archaebacterial enzymes is indicated (1) by the observation that components of comparable relative mobility resemble each other in appearance on (stained) SDS-polyacrylamide gels, e.g., component D is often a rather diffuse double band, and exhibits a distinct tinge after staining with Coomassie blue and (2) by their stoichiometry. In the RNA polymerases from *Sulfolobus* and the Thermoproteales, two small components (G and I) appear to be present twice per enzyme monomer, whereas another (F) is sometimes absent or occurs in substoichiometrical amounts.

B. HOMOLOGIES AMONG ARCHAEBACTERIAL RNA POLYMERASES

With the aim to find more certain criteria for the homology of components of different archaebacterial RNA polymerases, Western blots of their SDS-polyacrylamide gel electrophoretic patterns on nitrocellulose sheets (Towbin *et al.*, 1979; Burnette, 1981; Huet *et al.*, 1983) were challenged with rabbit antibodies against single components of the RNA polymerases of *Sulfolobus* and of *Methanobacterium* (Schnabel *et al.*, 1983a) (Fig. 2). Specific binding was visualized either with iodinated protein A or with peroxidase, coupled to a goat antirabbit antiserum plus *o*-dianisidine and H_2O_2.

The antiserum against the heaviest component of *Sulfolobus* polymerase reacted with the heaviest components of all other RNA polymerases from thermoacidophilic (including sulfur-dependent) archaebacteria, but reacted with the second and third components of the enzymes from the methanogens and with the second from *Halobacterium*. Conversely, the antiserum against the heaviest component of the *Methanobacterium* polymerase recognized the second component of the thermoacidophile polymerases (except for *Thermococcus*) but reacts with the heaviest component from other methanogens and the extreme halophiles.

As expected, antiserum against the second component of *Sulfolobus* reacts

---

FIG. 1.   Component patterns of DNA-dependent RNA polymerases from eubacteria (*L. curvatus* and *E. coli*), archaebacteria, and eukaryotes [*Saccharomyces* polymerases B(II) and A(I) and *Schizosaccharomyces* and *Candida* RNA polymerase A] obtained by SDS-polyacrylamide gradient slab gel electrophoresis.

TABLE I

ACTIVITIES OF DNA-DEPENDENT RNA POLYMERASES FROM *Sulfolobus*, *Thermoproteus*, AND *Desulfurococcus*[a]

Template	Inhibitor	RNA polymerase from						
		S. acidocaldarius	*T. tenax*	*D. mucosus*	*T. acidophilum*	*T. pendens*	*M. thermoautotrophicum*	*H. halobium*
—	—	0.2	0.7	0.6	0.3	0.2	0.01	0
ØH DNA	—	74.5	73.5	71.3	18.2	414	7.6	5.3
ØH DNA	Rifampicin	86.7	77.8	—	18.1			
ØH DNA	Streptolydigin	79.8	68.6	—	18.0			
ØH DNA	Actinomycin	0.2	0.5	—	0.3			
ØH DNA	Heparin	0.8	0.9	—	0.4			
ØH DNA	α-Amanitin	76.9	75.0	—	18.2			
T7 DNA	—	22.2	95.3	129.5	6.3	129	3.3	12.8
	Rifampicin	—	—	156.8	—	112		
	Streptolydigin	—	—	137.7	—	132		
	α-Amanitin	—	—	131.3	—	97		
T5 DNA	—	51.6	58.1	13.9	7.3	159	6.6	
T4 DNA	—	21.0	23.0	12.7	—	129		
P2 DNA	—	1.3	61.6	23.9	—		14	2.4
λ DNA	—	—	66.1	40.5	—			
P lac 5 DNA	—	—	74.6	54.4	—			
pBR322 DNA	—	—	—	127.8	25.1			
Calf thymus DNA (single-stranded)	—	27.7	22.0	23.9	—	58	4.4	36
poly[d(AT)·d(AT)]	—	64.3	99.7	2.7	455	88	110	114
PL1 DNA	—						26	
	Rifampicin						26	
	Streptolydigin						26	

[a] Expressed as nmoles of AMP incorporated per milligram enzyme per minute under optimized conditions on various templates without and with inhibitors.

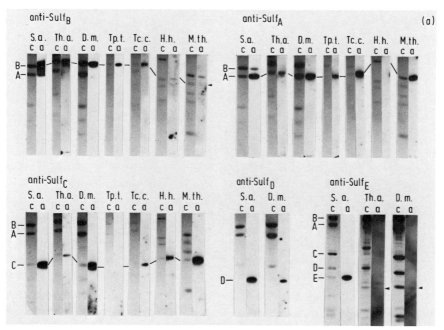

FIG. 2. Homologies of the components of the archaebacterial DNA-dependent RNA polymerases. Component patterns of the enzymes obtained by SDS-polyacrylamide gel electrophoresis, transferred to nitrocellulose filters [Western blots (Towbin *et al.*, 1979; Burnette, 1981)] were challenged with antibodies raised against the single components of two archaebacterial RNA polymerases (Schnabel *et al.*, 1983). (A) Cross-reactions with antibodies against the components of the RNA polymerase of *Sulfolobus acidocaldarius*. Bound antibodies were visualized by ^{125}I-labeled protein A, tracks labeled a. Tracks labeled c show the components directly visualized by Coomassie blue staining. Abbreviations: S.a., *Sulfolobus acidocaldarius;* Th.a., *Thermoplasma acidophilum;* D.m., *Desulfurococcus mucosus;* Tp.t., *Thermoproteus tenax;* Tc.c., *Thermococcus celer;* H.h., *Halobacterium halobium;* M.th., *Methanobacterium thermoautotrophicum.* (B) Cross-reactions with antibodies against the components of the RNA polymerase from *Methanobacterium thermoautotrophicum.* Bound antibodies were visualized by ^{125}I-labeled protein A. (C) Cross-reaction of antibodies against the components of the RNA polymerase from *Methanobacterium thermoautotrophicum* with components of the RNA polymerase of *Methanosarcina barkeri* and *Methanococcus thermolithotrophicus* and of antibodies against component B of the RNA polymerase of *Sulfolobus acidocaldarius* with the components B and B′ of the enzymes of methanogenic archaebacteria. The tracks labeled p show the bound antibodies (Towbin *et al.*, 1979) Abbreviations: Mg.th., *Methanobacterium thermoautotrophicum;* Mc.th., *Methanococcus thermolithotrophicus;* Ms.b., *Methanosarcina barkeri.* (See pp. 506–507.)

with the second component of other thermoacidophiles but reacts with the first component of the methanogen enzyme, whereas antibody against the second component of the methanogen polymerase reacts with the second component of other methanogen polymerases but reacts with the first component of the thermoacidophile polymerases. Again, as expected, antiserum against the *third* com-

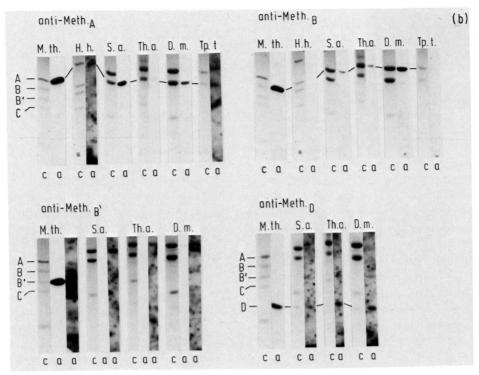

FIG. 2-B. (*Continued*)

ponent of the methanogen enzyme also reacts with the first component of the thermoacidophile enzyme, but reacts with the third component from other methanogens. Antiserum against the third component of *Sulfolobus* binds to the third components from other thermoacidophiles but binds to the fourth components from methanogens and halobacteria. Antibody against the fourth component from a methanogen reacts with the fourth component from other methanogens, but reacts with the third component of thermoacidophilic polymerases. The antibody against the fourth and fifth components from *Sulfolobus* react as expected with the components from the other thermoacidophiles. Antibodies against the fifth component from a methanogen react with the fourth component from the thermoacidophiles.

These cross-reactions prove the homology between (the heavier components of) the RNA polymerases of the two main branches of the archaebacteria. However, the two heaviest components appear in opposite sequence in the patterns of the thermoacidophiles as opposed to those of the methanogens and halophiles, and the heaviest components of the thermoacidophiles correspond to the second plus the third component from the methanogens. This is, however, not due to an immunochemical relation between these last two, because they do not cross-react

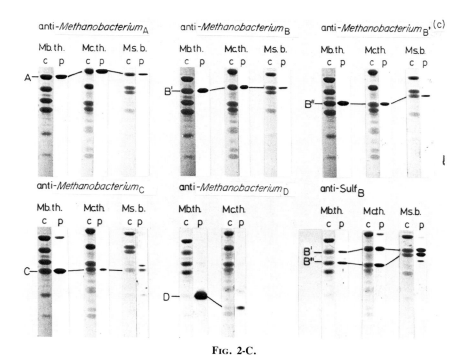

FIG. 2-C.

with each other. These two must, therefore, carry homologies to different parts of the (larger) first component of the thermoacidophiles.

In view of the fact that immunochemical cross-reactions reflect sequence and thus phylogenetic and probably functional homology, we propose to name homologous components of different RNA polymerases with the same large Roman letters.

Thus, the archaebacterial enzymes belong to either of two types (Fig. 3). All RNA polymerases from thermoacidophilic (and/or sulfur-dependent) archaebacteria, including *Thermoplasma,* have the composition BACDE(FG2HI2J) in which the sequence of symbols is in the order of the decreasing apparent size and the parentheses indicate lack of immunochemical evidence. All RNA polymerases from methanogens and extreme halophiles studied so far have, in contrast, the component pattern AB'B''CD. . . .

C. HOMOLOGY BETWEEN ARCHAEBACTERIAL AND THE NUCLEAR EUKARYOTIC RNA POLYMERASES

The same method of immunochemical analysis that has led to the establishment of the homologies between the components of archaebacterial RNA poly-

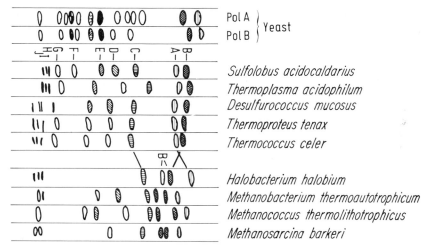

FIG. 3.   Schematic representation of the homologies shown in Figs. 2 and 5. For the homologies of the yeast RNA polymerases with each other, see Huet *et al.*, 1982.

merases had previously been used for the investigation of a possible correspondence between components of polymerases from nuclear eukaryotic and those from archaebacterial enzymes (Huet *et al.*, 1982) (Figs. 4 and 5).

Antibody against the heaviest single component A190 of yeast RNA polymerase I(A) reacted with spots of all but one of the archaebacterial RNA polymerases, but among eubacteria, reacted only with the polymerase of *E. coli*. Antibody against A135 bound to all archaebacterial enzymes tested. The antiserum against A49 bound to the *Thermoplasma* and *Halobacterium* polymerases; the antiserum against A40 bound to the enzymes from *Sulfolobus, Desulfurococcus, Halobacterium,* and *Methanobacterium,* and the antisera against A34.5, A25, and A14.5/14 bound to the *Halobacterium* polymerase.

The antisera against B220 or its proteolysis product B185 reacted with four of five archaebacterial enzymes but with none of the three eubacterial polymerases; the antiserum against B150 reacted with the enzymes from *Thermoproteus* and *Halobacterium*.

On Western blots, Anti A190 and Anti B185 reacted with the second component of the thermoacidophilic and (its counterpart) the first component of the *Halobacterium* enzyme, termed A (see previous paragraph). Anti A135 and Anti B150 were bound to the first component of the thermoacidophilic and to the second component of *Halobacterium* polymerase (now termed B). The other significant cross-reactions obtained with the spot tests could not be specified by challenging Western blots.

The cross-reaction between eukaryotic and eubacterial enzymes is limited to one or the other component. Quantitatively, it is much less pronounced.

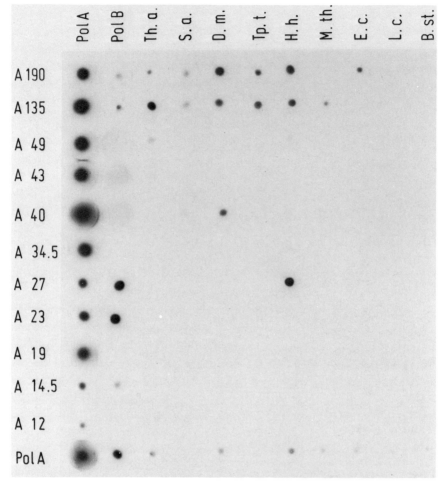

FIG. 4. Immunological spot test of eukaryotic, archaebacterial, and eubacterial RNA polymerases. RNA polymerases were spotted on nitrocellulose filters and challenged with antibodies raised against the native RNA polymerases A(I) (PolA) and B(II) (PolB) of yeast and against their single components as specified by their molecular weights. Bound antibodies were visualized by incubation with ^{125}I-labeled protein A. (For details, see Huet *et al.*, 1982, 1983.) Abbreviations: S.a., *Sulfolobus acidocaldarius;* Th.a., *Thermoplasma acidophilum;* D.m., *Desulfurococcus mucosus;* Tp.t., *Thermoproteus tenax;* H.h., *Halobacterium halobium;* M.th., *Methanobacterium thermoautotrophicum;* E.c., *Escherichia coli;* L.c., *Lactobacillus curvatus;* B.st., *Bacillus stearothermophilus.* (*continued*)

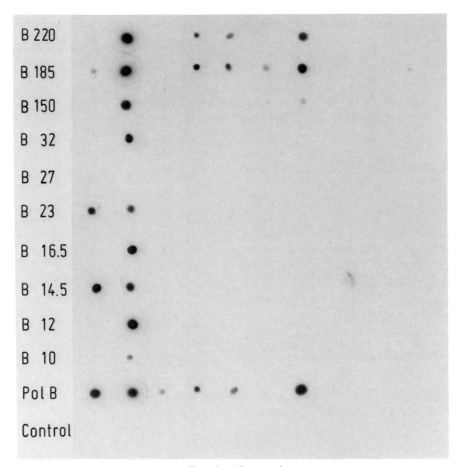

FIG. 4.   (*Continued*)

The cross-reaction between eukaryotic and archaebacterial enzymes and their components is sometimes stronger than between homologous components of RNA polymerases I and II from yeast.

1. THE FLAVONOLIGNANE DERIVATIVE SILYBIN STIMULATES TRANSCRIPTION BY ARCHAEBACTERIAL RNA POLYMERASES

Further evidence for the striking homology of archaebacterial and eukaryotic RNA polymerases is furnished by the finding that the flavonolignane derivative

FIG. 5.   Homologies of the large components of the RNA polymerases of some archaebacteria and yeast. Antibodies raised against the two largest components of polymerase A(I) (A190, A135) and B(II) (B185, B150) were used. (For details see Fig. 2 and Huet *et al.*, 1983.)

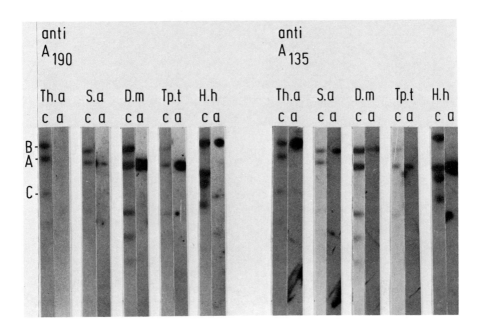

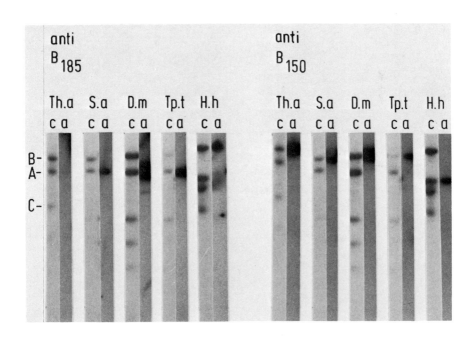

silybin, which is isolated from the thistle *Silybium marianum,* stimulates about twofold transcription by the RNA polymerases from the thermoacidophilic arch-archaebacteria (R. Schnabel *et al.,* 1982a) and eukaryotes, i.e., RNA polymerase I(A) (Machicao and Sonnenbichler, 1977). The stimulation affects the elongation, not the initiation, phase. The drug does not stimulate transcription either by eubacterial or by the other eukaryotic RNA polymerases (II and III), indicating (like the component patterns and the immunochemical data do) that, among eukaryotic polymerases, form I might be closest to the archaebacterial enzyme.

2. CONCLUSIONS

1. These data divide the archaebacterial RNA polymerases into two types: those from the thermoacidophilic (or sulfur-dependent) groups and those from the methanogenic (and extremely halophilic) archaebacterial groups. Except for *Thermoplasma,* this division also follows from comparative cataloging of 16 S rRNAs (Fox *et al.,* 1980; Woese *et al.,* 1984), from 16 S rRNA–DNA cross-hybridization (Tu *et al.,* 1982), and from 5 S rRNA (Fox *et al.,* 1982) and initiator tRNA (Kuchino *et al.,* 1982) sequences, indicating that the primary kingdom of the archaebacteria consists of two main branches (see review on thermoacidophiles including sulfur-dependent archaebacteria).

The exception, *Thermoplasma,* which has an RNA polymerase of the "thermoacidophilic type," appears closer to the methanogen branch (or in an isolated or bridging position) with respect to other phenotypic features (see Chapter 2 on *Thermoplasma* in review on thermoacidophiles).

2. At least in a few instances components of RNA polymerases of all three primary kingdoms show immunochemical homologies with each other, indicating that all RNA polymerases are derived from a common ancestor.

3. The strikingly extensive homology between archaebacterial and nuclear eukaryotic RNA polymerases points to a relationship between these two primary kingdoms that is closer than either has with the eubacteria. Compared to the polymerases of archaebacteria and eukaryotes, which are of the same basic type, the eubacterial enzymes appear to have evolved further, into a distinct type. In contrast, the eukaryotic RNA polymerases I and II and probably also III originate from an ancestral type resembling the archaebacterial enzyme. The separate evolution of the three types of eukaryotic enzymes should have followed the division of archaebacteria from "urkaryotes." Both branches of archaebacteria appear equally related to eukaryotes in this respect. In other features, e.g., 5 S rRNA and tRNA, the similarities between eukaryotes and *Sulfolobus* appear more extensive than those between eukaryotes and *Thermoplasma* and between the methanogens and halophiles.

An increasing body of evidence appears to support this view: Ribosomal A proteins (Matheson and Yaguchi, 1982), EFII (Kessel and Klink, 1982), some characteristics of 5 S rRNA (Fox *et al.*, 1982) and initiator tRNA (Kuchino *et al.*, 1982), the occurrence of glycoproteins, and the apparent absence of guanosine tetra- and pentaphosphates resemble the corresponding features in eukaryotes rather than eubacteria.

D. Purification of Archaebacterial RNA Polymerases

With the exception of the enzymes from *Halobacterium* (Zillig *et al.*, 1978; Madon and Zillig, 1983), *Halococcus* (Madon *et al.*, 1983), *Methanococcus*, and *Methanosarcina* (Thomm, 1983), polymerases of archaebacteria were purified by a variation of the polymin P method originally used for the isolation of *E. coli* RNA polymerase (Zillig *et al.*, 1979, 1980, 1983a,b; Sturm *et al.*, 1980; Prangishvilli *et al.*, 1982).

Steps involved were (1) preparation of a crude extract, often merely by suspension of cells in neutral buffer with the addition of a nonionic detergent, such as Triton X-100, for more sturdy organisms by means of a French press or sonication; (2) precipitation of nucleic acids plus bound and acidic proteins, including RNA polymerase with polymin P; (3) elution of proteins, including RNA polymerase from the washed precipitate at high ionic strength, followed by ammonium sulfate precipitation of the eluate; (4) DEAE chromatography of the redissolved and dialyzed precipitate. This step had to be omitted in the purification of the enzymes from *Thermoproteus* and *Thermofilum,* which did not adsorb; (5) heparin cellulose chromatography (Sternbach *et al.*, 1975) of the active fractions of the previous step; (6) DNA cellulose chromatography (according to Alberts *et al.*, 1968) and finally (7) sucrose glycerol gradient centrifugation. The buffer contained routinely $10^{-3}M$ EDTA and 40% v/v glycerol. If required, concentration was effected with a hollow fiber concentrator (Berghof, Tübingen) or miniconcentrator (Schulz, 1982).

After sucrose glycerol gradient centrifugation, the enzymes were essentially pure. Peptide chains were considered components, if they co-purified with each other and with the activity in at least three steps of the purification, e.g., heparin cellulose and DNA cellulose chromatography and gradient centrifugation. Antisera against single components of *Sulfolobus* polymerase co-precipitated the other components except H, demonstrating that 9 out of the 10 components are tightly linked (R. Schnabel, 1983). As long as the functions of the components remain unknown, the term *component* solely refers to this binding.

The RNA polymerases of *Methanococcus thermolithotrophicus* and *Methanosarcina barkeri* were inactivated in the polymin P step, possibly by dissociation. Therefore, the enzyme from *Methanococcus* was separated from the DNA by precipitation of the latter with polyethylene glycol (6000) in the presence of high salt (Thomm, 1983). The RNA polymerase from *Methanosarcina barkeri* was separated from the DNA by hydrophobic interaction chromatography (Thomm, 1983), a method applied recently to the isolation of proteins from rat liver chromatin (Schafer-Nielson and Rose, 1982).

The enzyme from *Methanosarcina barkeri* could not be eluted in active form from DNA-cellulose. Therefore, single-stranded DNA agarose was used in this case instead.

The purification of the enzyme from *M. barkeri* was further complicated by its irreversible dissociation during sucrose glycerol-gradient centrifugation. It remained stable during BioGel A 1.5m chromatography in the presence of 40% (v/v) glycerol. The enzymes of *Methanococcus* and *Methanosarcina* were both isolated in buffers containing 10^{-2} M $MgCl_2$.

The greatest difficulties were encountered in the purification of the RNA polymerase from *Halobacterium*. The use of buffers of low ionic strength, allowing the application of "normal" purification strategies, became possible by addition of 40–50% (v/v) glycerol and 0.025–0.05 M Mg^{2+}. The test had to be performed at low ionic strength, because initiation of transcription did not occur at high salt concentration. Precipitation by and elution off polymin P, which in most other cases led to the separation of RNA polymerase and DNA, did not work in this case. Initially, it was replaced by removal of DNA with DNase, followed by sizing (Zillig *et al.*, 1978). But the polymerase obtained after this step was neither completely DNA dependent nor able to transcribe native DNA, probably because it contained short pieces of bound residual DNA. This difficulty was surmounted by introducing phase partitioning steps for the separation of the enzyme from the DNA (Madon and Zillig, 1983). The enzyme purified in this manner contained an additional component required for transcription of DNA including native DNA, but the component was not absolutely required for that of poly[d(AT)]. However, polymerase prepared in this way is still unable to initiate transcription at high ionic strength, though elongation occurs at high salt. This problem, thus, remains to be solved.

E. Different Forms and Fragments of Archaebacterial RNA Polymerases

In several steps of the purification procedure, e.g., heparin cellulose and DNA cellulose chromatography, separate peaks of enzyme activity were observed. Sometimes, the component patterns of such forms appear indistinguishable, e.g., in the case of *Sulfolobus acidocaldarius* (Zillig *et al.*, 1979) where they separated in heparin cellulose chromatography and exhibited equal activities on poly[d(AT)] but different activities with native DNA templates. However, two forms of the RNA polymerase of *Desulfurococcus* were distinguished by the presence or absence of an additional component, in SDS-polyacrylamide gel electrophoresis migrating just in front of component C (Prangishvilli *et al.*, 1982) (Fig. 1).

A form of the polymerase of *Thermoplasma* has been distinguished from another by decreased affinity for its component D, which could be partially removed upon consecutive purification (Sturm *et al.*, 1980).

The RNA polymerases of *Halobacterium* (Madon and Zillig, 1983) and *Halococcus* (Madon *et al.*, 1983) exhibit a double band in place of component C in SDS-polyacrylamide gel electrophoresis (Fig. 6). Both C components sum up to one peptide chain per enzyme monomer. As revealed by cellogel electrophoresis, their charge densities differ (Zillig *et al.*, 1978). It appears possible that one is formed by modification of the other.

Analysis of fractions obtained in purification steps by SDS-polyacrylamide gel electrophoresis revealed the occurrence of inactive enzyme fragments. *Thermoplasma acidophilum* yielded a fragment devoid of components C and E (Sturm *et al.*, 1980), *Halobacterium* (Fig. 7), and *Methanobacterium* fragments lacking component A (Madon and Zillig, 1983; Stetter *et al.*, 1980). The corresponding released component A was eluted in a different position in heparin cellulose chromatography. In the case of *Halobacterium*, more than 80% of the enzyme components were found in the fragments eluting in front of the native enzyme peak.

RNA polymerase from *Methanococcus thermolithotrophicus* purified by a procedure in which the polyethylene glycol precipitation step was replaced by hydrophobic interaction chromatography contains an additional component migrating between components C and D (see Fig. 1). When the heparin cellulose chromatography step was omitted, three additional components, one between B′ and B″ and two more between C and D, were observed (Fig. 1). All of these did not cross-react immunochemically with any other component. In contrast, an additional band found in a preparation of *M. thermoautotrophicum* RNA poly-

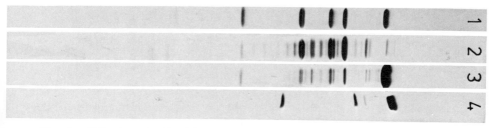

Fig. 6. SDS-polyacrylamide slab gel electrophoretogram of purified DNA-dependent RNA polymerases from *Halobacterium halobium* (track 1) and of fragment from heparin cellulose chromatography lacking component A and of the corresponding released fragment A from the same chromatography (tracks 2 and 3). RNA polymerase from *E. coli* (track 4).

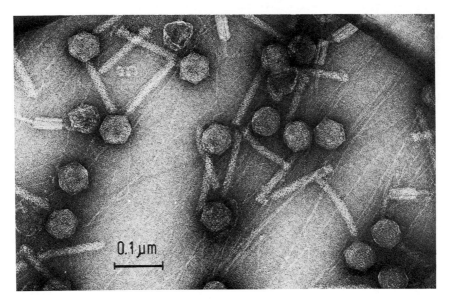

Fig. 7. Phage φH negatively stained with uranyl acetate. The bar represents 0.1 μm. (Electron micrograph courtesy of L. Hintermaier and E. Demm.)

merase was serologically related to component A and thus appeared to be a proteolytic cleavage product of the latter (Schnabel *et al.*, 1983). On poly[d(AT)] as template, no significant differences in the specific activities of all these different forms were observed.

It should be emphasized, that the term *form* in this context does not have the same meaning as for nuclear eukaryotic polymerases but merely indicates that different fractions of an enzyme can be distinguished either by being separated during purification and/or by their component patterns or other features. Some of these forms could be products of partial denaturation rather than having certain functions in *in vitro* transcription. The latter possibility has however not been excluded.

F. FUNCTION OF RNA POLYMERASE COMPONENTS

So far, only in a few instances is information pertaining to the role of polymerase components in the transcription process available.

Different preparations of *Sulfolobus* polymerase contain different, substoichiometric, sometimes insignificant amounts of component F. But no related difference in the specific activities was found. An inactive fraction of *Sulfolobus* polymerase, which was not bound to native DNA cellulose at 70°C, lacks all components smaller than E (Schnabel, 1983). Two fractions equally active on various templates were consecutively eluted in a salt gradient from a DNA-cellulose column at 70°C. In the less tightly bound fraction, component F was absent. The fraction eluting at higher salt concentration lacked components F and H, both of which do not appear to influence the *in vitro* overall activity of the enzyme (R. Schnabel, 1983). However, this does not exclude the possibility that components F and H are involved in transcription *in vivo*.

Component E of the RNA polymerase of *Thermoplasma* is required for, but released upon, tight binding of the enzyme to certain productive sites of the DNA (R. Schnabel *et al.*, 1982b). It is also required for the transcription of native DNA. The homologous component E of *Sulfolobus* polymerase is not released under comparable conditions, indicating considerable differences in the function of homologous proteins in different genera of archaebacteria.

A form of the RNA polymerase of *Halobacterium* lacking component ε (thus termed because its homology with the components of other archaebacterial polymerases has not yet been established) is unable to transcribe native and denatured DNA but shows significant activity on poly[d(A·T)] (Madon and Zillig, 1983).

Two nearly inactive fractions containing components of the RNA polymerase of *Thermoplasma* were separated by cellogel electrophoresis. The activity was restored by recombination, indicating reconstitution (Schnabel, 1983). However, the inactive fragments, A and enzyme without A, of the *Halobacterium* polymerase did not yield an active reconstitution product. Techniques for the complete dissociation and reconstitution, which could greatly facilitate the investigation of component function, have not yet been worked out for archaebacterial RNA polymerases.

G. *In vitro* Transcription by Archaebacterial RNA Polymerases

Though the optimal ionic strength and magnesium concentrations for the transcription by different archaebacterial RNA polymerases vary considerably, all enzymes isolated so far, including that from *Halobacterium,* work best at low ionic strength. The optimal temperatures for transcription by enzymes from extreme thermophiles vary for different enzymes and templates, but are usually around 80°–90°C. Several RNA polymerases of extreme thermophiles, most prominently that from *Thermoproteus,* are still entirely stable at temperatures not allowing transcription of naked DNA *in vitro* (at least 95°C in the case of the polymerase from *Thermoproteus*). Because such organisms grow at temperatures exceeding those optimal for *in vitro* transcription of naked native DNA, and even above the T_m's of the latter, it appears that stabilization of the template, e.g., by proteins or polyamines, must occur *in vivo*. Basic DNA binding proteins have indeed been found in archaebacteria (Searcy, 1975; Thomm *et al.,* 1982).

The abilities of different archaebacterial RNA polymerases to transcribe native DNAs and their overall transcription activities on different templates vary widely (Table I). The specific activities of the RNA polymerases of *Thermoplasma* and *Halobacterium* are much higher on poly[d(A·T)] than on native DNAs (3–20 times and more than 20 times, respectively). The specific activities of other enzymes are each near to that for *E. coli* RNA polymerase on certain templates, different for different polymerases, but low or insignificant on others, indicating genus-dependent template specificity, possibly in signal recognition.

The most striking example is that of *Desulfurococcus* RNA polymerase, for which phage T7 and pBR322 DNAs (both containing AT-rich sequences) are excellent but for which phage T4 and T5 DNAs are weak templates (in spite of their strong eubacterial promoters). This polymerase also transcribes T7 DNA asymmetrically, though not from its eubacterial promoters (Prangishvilli *et al.*, 1982).

H. Homologous Templates

The investigation of *in vitro* transcription by DNA-dependent RNA polymerases of archaebacteria requires homologous templates, possibly "chromatins." Such templates could be virus (phage) DNAs, plasmids, or cloned genes.

In the *Halobacterium halobium* system, a large satellite DNA (Moore and McCarthy, 1969) has been shown to contain a circular, superhelical plasmid of extreme genetic variability (Pfeifer *et al.*, 1981a,b), due to recombination in which many families of repetitive sequences are involved (Sapienza and Doolittle, 1982). However, genes have so far not been satisfactorily identified in this large genome.

The halobacteriophage φH (Fig. 7) is a temperate virus with linear DNA of 59,000 basepairs, which shows partial circular permutation (H. Schnabel *et al.*, 1982a). The DNA is a mixture of forms (Fig. 8) that are distinguished by inserts or deletions (Fig. 9), of which at least one insert is also present in the host plasmid (H. Schnabel *et al.*, 1982b). At least two of these forms (2 and 5) are interconvertible by the inversion of a sequence flanked by identical inserts. This genetic variability is probably of the same nature as that observed in the plasmid. Three types of clones from immune strains of *H. halobium* carry the phage genome or parts of it: (1) true lysogens, which contain the prophage in superhelical circular form and give rise to phage progeny frequently, (2) quasi-lysogens, which contain the whole, possibly slightly altered, phage genome and are distinguished by a low incidence of phage production, and (3) cells containing the invertible sequence of the phage forms 2 and 5 looped out as a plasmid (Schnabel and Zillig, 1984; Schnabel, 1984). The identification of genes, e.g., those governing immunity, and their transcription in the phage genome, appears

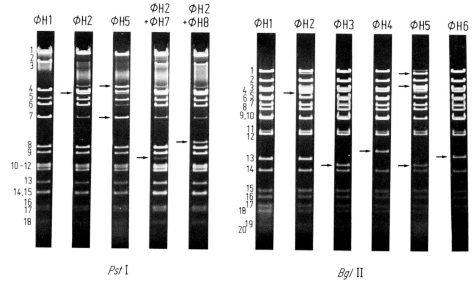

FIG. 8.   Restriction endonuclease fragments of phage φH DNA produced by the enzymes *Bgl*II or *Pst*I and separated by agarose gel electrophoresis. Phage φH exists as a mixture of variants, six of which have been isolated from single plaques. Their DNAs differ in structure, as shown by restriction fragments (arrows) that are not present in the DNA of the predominant variant φH1. φH7 and φH8 have not yet been isolated and are so far characterized by new minor fragments in mixtures of φH7 or φH8 with φH2.

straightforward. Unfortunately, the homologous RNA polymerase is yet unable to initiate *in vitro* transcription on native DNA at high ionic strength, possibly due to some deficiency, which could also involve the state of the template.

In the *Sulfolobus acidocaldarius* system, two plasmids have been identified and cloned, in parts or entirely, into *E. coli* vectors.

The plasmid pB12 of strain B12 of *S. acidocaldarius,* 15 kilobase pairs (kbp) long, exists integrated into a specific site of the host genome as well as in a free superhelical circular state (Yeats *et al.,* 1982). Ultraviolet induction leads to strong amplification accompanied by the breakdown of the host genome. The plasmid proved to be the genome of a virus-like particle (Martin *et al.,* 1984).

A plasmid pB6 of strain B6 of *S. acidocaldarius* (33 kbp) exists in different forms carried by different *Sulfolobus* clones, two of which are distinguished by an inversion, as well as other differences. The original isolate, from which the plasmid-carrying clone has been derived, contained virus-like particles in hexagonally packed crystalline arrays. The relationship between plasmid and virus is not understood. This plasmid has also been cloned into pBR322 and is the subject of genetic analysis (S. Yeats, P. McWilliam, and H. Neumann, unpublished).

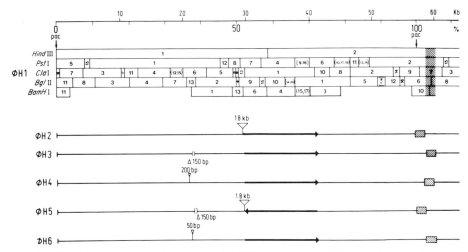

FIG. 9. Map of restriction fragments of φH DNA and structural differences between the variants. Compared to the predominant variant φH1, three types of changes are observed: (a) insertions of 1.8 kbp, 200 bp, or 50 bp; (b) a deletion of 150 bp; (c) an inversion of 11 kb (heavy arrow). The phage DNA is partially circularly permuted and terminally redundant. Phages contain linear DNA of 59 kb, corresponding to 103% of the φH1 genome. Insertions and deletions affect the ends of phage DNA molecules (shaded regions) because DNA length is determined by size of the phage head.

Of 15 isolates of the orders Methanomicrobiales and Methanococcales, only one, PL 12/M (belonging to the Methanomicrobiales), yielded a plasmid, pMP 1 of 7.0 kbp, which appears to exist in a free state as well as integrated in the chromosomal DNA (Thomm *et al.,* 1983). It was isolated from the pellet fraction rather than from the cleared lysate; it has been cloned in pBR322.

The structural genes for 16 S and 23 S rRNA of *Thermoplasma acidophilum,* which (as those for 5 S rRNA) are unlinked to other rRNA genes within the genome, have been cloned into pBR322 (Tu and Zillig, 1982) and into cosmids (R. Schnabel, unpublished).

These studies have not yet approached the main goal of such work—the identification of transcription signals and the establishment of a ''correct'' *in vitro* transcription system—but they appear to have opened the way.

I. POLYMERASE STRUCTURE AND TAXONOMY

The usefulness of RNA polymerase structure for taxonomy has been discussed previously (Zillig and Stetter, 1980). The insensitivity of archaebacterial RNA polymerases to rifampicin and streptolydigin (already in the completely DNA-dependent polymin P eluate) is an easily established criterion for the possible archaebacterial nature of a novel isolate.

Immunochemical cross-reactions between RNA polymerases of different archaebacteria, as revealed by the Western blotting technique that employs iodinated *Staphylococcus* protein A or antibody-coupled peroxidase, have been found throughout the archaebacteria among components of medium size or larger (see above). They form a strong basis for the establishment of the two main branches of this primary kingdom.

However, precipitation, which is the basis for Ouchterlony's in-gel immunodiffusion test, does not occur between a polymerase antiserum and a polymerase from a member of a different family (K. O. Stetter, unpublished). The presence or absence of a cross-reaction in the Ouchterlony's test can thus be used as criterion for phylogenetic distance. In a negative case, the difference is large enough to postulate different families, as was done for the taxonomic classification of the novel genera *Thermoproteus, Desulfurococcus, Thermofilum,* and *Thermococcus* within the order Thermoproteales (see review on this subject in this volume, Chapter 2) and of the family Methanothermaceae within the Methanobacteriales (Stetter *et al.*, 1980).

On the other hand, cross-reaction of the polymin P eluate of a novel isolate with an available antibody in the Ouchterlony test shows that the isolate is closely related to the source of the antigen. It also allows the immunochemical isolation of the RNA polymerase in question and thus its detailed comparison to those of other archaebacteria (Prangishvilli *et al.*, 1982).

So far, RNA polymerase structure belongs to the syntactic rather than semantophoretic features in a phylogenetic analysis (Balch, 1982). It appears, however, that RNA polymerase is a conservative molecule, probably as suitable for comparative sequence analysis as is ribosomal RNA.

NOTE ADDED IN PROOF

A later account of this topic containing further information is: Zillig, W., Schnabel, R., Stetter, K. O., Thomm, M., Gropp, F., and Reiter, W. D. (1985). The evolution of the transcription apparatus. *In* ''The Evolution of the Prokaryotes'' (K. H. Schleifer and E. Stackebrandt, eds.). Academic Press, Orlando.

REFERENCES

Alberts, B., Amodio, F., Jenkins, M., Goodman, E., and Ferris, F. (1968). Studies with DNA-cellulose chromatography. I. DNA binding proteins from *Escherichia coli. Cold Spring Harbor Symp. Quant. Biol.* **33,** 289–305.
Babinet, C. (1976). A new method for the purification of RNA-polymerase. *Biochem. Biophys. Res. Commun.* **26,** 639–644.
Balch, W. E. (1982). Methanogens: Their impact on our concept of procaryotic diversity. *Zentralbl. Bakteriol., Mikrobiol. Hyg., Abt. 1, Orig. C* **3,** 295–303.
Burgess, R. R. (1976). Purification and properties of *E. coli* polymerase. *In* ''RNA Polymerase'' (R. Losick and M. Chamberlin, eds.), pp. 69–100. Cold Spring Harbor Lab., Cold Spring Harbor, New York.

522   W. ZILLIG, K. O. STETTER, R. SCHNABEL AND M. THOMM

Burnette, W. N. (1981). "Western blotting": Electrophoretic transfer of protein from sodium
    dodecyl sulfate-polyacrylamide gels with antibody and radioiodinated protein A. *Anal. Bio-
    chem.* **112**, 195–203.
Fox, G. E., Stackebrandt, E., Hespell, R. B., Gibson, J., Maniloff, J., Dyer, T. A., Wolfe, R. S.,
    Balch, W. E., Tanner, R. S., Magrum, L. J., Zablen, L. B., Blakemore, R., Gupta, R., Bonen,
    L., Lewis, B. J., Stahl, D. A., Luehrsen, K. R., Chen, K. N., and Woese, C. R. (1980). The
    phylogeny of prokaryotes. *Science* **209**, 457–463.
Fox, G. E., Luehrsen, K. R., and Woese, C. R. (1982). Archaebacterial 5s ribosomal RNA.
    Zentralbl. Bacteriol., Mikrobiol. Hyg., Abt. 1, Orig. C **3**, 330–345.
Gierl, A., Zillig, W., and Stetter, K. O. (1982). The role of components and of the DNA-dependent
    RNA polymerase of *Lactobacillus curvatus* in promoter selection. *Eur. J. Biochem.* **125**, 41–
    47.
Halling, S. M., Burtis, K. C., and Doi, R. H. (1978). β' subunit of bacterial RNA polymerase is
    responsible for streptolydigin resistance in *Bacillus subtilis. Nature (London)* **272**, 837–839.
Heil, A., and Zillig, W. (1970). Reconstitution of bacterial DNA-dependent RNA polymerase from
    isolated subunits as a tool for the elucidation of the role of the subunits in transcription. *FEBS
    Lett.* **11**, 165.
Huet, J., Sentenac, A., and Fromageot, P. (1982). Spot-immunodetection of conserved determinants
    in eukaryotic RNA polymerases. *J. Biol. Chem.* **257**, 2613–2618.
Huet, J., Schnabel, R., Sentenac, A., and Zillig, W. (1983). Archaebacteria and eukaryotes posses
    DNA-dependent RNA polymerases of a common type. *EMBO J.* **2**, 1291–1294.
Ishihama, A., and Ito, K. (1972). Subunits of RNA polymerase in function and structure. II.
    Reconstitution of *E. coli* RNA polymerase from isolated subunits. *J. Mol. Biol.* **72**, 111–123.
Kessel, M., and Klink, F. (1982). Identification and comparison of 18 archaebacteria by means of
    the diphtheria toxin reaction. *Zentralbl. Bakteriol., Mikrobiol. Hyg., Abt. 1, Orig. C* **3**, 140–
    148.
Kuchino, Y., Ihara, M., Yabusaki, Y., and Nishimura, S. (1982). Initiator tRNAs from archaebac-
    teria show common unique sequence characteristics. *Nature (London)* **298**, 684–685.
Leib, C., Ernst, H., and Hartmann, G. R. (1980). Recognition of promoter sequences by RNA
    polymerase from different sources. *Mol. Biol., Biochem. Biophys.* **32**, 301–307.
Lill, U. I., Behrendt, E. M., and Hartmann, G. R. (1975). Hybridization in vitro of subunits of the
    DNA-dependent RNA polymerase from *E. coli* and *Micrococcus luteus. Eur. J. Biochem.* **52**,
    411–420.
Machicao, F., and Sonnenbichler, J. (1977). Mechanism of the stimulation of RNA synthesis in rat
    liver nuclei by silybin. *Hoppe-Seyler's Z. Physiol. Chem.* **358**, 141–147.
Madon, J., and Zillig, W. (1983). A form of the DNA-dependent RNA polymerase of *Halobac-
    terium halobium*, containing an additional component, is able to transcribe native DNA. *Eur. J.
    Biochem.* **133**, 471–474.
Madon, J., Leser, U., and Zillig, W. (1983). DNA-dependent RNA polymerase from the extremely
    halophilic archaebacterium *Halococcus morrhuae. Eur. J. Biochem.* **135**, 279–283.
Martin, A., Yeats, S., Janekovic, D., Reiter, W. D., Aicher, W., and Zillig, W. (1984). SAV 1, a
    temperate u.v.-inducible DNA virus-like particle from the archaebacterium *Sulfolobus acid-
    ocaldarius* isolate B12. *EMBO J.* **3**, 2165–2168.
Matheson, A. T., and Yaguchi, M. (1982). The evolution of the archaebacterial ribosome. *Zentralbl.
    Bakteriol., Mikrobiol. Hyg., Abt. 1, Orig. C* **3**, 192–199.
Moore, R. C., and McCarthy, B. J. (1969). Characterization of the deoxyribonucleic acid of various
    strains of halophilic bacteria. *J. Bacteriol.* **99**, 248–254.
Palm, P., Heil, A., Boyd, D., Grampp, B., and Zillig, W. (1975). The reconstitution of *E. coli*
    DNA-dependent RNA polymerase from its isolated subunits. *Eur. J. Biochem.* **53**, 283.
Pero, J., Nelson, J., and Fox, T. D. (1975). Highly asymmetric transcription by DNA polymerase

containing phage-SPOl-induced polypeptides and a new host protein. *Proc. Natl. Acad. Sci. U.S.A.* **72**, 1589–1593.

Pfeifer, F., Weidinger, G., and Goebel, W. (1981a). Characterization of plasmids in halobacteria. *J. Bacteriol.* **145**, 369–374.

Pfeifer, F., Weidinger, G., and Goebel, W. (1981b). Genetic Variability in *Halobacterium halobium. J. Bacteriol.* **145**, 375–381.

Prangishvilli, J., Zillig, W., Gierl, A., Biesert, L., and Holz, I. (1982). DNA-dependent RNA polymerase of thermoacidophilic archaebacteria. *Eur. J. Biochem.* **122**, 471–477.

Roeder, R. G. (1976). Eukaryotic nuclear RNA polymerase. *In* "RNA Polymerase" (R. Losick and M. Chamberlin, eds.), pp. 285–329. Cold Spring Harbor Lab., Cold Spring Harbor, New York.

Sapienza, C., and Doolittle, W. F. (1982). Unusual physical organization of the *Halobacterium genome. Nature (London)* **295**, 384–389.

Schafer-Nielson, C., and Rose, C. (1982). Separation of nucleic acids and chromatin proteins by hydrophobic interaction chromatography. *Biochim. Biophys. Acta* **696**, 323–331.

Schnabel, H. (1984). An immune strain of *Halobacterium halobium* carries the invertible L segment of phage ϕH as a plasmid. *Proc. Natl. Acad. Sci. U.S.A.* **81**, 1017–1020.

Schnabel, H., and Zillig, W. (1984). Circular structure of the genome of phage ϕH in a lysogenic *Halobacterium halobium. Mol. Gen. Genet.* **193**, 422–426.

Schnabel, H., Zillig, W., Pfäffle, M., Schnabel, R., Michel, H., and Delius, H. (1982a). *Halobacterium halobium phage* ϕH. *EMBO J.* **1**, 87–92.

Schnabel, H., Schramm, E., Schnabel, R., and Zillig, W. (1982b). Structural variability in the genome of phage ϕH of *Halobacterium halobium. Mol. Gen. Genet.* **188**, 370–377.

Schnabel, R. (1983). Untersuchungen zu Struktur und Funktion archaebakterieller RNA-Polymerasen. Ph.D. Thesis, Ludwig Maximilian University, Munich.

Schnabel, R., Sonnenbichler, J., and Zillig, W. (1982a). Stimulation by Silybin, a eukaryotic feature of archaebacterial RNA polymerases. *FEBS Lett.* **150**, 400–402.

Schnabel, R., Zillig, W., and Schnabel, H. (1982b). Component E of the DNA-dependent RNA polymerase of the archaebacterium *Thermoplasma acidophilum* is required for the transcription of native DNA. *Eur. J. Biochem.* **129**, 473–477.

Schnabel, R., Thomm, M., Gerardy-Schahn, R., Zillig, W., Stetter, K. O., and Huet, J. (1983). Structural homologies between different archaebacterial DNA-dependent RNA polymerases analysed by immunochemical comparison of their components. *EMBO J.* **2**, 751–755.

Schulz, W. (1982). Untersuchungen zur Initation der Transkription in *E. coli.* Ph.D. Thesis, Ludwig Maximilian University, Munich.

Searcy, D. G. (1975). Histone like protein in the prokaryote *Thermoplasma acidophilum. Biochim. Biophys. Acta* **395**, 535–547.

Sternbach, H., Engelhardt, R., and Lezius, A. G. (1975). Rapid isolation of highly active RNA polymerase from *E. coli* and its subunits by matrix bound heparin. *Eur. J. Biochem.* **60**, 51–55.

Stetter, K. O. (1977). DNA-dependent RNA polymerase from *Lactobacillus. Hoppe-Seyler's Z. Physiol. Chem.* **358**, 1093–1104.

Stetter, K. O., and Zillig, W. (1974). Transcription in Lactobacillus DNA-Dependent RNA polymerase from *Lactobacillus curvatus. Eur. J. Biochem.* **48**, 527–540.

Stetter, K. O., Winter, J., and Hartlieb, R. (1980). DNA-dependent RNA polymerase of the archaebacterium *Methanobacterium thermoautotrophicum. Zentralbl. Bakteriol., Mikrobiol. Hyg., Abt. 1, Orig. C* **1**, 201–218.

Stetter, K. O., Thomm, M., Winter, J., Wildgruber, G., Huber, H., Zillig, W., Janekovic, D., König, H., Palm, P., and Wunderl, S. (1981). *Methanothermus fervidus,* sp. nov, a novel extremely thermophilic methanogen isolated from an Icelandic hot spring. *Zentralbl. Bakteriol., Mikrobiol. Hyg., Abt. 1, Orig. C* **2**, 166–178.

Sturm, S., Schönefeld, V., Zillig, W., Janekovic, D., and Stetter, K. O. (1980). Structure and function of the DNA dependent RNA polymerase of the archaebacterium *Thermoplasma acidophilum. Zentralbl. Bakteriol., Mikrobiol. Hyg., Abt. 1, C* **1**, 12–25.

Thomm, M. (1983). Aufbau, Eigenschaften und immunologische Verwandtschaft der DNA-abhängigen RNA polymerasen methanogener Bakterien. Ph.D. Thesis, University of Regensburg.

Thomm, M., Stetter, K. O., and Zillig, W. (1982). Histone-like proteins in Eu- and Archaebacteria. *Zentralbl. Bakteriol., Mikrobiol. Hyg., Abt. 1, Orig. C* **3**, 128–139.

Thomm, M., Altenbuchner, J., and Stetter, K. O. (1983). Evidence for a plasmid in a methanogenic bacterium. *J. Bacteriol.* **153**, 1060–1062.

Towbin, H., Staehelin, T., and Gordon, J. (1979). Electrophoretic transfer of proteins from polyacrylamide gels to nitrocellulose sheets: Procedure and some applications. *Proc. Natl. Acad. Sci. U.S.A.* **76**, 4350–4354.

Tu, J., and Zillig, W. (1982). Organization of rRNA structural genes in the archaebacterium *Thermoplasma acidophilum. Nucleic Acids Res.* **10**, 7231–7245.

Tu, J., Prangishvilli, D., Huber, H., Wildgruber, G., Zillig, W., and Stetter, K. O. (1982). Taxonomic relations between archaebacteria including 6 novel genera examined by crosshybridization of DNAs and 16SrRNAs. *J. Mol. Evol.* **18**, 109–114.

Wiggo, J. L., Bush, J. W., and Chamberlin, M. (1979). Utilization of promoter and terminator sites on bacteriophage T7 DNA by RNA polymerases from a variety of bacterial orders. *Cell* **16**, 97–109.

Woese, C. R., Gupta, R., Hahn, C. M., Zillig, W., and Tu, J. (1984). The phylogenetic relationships of three sulfur-dependent archaebacteria. *Syst. Appli. Microbiol.* **5**, 97–105.

Yeats, S., McWilliam, P., and Zillig, W. (1982). A plasmid in the archaebacterium *Sulfolobus acidocaldarius. EMBO J.* **1**, 1035–1038.

Zillig, W., and Stetter, K. O. (1980). *In* "Genetics and Evolution of RNA Polymerase, t-RNA and Ribosomes" (S. Osawa, H. Ozeki, H. Uchida, and T. Yura, eds.), pp. 525–538. Univ. of Tokyo Press, Tokyo.

Zillig, W., Palm, P., and Heil, A. (1976). Function and reassembly of subunits of DNA-dependent RNA polymerase. *In* "RNA Polymerase" (R. Losick and M. Chamberlin, eds.), pp. 101–126. Cold Spring Harbor Lab., Cold Spring Harbor, New York.

Zillig, W., Stetter, K. O., and Tobien, M. (1978). DNA-dependent RNA polymerase from *Halobacterium halobium. Eur. J. Biochem.* **91**, 193–199.

Zillig, W., Stetter, K. O., and Janekovic, D. (1979). DNA dependent RNA polymerase from the archaebacterium *Sulfolobus acidocaldarius. Eur. J. Biochem.* **96**, 597–604.

Zillig, W., Stetter, K. O., Wunderl, S., Schulz, W., Priess, H., and Scholz, I. (1980). The *Sulfolobus* "*Caldariella*" group: Taxonomy on the basis of the structure of DNA-dependent RNA polymerases. *Arch. Microbiol.* **125**, 259–269.

Zillig, W., Stetter, K. O., Schnabel, R., Madon, J., and Gierl, A. (1982a). Transcription in Archaebacteria. *Zentralbl. Bakteriol., Mikrobiol. Hyg., Abt. 1, Orig. C* **3**, 218–227.

Zillig, W., Schnabel, R., Tu, J., and Stetter, K. O. (1982b). The phylogeny of archaebacteria, including novel anaerobic thermoacidophiles, in the light of RNA polymerase structure. *Naturwissenschaften* **69**, 197–204.

Zillig, W., Gierl, A., Schreiber, G., Wunderl, S., Janekovic, D., Stetter, K. O., and Klenk, H. P. (1983a). The archaebacterium *Thermofilum pendens* represents a novel genus of the thermophilic, anaerobic sulfur respiring *Thermoproteales. Syst. Appl. Microbiol.* **4**, 79–87.

Zillig, W., Holz, I., Janekovic, D., Schaefer, W., and Reiter, W. D. (1983b). The archaebacterium *Thermococcus* celer represents a novel genus within the thermophilic branch of the archaebacteria. *Syst. Appl. Microbiol.* **4**, 88–94.

CHAPTER 12

Antibiotic Sensitivity of Archaebacteria

AUGUST BÖCK AND OTTO KANDLER

I. Introduction

Antibiotics may block metabolic pathways, membrane transport, or macromolecular syntheses by interference with the function of essential high and low molecular weight compounds. Antibiotics are, therefore, important tools for elucidating biochemical and genetic mechanisms in the cell (Russel and Quessel, 1983). Because archaebacteria differ from eubacteria and eukaryotes (cf. Kandler, 1982a) in many basic biochemical characteristics, the sensitivity of archaebacteria toward antibiotics is likely to be quite different from that of the members of the other two kingdoms. First attempts to establish comparative surveys of the antibiotic sensitivity of members of the three kingdoms (Hammes et al., 1979; Hilpert et al., 1981; Pecher and Böck, 1981) have shown that the methanogens and halophiles tested are sensitive to a much smaller number of antibiotics than are the eubacteria (Table I). They are also not sensitive to cycloheximide, a typical inhibitor of eukaryotic cells, thus indicating that antibiotics may lack many of the target sites present in the eubacteria or the eukaryotes. It must, however, be emphasized that in vivo sensitivity or insensitivity toward an antibiotic does not provide decisive information on the absence or presence, respectively, of the specific target site. Lack of susceptibility to an antibiotic may be caused by the impermeability of the cell envelope or the cytoplasmic membrane, by the lack of a respective uptake system, or by the ability of the organism to inactivate the antibiotic. On the other hand, if susceptibility to an antibiotic is observed, it is necessary to demonstrate that the antibiotic

Copyright © 1985 by Academic Press, Inc.
All rights of reproduction in any form reserved.
ISBN 0-12-307208-5

TABLE I

EFFECT OF ANTIBIOTICS ON THE GROWTH OF METHANOGENIC AND HALOPHILIC BACTERIA[a,b]

Target	Antibiotic	Archaebacteria[c] — Halophiles								Archaebacteria[c] — Methanogens										Eubacteria[c]				Eukaryotes[c]	
		A	B	C	D	E	F	G	H	I	J	K	L	M	N	O	P	Q	R	S	T	U	V	W	X
Cell wall	Fosfomycin	—	—	—	—	—	—	—	—	—	—	—	—	—	—	—	—	—	—	10	12	4	18	—	—
	D-Cycloserine	—	—	—	—	—	—	—	—	—	—	—	—	—	—	—	9	—	—	13	16	15	13	—	—
	Vancomycin	—	—	—	—	—	—	—	—	—	—	—	—	—	—	—	—	—	—	7	10	14	4	—	—
	Penicillin G	—	—	—	—	—	14	—	—	—	—	—	—	—	—	—	—	—	—	2	26	11	29	—	—
	Cephalosporin C	—	—	—	—	18	—	18	—	—	—	—	—	—	—	—	—	—	—	2	22	11	9	—	—
	Nocardicin A	—	—	—	—	—	—	—	—	2	—	—	—	—	—	—	—	—	—	2	—	—	7	—	—
	Bacitracin	9	—	19	2	—	—	—	—	23	40	27	17	40	40	40	17	—	—	—	8	2	11	—	—
	Gardimycin	2	2	4	3	3	3	3	—	19	30	16	14	24	14	6	6	—	—	—	3	2	8	—	—
	Nisin	—	—	4	—	—	—	—	—	8	11	8	5	10	15	10	3	—	—	—	3	—	8	—	—
	Enduracidin	—	—	—	—	—	—	—	—	14	20	16	14	20	17	10	5	—	8	—	10	4	5	—	—
	Flavomycin	5	—	—	—	—	—	—	—	—	4	7	6	6	—	—	—	—	—	3	16	8	5	—	—
	Subtilin	—	—	8	—	—	—	—	—	—	2	—	—	—	—	5	10	—	—	—	6	2	6	—	—
RNA-polymerase	α-Amanitin	—	—	—	—	—	—	—	—	—	—	—	—	—	—	—	ND	—	ND	ND	ND	ND	ND	ND	ND
	Rifampicin	—	—	3	—	—	2	2	—	2	2	—	—	2	—	—	2	—	—	9	22	5	16	—	—

526

Protein biosynthesis / Membranes — Zone of inhibition table[a,b,c]

	A	B	C	D	E	F	G	G	H	I	J	K	L	M	N	O	P	Q	R	S	T	U	V	W	X
Protein biosynthesis																									
Cycloheximide	—	—	—	—	—	—	—	—	—	—	—	—	—	—	—	—	—	—	—	—	—	—	23	22
Chloramphenicol	—	—	—	—	—	—	—	—	11	—	25	13	40	—	12	—	ND	22	—	28	—	4	—	40
Virginiamycin	3	2	4	3	3	—	—	—	5	14	9	—	9	12	—	ND	—	12	15	—	10	5	19	15	—
Gentamicin	3	—	2	—	—	—	—	—	13	25	—	12	9	12	ND	—	8	8	—	15	9	13	18	10	—
Tetracycline	—	—	—	—	—	—	—	—	—	—	—	—	—	—	3	8	—	8	—	—	13	18	7	12	—
Oleandomycin	—	—	—	—	—	—	—	—	—	—	—	—	—	—	—	—	—	—	—	—	13	8	15	—	—
Erythromycin	—	—	—	—	—	—	—	—	—	—	—	—	—	—	—	3	—	3	—	—	3	17	11	15	—
Kanamycin	—	—	—	—	—	—	—	—	—	—	—	—	—	—	—	—	—	—	—	—	9	13	7	8	—
Membranes																									
Gramicidin S	—	—	—	—	—	—	—	—	5	7	7	5	—	11	15	10	4	16	—	—	4	4	5	3	4
Gramicidin D	—	—	—	—	—	—	—	—	—	—	—	—	—	16	11	—	—	—	—	—	4	4	—	4	3
Polymixin	—	4	—	—	—	—	—	—	4	—	—	—	—	8	8	—	—	—	—	—	5	—	—	4	3
Amphotericin B	—	—	—	—	—	—	—	—	—	—	—	—	—	—	—	—	—	—	—	—	—	—	3	6	5
Valinomycin	5	7	—	—	—	—	—	—	—	—	—	—	—	—	—	—	—	—	—	—	2	2	—	—	—
Nonactin	5	—	7	—	—	—	—	—	—	—	—	—	—	—	—	—	—	—	—	—	2	5	—	—	—
Monensin	2	4	6	2	3	4	3	2	—	5	10	4	35	25	12	15	4	25	—	—	6	3	10	—	—
Lasalocid	—	6	2	2	4	3	2	2	19	22	25	30	40	35	42	21	40	40	—	—	15	8	15	—	—

[a] Zone of inhibition in millimeters (Hilpert et al., 1981).

[b] —, no inhibition; ND, not determined.

[c] A, Halobacterium halobium; B, Halobacterium saccharovorum; C, Halobacterium salinarium; D, Halococcus morrhuae; E, H. morrhuae CCM 537; F, H. morrhuae CCM 859; F, H. morrhuae CCM 2226; G, H. morrhuae CCM 2526; H, Halobacterium sp. r-4; I, Methanobrevibacter arboriphilus AZ; J, Methanobrevibacter smithii; K, Methanobacterium bryantii; L, M. bryantii M.o.H.G.; M, Methanobacterium formicicum; N, Methanobacterium thermoautotrophicum ΔH; O, M. thermoautotrophicum 1; P, Methanococcus vannielii; Q, Methanospirillum hungatei; R, Methanosarcina barkeri; S, Escherichia coli; T, Staphylococcus aureus; U, Bacillus subtilis; V, Mic. luteus; W, Sacch. cerevisiae; X, Hansenula sp.

acts on its usual target and that this is the sole site of inhibition. As will be shown, there are several examples where antibiotics, which are inhibitory to particular groups of archaebacteria, act on some unusual "nonspecific" cellular sites.

In some cases where the specific target of the antibiotic is an enzyme, the addition of the natural product of that enzyme may release the inhibition by the antibiotic, thus proving the specificity of the target. If radioactively labeled antibiotics are available, the presence of a specific target site may be also demonstrated by *in vivo* experiments that are followed by the isolation of the resulting labeled target. In all other cases, *in vitro* assays are required to identify the target site.

Apart from using antibiotics as tools in studies on archaebacterial biochemistry, knowledge of the *in vivo* susceptibility of the archaebacteria to antibiotics may aid in devising selective media for the isolation and enrichment of new organisms (Weisburg and Tanner, 1982) and in maintaining the purity of cultures.

II. Methods for Susceptibility Tests

Both plate diffusion (Hammes *et al.,* 1979; Hilpert *et al.,* 1981) and tube dilution methods (Pecher and Böck, 1981) have been employed in testing *in vivo* susceptibility of archaebacteria. Generally, tube dilution tests are more sensitive, easier to standardize, and give a rather precise value for the minimal inhibitor concentration. Plate diffusion assays are very convenient for screening, but they are limited by low diffusion rates, especially in the case of lipophilic compounds, e.g., in the high-salt growth medium of extreme halophiles. Thus, at the same time, a negative result may be obtained on plates while the tube test indicates susceptibility.

In vitro testing of antibiotic sensitivity may be performed by employing homogenates or more or less purified enzyme preparations when the supposed target site is an enzyme. However, in the case of complex pathways involving multienzyme complexes, e.g., the synthesis of nucleic acid, proteins, etc., a test system has to be chosen that allows the *in vitro* determination of the partial reaction with which the antibiotic normally interferes. This requires, however, information on signal structures in archaebacteria that determine, for instance, initiation of macromolecule syntheses. Because suitable test systems are not yet available, our present knowledge of the details of the action of antibiotics in the archaebacteria is scarce and limited to a few compounds.

III. Target Sites

A. CELL ENVELOPE

Cell envelopes of the various lineages of the archaebacteria exhibit a distinct chemical and structural diversity (Kandler, 1982b; see Chapter 9 by Kandler and König). Because there is no common cell wall polymer, such as the eubacterial murein (peptidoglycan), no common target for any group of antibiotics can be expected. In fact, antibiotics that interfere with the synthesis of muramic acid (fosfomycin; Kahan *et al.*, 1974), of the UDP-MurNAc-pentapeptide precursor of murein (D-cycloserine, vancomycin), or the cross-linking reaction (penicillin, cephalosporin, nocardicin A), generally do not inhibit archaebacteria. The only exception to this is *Methanococcus vannielii* (Table I), which is susceptible to D-cycloserine (Jones *et al.*, 1977; Hilpert *et al.*, 1981), even though this organism does not contain murein (Jones *et al.*, 1977; Kandler and König, 1978) or any other known D-alanine-containing structure. The inhibition cannot be released by the addition of a large excess of D-alanine (Hilpert *et al.*, 1981), a powerful competitor of D-cycloserine in eubacteria (Neuhaus, 1967). The target of D-cycloserine in *M. vannielii* remains to be elucidated. The target site is certainly not the synthesis of D-alanine, and it may be unique to *M. vannielii*, because none of the numerous other methanogens and halobacteria tested were sensitive to D-cycloserine.

Bacitracin, the best known inhibitor of the lipid cycle of murein biosynthesis, prevents the dephosphorylation of isoprenoyl pyrophosphate to the respective monophosphate, an essential reaction in the biosynthesis of murein (Stone and Strominger, 1971). Surprisingly, bacitracin inhibits not only those methanogens containing pseudomurein, whose biosynthesis may also involve lipid intermediates, but also *M. vannielii* and the halophiles, which do not contain pseudomurein. Thus, bacitracin may act on archaebacteria "through inhibition of de-phosphorylation of the isoprenoyl pyrophosphate precursor of this organism's diether lipids" as pointed out by Basinger and Oliver (1979).

Bacitracin, often used as a feed additive, also inhibits biogas formation from agricultural wastes (Hilpert *et al.*, 1984). However, it was not shown whether the decrease of methane production from complex substrates was caused by a direct inhibition of the methanogens or by the inhibition of the fermentative population, which forms organic acids and H_2.

Of the antibiotics that prevent the incorporation of the lipid-bound precursor into the glycan strands, gardimycin (Somma *et al.*, 1977) inhibits not only the methanogens but also the halophiles, whereas inhibition by enduracin (Matsuhashi *et al.*, 1970) and nisin (Reisinger *et al.*, 1980) is almost solely restricted to the pseudomurein-containing methanogens (Table I). Subtilin, which is chemically related to nisin (Gross et al., 1973), exhibits a similar but less complete inhibition pattern. Flavomycin, which is also reported to prevent the formation of glycan strands of murein (Lughtenberger et al., 1971; Hammes and Neuhaus, 1974) by an as yet unknown mechanism, does not inhibit archaebacteria (Table I). Also, the addition of high levels of flavomycin (5.0 mg/liter) to biogas fermenters operated with municipal sludge had no effect on methane production (Hilpert et al., 1984). However, there are two exceptions, Methanobacterium bryantii and Halobacterium halobium (Table I). The sensitivity of H. halobium toward flavomycin was also reported by Mescher and Strominger (1975). The reason for such an erratic inhibition pattern remains to be elucidated.

In conclusion, the antiarchaebacterial spectrum of the antibiotics directed against the lipid-bound precursors of murein indicates that these antibiotics inhibit archaebacteria by different mechanisms. They may interfere with different lipid-bound precursors of various carbohydrate-containing polymers (pseudomurein, heteropolysaccharides, glycoprotein) or with the biosynthesis of isoprenoid diether lipids, which is typical of archaebacteria (Langworthy et al., 1982). The latter may be the sole target in halobacteria and non-pseudomurein-containing methanogens, whereas pseudomurein precursor synthesis may be the main target in the pseudomurein-containing members of the Methanobacteriales.

The actions of antibiotics directed against cell wall polymers have only rarely been studied in thermoacidophilic archaebacteria because they possess no cell envelope (Thermoplasma) or an envelope consisting of glycoprotein subunits (Sulfolobus, Thermoproteales) (see this volume, Chapter 9). However, vancomycin was employed to demonstrate the absence of a murein-containing cell wall in Thermoplasma and Sulfolobus (Brock, 1978). Although Bacillus acidocaldarius, a murein-containing control organism, was sensitive to vancomycin under the severe conditions (pH 3 and 56°C) necessary to grow the thermoacidophiles, Thermoplasma and Sulfolobus resisted a concentration as high as 5 mg vancomycin/ml.

B. CELL MEMBRANES

Of the membrane-active antibiotics tested (Table I), the Na^+-ionophor monensin and the K^+-ionophor lasalocid are the most effective in vivo inhibitors (Table I). They are of practical use as feed additives for improvement of feed efficiency (Chen and Wolin, 1979).

On the basis of experiments with the complex flora in rumen fluid (Chen and Wolin, 1979; Slyter, 1979) and with washed suspensions of rumen microorganisms in buffered solution (Nevel and Demeyer, 1977), it is doubtful that the two antibiotics inhibit methane formation by a primary attack on the methanogenic component of the rumen flora. The accumulation of propionate and the decrease of the formation of acetic acid and H_2 (caused by monensin added to rumen fluid) indicate a strong inhibition of the fermentative rather than the methanogenic flora. Addition of monensin to fermenting municipal sludge, however, showed the opposite effect. Here, acetic acid was the first accumulated organic acid, methane production was severely inhibited, and total counts of methanogens were markedly decreased (Hilpert *et al.*, 1983, 1984), thus indicating a direct inhibition of the methanogens. The discrepancy between the effects of monensin on the fermentation in the rumen and on sewage fermentation may be due to the fact that the Na^+ concentration in rumen fluid is higher than in municipal sludge. As shown by Perski *et al.* (1982), the inhibition of methane formation by monensin is almost completely released by high Na^+ concentrations.

Due to the unusual membrane lipids of the archaebacteria, it is not justified, *a priori,* to assume that antibiotics that are shown to be directed against membrane functions in eubacteria will act in the same way in archaebacteria. Therefore, the effect of such antibiotics on cation transport and ATP levels have been studied on pure cultures of methanogens (Jarrell and Sprott, 1983; Hilpert *et al.*, 1984).

The addition of monensin or lasalocid to cell suspensions of *Methanobacterium bryantii* containing a very low concentration of K^+ (0.2 mmole/liter) lead to an immediate release of all the intracellular K^+ (Hilpert *et al.*, 1984). Similar experiments carried out with the same organisms but at much higher K^+ concentration in the suspension buffer (5.7 mmole/liter) resulted in a partial release only of the intracellular K^+ leading to a decrease of the ratio of K^+ inside and outside the cell by more than 50% (Jarrell and Sprott, 1983). Further studies have shown that monensin, nigericin, and gramicidin also have distinct effects on the pH gradient across the cytoplasmic membrane, the intracellular ATP level, the electrical potential, and the proton motive force. Hence, these antibiotics are also directed against the membranes of archaebacteria. Valinomycin, however, was found to be much less effective (Jarrell and Sprott, 1983), which is in agreement with its inactivity against methanogens and halophiles in *in vivo* tests (Table I). The effect of membrane-active antibiotics on thermoacidophilic archaebacteria has not yet been studied.

C. DNA Structure and Synthesis

Only a few compounds that complex with DNA or that specifically inhibit enzymes involved in DNA replication have been tested so far. They include actinomycin, adriamycin, bleomycin sulfate (Hook *et al.*, 1984), and nalidixic acid (Pecher and Böck, 1981). Only *in vivo* susceptibility studies were carried out. Inhibition was obtained with adriamycin in the case of a few methanogens, but the other compounds proved to be completely inactive at the highest concentrations tested.

D. DNA-Dependent RNA Polymerase

The DNA-dependent RNA polymerase from archaebacteria shares considerable structural homologies with the RNA polymerase A from eukaryotes (Zillig *et al.*, 1982). This similarity is nicely reflected by the sensitivity of the archaebacterial enzyme to transcriptional inhibitors.

The rifampicins, which inhibit all eubacterial polymerases tested at concentrations below 1 µg/ml, by forming a 1:1 stoichiometric complex (Wehrli *et al.*, 1968), are completely inactive against the enzymes from extreme halophiles, methanogens, and thermoacidophiles (Zillig *et al.*, 1978, 1979; Stetter *et al.*, 1980; Prangishvilli *et al.*, 1982). Concentrations up to 200 µg/ml were tested. Streptolydigin was inactive as well.

α-Amanitin, a toxin from *Amanita phalloides*, is known to interfere with the activity of the eukaryotic polymerases B and C but not with polymerase A (Wieland, 1972). The DNA-dependent RNA polymerases of archaebacteria were not affected by it (Zillig *et al.*, 1982). Conversely, however, the alkaloid silybin, which stimulates the rate of transcription by eukaryotic polymerase A, was found to exert the same effect on the archaebacterial enzymes (Zillig *et al.*, 1982).

In conclusion, archaebacterial RNA polymerases resemble the type A of eukaryotic enzymes in their antibiotic sensitivity pattern. The polymerases are resistant to the eubacterial inhibitors rifampicin and streptolydigin; they are not inhibited by α-amanitin, but are stimulated by silybin.

In contrast to these *in vitro* data, *Halobacterium halobium* and *Halobacterium cutirubrum* are highly susceptible to rifampicin *in vivo* (Zillig *et al.*, 1978; Pecher and Böck, 1981). Minimal inhibitor concentrations as low as 4 µg/ml have been determined (Pecher and Böck, 1981). This strong inhibitor effect, however, has been ascribed to the interference of rifampicin, with the integrity of the cytoplasmic membrane causing rapid lysis at a concentration of 100 µg/ml and slow lysis at 10 µg/ml (Zillig *et al.*, 1982).

E. Protein Synthesis

1. Aminoacyl-tRNA Synthetases

Aminoacyl-tRNA synthetases play important roles in the control of fidelity and the regulation of protein synthesis. They are not only responsible for charging the individual tRNA species with the cognate amino acids but in addition are key enzymes for the control of the formation of amino acid biosynthetic enzymes and of ribosomes (Neidhardt *et al.*, 1975). *In vivo* investigation of these regulatory processes in archaebacteria requires the availability of inhibitors that interfere with the charging of tRNA.

Pseudomonic acid, a metabolite of *Pseudomonas fluorescens* strain NCIB 10586 (Banks *et al.*, 1971) was found to be a powerful inhibitor of the isoleucyl-tRNA synthetase of *Halobacterium halobium* and two methanogens tested, *Methanobacterium formicicum* and *Methanococcus vannielii* (A. Böck and U. Bär, unpublished results). Figure 1 shows the growth response of *M. vannielii* to different concentrations of the antibiotic and demonstrates that inhibition is released by high concentrations of L-isoleucine in the medium. This and the *in vitro* action on isoleucyl-tRNA synthetase activity (Fig. 1) suggests that pseudomonic acid acts at a single site, namely isoleucyl-tRNA synthetase and competes with the natural substrate L-isoleucine. Its mechanism of action in archaebacteria is therefore identical to that reported for eubacteria and eukaryotes (Hughes and Mellows, 1980).

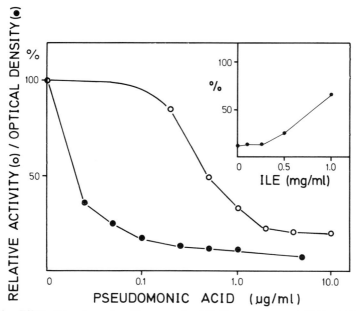

FIG. 1. Effect of pseudomonic acid on growth of *Methanococcus vannielii* (A_{420}) (●) and on isoleucyl-tRNA synthetase activity *in vitro* (○). The insert figure shows the reversal of growth inhibition (in presence of 0.1 μg pseudomonic acid per milliliter) by different concentrations of L-isoleucine in the medium. The ordinate of the insert figure gives the optical density values relative to that from a culture without pseudomonic acid. Isoleucyl-tRNA synthetase activity was measured by the standard tRNA aminoacylation reaction (Comer and Böck, 1976) employing total tRNA from *E. coli* and L-isoleucine at 20 μM.

2. ELONGATION FACTORS

There are a few antibiotics known to interfere with the activity of the elongation factor EF-Tu in eubacteria; kirromycin binds in a 1:1 ratio to EF-Tu and prevents the dissociation of EF-Tu from the ribosome (Wolf *et al.*, 1977). Pulvomycin, on the other hand, blocks the formation of the ternary aminoacyl-tRNA·EF-Tu·GTP complex (Wolf *et al.*, 1978). In the archaebacterial systems tested, namely *Halobacterium halobium* (Kessel and Klink, 1981) and *Sulfolobus solfataricus* (Cammarano *et al.*, 1982), both antibiotics were ineffective. This was quoted by the latter authors as a typical eukaryotic characteristic of the EF-Tu equivalent from archaebacteria. In view of the recent detection of a completely kirromycin-resistant eubacterium, *Lactobacillus brevis* (Wörner and Wolf, 1982), the validity of this statement must await the investigation of more archaebacterial species, that represent different groups. In this respect, we want to emphasize that it is important to support data obtained on halophilic or thermophilic systems by those from mesophilic and nonhalophilic organisms in order to gain information on whether high salt environment or high temperature interfere with antibiotic binding and whether loss of antibiotic sensitivity may be a consequence of the adaptation to these extreme environments.

Fusidic acid interferes with the activity of elongation factor G from eubacterial and eukaryotic organisms (Tanaka, 1975). It also blocks polyphenylalanine synthesis by *Methanococcus vannielii* ribosomes (A. Böck and U. Bär, unpublished results).

3. THE RIBOSOMES

Archaebacterial ribosomes, although originally thought to be of the normal 70 S prokaryotic type, seem to possess a mixture of eubacterial, eukaryotic, and unique features (Chapter 7 by Matheson). The antibiotic sensitivity pattern strengthens this view. In addition, it reveals that the considerable heterogeneity in protein composition (Schmid and Böck, 1982a,b; Londei *et al.*, 1982), particle size (Londei *et al.*, 1982), and 5 S rRNA primary structure (Fox *et al.*, 1982) among archaebacteria is paralleled by comparable variety of response to protein synthesis inhibitors.

a. Aminoglycoside–Aminocyclitol Antibiotics and Viomycin. From their mode of action aminoglycoside–aminocyclitol antibiotics can be divided into two main groups, namely the streptidine and the 2-deoxystreptamine compounds (Zierhut *et al.*, 1979). Members of both groups have been tested for their *in vivo* (Pecher and Böck, 1981; Hilpert *et al.*, 1981; Weisburg and Tanner, 1982; Hook *et al.*, 1984) and *in vitro* (Elhardt and Böck, 1982; Böck *et al.*, 1983; Londei, 1983; Sanz and Amils, 1983) activity against archaebacteria. A comprehensive *in vitro* study on 20 aminoglycosides was carried out employing the *Methanococcus in vitro* poly(U)-dependent system (Böck *et al.*, 1983). Compounds belonging to the streptidine class of aminoglycosides (streptomycin, bluensomycin) were completely ineffective in inhibiting polyphenylalanine synthesis or in promoting translational misreading. The results correlated well with the lack of binding of radioactively labeled dihydrostreptomycin to ribosomes from *Halobacterium halobium, Methanobacterium bryantii,* or *Methanosarcina barkeri* (Schmid *et al.*, 1982).

A rather clearcut structure–activity relationship, on the other hand, was observed when 2-deoxystreptamine compounds were tested. Those containing the 4,5-disubstituted cyclohexitol (neomycin group) were much more active than the 4,6-disubstituted (gentamicin group) ones (Böck *et al.*, 1983). The degree to which polyphenylalanine formation and translational misreading was affected was closely paralleled by the growth-inhibitory effects, with neomycin being the most potent compound. As in eubacteria, neomycin and related compounds were bactericidal (Böck *et al.*, 1983). Londei (1983) and Sanz and Amils (1983) studied the effect of aminoglycosides on polyphenylalanine synthesis by *Sulfolobus solfataricus* ribosomes. The only compound exhibiting activity, and that only at high drug-to-ribosome ratios, was neomycin.

Other members of the archaebacteria were only tested for their *in vivo* susceptibility to aminoglycosides. Slight to negligible effects were seen with extreme halophiles (Hilpert *et al.*, 1981; Pecher and Böck, 1981; Weisburg and Tanner, 1982), which might be a consequence of the interference of high salt concentrations with any electrostatic interaction of the highly basic compounds with their binding site. Within the methanogens, on the other hand, a variable response was observed, depending on the compound tested and on the method used (Pecher and Böck, 1981; Hilpert *et al.*, 1981; Weisburg and Tanner, 1982; Hook *et al.*, 1984). It is, however, problematic to conclude from negative *in vivo* results an absence of cellular binding site for these compounds, since the insusceptibility may merely reflect an inability for aminoglycoside uptake. It is well known that aminoglycosides are accumulated via a still-unknown carrier with electron transport-generated membrane potential as the driving force (Bryan and Van den Elzen, 1977). In general, therefore, anaerobic organisms possess a high intrinsic resistance to these compounds (Bryan and Van den Elzen, 1977).

Viomycin, a peptide antibiotic, appears to share an identical binding site and a similar mechanism of action with the 2-deoxystreptamine aminoglycosides (Tanaka, 1982). It strongly inhibits initiation of protein synthesis and the ribosomal "translocation reaction." In view of the inefficiency of many 2-deoxystreptamine compounds to inhibit polyphenylalanine synthesis (a measure of the translocation of peptidyl-tRNA; Cabañas *et al.*, 1978) at archaebacterial ribosomes, it was of interest to study the effect of viomycin. *In vivo* growth of *Methanococcus* was not influenced by concentrations as high as 500 μg/ml; also, no inhibition of polyphenylalanine formation by the ribosomes of this organism was observed. Rather, a 50% stimulation of phenylalanine incorporation occurred, indicating that the drug interacts with the ribosome but in a manner not influencing peptidyl-tRNA translocation (Böck et al., 1983).

Spectinomycin and kasugamycin are aminoglycosides that rather specifically interfere with protein chain initiation and that, in constrast to the classical aminoglycosides, have no effect on translational misreading (Tanaka, 1982). They did not inhibit growth of the archaebacteria tested in concentrations as high as 500 μg/ml (Pecher and Böck, 1981). Due to the lack of a suitable test system, their in vitro activity has not yet been analyzed.

   b. Tetracyclines. Tetracyclines inhibit protein synthesis on 70 S eubacterial ribosomes and—to a somewhat lesser degree—also on 80 S ribosomes. Their specific effect on bacterial growth is based on the fact that eukaryotic cells in contrast to eubacteria lack the capacity for drug accumulation (Kaji and Ryoji, 1979). Tetracycline, chlorotetracycline, and minocycline were tested in the poly(U)-dependent system from *Methanococcus vannielii, Methanosarcina barkeri* (Elhardt and Böck, 1982; A. Böck and U. Bär, unpublished), and *Sulfolobus solfataricus* (Londei, 1983; Sanz and Amils, 1983). Polyphenylalanine formation (which is normally strongly inhibited by tetracycline in the case of both eubacterial and eukaryotic ribosomes) was refractory to inhibition by tetracycline and chlorotetracycline in the *Methanococcus* and *Sulfolobus* systems, but was moderately affected in the *Methanosarcina* system. Minocycline caused a 20% inhibition at high concentration with *Methanococcus* ribosomes. The apparent lack of susceptibility of *Methanococcus* polypeptide synthesis contrasted with the growth inhibitory effect (Pecher and Böck, 1981) of tetracycline antibiotics. This discrepancy was resolved by the finding that tetracycline compounds, at the minimal inhibitory concentrations, caused lysis of *Methanococcus* cells. This is reminiscent of the in vivo effect of rifampicin on *Halobacterium* (Zillig et al., 1982) and may reflect a perturbance of the ether lipid-containing membranes of archaebacteria by "intercalation" of hydrophobic, planar molecules.

   c. The Macrolide Antibiotics. With very few exceptions, these 50 S targeted inhibitors did not affect the growth of extreme halophiles and methanogens (Hilpert et al., 1981; Pecher and Böck, 1981), even at concentrations above 100 μg/ml. However, because they generally do not inhibit poly(U)-dependent polyphenylalanine formation, conclusive experimental evidence for their effect on archaebacterial ribosomes must await the development of other test systems. With *Methanococcus* and *Sulfolobus* ribosomes and poly(U,G) (base ratio 4:1) as message, erythromycin was ineffective (Elhardt and Böck, 1982; Londei, 1983; Sanz and Amils, 1983). Under this condition phenylalanine polymerization by *E. coli* ribosomes was slightly affected.

d. Chloramphenicol and Tiamulin. Chloramphenicol inhibits growth of all methanogens tested at low concentrations (McKellar and Sprott, 1979; Hilpert *et al.*, 1981; Pecher and Böck, 1981). It is not inhibitory for multiplication of Thermoproteales species (Zillig *et al.*, 1981). Extreme halophiles were not susceptible when tested by the plate diffusion assay (Hilpert *et al.*, 1981), but two species showed moderate sensitivity when assayed in liquid culture. Analysis of the addition of chloramphenicol to cultures of *Halobacterium halobium* revealed that growth rate is reduced to less than one-half of the control in presence of 200 μg chloramphenicol per milliliter (border of solubility). No complete bacteriostasis was observed (Schmid *et al.*, 1982).

In vitro measurement of inhibition by chloramphenicol is complicated by the fact that it cannot be assessed in the poly(U) system (cf. Pestka, 1975). With poly(U,G) as message, however, phenylalanine incorporation is sensitive to the antibiotic (Kucan and Lipmann, 1964); no inhibition of polypeptide synthesis by chloramphenicol on *Methanococcus vannielii* and *Sulfolobus solfataricus* ribosomes was obtained under this condition (Elhardt and Böck, 1982; Londei, 1983; Sanz and Amils, 1983). This is corroborated by the lack of binding of radioactively labeled antibiotic to ribosomes from two methanogenic and one halophilic organism (Schmid *et al.*, 1982). These findings suggest that archaebacterial ribosomes in general, including those from methanogens, may possess intrinsic resistance to chloramphenicol and that the *in vivo* inhibitory effect observed occurs at some other site. Inhibition of energy metabolism of methanogens may be the actual target (Gunsalus and Wolfe, 1978; McKellar and Sprott, 1979).

Tiamulin, a 50 S ribosomal subunit targeted antibiotic (Högenauer, 1979), was moderately effective in inhibiting the growth of *Methanococcus vannielii* but almost ineffective on other methanogens and extreme halophiles (Pecher and Böck, 1981). Pleuromutilin, the natural product from which the semisynthetic tiamulin was derived, also showed growth inhibitory effects on two species of *Methanococcus* and of *Methanobacterium bryantii* (Hook *et al.*, 1984). In the *in vitro* poly(U) system derived from *Methanococcus* and *Methanosarcina*, tiamulin was completely ineffective under the conditions employed.

e. Virginiamycin and Thiostrepton. Hilpert *et al.* (1981) discovered that virginiamycin was a powerful inhibitor of the growth of most of the halophiles and the methanogens tested. Hook *et al.* (1984) reported growth inhibition of several other methanogens by A2315, a compound also belonging to the virginiamycin family of antibiotics (Chamberlin and Chen, 1977). *In vivo* results, in this case, were supported by the strong inhibition of polyphenylalanine synthesis by *Methanococcus* and *Methanobacterium* ribosomes *in vitro* (Elhardt and Böck, 1982). As reported for eubacterial ribosomes (Cocito, 1979), virginiamycin M and S compounds act synergistically *in vivo* and *in vitro* in the *Methanococcus* system

(C. Cocito and A. Böck, unpublished results). No effect was observed on poly-phenylalanine synthesis by *Sulfolobus* ribosomes (Londei, 1983; Sanz and Amils, 1983).

Thiostrepton is one of the few antibiotics for which detailed knowledge of its binding to the 70 S ribosome and its mechanism of action is available (Cundliffe, 1983). Because it is rapidly inactivated in the growth medium used to cultivate methanogens, its effect on archaebacteria was only studied *in vitro*. Of all the antibiotics tested, thiostrepton showed the most pronounced inhibitory effect on polyphenylalanine synthesis by *Methanococcus* and *Methanobacterium* ribosomes (Elhardt and Böck, 1982). Again, *Sulfolobus* ribosomes were not affected (Londei, 1983; Sanz and Amils, 1983).

f. Inhibitors of Protein Synthesis on 80 S Ribosomes. In view of the phy-logenetic status of archaebacteria, the sensitivity of ribosomes from these orga-nisms to inhibitors of protein synthesis on 80 S ribosomes was of particular interest. Trichodermin was growth inhibitory to the two species of halophiles tested (Pecher and Böck, 1981), but the specificity of this inhibition has not yet been supported by *in vitro* data. Cycloheximide and emetine were inactive both *in vivo* (Pecher and Böck, 1981; Hilpert *et al.*, 1981) and in the methanogen *in vitro* polypeptide synthesis system (Elhardt and Böck, 1982). Neither did ricin inactivate ribosomes from the two methanogens.

Apparent exceptions to this pattern were anisomycin, verrucarin A, and tri-chodermin, all specific inhibitors of 80 S ribosomes (Cundliffe *et al.*, 1974; Jimenez and Vazquez, 1979). Anisomycin prevented growth of halophiles and Methanobacteriales species at low concentrations (Pecher and Böck, 1981); con-sistent with these *in vivo* results, strong interference with polypeptide synthesis by ribosomes from *Methanobacterium formicicum* (Elhardt and Böck, 1982) and *Halobacterium cutirubrum* (Kessel and Klink, 1981) was found. *Methanococcus vannielii* and *Methanosarcina barkeri* exhibited a rather high intrinsic *in vivo* resistance to anisomycin, which was paralleled by higher resistance of their ribosomes *in vitro*. The finding that ribosomes from *Thermoplasma* are also sensitive to anisomycin (see Chapter 8 by Klink) suggests that the presence of binding site(s) for this antieukaryotic compound may be a general feature of archaebacterial ribosomes.

Definite proof that anisomycin may act at a similar site on archaebacterial ribosomes as it does on 80 S eukaryotic ones was provided by mutant studies. Isolated mutants from *Methanobacterium formicicum* were resistant to 50 μg anisomycin per milliliter. The mutations conferred high level *in vitro* resistance to polyphenylalanine synthesis; altered 50 S subunits were responsible for the resistance property (H. Hummel and A. Böck, unpublished results).

In addition to anisomycin, two other antieukaryotic protein synthesis inhib-itors were found to interfere with polypeptide formation by ribosomes from

Methanobacterium formicicum, namely verrucarin A and trichodermin. Whereas trichodermin was only effective in a poly(U,G) (4:1) primed system, verrucarin A inhibited poly(U)-dependent incorporation as well. The mutants of *M. formicicum* originally selected for resistance to anisomycin (see above) showed high cross-resistance both to verrucarin A and to trichodermin. This is exactly what was found for 80 S ribosomes from anisomycin-resistant yeast mutants (Jimenez and Vazquez, 1979); it provides evidence that the sites of interaction for these three peptidyl-transfer inhibitors, which are similar to those reported for the yeast ribosome, exist in *Methanobacterium* ribosomes.

IV. Conclusions

Antibiotics that act specifically are powerful tools for physiological, genetic, biochemical, and evolutionary studies. In physiological analysis, they permit the investigation of the *in vivo* functional role of an inhibited reaction and its regulatory consequences. For example, pseudomonic acid can be used to study the role of isoleucyl-tRNA in the regulation of isoleucine biosynthesis in archaebacteria. Furthermore, the selection of mutants resistant to antibiotics allows the extension of the classical mutational analysis to archaebacterial systems and may provide suitable selective markers for beginning genetic studies.

Equally important, however, is the fact that antibiotics, under certain conditions, may be used as indicators of the phylogenetic status of an organism. This requires, however, a detailed biochemical investigation of the specificity of their action. From the results summarized in this review, it is clear that we are only at the beginning—the data-collecting stage—of this analysis. New test systems, especially those in which initiation of RNA and protein synthesis can be studied, have to be developed in order to extend the investigation to those compounds interfering with these steps. Such detailed studies are especially important for the inhibitors of protein synthesis, because, as outlined, the translational apparatus appears to be more diverse than the transcriptional one within the archaebacterial lineage. Furthermore, the analysis, which has concentrated primarily on *Methanococcus* and *Sulfolobus,* must be extended to other archaebacteria in order to allow general conclusions. Despite the scarcity of our information, the following facts have already emerged.

(1) Ribosomes of methanogens (*Methanococcus, Methanobacterium*) differ greatly in their antibiotic sensitivity pattern from those of the thermoacidophilic *Sulfolobus.* It is important to analyze whether the insensitivity of *Sulfolobus* ribosomes to almost all of the 50 antieubacterial and antieukaryotic compounds tested (Londei, 1983; Sanz and Amils, 1983) is a general feature of thermophilic archaebacteria or whether it is a specific property of *Sulfolobus.* In addition, it is

important to study the biochemical basis of the insensitivity of *Sulfolobus* to those antibiotics that are inhibitory to methanogens. Thiostrepton might be an excellent model compound for this purpose because its binding site can be reconstituted from purified 23 S ribosomal RNA plus a single large subunit ribosomal protein (Cundliffe, 1983). This should answer the question of whether it is modification or sequence variation of RNA or some protein component that renders ribosomes from *Sulfolobus* resistant.

(2) A number of antieubacterial protein synthesis inhibitors are inactive in all of the archaebacterial systems tested. Streptomycin and, especially, the macrolides and chloramphenicol are interesting examples. These compounds also have no effect on protein synthesis at 80 S ribosomes. Streptomycin seems to exert its bactericidal effect in eubacteria by interfering with the initiation step (Wallace and Davis, 1973); chloramphenicol and erythromycin, with the peptidyl transfer and the translocation reactions, respectively (Pestka, 1975). It is interesting to note that other anti-70 S inhibitors—for initiation (spectinomycin, viomycin, gentamicin group aminoglycosides), for peptidyl transfer (lincomycin, tiamulin), and for translocation (viomycin)—are also inactive in *Methanococcus* and *Sulfolobus*. This suggests that ribosomes from these archaebacteria in comparison to those from eubacteria possess differently structured sites for initiation, peptidyl transfer, and translocation. Inhibition by anisomycin, which affects peptidyl transfer in 80 S ribosomes, indicates instead that the respective sites in archaebacterial ribosomes are structurally related to those from eukaryotic organelles. It is important to investigate whether this correlated with sequence changes in those parts of the eubacterial 23 S rRNA molecule implicated recently in the peptidyl transfer and translocation reactions (for review, see Skinner *et al.*, 1983).

On the other hand, the archaebacterial 50 S subunit domain connected with the GTPase activity linked to elongation factor G appears to bear eubacterial features as judged by the strong interaction of *Methanococcus* ribosomes with thiostrepton (Elhardt and Böck, 1982).

V. Summary

Whereas the transcriptional apparatus of archaebacteria shows a uniform and "eukaryotic-type" antibiotic sensitivity, that of the translational system is heterogeneous. Protein synthesis of ribosomes from *Sulfolobus* is insensitive to almost all antieubacterial and antieukaryotic compounds; ribosomes from methanogenic and halophilic organisms also lack binding sites for many classical eubacterial ribosome inhibitors, but at the same time they possess sites for others, as well as for some inhibitors of 80 S ribosomes.

REFERENCES

Banks, G. T., Barrow, K. D., Chain, E. B., Fuller, A. T., Mellows, G., and Woolford, M. (1971). Pseudomonic acid: An antibiotic produced by *Pseudomonas fluorescens*. *Nature (London)* **234**, 416–417.

Basinger, G. W., and Oliver, J. D. (1979). Inhibition of *Halobacterium cutirubrum* lipid biosynthesis by bacitracin. *J. Gen. Microbiol.* **111**, 423–427.

Böck, A., Bär, U., Schmid, G., and Hummel, H. (1983). Aminoglycoside sensitivity of ribosomes from the archaebacterium *Methanococcus vannielii:* Structure-activity relationship. *FEMS Microbiol. Lett.* **20**, 435–438.

Brock, T. D. (1978). "Thermophilic Microorganisms and Life at High Temperatures." Springer-Verlag, Berlin and New York.

Bryan, L. E., and Van den Elzen, H. M. (1977). Effects of membrane energy mutations and cations on streptomycin and gentamicin accumulation by bacteria: A model for entry of streptomycin and gentamicin in susceptible and resistant bacteria. *Antimicrob. Agents Chemother.* **12**, 163–177.

Cabañas, M. J., Vazquez, D., and Modolell, J. (1978). Dual interference of hygromycin B with ribosomal translocation and with aminoacyl-tRNA recognition. *Eur. J. Biochem.* **87**, 21–27.

Cammarano, P., Teichner, A., Chinali, G., Londei, P., De Rosa, M., Gambacorta, A., and Nicolaus, B. (1982). Archaebacterial elongation factor Tu insensitive to pulvomycin and kirromycin. *FEBS Lett.* **148**, 255–259.

Chamberlin, J. W., and Chen, S. (1977). A 2315, New antibiotics produced by *Actinoplanes philippinensis*. 2. Structure of A 2315. *J. Antibiot.* **30**, 197–201.

Chen, M., and Wolin, M. J. (1979). Effect of monensin and lasalocid sodium on the growth of methanogenic and rumen saccharolytic bacteria. *Appl. Environ. Microbiol.* **38**, 72–77.

Cocito, C. (1979). Antibiotics of the virginiamycin family, inhibitors which contain synergistic components. *Microbiol. Rev.* **43**, 145–198.

Comer, M. M., and Böck, A. (1976). Genes for the α and β subunits of the phenylalanyl-transfer ribonucleic acid synthetase of *Escherichia coli*. *J. Bacteriol.* **127**, 923–933.

Cundliffe, E. (1983). Antibiotics as probes of ribosomal structure and function. *In* "Chemotherapeutic Strategy" (D. I. Edwards and D. R. Hiscock, eds.), pp. 65–78. MacMillan Press, London.

Cundliffe, E., Cannon, M., and Davies, J. (1974). Mechanism of inhibition of eukaryotic protein synthesis by trichothecene fungal toxins. *Proc. Natl. Acad. Sci. U.S.A.* **71**, 30–34.

Elhardt, D., and Böck, A. (1982). An in vitro polypeptide synthesizing system from methanogenic bacteria: Sensitivity to antibiotics. *Mol. Gen. Genet.* **188**, 128–134.

Fox, G. E., Luehrsen, K. R., and Woese, C. R. (1982). Archaebacterial 5S RNA. *Zentralbl. Bakteriol. Hyg., Abt. 1, Orig. C* **3**, 330–345.

Gross, H., Kiltz, H. H., and Nebelin, E. (1973). Subtilin. VI. Die Struktur des Subtilins. *Hoppe-Seyler's Z. Physiol. Chem.* **354**, 810–812.

Gunsalus, R. P., and Wolfe, R. S. (1978). ATP activation and properties of the methyl coenzyme M reductase system in *Methanobacterium thermoautotrophicum*. *J. Bacteriol.* **135**, 851–859.

Hammes, W. P., and Neuhaus, F. C. (1974). On the mechanism of action of vancomycin inhibition of peptidoglycan synthesis. *Antimicrob. Agents Chemother.* **6**, 722–728.

Hammes, W. P., Winter, J., and Kandler, O. (1979). The sensitivity of the pseudomurein-containing genus *Methanobacterium* to inhibitors of murein synthesis. *Arch. Microbiol.* **123**, 275–279.

Hilpert, R., Winter, J., Hammes, W., and Kandler, O. (1981). The sensitivity of archaebacteria to antibiotics. *Zentralbl. Bakteriol. Mikrobiol. Hyg., Abt. 1, Orig. C* **2**, 11–20.

Hilpert, R., Winter, J., and Kandler, O. (1983). Fütterungszusätze und Desinfektionsmittel als Störfaktoren bei der anaeroben Faulung landwirtschaftlicher Abfälle. *Münch. Beitr. Abwasser-, Fisch.- Flussbiol.* **36**, 162–176.

Hilpert, R., Winter, J., and Kandler, O. (1984). Feed additives and disinfectants as inhibitory factors in anaerobic digestion of agricultural wastes. *Agric. Wastes* **10**, 103–116.

Högenauer, G. (1979). Tiamulin and pleuromutilin. *In* "Antibiotics" (F. E. Hahn, ed.), Vol. 1, pp. 344–360. Springer-Verlag, Berlin and New York.

Hook, L. A., Corder, R. E., Hamilton, P. T., Frea, J. I., and Reeve, J. N. (1984). Modification and use of an ultra-low oxygen chamber for the growth of methanogens. *In* "Microbial Chemoautotrophy" (W. R. Strohl and O. H. Tuovinen, eds.), Ohio State Univ. Biosci. Colloq. 8, pp. 275–289. Ohio State Univ. Press, Columbus.

Hughes, J., and Mellows, G. (1980). Inhibition of isoleucyl-transfer ribonucleic acid synthetase in *Escherichia coli* by pseudomonic acid. *Biochem. J.* **176**, 305–318.

Jarrell, K. F., and Sprott, G. D. (1983). The effects of ionophores and metabolic inhibitors on methanogenesis and energy-related properties of *Methanobacterium bryantii*. *Arch. Biochem. Biophys.* **225**, 33–41.

Jimenez, A., and Vazquez, D. (1979). Anisomycin and related antibiotics. *In* "Antibiotics" (F. E. Hahn, ed.), Vol. 2, pp. 1–19. Springer-Verlag, Berlin and New York.

Jones, J. B., Bowers, B., and Stadtman, T. C. (1977). *Methanococcus vannielii:* Ultrastructure and sensitivity to detergents and antibiotics. *J. Bacteriol.* **130**, 1357–1363.

Kahan, F. M., Kahan, J. S., Cassidy, P. J., and Kropp, H. (1974). The mechanism of action of fosfomycin (Phosphomycin). *Ann. N.Y. Acad. Sci.* **235**, 364–386.

Kaji, A., and Ryoji, M. (1979). Tetracycline. "Antibiotics" (F. E. Hahn, ed.), Vol. 1, pp. 344–360. Springer-Verlag, Berlin and New York.

Kandler, O., ed. (1982a). "Archaebacteria." Fischer, Stuttgart.

Kandler, O. (1982b). Cell wall structures and their phylogenetic implications. *Zentralbl. Bakteriol., Mikrobiol. Hyg., Abt. 1, Orig. C* **3**, 149–160.

Kandler, O., and König, H. (1978). Chemical composition of the peptidoglycan-free cell walls of methanogenic bacteria. *Arch. Microbiol.* **118**, 141–152.

Kessel, M., and Klink, F. (1981). Elongation factors of the extremely halophilic archaebacterium *Halobacterium cutirubrum*. *Eur. J. Biochem.* **114**, 481–486.

Kucan, Z., and Lipmann, F. (1964). Differences in chloramphenicol sensitivity of cell-free amino acid polymerization systems. *J. Biol. Chem.* **239**, 516–520.

Langworthy, T. A., Tornabene, T. G., and Holzer, G. (1982). Lipids of archaebacteria. *Zentralbl. Bakteriol., Mikrobiol. Hyg., Abt. 1, Orig. C* **3**, 228–244.

Londei, P. (1983). Action of inhibitors of protein synthesis on archaebacterial (*Caldariella acidophila*) ribosomes. *Abstr. Meet. "Antibiot. 83."*

Londei, P., Teichner, A., and Cammarano, P. (1982). Particle weights and protein composition of the ribosomal subunits of the extremely thermoacidophilic archaebacterium *Caldariella acidophila*. *Biochem. J.* **209**, 461–470.

Lughtenberg, E. J. J., van Schijndee-van Dam, A., and van Bellegem, T. H. M. (1971). In vivo and in vitro action of new antibiotics interfering with the utilization of N-acetylglucosamine-N-acetylmuramyl-pentapeptide. *J. Bacteriol.* **108**, 20–29.

McKellar, R. C., and Sprott, G. D. (1979). Solubilization and properties of a particulate hydrogenase from *Methanobacterium* strain G2R. *J. Bacteriol.* **139**, 231–238.

Matsuhashi, M., Ohara, J., and Yoshiyama, Y. (1970). Inhibition of bacterial cell-wall synthesis in vitro by enduracidin, a new polypeptide antibiotic. *Prog. Antimicrob. Anticancer Chemother., Proc. Int. Congr. Chemother., 6th, 1969* Vol. 1, p. 226.

Mescher, M. F., and Strominger, J. L. (1975). Bacitracin induces sphere formation in *Halobacterium* species which lack a wall peptidoglycan. *J. Gen. Microbiol.* **89**, 375–378.

Neidhardt, F. C., Parker, J., and McKeever, W. G. (1975). Function and regulation of aminoacyl-tRNA synthetases in procaryotic and eucaryotic cells. *Annu. Rev. Microbiol.* **29**, 215–250.

Neuhaus, F. (1967). D-cycloserine, as it is observed with eubacteria. *In* "Antibiotics" (D. Gottlieb and P. D. Shaw, eds.), Vol. 1, pp. 40–83. Springer-Verlag, Berlin and New York.

Nevel van, C. J., and Demeyer, D. I. (1977). Effect of monensin on rumen metabolism in vitro. *Appl. Environ. Microbiol.* **34**, 251–257.

Pecher, T., and Böck, A. (1981). In vivo susceptibility of halophilic and methanogenic organisms to protein synthesis inhibitors. *FEMS Microbiol. Lett.* **10**, 295–297.

Perski, H. J., Schönheit, P., and Thauer, R. K. (1982). Sodium dependence of methane formation in methanogenic bacteria. *FEBS Lett.* **143**, 323–326.

Pestka, S. (1975). Chloramphenicol. "Antibiotics" (J. W. Corcoran and F. E. Hahn, eds.), Vol. 3, pp. 370–395. Springer-Verlag, Berlin and New York.

Prangishvilli, D., Zillig, W., Gierl, A., Biesert, L., and Holz, J. (1982). DNA-dependent RNA-polymerases of thermoacidophilic archaebacteria. *Eur. J. Biochem.* **122**, 471–477.

Reisinger, P., Seidel, H., Tschesche, H., and Hammes, W. P. (1980). The effect of nisin on murein synthesis. *Arch. Microbiol.* **127**, 187–193.

Russel, A. D., and Quessel, L. B., eds. (1983). "Antibiotics: Assessment of Antimicrobial Activity and Resistance." Academic Press, London.

Sanz, J. L., and Amils, R. (1983). Sensitivity of thermoacidophilic archaebacteria to protein synthesis inhibitors. *Abstr. Meet. "Antibiot. 83."*

Schmid, G., and Böck, A. (1982a). The ribosomal protein composition of the archaebacterium *Sulfolobus. Mol. Gen. Genet.* **185**, 498–501.

Schmid, G., and Böck, A. (1982b). The ribosomal protein composition of five methanogenic bacteria. *Zentralbl. Bakteriol., Mikrobiol. Hyg., Abt. 1, Orig. C* **3**, 347–353.

Schmid, G., Pecher, T., and Böck, A. (1982). Properties of the translational apparatus of archaebacteria. *Zentralbl. Bakteriol., Mikrobiol. Hyg., Abt. 1, Orig. C* **3**, 209–217.

Skinner, R., Cundliffe, E., and Schmidt, F. J. (1983). Site of action of a ribosomal RNA methylase responsible for resistance to erythromycin and other antibiotics. *J. Biol. Chem.* **258**, 12702–12706.

Slyter, L. L. (1979). Monensin and dichloroacetamide influences on methane and volatile fatty acid production by rumen bacteria in vitro. *Appl. Environ. Microbiol.* **37**, 283–288.

Somma, S., Merati, W., and Parenti, F. (1977). Gardimycin, a new antibiotic inhibiting peptidoglycan synthesis. *Antimicrob. Agents Chemother.* **11**, 396–401.

Stetter, K. O., Winter, J., and Hartlieb, R. (1980). DNA-dependent RNA polymerase of the archaebacterium *Methanobacterium thermoautotrophicum. Zentralbl. Bakteriol., Mikrobiol. Hyg., Abt. 1, Orig. C* **1**, 201–218.

Stone, K. J., and Strominger, J. L. (1971). Mechanisms of action of bacitracin: Complexation with metal ion and C_{55}-isoprenyl pyrophosphate. *Proc. Natl. Acad. Sci. U.S.A.* **68**, 3223–3227.

Tanaka, N. (1975). Fusidic acid. "Antibiotics" (J. W. Corcoran and F. E. Hahn, eds.), Vol. 3, 436–447. Springer-Verlag, Berlin and New York.

Tanaka, N. (1982). Mechanism of action of aminoglycoside antibiotics. *In* "Aminoglycoside Antibiotics" (H. Umezawa and I. R. Hooper, eds.), pp. 221–266. Springer-Verlag, Berlin and New York.

Wallace, B. J., and Davis, B. D. (1973). Cyclic blockade of initiation sites by streptomycin-damaged ribosomes in *Escherichia coli:* An explanation for dominance of sensitivity. *J. Mol. Biol.* **75**, 377–390.

Wehrli, W., Knüsel, F., and Staehelin, M. (1968). Interaction of rifamycin with bacterial RNA polymerase. *Proc. Natl. Acad. Sci. U.S.A.* **61**, 667–671.

Weisburg, W. G., and Tanner, R. S. (1982). Aminoglycoside sensitivity of archaebacteria. *FEMS Microbiol. Lett.* **14**, 307–310.

Wieland, T. (1972). Struktur und Wirkung der Amatoxine. *Naturwissenschaften* **59**, 225–231.

Wolf, H., Chinali, G., and Parmeggiani, A. (1977). Mechanism of the inhibition of protein synthesis by kirromycin. Role of EF-Tu and ribosomes. *Eur. J. Biochem.* **75**, 67–75.

Wolf, H., Assmann, D., and Fischer, E. (1978). Pulvomycin, an inhibitor of protein biosynthesis preventing ternary complex formation between elongation factor Tu, GTP and aminoacyl-tRNA. *Proc. Natl. Acad. Sci. U.S.A.* **75**, 5324–5328.

Wörner, W., and Wolf, H. (1982). Kirromycin-resistant elongation factor Tu from wild-type of *Lactobacillus brevis*. *FEBS Lett.* **146**, 322–326.

Zierhut, G., Piepersberg, W., and Böck, A. (1979). Comparative analysis of the effect of aminoglycosides on bacterial protein synthesis in vitro. *Eur. J. Biochem.* **98**, 577–583.

Zillig, W., Stetter, K. O., and Tobien, M. (1978). DNA-dependent RNA polymerase from *Halobacterium halobium*. *Eur. J. Biochem.* **91**, 193–199.

Zillig, W., Stetter, K. O., and Janekovic, D. (1979). DNA-dependent RNA polymerase from the archaebacterium *Sulfolobus acidocaldarius*. *Eur. J. Biochem.* **96**, 597–604.

Zillig, W., Stetter, K. O., Schäfer, W., Janekovic, D., Wunderl, S., Holz, I., and Palm, P. (1981). *Thermoproteales:* A novel type of extremely thermoacidophilic anaerobic archaebacteria isolated from Icelandic Solfataras. *Zentralbl. Bakteriol., Mikrobiol. Hyg., Abt. 1, Orig. C* **2**, 205–227.

Zillig, W., Stetter, K. O., Schnabel, R., Madon, J., and Gierl, A. (1982). Transcription in archaebacteria. *Zentralbl. Bakteriol. Mikrobiol. Hyg., Abt. 1, Orig. C* **3**, 218–227.

CHAPTER 13

Genome Structure in Archaebacteria

W. FORD DOOLITTLE

I. Introduction

Arguments for an early divergence of archaebacterial genomic lineages from eubacterial genomic lineages, and of either from that genomic lineage now resident within eukaryotic nuclei, are many and compelling (Woese, 1981, 1982; Woese and Fox, 1977; Fox et al., 1980; Doolittle, 1980). We may never be able to decide (1) whether archaebacterial and eukaryotic nuclear genomic lineages share a common ancestor more recent than that which either shares with eubacterial genomic lineages, (2) whether all divergences are so ancient as to preclude construction of an experimentally verifiable cladogram for these three "primary kingdoms," or (3) whether frequent genetic exchanges early in the divergence of these kingdoms render questions of branching order meaningless (Woese, 1982; Bayley, 1982). We can, nevertheless, be sure that many of the features that now distinguish eubacterial, eukaryotic nuclear, and archaebacterial lineages represent independently achieved solutions to evolutionary problems as yet unsolved by their common "progenotic" ancestor (Doolittle, 1980; Woese, 1982). Many of these distinguishing features involve the structure and function of the machineries of transcription and translation, and may be interpreted to mean that this

Copyright © 1985 by Academic Press, Inc.
All rights of reproduction in any form reserved.
ISBN 0-12-307208-5

progenotic ancestor was, at the time of the divergence of the primary kingdoms, still under strong selection pressure to increase the accuracy and speed with which hereditary information is transferred from gene to protein.

Thus, in determining how archaebacterial genomes and genes are organized and expressed and how that expression is regulated, we can expect to learn not only something about the variety of possible evolutionary solutions to problems of coupling genotype to phenotype, but also, through comparisons of the molecular biologies of the three kingdoms, something about the more primitive molecular biology of their ancient common ancestor. We can also expect to find much that we cannot readily understand. We can hope that, in our attempts to understand, we will develop new and more general ways of thinking about evolutionary molecular biology.

The development of archaebacterial molecular biology as a discipline may follow a somewhat idiosyncratic course for three reasons. First, there are as yet no known routes for "classical" genetic analysis. Information on gene structure and genome organization will come first, if not exclusively, through the application of recombinant DNA techniques. Second, much of this information is likely to come first from the halobacteria, because these are the most easily manipulated of the archaebacteria. The extent to which it can be generalized to more remote archaebacterial taxa, in particular to the phylogenetically diverse thermoacidophiles, will remain for some time a vexing and unsolved problem. Third, our attempts to interpret the emerging data within existing intellectual frameworks provided by our knowledge of eubacterial and eukaryotic molecular biologies will almost certainly lead to basic misunderstandings. These will only be cleared up when we gain a broader appreciation for just how different the archaebacteria really are.

II. Overall Genome Structure

A. Genome Size

Results of renaturation kinetic analysis have been reported for only a limited number of archaebacteria. *Halobacterium* species (Moore and McCarthy, 1969a) appear to have sequence complexities comparable to that of *Escherichia coli* [2.5 $\times 10^9$ daltons, or approximately 4×10^3 kilobase pairs (kbp)]. *Thermoplasma acidophilum* shows an apparent genome size of 0.8×10^9 daltons, within the range exhibited by the "degenerate" eubacterial mycoplasmas, with which it was earlier and erroneously grouped (Searcy and Doyle, 1975). The genome of *Methanobacterium thermoautotrophicum* appears a bit larger than this (1.1×10^9 daltons) and is the only archaebacterial genome that shows, in simple renaturation kinetic analyses, any substantial fraction of repetitive DNA (6%) (Mitchell *et al.*, 1979).

B. BASE COMPOSITION

As expected for an ancient group with deep internal phylogenetic divergences, archaebacteria exhibit a broad spectrum of chromosomal G + C contents, with the methanogens themselves spanning the range from 21 to 61 mol% G + C (Balch *et al.*, 1979). If we include halobacteria within the methanobacterial lineage, as 16 S rRNA catalog analyses indicate that we should (Woese and Fox, 1977; Fox *et al.*, 1980), then the higher value is extended further (up to 64–68 mol%) (Moore and McCarthy, 1969a,b; Soliman and Trüper, 1982). Thus, the diversity in G + C contents shown by this apparently monophyletic assemblage is nearly as great as that shown by all the prokaryotes (Stanier *et al.*, 1976). Other major groups of archaebacteria so far recognized—the Thermoplasmales (sole member *Thermoplasma*), the Sulfolobales (*Sulfolobus*), and the Thermoproteales (*Thermoproteus* and *Desulfurococcus*)—show G + C contents between 46 and 60 mol%, although this more limited range may reflect nothing more than the as yet limited number of studied members of these groups (Searcy and Doyle, 1975; Zillig *et al.*, 1982).

C. "SATELLITES" AND PLASMIDS

To my knowledge, only halobacterial DNA has been shown to be resolvable, upon CsCl equilibrium density-gradient centrifugation, into components of substantially different G + C content. Moore and McCarthy (1969a,b) identified, in species of *Halobacterium* and *Halococcus,* both a "chromosomal" component of 66 to 68 mol% G + C and a "satellite" component of 57 to 60 mol% G + C. This latter fraction comprised from 11 to 36% of the total DNA of the strains examined. Because halobacterial DNA behaved as single copy in renaturation kinetic experiments, the authors concluded that the satellite fraction did not represent the DNA of a multicopy plasmid. Instead, they conjectured that halobacterial chromosomes might contain one or more regions of relatively low G + C content, which were the products of whole or piecemeal integration of a foreign plasmid introduced into halobacterial cells early in their evolutionary history. Ironically, subsequent work has shown that halobacteria *do* contain extrachromasomal, covalently closed, circular DNAs of relatively low G + C content, which we must perforce call plasmids, and yet it has also shown that chromosomal integration occurs quite readily now and, presumably, has been common throughout the evolution of the halobacteria.

Although Lou (cited in Bayley and Morton, 1978) detected covalently closed, circular DNAs in halobacteria, the first published characterization of halobacterial (or indeed, archaebacterial) plasmids was provided by Simon in 1978. Simon identified, in *Halobacterium salinarium,* three distinct covalently closed, circular DNAs. The spontaneous loss of one of those was (in four of four instances) associated with the loss of the ability to produce gas vacuoles. Weidinger *et al.* (1979) described in some detail the major plasmid (pHH1) of *Halobacterium halobium.* This 150 kbp, covalently closed, circular DNA is, as expected, of relatively low (approximately 59 mol%) G + C content but does not account for all of the low G + C satellite DNA. A fraction of the rest appears to be contained in a collection of small heterogeneous circular DNAs, but much of it must also represent regions of relatively low G + C content within the "chromosome" (or, at least, within that fraction of DNA not easily isolated as covalently closed circles; Weidinger *et al.,* 1979; Pfeifer *et al.,* 1982).

The plasmids of "wild-type" *Halobacterium halobium* and plasmids borne by various spontaneous mutants of this strain have been relatively well characterized physically, although it is our view that the high levels of spontaneous rearrangement which these plasmids (and the halobacterial genome in general) suffer ensure that plasmid restriction maps constructed by different laboratories, or by the same laboratory with different plasmid preparations, will not prove identical. Pfeifer *et al.* (1981a, 1982) mapped pHH1 from the "wild-type" *H. halobium* strain NRC817 (from the collection of the National Research Council in Ottawa), with *Pst*I, *Eco*RI, and *Hin*dIII, and showed it to differ from the similarly sized (about 150 kbp) plasmid pHH2 of a wild-type *H. halobium* strain DSM 670 (from the Deutsch Sammlung for Mikroorganismen in Gottingen) by four insertions or deletions ranging in size from <0.8 to >6 kbp. The gas vacuole-deficient *H. halobium* variant DSM 671 bears a smaller (75 kbp) plasmid—pHH3.

Halobacterium halobium strain R1, obtained some 14 years ago as a spontaneous gas vacuole-deficient variant of the Ottawa "wild-type" strain NRC-1 (whose relationship to NRC817 is unclear) bears an extensively rearranged 50–55 kbp plasmid. *Hin*dIII digestion of this plasmid yields fragments of 40, 8.0, and 7.1 kbp (C. Sapienza, unpublished). Restriction endonuclease digestions of the latter two fragments show them to be identical except for an internal region of some 0.9 kbp, which represents a copy of a repetitive element (1SH50) with typical transposon-like structure (see below). One could assume that during the derivation of the contemporary R1 plasmid from that of its parent NRC-1, there has been both a duplication of a 7.1 (or 8.0)-kbp fragment and the insertion (or deletion) of a copy of 1SH50 into (or from) one of the duplicates. The order of these events, their relationship to other rearrangements, and the role of such rearrangements in the apparently permanent loss of genetic determinants for gas vacuole production remain mysterious.

Pfeifer and collaborators (1981a, 1982) also find an approximately 120 kbp plasmid-bearing sequence homologous to pHH1 in *H. cutirubrium,* and two plasmids each (120 and 6 kbp, and 90 and 6 kbp, respectively) in strains of *H. trapanicum* and *H. volcanii.* The latter exhibit no detectable homology to plasmids in any others strains. All plasmid-bearing strains with the exception of *H. trapanicum* contain a relatively low G + C "satellite" DNA component, which can be separated from the majority of the DNA on malachite green-bisacrylamide columns (Weidinger *et al.,* 1979; Pfeifer *et al., 1982).*

Although some isolates of *H. salinarium* and *H. tunisiensis* examined by these workers contained no detectable covalently closed, circular DNA, sequences homologous to pHH1 were found in total DNA by Southern blot analysis. It is of interest that DSM 670, as first obtained from the Deutsch Sammlung for Mikroorganismen, was devoid of plasmid. However, various subclones derived from this strain contained a variety of covalently closed, circular DNAs with homology to pHH1. Integration and excision of plasmid-related low G + C DNA, either intact or in pieces, seems, as predicted by Moore and McCarthy, a not uncommon event.

Even more intriguing in this context is the situation recently described for *Sulfolobus acidocaldarius* (strain B12) by Yeats *et al.* (1982). DNAs prepared from batch cultures contain covalently closed, circular DNAs of some 15 kbp, but these are normally recoverable only in very low and variable quantities. Ultraviolet irradiation results, after a 2–3 hr lag, in substantial increases in plasmid yield. When only a single ultraviolet light induction is used, growth and total DNA synthesis proceed normally. If, however, the same dose of ultraviolet light is given at three successive intervals separated by 90 min, cell viability is reduced to 1%, chromosomal DNA is degraded, and plasmid DNA recovery is greatly enhanced. Similar irradiation of *Sulfolobus* strain DSM 639 produces no plasmid and lacks lethal effect. Yeats *et al.* (1982) interpret this to mean that (1) plasmid DNA is normally integrated within the chromosome of *Sulfolobus acidocaldarius* B12, (2) the low spontaneous yields of plasmid reflect low rates of spontaneous excision and autonomous replication, (3) excision and replication are ultraviolet light-inducible and result directly or indirectly in the destruction of chromosomal DNA, but (4) susceptibility to induction depends upon cell cycle events, so that only a fraction of an asynchronously growing population is inducible at any one time. Southern blot experiments using labeled plasmid and total DNA indicate that the entire plasmid is normally integrated in a unique orientation at a unique chromosomal site. Yeats *et al.* (1982) also suggest that the extreme variation in colony site exhibited by *S. acidocaldarius* plated on 10% starch gel medium may be due to spontaneous plasmid "induction" in some but not all clones.

All this sounds very much like prophage induction. However, there is no substantial qualitative change in RNA and protein synthesis after induction, no release of plasmid DNA into the medium, and no production of particles capable of infecting either uninduced B12 or DSM639. This plasmid may be the genome of a defective phage, or a phage for which there is no available sensitive host. However, there seems as yet not strong reason to adopt this interpretation, and the behavior of presumed nonviral plasmid DNAs in halobacteria makes it seem less attractive.

Extrachromosomal DNAs of as yet unknown nature have been described in two methanogens (G. Howard, cited in Reeve *et al.*, 1982). It seems likely that the possession of plasmids, and a variety of intimate relationships between plasmid and chromosomal DNAs, will prove a general characteristic of archae-bacteria.

D. CHROMATIN

Searcy and collaborators have isolated and sequenced a histone-like DNA-binding protein (HTa) from *Thermoplasma acidophilum* (Stein and Searcy, 1978; Searcy and Stein, 1980; DeLange *et al.*, 1981; Searcy, 1982). This protein shows statistically significant homology to eukaryotic nuclear histones, as well as to the DNA-binding protein HU of *E. coli*. Searcy suggests that the *Thermoplasma* protein is more like eukaryotic nuclear than eubacterial DNA-binding proteins, because it condenses DNA into globular particles containing a core of four HTa subunits and a 40-bp loop of DNA, whereas *E. coli* HU forms much larger complexes, with some 16–20 protein molecules associated with about 275 bp of DNA. Perhaps more significant is the observation that HTa binds DNA under conditions similar to those found *in vivo* and protects it from nuclease digestion, whereas HU does neither (Stein and Searcy, 1978).

Thomm *et al.* (1982) have described DNA-binding proteins from *Methanobacterium thermoautotrophicum* and several species of *Sulfolobus*. None of these showed immunological cross-reactivity with *Thermoplasma* HTa, whereas the DNA-binding proteins of a variety of gram-negative and gram-positive eubacteria do react with antisera prepared against *E. coli* HU. Although HTa and the nucleosome-like structures that it presumably forms *in vivo* are, as Searcy (1982) states, "suggestively eukaryotic," it remains unclear whether this reflects homology or convergence. It is even less clear to what extent archaebacterial (or indeed eubacterial) genomes resemble, *in vivo*, the chromosomes of eukaryotic cells.

III. Genome Instability

A. MUTABILITY AND VARIATION IN HALOBACTERIA

Spontaneous mutations affecting the production of gas vacuoles, bacterioruberin and (under some conditions) bacteriorhodopsin result in striking alterations in the appearance of colonies of *Halobacterium halobium*. Such mutations occur at frequencies of between one in 10^4 and one in 10^2 cells plated. Spontaneous reversions at nearly the same frequencies are often observed with such "mutants." Weber and Leighton (1982) indicated that mutations to a variety of auxotrophic characters and to temperature sensitivity occur at similarly high frequencies and are also reversible. Even more surprising is the observations by these authors (Pfeifer *et al.* (1981b) and by C. Sapienza (unpublished) of occasional sectored colonies in which the frequency of spontaneous pigment mutants is increased perhaps 100-fold. Such "hypervariable mutants" represent about 1 in 10^4 of all cells plated, and the hypervariability is retained in at least some subclones, although it, too, appears unstable. Since available evidence (see below) suggests that many or most spontaneous mutations in *H. halobium* result from the insertion and/or excision of mobile repetitive DNA sequence elements, the existence of such hypervariable mutants may suggest that mobility itself is under genetic control.

B. PHYSICAL REARRANGEMENTS IN HALOBACTERIAL PLASMID DNAs

Although earlier work from Goebel's laboratory (Weidinger *et al.*, 1979) provided evidence for a correlation between loss of the ability to produce gas vacuole protein and specific insertions into the DNA of pHH1, further work from this group showed that plasmid rearrangements are frequent and complex, so that it is difficult to establish any simple relationship between them and any single phenotypically expressed mutation. In an analysis of the covalently closed, circular DNAs of a number of gas vacuole and/or pigment-deficient strains, Pfeifer *et al.* (1981b) found (1) that all pigment-deficient strains except those unable to produce retinal exhibited changes in plasmid DNA, (2) that most changes represented insertions of segments of DNA ranging in size from 0.35 to 2.1 kbp, with those of the smallest size being the most frequent, (3) that many plasmids had suffered multiple insertions, and (4) that revertants to wild-type had either lost the initial insertion or experienced additional insertions and/or deletions. These authors found no rearrangements in the plasmids of 10 single colonies of apparent "wild" phenotype, although we (Sapienza and Doolittle, 1982; Sapienza *et al.*, 1982) have shown that phenotypically silent rearrangements of both plasmid and chromosomal DNA do indeed occur.

C. PHYSICAL REARRANGEMENTS IN THE DNA OF PHAGE ϕH

Schnabel and co-workers (1982) have shown that the bacteriophage ϕH suffers a variety of structural rearrangements during lytic growth in *H. halobium*. In their complexity and high frequency of occurence, these rival those experienced by normally resident extrachromosomal (plasmid) DNAs. The genome of this phage is a partially circularly permuted, terminally redundant, double-stranded linear, 39 kbp DNA of 65 mol% G + C. Regions of this genome show strong sequence homology to a limited number of restriction endonuclease-generated fragments of both plasmid and chromosomal DNAs of *H. halobium,* which may indicate genetic exchange early in the evolution of the phage–host relationship.

It is clear that such exchange still occurs during normal lytic growth. Phage DNA prepared from mass lysates produces, upon restriction endonuclease digestion, both major and minor (substoichiometric) fragments. DNA from phage purified from single plaques shows few if any minor fragments, and yet DNA purified from different plaques may show different patterns of major fragments.

All these observations can be explained as the consequence of rearrangements and/or insertions and deletions that occur during prolonged lytic growth. In further characterizing this variability, Schnabel *et al.* (1983) examined the DNAs from some 71 single plaque isolates, and they found among them six distinct classes of variant genome structure. One variant, ϕH2, appeared to have suffered an insertion of 1.8 kbp at position 51 (of a nominal 100) on the ϕH1 restriction map. A second, ϕH3, had experienced a 150-bp deletion at position 35. A third, ϕH5, showed both this small deletion, the 1.8-kbp insertion characteristic of ϕH2, and an inversion of 11.2 kbp of DNA lying immediately beyond (positions 52–71) this insertion. Variants ϕH4 and ϕH6 had suffered 200- and 50-bp insertions, respectively, at similar sites around position 32. Southern hybridization experiments showed that the 1.8-kbp insertion of variants ϕH2 and ϕH5 is present in a single copy at position 72 in wild-type ϕH1 and is also represented twice in the genome of uninfected *H. halobium.* Restriction endonuclease and electron microscopic studies indicated that, in ϕH5, the two 1.8 kbp regions are in inverted orientation, flanking the invertible 11.2 kbp sequence.

The ease with which such inversions and deletions are detected in randomly selected single plaque isolates argues for their high frequency of occurrence and makes it seem likely that many other events of this sort occur but escape detection because they disrupt the lytic cycle. It is interesting in this context that some phage-resistant strains surviving ϕH infection appear to bear even more extensive and unstable modifications of phage genetic information (Schnabel *et al.,* 1982, 1983; W. Zillig, personal communication).

D. PHYSICAL REARRANGEMENTS IN THE BACTERIORHODOPSIN GENE

Genetic rearrangements are more easily detected in relatively small autonomously replicating DNAs, which can be readily isolated from chromosomal

DNA. However, there is both physical and genetic evidence that insertions, deletions, and perhaps other rearrangements affect chromosomal sequences. Most of this evidence comes from the analysis of mutants of the only characterized chromosomal gene—that for the apoprotein of *H. halobium* bacteriorhodopsin. This gene has been cloned into pBR322 by Chang *et al.* (1981) and by Betlach *et al.* (1983), and both protein-coding and noncoding flanking regions have been sequenced by Dunn *et al.* (1981).

Spontaneous (and revertible) mutants that fail to produce bacteriorhodopsin have been characterized by both groups. The most thoroughly studied are those that arose, on at least two independent occasions, from the insertion of a now fully sequenced transposable element ISH1 (see below). Even though ISH1 is oppositely oriented in the two mutants, insertion occurred at the same site. The target sequence, AGTTATTG, comprises positions 8–15 of the bacteriorhodopsin coding sequence and was duplicated (to give "flanking direct repeats") in both cases. It is of interest that the transcript of this octameric sequence is precisely looped out in the secondary structural model for bacteriorhodopsin mRNA proposed by Dunn *et al.* (1981). It is also of interest that neither insertion mutant produces bacteriorhodopsin mRNA detectable by Northern hybridization with the labeled bacteriorhodopsin coding sequence. ISH1 must, in either orientation, interrupt bacteriorhodopsin gene transcription or destabilize the bacteriorhodopsin transcript (U. L. RajBhandary, personal communication). These workers have also found, but have not yet described, mutants resulting from the insertion of an approximately 0.5 kbp element into the middle of the bacteriorhodopsin coding sequence.

Betlach *et al.* (1983) described a greater variety of bacteriorhodopsin mutants, using a cloned 5.1-kbp *Pst*I fragment containing the bacteriorhodopsin coding sequence as a Southern hybridization probe. Of 12 mutants examined, only one bore no detectable insertion or deletion. One showed a 350-bp insertion near the middle of the bacteriorhodopsin coding region, seven bore insertions comparable in size and probably in position to the ISH1 insertions characterized by Dunn *et al.* (1981), and an eighth carried a 900-bp insertion in this region. Another mutant bore a 450-bp insertion at least 300 bp transcriptionally upstream from the bacteriorhodopsin-coding region. More surprisingly, two had experienced insertions (of 600 or 3000 bp) at least 1400 bp transcriptionally upstream. Phenotypic revertants of these latter retained the original insertion and gained additional insertions within the same region.

The fact that insertions in regions so remote from the bacteriorhodopsin coding region can result in failure to produce this protein may mean, as Betlach *et al.* (1983) suggest, that this region is transcribed as part of a much larger mRNA, from an as yet unidentified promoter. Dunn *et al.* (1981) found no sequences bearing obvious homology to either eubacterial or eukaryotic promoters within the first 340 bp upstream from the bacteriorhodopsin coding region. They did find, by S1 nuclease mapping, that the 5′ terminus of what would appear to be mature bacteriorhodopsin mRNA *in vivo* maps only three bp 5′ to the translation-initiating ATG. This terminus may, however, be the product of rapid *in vivo* processing of a much larger primary transcript, rather than the product of an as yet unidentified nearby promoter.

E. Agents of Genomic Instability

It appears that much of the genetic and physical instability of halobacterial extrachromosomal and chromosomal DNA reflects specific and/or nonspecific insertions and deletions occuring at high frequences. In eubacteria and eukaryotes, such events most often reflect the activities of transposable elements (Kleckner, 1981). In both, such elements are often members of families of dispersed repetitive sequences, often show characteristic structural features (the possession of terminal inverted repeats, which may be part of longer direct or inverted repeats) and characteristically generate, upon insertion, duplications of sequences present only once at the target of insertion.

From the little information already available, it is possible to conclude that halobacterial genomes do indeed owe much of their instability to the presence of repetitive sequences, some of which look and behave like transposable elements, many or all of which effect or are affected by genomic rearrangements of astonishingly high frequency, and some of which may possess uniquely archaebacterial features.

1. The Presence of Mobile Repetitive Sequences in Halobacteria

Through the simple expedient of probing total *H. halobium* DNA with small, randomly cloned genomic fragments, we were able to show that the genome of *H. halobium* contains at least 500 repetitive elements. Comparisons of hybridization patterns indicated that these comprise at least 50 different families, with from 2 to about 20 members each. Individual members of different families are often clustered, and repeat sequences seem in general to be confined to regions of the genome of relatively low G + C content. Nevertheless, it is possible to show that several families have members present on both chromosome and plasmid. It can also be demonstrated that repetitive elements, even if often clustered, are distributed (either randomly or regularly) over at least one-fourth to one-half of the entire length of the *H. halobium* genome (Sapienza and Doolittle, 1982).

Similar experiments with cloned DNAs from *H. volcanii* show that this genome also contains multiple repetitive sequences, although the number of different families in this genome may be somewhat smaller and the number of members in each family may be somewhat larger. Almost all families represented in *H. volcanii* have homologs detectable by Southern hybridization in *H. halobium,* although the converse is not true. In a broader survey, C. Sapienza (unpublished) found that many (although probably in no case all) of the repetitive sequence families in the genome of *H. halobium* are homologous to repetitive sequence families in the genomes of *H. volcanii, H. salinarium, H. saccharovorum, H. trapanicum,* and *H. vallismortis.*

It is possible to demonstrate that repetitive elements identified in this way effect or are affected by genomic rearrangements and to measure the frequency of such rearrangements, again through Southern blot hybridization with cloned members of different families. Total *Eco*RI-digested DNAs from 19 isolates derived from a single *H. halobium* cell and maintained separately for 34 (or in four cases, 215) generations were individually probed with cloned members of seven distinct repetitive sequence families. "Events" (losses or gains, in single isolates, of hybridizing genomic fragments) were detected with all probes, at an average frequency of 0.004 events/family/cell generation (Sapienza *et al.,* 1982). It is impossible to determine what fraction of these "events" represents bona fide transposition, although it seems likely, from the results discussed above, that many of them do. If this is so, then frequencies of transposition in *H. halobium* are as high or higher than those reported for any eubacterial transposable element (Kleckner, 1981). This apparent high frequency, together with the sheer number of such mobile elements in the halobacterial genome, implies an extraordinary degree of genomic instability. At these frequencies, two daughter cells have only an 80% chance of bearing physically identical genomes.

2. SEQUENCED ELEMENTS

The 1118 bp insert, ISH1, that was found responsible on at least two independent occasions for inactivation of the *H. halobium* bacteriorhodopsin gene, has been completely sequenced (Simsek *et al.,* 1982). It bears imperfect inverted terminal repeats of the form TGCCTTGTT- - -(1110 bp)- - -CAACGAGGCA. It contains an open reading frame of 810 nucleotides on one strand and an open reading frame of 402 nucleotides on the other. At least the first appears to be transcribed, because an RNA of 900 nucleotides can be detected in total cellular RNA by Northern hybridization with cloned ISH1 sequences as probes. Different *H. halobium* strains bear different numbers (from one to five) of copies of ISH1, and the amount of this RNA varies proportionately with copy number. Its size does not vary with either copy number or position, suggesting that transcription may be both initiated and terminated at sites within the element.

W.-L. Xu (unpublished) has completed 80–90% of the sequence of the 900-bp

element ISH50 found in one of the two otherwise identical HindIII fragments of the 50–55 kbp plasmid of H. halobium strain R1 referred to above. This element bears 10 bp imperfect inverted terminal repeats of the form CGCTCTTGGG- -
-(about 900 bp)- - -CACAAGAGCG, and it is flanked by the directly repeated octamer TTGTGGAAT. This "target" sequence is present only once in the HindIII fragment lacking the insertion. Since some 100–200 bp of internal sequence remain to be determined, it is premature to scan for open reading frames. Nevertheless, this element will probably prove similar in structure (though not in sequence) to ISH1.

IV. Approaches to the Analysis of Genome Organization and Gene Structure

In the absence of known "natural" mechanisms for genetic exchange between archaebacteria, genetic analysis must proceed via indirect routes. Some of these are, of course, quite powerful, and several approaches to physical characterization, sequence analysis, and functional dissection of archaebacterial genes have already proven their utility.

A. BACTERIORHODOPSIN

The bacteriorhodopsin apoprotein structural gene was cloned, on two independent occasions, through the use of labeled synthetic oligonucleotides as probes for colony hybridization with E. coli "shotgun" pBR322 transformants. The sequences of the potentially useful oligonucleotide probes were dictated in the first instance by the known amino acid sequence of the protein (Chang et al., 1981; Dunn et al., 1981).

B. RIBOSOMAL RNA

Ribosomal RNA (rRNA) structural genes have been physically mapped in several species and were cloned from Thermoplasma acidophilum (Tu and Zillig, 1982), Sulfolobus solfataricus (M. Nuell, unpublished), and Halobacterium volcanii (A. Yuki, unpublished) using the labeled RNAs as probes in genomic Southern blot or colony hybridization experiments. Results to date show an unexpected mix of similarities and differences in rRNA gene organization within and between archaebacteria and eubacteria. Halobacterium halobium has but a single rRNA cluster, with rRNA genes arrayed and probably co-transcribed in the order 5'-16S-23S-5S-3' (Hofman *et al.*, 1979). Preliminary mapping indicates that the total length of the cluster is no more than 14 kbp, and more likely between 5 and 6 kbp.

Halobacterium volcanii appears to have two such rRNA gene clusters. C. R. Woese and collaborators (personal communication) have cloned two *Eco*RI fragments (total length 13.5 kbp) within which genes for 16 S, 23 S, and 5 S rRNAs genes are clustered, in that order, within 6 to 7 kbp (J. D. Hofman, unpublished). *Eco*RI digested genomic DNA shows two fragments hybridizing to the 3'-terminal portion of 23S and to 5S rRNA, but only one hybridizing to 16 S rRNA. However, genomic DNA digested with other restriction endonucleases shows the expected doublets hybridizing to 16 S. C. J. Daniels (unpublished) has cloned a *Pst*I fragment that contains regions hybridizing with 23 S and 5 S rRNAs. This fragment has, within these coding regions, a restriction map very similar to that of Woese's cloned fragment, but it shows no restriction site homology to it in regions 3' to the 5 S coding region; presumably it derives from the "second" rRNA gene cluster.

In contrast to these typically "eubacterial" patterns of rRNA gene organization, *Thermoplasma acidophilum* shows no close linkage of genes for 16 S, 23 S, and 5 S rRNA (Tu and Zillig, 1982). Quantitative hybridization experiments indicate that there is (as in *H. halobium* and the eubacterial mycoplasmas) but a single copy of each rRNA gene. No fragments of genomic DNA generated by any of six restriction endonucleases with 6-bp recognition sites hybridize to more than a single rRNA. If *Thermoplasma* rRNA genes form a single cluster of the order 16S-23S-5S, then the intergenic spacers must be *at least* 7.5 and 6 kbp, respectively. Thus the total cluster, if it exists, must be at least 18 kbp in length. This minimum estimate is based on the assumption of misleadingly nonrandom distribution of restriction endonuclease sites; there is in fact no evidence for any clustering of rRNA genes in *Thermoplasma*. Equally "atypical" dispositions of rRNA genes many be found among the methanogens and Thermoproteales (W. Zillig, personal communication).

C. Genes for Unknown Proteins

A more courageous approach to the development of a surrogate archaebacterial genetics has been taken by Bollschweiler and Klein (1982) and by Hamilton and Reeve (1983). The former have cloned random *Hin*dIII fragments of *Methanobrevibacter arboriphilus* into the *Hin*ddIII site of a pBR322 derivative, and looked for synthesis of novel polypeptides following introduction of the recombinant plasmids into *E. coli*. In two cases, such polypeptides were found, and their syntheses and molecular weights were independent of the orientation of the cloned fragment, when this was reversed by recloning. This, together with the fact that the novel polypeptides could easily be encoded within fragments of the sizes cloned and the fact that their expression was not coupled with the activity of *lac* and *tet* promoters resident on the cloning vectors, suggests that transcription and translation were both initiated and terminated from sites within the cloned archaebacterial DNA.

In a more extensive study, Hamilton and Reeve (1984) have obtained similar results with DNA from *Methanosarcina barkeri* and *Methanobrevibacterium smithii*. In two instances, recombinant clones producing novel polypeptides (in minicells) were selected from among those that could complement, *in trans*, mutants at the *argG* and *purE* loci of *E. coli*.

It is difficult to escape the conclusion that archaebacterial genes can be transcribed and translated into functional protein by *E. coli* RNA polymerase and the *E. coli* translation apparatus. Given the possession by archaebacterial 16S rRNAs of a properly positioned Shine–Dalgarno 3′-terminal sequence (Woese and Fox, 1977) and the near certainty that archaebacteria and eubacteria share a common genetic code, the latter is not completely unexpected (Bayley and Morton, 1978). However, the absence of anything resembling a eubacterial promoter in either ISH1 [which appears to be transcribed from an element-internal site (Simsek *et al.*, 1982)] or in the anticipated region 5′ to the bacteriorhodopsin apoprotein structural gene make the observed apparent transcription by *E. coli* RNA polymerase a surprise. It is clearly essential to identify, sequence, and compare promoter sites recognized by archaebacterial and eubacterial RNA polymerase on both heterologous and homologous DNAs.

ACKNOWLEDGMENTS

We are grateful to W. Zillig, U. L. RajBhandary, M. Betlach, J. Reeve, and C. R. Woese for communication of unpublished results, and to the Medical Research Council of Canada for support.

REFERENCES

Balch, W. E., Fox, G. E., Magrum, L. J., Woese, C. R., and Wolfe, R. S. (1979). Methanogens: Reevaluation of a unique biological group. *Microbiol. Rev.* **43**, 260–296.

Bayley, S. T. (1982). Problems in tracing the early evolution of cells as illustrated by the archaebacteria and particularly by the halobacteria. *In* "Archaebacteria" (O. Kandler, ed.), pp. 65–68. Fischer, Stuttgart.

Bayley, S. T., and Morton, R. A. (1978). Recent developments in the molecular biology of extremely halophilic bacteria. *CRC Crit. Rev. Biochem.* **6**, 151–205.

Betlach, M., Pfeifer, F., Friedman, J., and Boyer, H. W. (1983). Bacterio-opsin mutants of *H. halobium*. *Proc. Natl. Acad. Sci. U.S.A.* **80**, 1416–1420.

Bollschweiler, C., and Klein, A. (1982). Polypeptide synthesis in *Escherichia coli* directed by cloned *Methanobrevibacter arboriphilus* DNA. *In* "Archaebacteria" (O. Kandler, ed.), pp. 101–109. Fischer, Stuttgart.

Chang, S. H., Majundar, A., Dunn, R., Makabe, O., RajBhandary, U. L., Khorana, H. G., Ohtsuka, E., Tanaka, T., Taniyama, Y. O., and Ikehara, M. (1981). Bacteriorhodopsin: Partial sequence of mRNA provides amino acid sequence in the precursor region. *Proc. Natl. Acad. Sci. U.S.A.* **78**, 3398–3402.

DeLange, R. J., Williams, L. C., and Searcy, D. G. (1981). A histonelike protein (HTa) from *Thermoplasma acidophilum* II. Complete amino acid sequence. *J. Biol. Chem.* **256**, 905–911.

Doolittle, W. F. (1980). Revolutionary concepts in evolutionary cell biology. *Trends Biochem. Sci.* **5**, 146–149.

Dunn, R., McCoy, J., Simsek, M., Majundar, A., Chang, S. H., Rajbhandary, U. L., and Khorana, H. G. (1981). The bacteriorhodopsin gene. *Proc. Natl. Acad. Sci. U.S.A.* **78**, 6744–6748.

Fox, G. E., Stackebrandt, E., Hespell, R. B., Gibson, J., Maniloff, J., Dyer, T. A., Wolfe, R. S., Balch, W. E., Tanner, R. S., Magrum, L. J., Zablen, L. B., Blakemore, R., Gupta, R., Bonen, L., Lewis, B. J., Stahl, D. A., Luehrsen, K. R., Chen, K. N., and Woese, C. R. (1980). The phylogeny of prokaryotes. *Science* **209**, 457–463.

Hamilton, P. T., and Reeve, J. N. (1984). Cloning and expression of archaebacterial DNA from methanogens in *Escherichia coli*. *In* "Microbial Chemoautotrophy" (W. R. Strohl and O. H. Tuovinen, eds.), pp. 291–308. Ohio State Univ. Press, Columbus.

Hofman, J. D., Lau, R. H., and Doolittle, W. F. (1979). The number, physical organization and transcription of ribosomal RNA cistrons in an archaebacterium: *Halobacterium halobium*. *Nucleic Acids Res.* **7**, 1321–1333.

Kleckner, N. (1981). Transposable elements in prokaryotes. *Ann. Rev. Genet.* **15**, 341–404.

Mitchell, R. M., Loeblich, L. A., Klotz, L. C., and Loeblich, A. R. (1979). DNA organization of *Methanobacterium thermoautotrophicum*. *Science* **204**, 1082–1084.

Moore, R. L., and McCarthy, B. J. (1969a). Characterization of the deoxyribonucleic acid of various strains of halophilic bacteria. *J. Bacteriol.* **99**, 248–254.

Moore, R. L., and McCarthy, B. J. (1969b). Base sequence homology and renaturation studies of the deoxyribonucleic acid of various strains of extremely halophilic bacteria. *J. Bacteriol.* **99**, 255–262.

Pfeifer, F. G., Weidinger, G., and Goebel, W. (1981a). Characterization of plasmids in halobacteria. *J. Bacteriol.* **145**, 369–374.

Pfeifer, F. G., Weidinger, G., and Goebel, W. (1981b). Genetic variability in *Halobacterium halobium*. *J. Bacteriol.* **145**, 375–381.

Pfeifer, F., Ebert, K., Weidinger, G., and Goebel, W. (1982). Structure and function of chromosomal and extrachromosomal DNA in halobacteria. *In* "Archaebacteria" (O. Kandler, ed.), pp. 110–119. Fischer, Stuttgart.

Reeve, J. N., Tron, N. J., and Hamilton, P. T. (1982). Beginning genetics of methanogens. *In* "Genetic Engineering of Microorganisms for Chemicals" (A. Hollaender, R. D. DeMoss, S. Kaplan, J. Konisky, D. Savage, and R. S. Wolfe, eds.), pp. 233–244. Plenum, New York.

Sapienza, C., and Doolittle, W. F. (1982). Unusual physical organization of the halobacterial genome. *Nature (London)* **395**, 384–389.

Sapienza, C., Rose, M. R., and Doolittle, W. F. (1982). High frequency genomic rearrangement involving archaebacterial repeat sequence elements. *Nature (London)* **299**, 182–185.

Schnabel, H., Zillig, W., Pfaffle, M., Schnabel, R., Michel, H., and Delius, H. (1982). *Halobacterium halobium* phage φH. *EMBO J.* **1**, 87–92.

Schnabel, H., Schramm, E., Schnabel, R., and Zillig, W. (1983). Structural variability in the genome of phage φH of *Halobacterium halobium*. *Mol. Gen. Genet.* **188**, 370–377.

Searcy, D. G. (1982). Thermoplasma: A primordial cell from a refuse pile. *Trends Biochem. Sci.* **7**, 183–184.

Searcy, D. G., and Doyle, E. K. (1975). Characterization of *Thermoplasma acidophilum* deoxyribonucleic acid. *Int. J. Syst. Bacteriol.* **25**, 286–289.

Searcy, D. G., and Stein, D. B. (1980). Nucleoprotein subunit structure in an unusual prokarotic organism: *Thermoplasma acidophilum. Biochim. Biophys. Acta* **609**, 180–185.

Simon, R. D. (1978). Halobacterium strain 5 contains a plasmid which is correlated with the presence of gas vacuoles. *Nature (London)* **273**, 314–317.

Simsek, M., DasSarma, S., RajBhandary, U. L., and Khorana, H. G. (1982). A transposable element from *Halobacterium halobium* which inactivates the bacteriorhodopsin gene. *Proc. Natl. Acad. Sci. U.S.A.* **79**, 7268–7272.

Soliman, G. S. H., and Trüper, H. G. (1982). *Halobacterium pharaonis* sp. nov., a new, extremely haloalkiliphilic archaebacterium with low magnesium requirement. *In* "Archaebacteria" (O. Kandler, ed.), p. 318. Fischer, Stuttgart.

Stanier, R. Y., Adelberg, E. A., and Ingraham, J. (1976). "The Microbial World." Prentice-Hall, Englewood Cliffs, New Jersey.

Stein, D. B., and Searcy, D. G. (1978). Physiologically important stabilization of DNA by a prokarotic histone-like protein. *Science* **202**, 219–221.

Thomm, M., Stetter, O., and Zillig, W. (1982). Histone-like proteins in eu- and archaebacteria. *In* "Archaebacteria" (O. Kandler, ed.), pp. 128–139. Fischer, Stuttgart.

Tu, J., and Zillig, W. (1982). Organization of rRNA structural genes in the archaebacterium *Thermoplasma acidophilum. Nucleic Acids Res.* **10**, 7231–7245.

Weber, H. J., and Leighton, T. J. (1982). Genetic instability of *Halobacterium halobium*. In "Archaebacteria" (O. Kandler, ed.), p. 350. Fischer, Stuttgart.

Weidinger, G., Klotz, G., and Goebel, W. (1979). A large plasmid from *Halobacterium halobium* carrying genetic information for gas vacuole formation. *Plasmid* **2**, 377–386.

Woese, C. R. (1981). Archaebacteria. *Sci. Am.* **244**, 98–122.

Woese, C. R. (1982). Archaebacteria and cellular origins: An overview. *In* "Archaebacteria" (O. Kandler, ed.), pp. 1–17. Fischer, Stuttgart.

Woese, C. R., and Fox, G. E. (1977). Phylogenetic structure of the prokaryotic domain: The primary kingdoms. *Proc. Natl. Acad. Sci. U.S.A.* **74**, 5088–5090.

Yeats, S., McWilliam, P., and Zillig, W. (1982). A plasmid in the archaebacterium *Sulfolobus acidocaldarius. EMBO J.* **9**, 1035–1038.

Zillig, W., Stetter, K. O., Prangishvilli, D., Schafer, W., Wunderl, S., Janekovic, D., Holz, I., and Palm, P. (1982). *Desulfurococcaceae,* the second family of the extremely thermophilic, anaerobic, sulfur-respiring *Thermoproteales. In* "Archaebacteria" (O. Kandler, ed.), pp. 304–317. Fischer, Stuttgart.

Epilogue

Archaebacteria: The Urkingdom

CARL R. WOESE AND RALPH S. WOLFE

Archaebacteria are viewed by most biologists today as a very special type of bacteria. A smaller number of biologists have even come to see them as a unique class of organisms that are interesting in their own right. However, their real value to us will lie not in their being some "third form of life" per se. Rather, they will be known and valued for the perspective that they provide regarding the evolution of the cell in general and the relationship of bacteria to the evolution of the eukaryotic cell in particular. They, in effect, provide a conceptual bridge between the bacteria and the eukaryotes.

Archaebacteria would seem to be the most diverse of the three major groups of organisms. This can be seen by genotypic (i.e., sequence) measure—their ribosomal RNA sequences, for example, can be more distant from one another than are the extreme distances among eubacterial rRNAs. Similarly, by phenotypic measure, archaebacteria show comparably extreme variation. For example, archaebacterial 5 S rRNA secondary structures define a spectrum of types that run from a somewhat eubacterial type near the one extreme to a eukaryotic-like type near the other (see Chapter 5). The sulfur-dependent archaebacteria appear so unlike the methanogens and halophiles that a few biologists would split the group into two kingdoms. The reason for the genotypic and phenotypic depth to the archaebacterial group is unknown at present. It could merely reflect a faster evolutionary clock in the archaebacterial line of descent. However, it could also mean that the archaebacterial group is older than the other two or that the ancestor common to all archaebacteria was a simpler, more primitive (less evolved) entity than is either the eubacterial or the eukaryotic common ancestors.

Archaebacteria tend to be found in "extreme" habitats. The least extreme of these are some of the methanogen niches, which are extremely anaerobic but otherwise not unusual. The ease with which archaebacteria are recovered from thermophilic niches is remarkable. Seemingly, all hot springs contain archaebacteria, often as their dominant species. (However it is likely but not yet certain that they occur in alkaline hot springs.) They have even been isolated from the vicinity of deep sea hot springs (in the few attempts at isolations therefrom), and the presence of methane in vent exudates suggests their widespread presence there. The highest recorded optimum temperatures for an isolated species of archaebacteria is 105°C. (This, however, is probably a temporary record.) In

Copyright © 1985 by Academic Press, Inc.
All rights of reproduction in any form reserved.
ISBN 0-12-307208-5

addition to the many thermophilic archaebacterial species, a number of others will withstand boiling—but cannot grow at or near that temperature. Although some of the archaebacterial high temperature niches are also occupied by eubacterial species, the latter, by their phylogenetic relationships, appear to be true adaptations from mesophilic ancestral forms. It is not clear that this is the case for archaebacteria, however. The majority of the Thermoproteales appear to be basically if not exclusively thermophilic, which is never the case for eubacteria, wherein only phylogenetically restricted groups are predominantly thermophilic.

Although classically the archaebacteria growing in extreme habitats were perceived only as examples of adaptations, this interpretation must now be questioned. ''Extreme,'' after all, is a somewhat relative term. What we think of as extreme for eukaryotes is not necessarily extreme for bacteria—anaerobic growth, for example. And so it may be with thermophilia and the archaebacteria. If so, one might picture ancestral archaebacteria inhabiting niches of the type now defined by the various (volcanically active) rifts in the crust of the earth. Indeed, it is perhaps constructive to consider archaebacteria as being the organisms of the earth's crustal rifts.

Geologists and others now feel that during the first billion or so years of its existence, the surface of the earth was far hotter than at present. If so, one could argue that ancestral archaebacterial phenotypes were established during this era—which makes the term *archaebacteria* a particularly appropriate one.

What are the evolutionary relationships among the three major groups of organisms that inhabit this planet? What is the nature of their common ancestor? These interrelated questions are not now answerable, but we do have some interesting clues as to what might be. As biologists began to get over the notion that all prokaryotes have to be specifically related to one another, they began to see certain specific similarities between archaebacteria and eukaryotes, e.g., the fact that archaebacterial RNA polymerases resemble their eukaryotic more than their eubacterial counterparts (see Chapter 11). At first, the evidence was not impressive, especially since there also existed some specific similarities between eubacteria and eukaryotes, e.g., the possession of ribothymidine in transfer RNAs. However, now that more evidence exists, the case has become stronger. It is based for the most part on phenotypic evidence. The genotypic evidence that now exists seems contradictory. For example, by ribosomal A protein sequences, archaebacteria appear closer to eukaryotes, while in terms of 16 S rRNA sequences, they appear, if anything, closer to eubacteria.

The eukaryotic cell is known to be, at least to some extent, a chimera. It's organelles are of eubacterial origin as are some of its nuclear genes. We now need to determine the extent to which and the ways in which, archaebacteria have contributed to the chimeric eukaryotic cell. The question is one of whether the eukaryotic cell is completely chimeric, whether all of its structures are of eubacterial or archaebacterial origin, or whether there exists a line of descent (represented in the major part of the eukaryotic cell today, and only in eukaryotic cells) that can be called a truely eukaryotic line. A number of biologists feel that the original eukaryotic cell (the "urkaryote") that served as a host for the endosymbionts that would become organelles was an archaebacterium. If so, then the two prokaryotic lines are the basic life forms and their combination has produced eukaryotic life.

However, recent developments in archaebacterial research have given rise to a more radical suggestion regarding the basic phylogenetic categories, i.e., that the archaebacteria have given rise to one or both of the other two kingdoms. The resemblance between archaebacteria and eukaryotes is most pronounced in the case of the sulfur-dependent archaebacteria. For example, it is only the sulfur-dependent archaebacteria and the eukaryotes that have highly modified ribosomal and transfer RNAs. And only in these two groups have 5 S rRNAs with triphosphorylated 5' termini been found.

On the other hand, a specific phenotypic resemblance between the methanogen halophile side of the archaebacteria and the eubacteria can also be claimed. Specific resemblances include: (1) relatively low levels of modified nucleotides in ribosomal and transfer RNAs, (2) formation of cross walls in cell division, (3) possession of viruses with heads and tails, i.e., bacteriophages, (4) closely linked rRNA genes (same DNA sense strand) whose order is 16 S rRNA-(tRNA)-23 S rRNA-5 S rRNA, and (5) 5 S rRNAs that have a 5' terminal monophosphate only.

There are several problems with such a view at this point in time. The main one is that it is supported almost solely by (rather weak) phenotypic evidence, and so it is easily open to the criticism that the few (in some cases, ill-defined) similarities between eukaryotes and sulfur-dependent archaebacteria or eubacteria and methanogen halophiles merely reflect convergence. Also, there exist ostensible counter examples, e.g., rhodopsin is a eukaryotic characteristic found in the extreme halophiles. The matter will be resolved when sufficient genotypic (i.e., sequence) evidence is brought to bear.

What we are saying should not be construed to mean that the archaebacteria are not a phylogenetically coherent unit, however. The genotypic evidence (e.g., complete or partial sequences of 16 S and 5 S rRNA, of tRNA and of a few ribosomal proteins) is clear on this point. All archaebacteria are closer to one another than to any organisms from either of the other two kingdoms. However, archaebacteria may exist in a special, primary relationship to the other two kingdoms. It is now reasonable to consider that archaebacteria are not merely a "third form of life," one of three phylogenetically equivalent primary kingdoms, but that they have higher phylogenetic status than the other two primary kingdoms. In other words, they could represent the evolutionary stage immediately succeeding that of the universal ancestor. Archaebacteria evolved as a group before either eubacteria or eukaryotes did. The latter two kingdoms then arose as (temporary fast clock) offshoots of the archaebacteria—the eukaryotic line of descent arising from within the sulfur-dependent archaebacteria, the eubacterial line from within the methanogen halophile group of archaebacteria. Thus, at base, there may be only one primary group of organisms, the archaebacteria—for which we would then reserve the term *Urkingdom*—and the other two major kingdoms, the eubacteria and the eukaryotes, have arisen (independently) from it.

Because the eukaryotic cell is cytologically so different from the prokaryotes, it is useful to try to span conceptually the evolutionary gap between them. The eukaryotic organelles do not cause a problem in this regard. They have clearly arisen as bacterial endosybionts within the basic eukaryotic cytoplasm. It is the eukaryotic nucleus that seems to raise questions. Its origin is usually seen in terms of endosymbioses, which has led to some largely unsatisfactory proposals. Perhaps we are asking the wrong question. Is it possible that it is not the nucleus that has arisen within the cytoplasm, but the cytoplasm that has arisen around the nucleus? In this way, one could picture the transition from an ancient archaebacterium (of the sulfur-dependent type, perhaps) to a true urkaryote, by an expansion of its periplasmic space, which as it develops, acquires more and more of the cellular functions initially confined to the protonucleus. In this way, the translation function would have moved from the original cell, the protonucleus, into the expanded periplasmic space, the evolving cytoplasm—in which case it is conceivable that translation may still exist within nuclei of some sufficiently primitive eukaryotes.

Thus, we come to see that archaebacteria are not "merely bacteria." Their discovery does not represent the mere finding of another novel group of organisms. Archaebacteria truely are a new kingdom of organisms, in every sense of the word. They present us with biochemical novelty (only beginning to be elucidated) that staggers the imagination. They introduce us to a world of thermophilic organisms and, in so doing, cause us to reexamine our concept of what "normal" and "extreme" are as regards bacterial environments. They radically alter our (poorly defined) ideas as to what prokaryotes are and what relationships they bear to eukaryotes—and they do away forever with the mistaken notion that eukaryote–prokaryote represents a phylogenetic dichotomy. They give us a new and more refined perspective on the origin of the eukaryotic cell. And they may represent an evolutionary stage that is older than one or both of the eubacteria or the eukaryotes. They may be *The Urkingdom*.

Index